New Insights into Geoscience

New Insights into Geoscience

Editor: Luther Doyle

Callisto Reference

www.callistoreference.com

Callisto Reference,
118-35 Queens Blvd., Suite 400,
Forest Hills, NY 11375, USA

Visit us on the World Wide Web at:
www.callistoreference.com

ISBN: 978-1-64116-285-2 (Hardback)

Cataloging-in-Publication Data

New insights into geoscience / edited by Luther Doyle.
p. cm.
Includes bibliographical references and index.
ISBN 978-1-64116-285-2
1. Earth sciences. 2. Geology. I. Doyle, Luther.
QE26.3 .N49 2020
550--dc23

Table of Contents

Preface

This book has been an outcome of determined endeavour from a group of educationists in the field. The primary objective was to involve a broad spectrum of professionals from diverse cultural background involved in the field for developing new researches. The book not only targets students but also scholars pursuing higher research for further enhancement of the theoretical and practical applications of the subject.

Geoscience is the discipline concerned with all the fields of natural science that are related to planet Earth. It studies the constitution of the Earth and its atmosphere. Geoscience, also known as Earth science, studies the physical characteristics and processes of the Earth such as earthquakes, rain, flood, fossils, etc. It is classified into four main branches of study, which are the atmosphere, biosphere, hydrosphere and lithosphere. It is the study of Earth as well as its neighbours in space. It focuses on locating or developing mineral and energy resources available on or in the Earth. It analyzes the impacts of human activities on the Earth's environment and makes strategies to protect the planet. Tools from other fields such as chemistry, biology, physics, mathematics, geography, etc., are used in geoscience to understand the Earth's working and evolution. It also focuses on predicting the natural disasters like earthquakes, floods or volcanoes. This book contains some path-breaking studies in the field of geoscience. It provides significant information of this discipline and helps in developing a good understanding of it and related fields. This book will serve as a valuable source of reference for graduate and post graduate students.

It was an honour to edit such a profound book and also a challenging task to compile and examine all the relevant data for accuracy and originality. I wish to acknowledge the efforts of the contributors for submitting such brilliant and diverse chapters in the field and for endlessly working for the completion of the book. Last, but not the least; I thank my family for being a constant source of support in all my research endeavours.

Editor

Preface

1

Coastline evolution of Portuguese low-lying sandy coast in the last 50 years: an integrated approach

Cristina Ponte Lira, Ana Nobre Silva, Rui Taborda, and Cesar Freire de Andrade

IDL – Instituto Dom Luiz, Faculdade de Ciências da Universidade de Lisboa, Campo Grande, 1749-016 Lisbon, Portugal

Correspondence to: Cristina Ponte Lira (fclira@fc.ul.pt)

Abstract. Regional/national-scale information on coastline rates of change and trends is extremely valuable, but these studies are scarce. A widely accepted standardized methodology for analysing long-term coastline change has been difficult to achieve, but it is essential to conduct an integrated and holistic approach to coastline evolution and hence support coastal management actions. Additionally, databases providing knowledge on coastline evolution are of key importance to support both coastal management experts and users.

The main objective of this work is to present the first systematic, national-scale and consistent long-term coastline evolution data of Portuguese mainland low-lying sandy coasts.

The methodology used quantifies coastline evolution using a unique and robust coastline indicator (the foredune toe), which is independent of short-term changes.

The dataset presented comprises (1) two polyline sets, mapping the 1958 and 2010 sandy beach–dune system coastline, both optimized for working at 1 : 50 000 scale or smaller; (2) one polyline set representing long-term change rates between 1958 and 2010, each estimated at 250 m; and (3) a table with minimum, maximum and mean of evolution rates for sandy beach–dune system coastline. All science data produced here are openly accessible at https://doi.pangaea.de/10.1594/PANGAEA.859136 and can be used in other studies.

Results show beach erosion as the dominant trend, with a mean change rate of $-0.24 \pm 0.01\ \mathrm{m\,year^{-1}}$ for all mainland Portuguese beach–dune systems. Although erosion is dominant, this evolution is variable in signal and magnitude in different coastal sediment cells and also within each cell. The most relevant beach erosion issues were found in the coastal stretches of Espinho–Torreira and Costa Nova–Praia de Mira, Cova da Gala–Leirosa, and Cova do Vapor–Costa da Caparica. The coastal segments Minho River–Nazaré and Costa da Caparica adjacent to the coast exhibit a history of major human interventions interfering with the coastal system, many of which originated and maintained a sediment deficit. In contrast, the coastal segments Troia–Sines and Sines–Cape S. Vicente have experienced less intervention and show stable or moderate accretion behaviour.

1 Introduction

The coastal zone is intrinsically dynamic, where changes occur at different temporal and spatial scales. Along any sandy coastal stretch, changes in coastline position (erosion or accretion) are expected and occur in response to variations in sea level, sediment budget and hydrodynamic conditions. As human activities and occupation grow in coastal areas, management problems and challenges also increase as a result of needing to accommodate both coastal change and occupation. Beach erosion is one of the leading management problems that coastal regions have to face worldwide (Phillips and Jones, 2006) and accurate information on coastline movement rates and trends is essential to support sustainable management strategies. This objective is particularly relevant in a climate change scenario, which may determine an increase in the intensity of coastal response (e.g. excess erosion induced by sea-level rise).

The idea that coastlines are under threat by climate change acceleration is relatively recent (Carter and Woodroffe, 1994) and has increased the demand for accurate knowledge on coastal evolution. This encouraged the development of numerous long-term (> 30 years according to Fletcher et al., 2012) studies worldwide which, in most cases, lack the broader scale perspective, instead evaluating change at micro/meso spatial scales and generally focusing only on known problematic areas.

Pioneer studies considering a national-scale approach to coastline change were promoted in the USA and Europe in the early 21st century by the USGS (United States Geological Survey) (USGS, 2004) and the European Commission (European Commission, 2004). The United States launched the National Assessment of Coastline Change Project, which delivered a number of reports organized by coastal regions (e.g. Fletcher et al., 2012). The European counterpart is the EUROSION project framework (European Commission, 2004), which conducted a country-by-country evaluation of coastal erosion at the European scale and delivered three main final products: (1) online reports, including major findings and recommendations; (2) a GIS database, which includes Europe's coastline vectors at 1 : 100 000 scale; and (3) a coastline management guide.

Despite this effort, there is still no consensual coastline definition for reporting coastal evolution at national to global scale or to ensure comparability between datasets. Therefore, most of the coastline evolution works are still not adequate for inter-comparison. This fact is even more striking as the 2012 European Environment Agency (EEA) report on climate change (EEA, 2012) does not include coastline position as an EEA indicator on threats to the coastal zone, because regular updates of this type of information are not expected.

In Portugal, coastline evolution studies are usually local to mesoscale in their spatial scope and essentially focus on problematic areas (e.g. Abecassis et al., 1970; Barceló, 1971; Castanho et al., 1974; Bettencourt and Ângelo, 1992; Ferreira, 1992; Oliveira, 2005; Pinto and Teixeira, 2005; Veloso-Gomes et al., 2009; Taveira Pinto et al., 2009; Rebêlo et al., 2011; Silva et al., 2013), and different coastline proxies are used (e.g. European Commission, 2004; Oliveira, 2005; Ferreira et al., 2006; Rebêlo et al., 2011; Cenci et al., 2013). Thus, these analyses can not provide a complete and unified approach for all of the Portuguese territory.

From the above, one may conclude that an approach using metrics that are consistent from one coastal zone to another is essential to address coastal evolution in an integrated and holistic manner. A single and robust coastal indicator capable of representing long-term change is indispensable and should be evaluated at a national and broader scale.

This work aims at presenting the first systematic, national-scale and consistent long-term coastline evolution study of Portuguese mainland low-lying sandy coast. Coastline evolution is quantified using a single robust coastline indicator (foredune toe), which is, to a larger extent, independent of short-term morphodynamic changes. The coastline and evolution data are fully and freely available at http://doi.pangaea.de/10.1594/PANGAEA.853654 and can be used in other studies. This dataset comprises (1) two polyline sets, mapping the 1958 and 2010 sandy beach–dune system coastline, both optimized for working at 1 : 50 000 scale or smaller, and (2) one polyline set representing long-term change rates between 1958 and 2010, each estimated at 250 m.

2 Study area

The Portuguese mainland coastline extends south from the Minho River to west of the Guadiana river mouth (Fig. 1) along ca. 900 km (variable with the scale of representation). The coast includes a wide range of morpho-sedimentary environments, such as beaches, cliffs, estuaries, lagoons and barrier islands (Andrade and Freitas, 2002; Ferreira and Matias, 2013). Coastal districts house around three-quarters of the population and generate around 80 % of the gross national product (Duarte Santos et al., 2014b), with increasing economic activity in these areas, especially from tourism.

Tides along the Portuguese coast are semi-diurnal and meso-tidal and are unimportant to coastal processes except in the vicinity of inlets and tide-dominant basins (Andrade and Freitas, 2002). According to Andrade and Freitas (2002), Portuguese western- and southern-facing coasts present asymmetric characteristics regarding wave energy and wave climate. The western coast is fully exposed to the dominant NW high-energy swells (Andrade and Freitas, 2002). The south-facing coastal stretches are sheltered from the dominant waves, resulting in a milder wave regime (Ferreira and Matias, 2013). The high-energetic wave regime at the western coast induces a net potential littoral drift directed southward that can reach up to 10^6 m^3 year^{-1}. This potential falls one or more orders of magnitude in linear coastal stretches that are rotated somewhat clockwise, along equilibrium arcuate embayments, and in sections sheltered from the prevailing NW waves; in these cases the residual drift may be annulled or even reversed in direction (Andrade and Freitas, 2002). At the south coast of Portugal, dominant waves drive an easterly directed littoral drift of ca. 10^4 to 10^5 m^3 year^{-1}, depending upon coastal orientation (Ferreira and Matias, 2013); exceptions are NW/SW coastal ribbons that face the prevailing swell and show almost null residues of longshore power and drift (Andrade and Freitas, 2002).

2.1 Sediment cells

According to EUROSION resolution 1 (European Commission, 2004), the sediment status must be accounted for each coastal sediment cell. Sediment cells are independent of each other in terms of sediment transfers, which means actions taken within a specific cell may impact sections of the same cell but will not significantly affect adjacent cells.

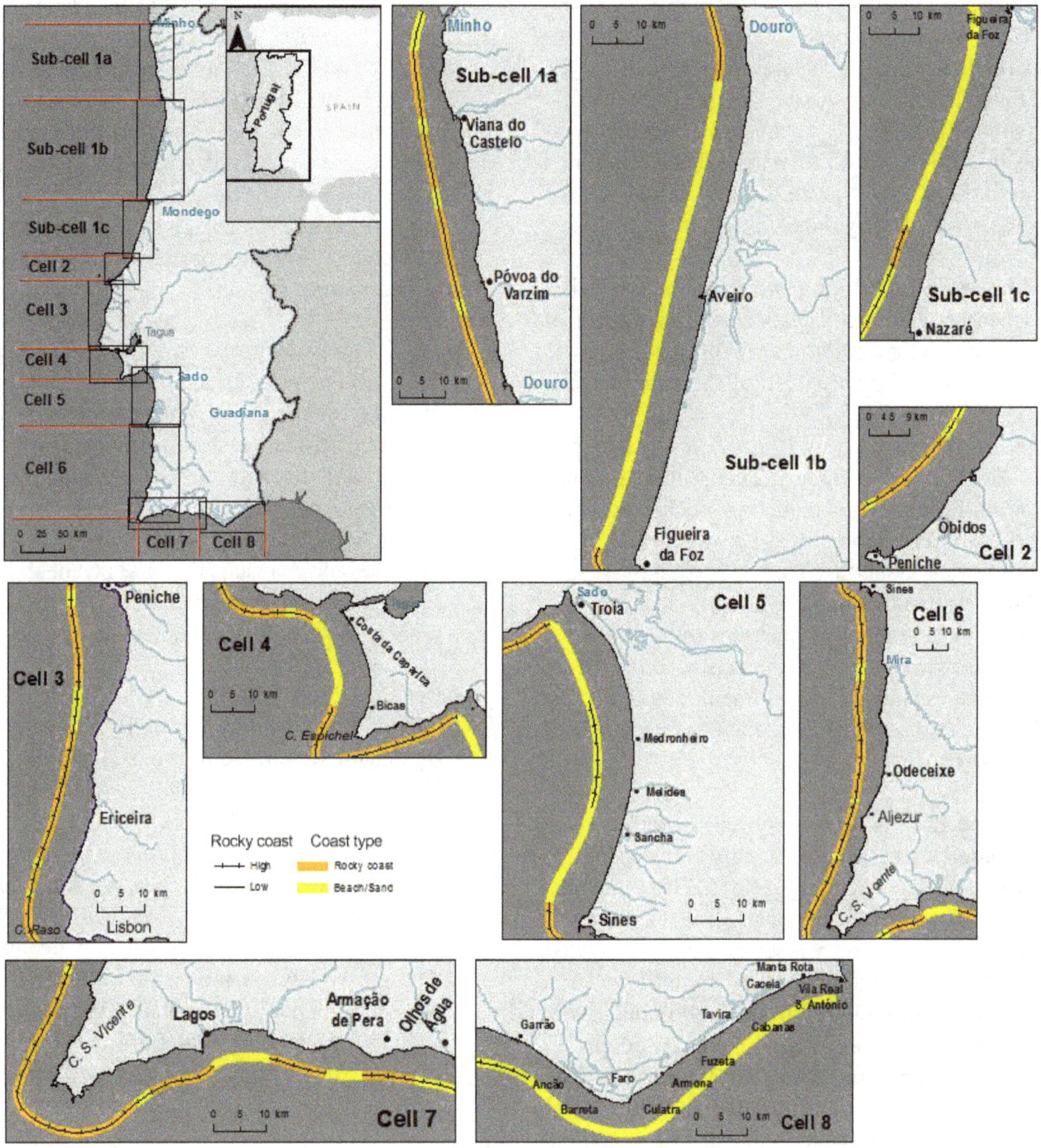

Figure 1. Sediment cells for the Portuguese mainland coast (cell limits in red) and coastal characterization.

The works of Andrade and Freitas (2002) and later of Duarte Santos et al. (2014a) defined and characterized eight sediment cells (Fig. 1) along the Portuguese mainland littoral, according to its geomorphological characteristics and sedimentary dynamics:

- *Cell 1* – Minho river mouth to Nazaré. This cell is divided into three sub-cells: from the Minho to Douro rivers (sub-cell 1a), from Douro River to Cape Mondego (1b) and from Cape Mondego to Nazaré (sub-cell 1c).

- *Sub-cell 1a* is characterized by a low rocky coast, with numerous small sandy and gravel beaches (which are sometimes extensive), whose development is strongly associated with local sheltering effects in the first case and river estuaries in the latter. The low-lying sandy coast accounts for 50 % of the total coastal stretch.

- *Sub-cell 1b* coastline presents geomorphological characteristics similar to cell 1a at the north, and further south it essentially consists of an extensive linear beach, backed by coastal dune systems and only interrupted by the Aveiro inlet. The southernmost section is mainly constituted by cliffs and sandy beaches. Beach–dune systems represent 75 % of this sector's total length.

- *Sub-cell 1c* represents a rocky coast to the north, which becomes a highly developed sandy beach at Figueira da Foz that is made by the retention effect of the port of

Figueira da Foz's northern jetty. This sediment retention induced downdrift coastline retreat, which led to the construction of rigid structures, to the south of the port of Figueira da Foz (Duarte Santos et al., 2014a). Further south the coast is low, sandy and straight, culminating in cliffs bordered by narrow beaches, which extend until the area of Nazaré. In this sector, 72 % of the beaches are backed by dunes.

- *Cell 2* – Nazaré to Peniche, which mainly consists of cliffs bordered by rocky platforms to the north of Óbidos lagoon and generally narrow linear beaches to the south, accounts for only 25 % of the entire coastal sector.

- *Cell 3* – Peniche to Cape Raso. This section is dominated by cliffs presenting numerous pocket beaches, with very different geometry. The wider and short beaches are often limited by small dune fields and developed next to stream mouths, whereas linear and narrow beaches lacking dunes, sometimes to the extent of kilometres, are associated with the existence of natural headlands that provide limited sediment retention. This sector has one of the lowest percentages of beach–dune systems, a mere 6 %.

- *Cell 4* – Cape Raso to Cape Espichel. In terms of its geomorphological characteristics, this cell can be divided into two distinct sectors, separated by the Tagus river mouth. Northwest of the Tagus estuary the coast is dominated by cliffs, with a set of small embedded beaches, which are limited landward by cliffs or artificial structures and sheltered from the dominant NW wave regime. South of the Tagus, the coast adopts an arched configuration, suggesting an equilibrium bay, forming a continuous sandy coast from Costa da Caparica to Bicas. Further south, until Cape Espichel, the coastline mainly develops in cliffs, occasionally interrupted by small sandy or gravelly pocket beaches. Half of the sector (50 %) is represented by beach–dune systems.

- *Cell 5* – Cape Espichel to Sines. The coast between Cape Espichel and Sado inlet is sheltered from the prevailing wave regime and only a few pocket beaches exist in indentations affecting the rocky plunging cliffs of the Arrábida chain. Between the Troia peninsula and Sines, it is a continuous sandy coast, with an arched configuration, similar to that observed in the Caparica–Espichel section. The beach is bordered by dunes in most of the sector, mainly north of Medronheiro; south of this point, cliffs occur until the Melides lagoon and, occasionally, between the Sancha lagoon and Moinhos stream. In this sector, 54 % of the coast is represented by low-lying sandy beaches.

- *Cell 6* – Sines to Cape S. Vicente. This stretch is dominated by cliffs (generally high) carved into resistant Palaeozoic and Mesozoic rocks. These cliffs are disrupted by the Mira, Odeceixe and Aljezur estuaries and lowlands, which, together with cliff indentations, accommodate sandy and gravelly beaches, usually with reduced width. The beaches developed on larger stream mouths are often backed by dune systems. Together with cell 3, this sector presents a small percentage (3 %) of beach–dune systems.

- *Cell 7* – between Cape S. Vicente and Olhos de Água. This cell is characterized by an extremely varied morphology, where cliff segments alternate with beaches confined between resistant headlands or developed at stream mouths and in barriers separating small lagoons/estuaries from the open ocean. From Lagos to Olhos de Água the coast is extremely crenulated. The bays of Lagos and Armação de Pera include dune systems of significant size. In this cell, 12 % of the coastline belongs to the beach–dune system geomorphological type.

- *Cell 8* – from Olhos de Água to the Guadiana river mouth. The coastline is mainly dominated by a barrier island system (Ria Formosa) and the Manta Rota–Vila Real de Santo António coastal plain. The western end, between Olhos de Água and Garrão, is comprised of a sandy beach bordered by cliffs. The barrier island system, which separates the sea from the second, larger Portuguese lagoon, includes (from west to east) two peninsulas (Ancão and Cacela), five barrier islands (Barreta, Culatra, Armona, Tavira and Cabanas) and six tidal inlets (Ancão, Faro-Olhão, Armona, Fuseta, Tavira and Cacela inlets). In total, 66 % of this sector's beaches are backed by dunes.

3 Methods

3.1 Beach coastline indicator

In coastal evolution studies the selection of a suitable coastline definition is a primary issue. Long-term studies, such as the one presented, are based on discrete measurements that represent snapshots of the coastal system, usually separated by a large temporal gap. To obtain an unbiased measure of coastal evolution, the measurements should, therefore, be as independent as possible of both high-frequency water level changes and the seasonal beach morphological cycle.

In this sense, we have considered the coastline concept (as given in Carapuço et al., 2016) over the use of the shoreline (the physical interface of land and water) since is more conservative in spatial location. Nevertheless, due to the extremely dynamic nature of this feature, coastline mapping is generally based on an indicator which, according to Boak and Turner (2005), is a feature used as a proxy to represent the "true" coastline position. Coastline indicators may be physical coastal features such as the bluff top/cliff top, landward edge of shore protection structure, foredune toe and seaward vegetation line. Considering that this study targets low-lying sandy coasts backed by dunes, the coastline indicator used in this study was the foredune toe, and recognized as the morphological feature less affected by short-term (tidal) and medium-term (seasonal) changes (e.g. Komar et al., 2001; Ferreira et al., 2006; Del Río and Gracia, 2013).

The foredune toe indicator is described by either a slope break or the seaward limit of vegetation. The latter criterion was only used when the first was not clearly recognized. In cases where the beach is not backed by a dune (e.g. overwashed sections of sandy barriers, extensive blowouts in the back beach) the aforementioned indicator could not be mapped. These situations constitute a limitation of the proposed methodology, and other methods should be used when analysing other geomorphological coastal types. Nevertheless, using this method it was possible to map 92 and 95 % of all the low-lying sandy coast of mainland Portugal for the years 1958 and 2010, respectively. The shoreline evolution assessment was conducted for 84 % of the sandy coast, where the information from both coastlines overlaps.

3.2 Coastline mapping

3.2.1 Data sources

Two sets of data were used in the coastline extraction procedure:

1. digital aerial photographs from the USAF 1958 flight (photos taken on different days and no flight plan was available to the authors) with 0.5 m pixel resolution;

2. digital orthophotomaps of the year 2010 (IGEO, 2010) (photos taken on different days of 2010) with 0.2 m pixel resolution.

3.2.2 Mapping procedures

Due to the unrectified nature of the 1958 aerial photographs, images were georeferenced using the 2010 orthophotomaps. The georeferencing procedure, consisting in the coordination of points on the image to be georeferenced with points on a geographically referenced dataset, was conducted according to Fig. 2 in a GIS environment. The individual photos were grouped into mosaics covering continuous sandy coastal areas. A total of 158 photos were used to produce 56 mosaics, using the Photomerge capability of Photoshop®. This procedure finds the optimal borders between the images and creates seams based on those borders, requiring an image overlapping of 40 to 70 %. Mosaicking minimizes misalignments, which are especially visible in borders of two contiguous images (Wan et al., 2013); the main advantages of this include image spatial continuity, allowing for coastline location on longer coastal stretches and facilitating the search for common ground control points (GPCs) (Verhoeven et al., 2012).

Mosaics were georeferenced using GCPs that corresponded to common elements in both datasets, evenly distributed along the coastline. Between 5 and 10 GCPs per kilometre of coastline were used, accounting for availability of common elements, mosaicking distortions and terrain elevation. The georeferencing was based on a *spline* adjustment method (a piecewise polynomial that maintains continuity and smoothness between adjacent polynomials).

The work of Rocchini and Di Rita (2005) showed that georeferencing by polynomial fit is a very robust tool in flat areas, with performance similar to that obtained by orthorectification, reporting significant increases in the error as terrain becomes more rugged. In this present work an independent study for accuracy assessment related to the georeferencing process and coastline vectorization was also conducted. Results from this evaluation will be used in the uncertainty assessment of coastline rates and are presented in the next section.

The coastline digitization process, which consisted in manually recognizing and digitizing the coastline indicator on the images, was conducted using a ranging visualization scale between 1 : 5000 and 1 : 8000. The coastline final position segments, for both analysed dates, were optimized for working at a 1 : 50 000 scale or smaller.

3.3 Coastline evolution

Coastline evolution rates of change (R) were calculated in a GIS environment using the Digital Shoreline Analysis System (Thieler et al., 2009) and the Coastline Change Mapper (Psuty et al., 2010), both ArcGIS extensions. Transects were placed every 250 m, roughly perpendicular to the coastline trend. Rate-of-change results were grouped together along the same littoral cell and maximum ($\mathrm{Max}(R)$), mean ($\bar{R}$) and minimum ($\mathrm{Min}(R)$) rates were assessed for each one. Results are presented for the overall sandy beach–dune system coastline as well.

3.3.1 Position and rate uncertainty

Several sources of error affect the accuracy of coastline position and consequently coastline change rates. Measurement uncertainties include errors related to coastline digitization (E_d) (Fletcher et al., 2003), image resolution (E_ir) (Coyne et al., 1999; Catalão et al., 2002) and image rectification (E_r) (Shoshany et al., 1996; Fletcher et al., 2003).

According to Fletcher et al. (2003) these errors are random and uncorrelated and can be represented by a single measure (U_p) calculated as the square root of the sum of the squares (Coyne et al., 1999; Fletcher et al., 2003):

$$U_p = \sqrt{E_\mathrm{d}^2 + E_\mathrm{r}^2 + E_\mathrm{ir}^2}. \qquad (1)$$

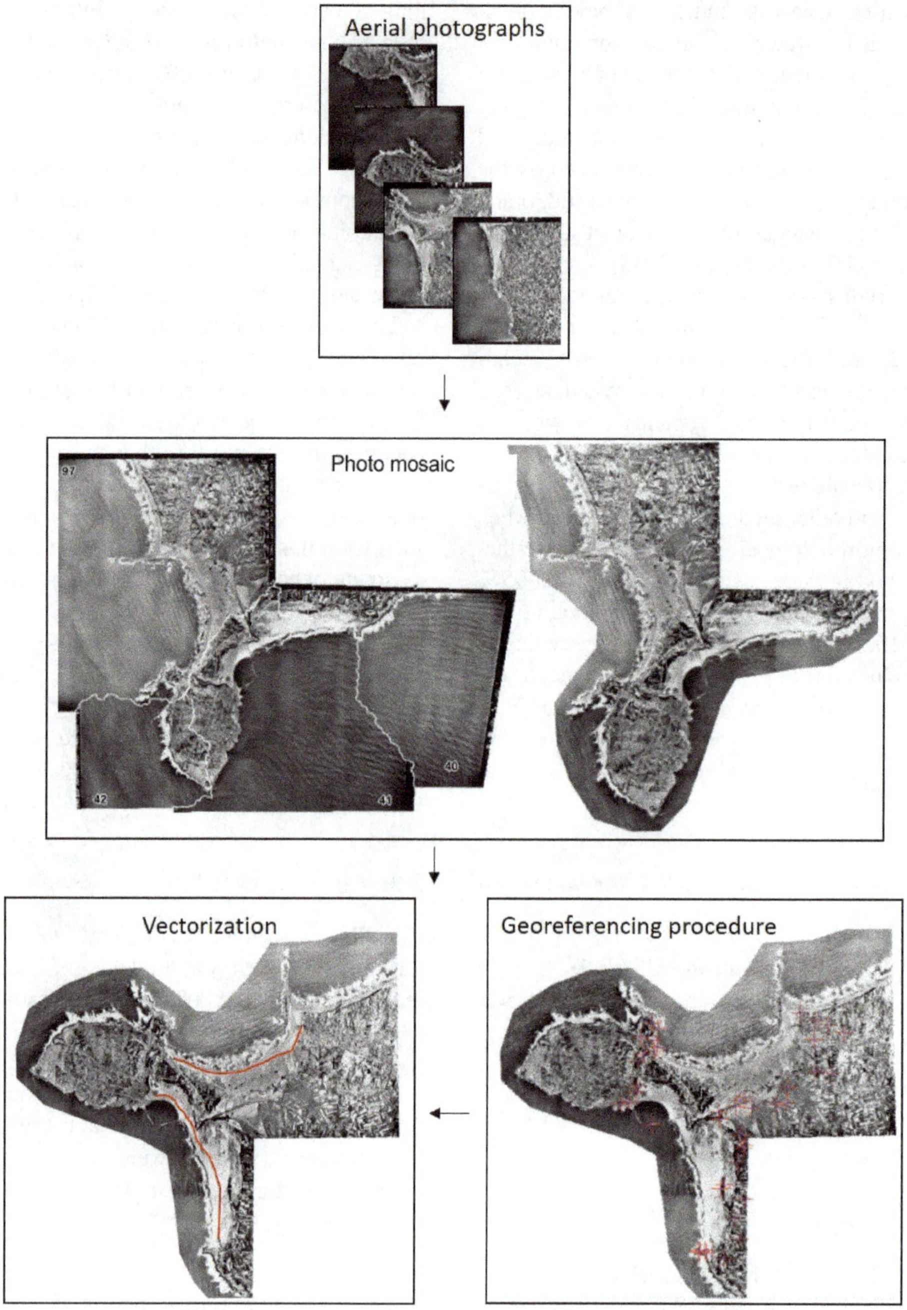

Figure 2. Flowchart procedures for using aerial photographs to extract coastline indicators.

Fletcher et al. (2003) state another physical component of error, representing short-term variability in coastline position. In this study, these sources of error were eliminated by adopting the foredune shoreward edge approach, in accordance with other works (e.g. Del Río and Gracia, 2013).

The uncertainty related to the georeferencing procedures (E_r) of 1958 aerial photos was evaluated using three mosaics, depicting different geomorphological settings. Validation with an independent subset of control points (10 % of the original GCPs) was used to assess the root-mean-square error (RMSE) for each mosaic. This process ensures an independent evaluation of the georeferencing errors that was considered representative of all other mosaics. Results show a maximum RMSE of 4.79 m for all three mosaics; hence, E_r was considered to be 5 m. Uncertainties related to the vectorization process of the coastline indicator (E_d), which were digitized by three different operators, resulted in a 7 m error. Image resolution is 0.5 m (E_{ir}). Thus, U_{p1958} was estimated as 8.6 m.

In a similar way to the aerial photos, the uncertainties related to the use of orthophotomaps were estimated as $E_r = 0$, due to its rectified nature, and $E_d = 5$ m and $E_{ir} = 0.2$ m. Therefore, U_{p2010} was estimated as 5.0 m.

The coastline change rate uncertainty (U_R) is calculated, according to Fletcher et al. (2012), by using the uncertainty values for both coastlines (U_{p1958} and U_{p2010}) and the time frame (t) between them:

$$U_R = \frac{\sqrt{U_{p1958}^2 + U_{p2010}^2}}{t}. \qquad (2)$$

In this work, the uncertainty affecting coastline change rate was calculated as ± 0.2 m year^{-1}, for the 52-year period of analysis.

The uncertainty of a mean rate can be calculated, according to Fletcher et al. (2012), as the root sum of squares of rate uncertainties (U_R) at all transects divided by their total number (n):

$$U_{\bar{R}} = \frac{\sqrt{\sum_n^{i=1} U_{R,i}^2}}{n}. \qquad (3)$$

$U_{\bar{R}}$ was evaluated for each individual sediment cell taking into account the number of transects in each one, and the same procedure was applied to all low-lying sandy coastline of mainland Portugal. $U_{\bar{R}}$ values are shown in Table 1.

4 Results and discussion

The coastline evolution dataset comprises (1) two coastlines, an older one dated 1958 and a modern one dated 2010, and (2) long-term change rates between 1958 and 2010, which were estimated at a 250 m interval assuming a linear rate of change. All science data are available as polylines and is openly accessible at http://doi.pangaea.de/10.1594/PANGAEA.853654. These results have also been incorporated as an update of the EUROSION project information on low-lying sandy coastal ribbons of mainland Portugal, in collaboration with the IPMA (Instituto do Mar e da Atmosfera) agency and the EMODnet project Portuguese partner for the geology section (EMODnet, 2009). The dataset was incorporated into the "Coastal Migration Map – Coastal Behaviour for all European countries" dataset and is currently available for visualization at http://194.66.252.216/geonetwork/srv/por/catalog.map (last access: May 2016).

Table 1 shows maximum, minimum and mean coastline evolution rates found in each sediment cell and for the total coastline of mainland Portugal and is openly accessible at https://doi.pangaea.de/10.1594/PANGAEA.859135. The total length occupied by sandy beach–dune systems corresponds to ca. 40 % of the total mainland Portuguese coastline.

Sub-cell 1a presents an erosion maximum of -1.84 ± 0.2 m year^{-1}, mainly to the south of the Lima estuary. An accretion maximum of $+5.83 \pm 0.2$ m year^{-1} is reached next to the same river mouth, due to interference of both harbour jetties in sediment transport patterns. The mean rate of change is -0.29 ± 0.02 m year^{-1}, showing an erosive trend (Fig. 3). According to Duarte Santos et al. (2014a) this pattern is related to the reduction in river sediment supply and harbour dredging.

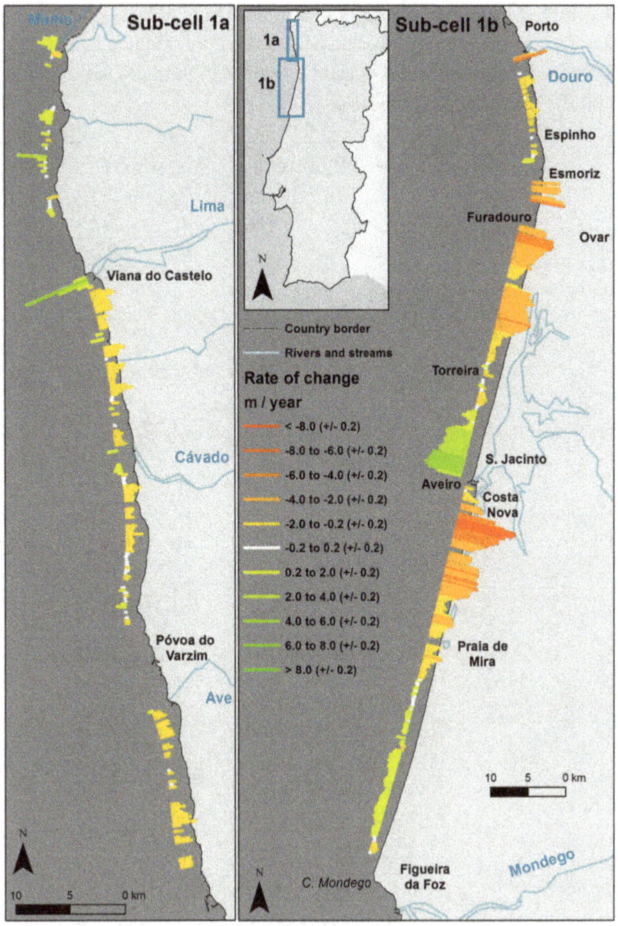

Figure 3. Long-term coastline rates of change in metres per year for sediment sub-cells 1a and 1b.

Sub-cell 1b is dominated by an erosional trend with an average retreat rate of -0.91 ± 0.01 m year^{-1}; this figure exceeds by a factor of 3 to 5 the average recession observed elsewhere along the same coast and the global average. Erosional trends extend almost continuously from the Douro river mouth to Torreira and also further south, from the Aveiro inlet to Praia de Mira, where a maximum rate of change (-7.38 ± 0.2 m year^{-1}) is observed. Accretional trends are observed updrift of significant obstacles to littoral drift such as Cape Mondego and Aveiro jetties (Fig. 3). Similarly to sub-cell 1a, the main factor driving erosion is the loss of fluvial sediment supply and sediment retention in artificial structures. This sub-cell contains the most vulnerable areas

Table 1. Minimum (Min(R)), maximum (Max(R)) and mean ($\bar{R}$) of evolution rates (R) for sandy beach–dune systems (SBDS) coastline. Total coastline length (km) and SBDS coastline length (in km and in percentage). All values are presented for each sediment cell and for all mainland Portugal SBDS coastline. Values in parentheses depict (a) % of transects with $R \leqslant -0.2$, (b) % of transects with $-0.2 < R < 0.2$ and (c) % of transects with $R \geqslant -0.2$.

Cell	Location[1] (start, end)	Min(R) $\pm U_R$ (m year^{-1})	Max(R) $\pm U_R$ (m year^{-1})	$\bar{R} \pm U_{\bar{R}}$ (m year^{-1}) (a) (b) (c)	Total[2] coastline length (km)	SBDS[3] coastline length (km)	SBDS[3] coastline length (%)
1a	41°51′47 N, 8°52′13 W 41°8′51 N, 8°40′28 W	−1.94± 0.2	5.63 ± 0.2	−0.29 ± 0.02 (60) (19) (21)	95	48	50
1b	41°8′51 N, 8°40′28 W 40°11′11 N, 8°54′31 W	−7.38 ± 0.2	4.67 ± 0.2	−0.91 ± 0.01 (60) (10) (30)	120	91	75
1c	40°11′11 N, 8°54′31 W 39°36′16 N, 9°5′6 W	−3.77 ± 0.2	2.75 ± 0.2	−0.19 ± 0.01 (39) (26) (35)	69	49	72
2	39°36′16 N, 9°5′6 W 39°21′30 N, 9°24′27 W	-1.07 ± 0.2	1.63 ± 0.2	−0.17 ± 0.03 (59) (31) (10)	54	13	24
3	39°21′30 N, 9°24′27 W 38°42′29 N, 9°29′4 W	−0.65 ± 0.2	1.06 ± 0.2	0.02 ± 0.04 (30) (48) (22)	107	6	6
4	38°42′29 N, 9°29′4 W 38°24′50 N, 9°13′19 W	−4.57 ± 0.2	1.20 ± 0.2	−0.04 ± 0.03 (11) (29) (60)	32	16	50
5	38°24′50 N, 9°13′19 W 37°57′14 N, 8°53′14 W	−1.09 ± 0.2	8.18 ± 0.2	0.45 ± 0.01 (28) (28) (44)	99	53	54
6	37°57′14 N, 8°53′14 W 37°1′19 N, 8°59′44 W	−0.95 ± 0.2	1.32 ± 0.2	0.28 ± 0.06 (36) (9) (55)	133	8	6
7	37°1′19 N, 8°59′44 W 37°5′23 N, 8°11′4 W	−0.66 ± 0.2	1.96 ± 0.2	0.23 ± 0.03 (12) (56) (32)	94	11	12
8	37°5′23 N, 8°11′4 W 37°9′57 N, 7°23′38 W	−8.96 ± 0.2	12.99 ± 0.2	0.01 ± 0.01 (45) (17) (38)	84	56	66
All	−60 518, 244 795 64 870, −277 252	−8.96 ± 0.2	12.99 ± 0.2	−0.24 ± 0.01 (46) (20) (34)	887	350	40

[1] Latitude, longitude in ETRS 89. [2] Total coastline length was estimated using the available EUROSION coastline (European Commission, 2004) for the Portuguese mainland littoral. [3] SBDS coastline length was calculated using the proposed 2010 coastline.

concerning the erosion risk in Portugal (Duarte Santos et al., 2014a).

In sub-cell 1c, erosion is mainly concentrated in its northern section, up to about 20 km to the south of the Mondego river mouth, in relation to harbour development works, including dredging and jetties. Here the retreat rates locally reach a maximum of -3.77 ± 0.2 m year^{-1} at Lavos. Further south, accretion dominates, reaching a maximum of $+1.39 \pm 0.2$ m year^{-1} in the vicinity and updrift of Nazaré, related with sediment retention by a rocky headland. Mean coastline rates of change indicates a general erosional trend, with a value of -0.34 ± 0.01 m year^{-1} (Fig. 4).

Cell 2 presents a set of five discontinuous sandy beach–dune systems with a mean evolution rate of -0.17 ± 0.03 m year^{-1}. Erosional coastal stretches are found at Nazaré–Salgados, Cova da Areia and the Baleal embayment, east of Peniche. Accretion was detected in São Martinho do Porto bay and the Óbidos sand barrier, with a maximum of $+1.63 \pm 0.2$ m year^{-1} in the former case (Fig. 4).

Cell 3 shows rare beach–dune systems confined to coastline indentations in relation to small river mouths. The signal and rate of change are site-specific, with erosion at Consolação, Ponta da Lamporeira and Guincho (north of Cape Raso) (-0.49, -0.22 and -0.65 ± 0.2 m year^{-1}, respectively), and accretion at Areia Branca, Santa Rita and Ponta da Lamporeira ($+0.74$, $+1.06$ and $+0.23 \pm 0.2$ m year^{-1}, respectively) (Fig. 4). The mean coastline rate of change ($+0.02 \pm 0.04$ m year^{-1}) is within the method uncertainty, indicating non-significant long-term evolution. This result is compatible with negligible anthropogenic influence in sediment supply; thus, the measured changes essentially reflect the coastal system's natural variability.

Cell 4 northern sector, trending east–west, contains only small pocket beaches limited by either cliffs or artificial structures; these were not considered in this study. The southern sector (extending from Cova do Vapor to Bicas) is dominated by one continuous sandy beach–dune system, north of Lagoa de Albufeira, and by beaches backed by cliffs to the south. Coastal change was not measured at Costa da Caparica due to the existence of a seawall. Likewise, in places between Costa da Caparica and Lagoa de Albufeira, where the lack of dune vegetation and overexposure of the

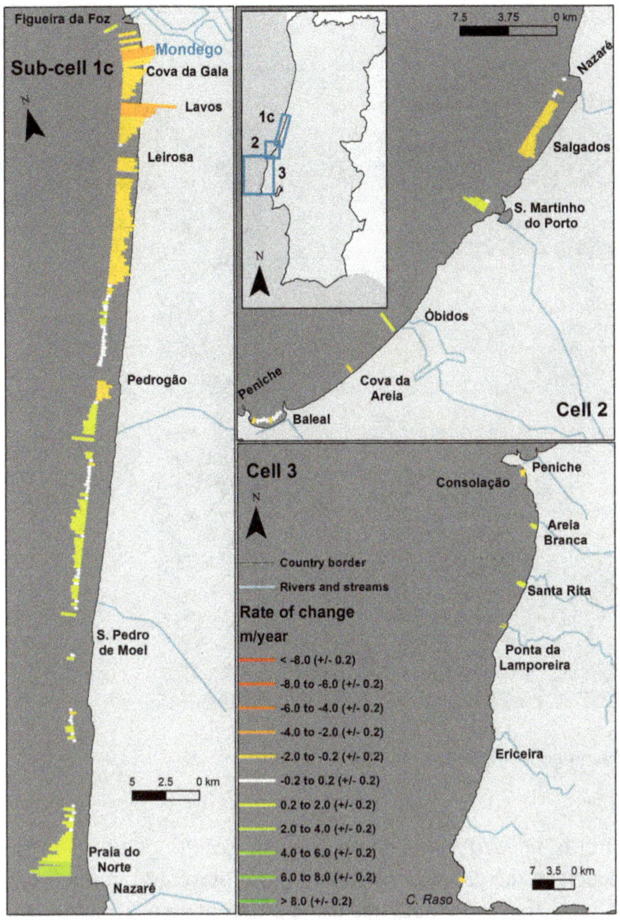

Figure 4. Long-term coastline rates of change in metres per year for sediment sub-cell 1c and cells 2 and 3. Note: the sub-cell 1c coast is shown with a 15° rotation.

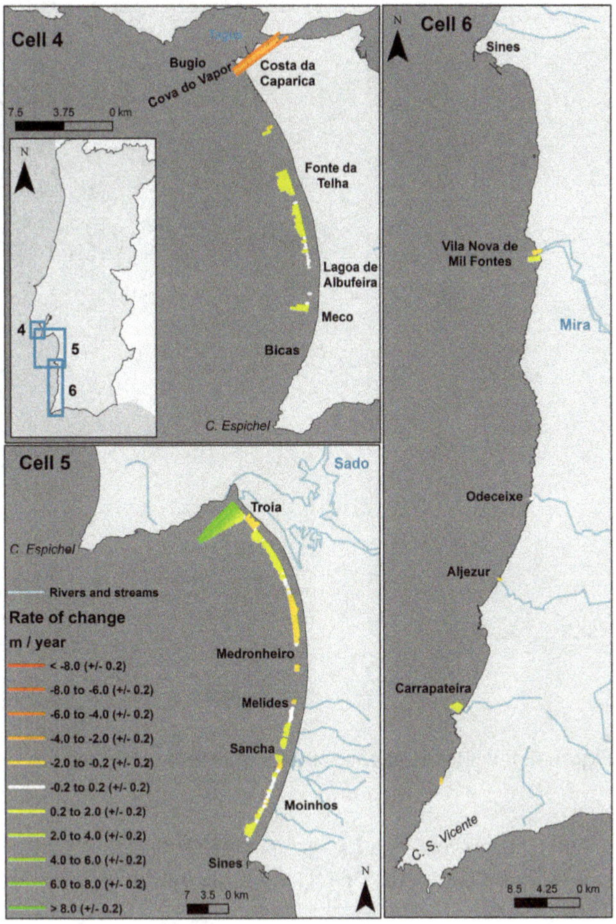

Figure 5. Long-term coastline rates of change in metres per year for sediment cells 4, 5 and 6.

1958 photographs precluded the coastline indicators identification, coastal change was not measured. In this sector, two contrasting trends were found: (1) an intense erosional trend with a maximum of $-4.57 \pm 0.2\,\mathrm{m\,year^{-1}}$, to the south and next to the Tagus river mouth, north of Costa da Caparica, and (2) south of Costa da Caparica, a mild accretional trend alternating with stable stretches, reaching a maximum of $+1.20 \pm 0.2\,\mathrm{m\,year^{-1}}$ at Meco (Fig. 5). Globally, this sector presents a slight erosive trend with a mean rate of $-0.04 \pm 0.03\,\mathrm{m\,year^{-1}}$, despite the strong erosive tendency north of Costa da Caparica. The erosional process north of Costa da Caparica was aggravated in the second quarter of the 20th century due to sand deficit in relation to intensive dredging; in response, a seawall and a groyne field were built (Appendix I – Duarte Santos et al., 2014a). More recently the beaches were the object of artificial nourishment (Appendix VI – Duarte Santos et al., 2014a). Coastal retreat rates reported here would have been higher in the absence of beach nourishment.

The geomorphological settings of cells 5 and 4 are broadly homothetic, but the intensity and spatial distribution of erosion/accretion rates are different. In cell 5, the northern tip of the Troia sand spit presents the largest accretion trend ($+8.18 \pm 0.2\,\mathrm{m\,year^{-1}}$). The erosion maximum, located immediately to the south of the accretion area, is $-1.09 \pm 0.2\,\mathrm{m\,year^{-1}}$ (Fig. 5). In this sector, mean coastline rates of change reveals the highest accretional trend for all sediment cells in mainland Portugal ($+0.45 \pm 0.01\,\mathrm{m\,year^{-1}}$).

Cell 6, which presents geomorphological similarities with cell 3, shows both the maximum erosive trend ($-0.95 \pm 0.2\,\mathrm{m\,year^{-1}}$) and the maximum accretion trend ($+1.32 \pm 0.2\,\mathrm{m\,year^{-1}}$) close to Vila Nova de Milfontes. The mean rate of change in beach–dune systems is $+0.28 \pm 0.06\,\mathrm{m\,year^{-1}}$, translating into an overall accretional trend of the few beach–dune systems' stretches (Fig. 5).

In the western south-facing Algarve coast, cell 7 shows a maximum erosion rate of $-0.66 \pm 0.2\,\mathrm{m\,year^{-1}}$ at Armação de Pera bay and a maximum rate of accretion of

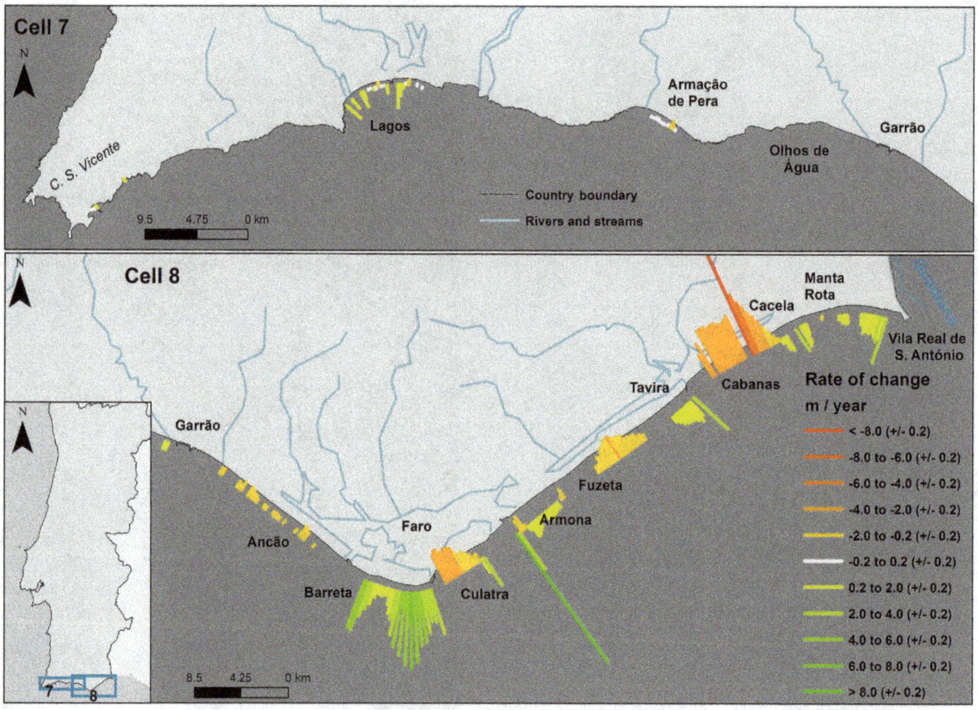

Figure 6. Long-term coastline rates of change in metres per year for sediment cells 7 and 8.

$+1.96 \pm 0.2\,\text{m year}^{-1}$ at Lagos bay. The mean rate of change is $+0.28 \pm 0.03\,\text{m year}^{-1}$, revealing an overall accretional trend (Fig. 6).

Cell 8 presents a complex evolution pattern, with sections in erosion alternating with sections in accretion. This pattern is mostly related with the morphological evolution of Ria Formosa tidal inlets, determined by either natural or artificial causes (Fig. 7). The maximum and minimum values should be interpreted with caution because they represent localized effects of inlet-barrier tip morphological readjustments. Moreover, retention effects induced by jetties also contributed to high values of change (e.g. Faro and Guadiana inlets). The erosion maximum of $-8.96 \pm 0.2\,\text{m year}^{-1}$ is observed at the western tip of the Cacela barrier and the accretion maximum of $+12.99 \pm 0.2\,\text{m year}^{-1}$ at the eastern boundary of the Armona inlet (Fig. 6). The mean rate of change for this section falls within the uncertainty range ($+0.01 \pm 0.01\,\text{m year}^{-1}$), which indicates an overall balanced coastal sediment budget. Additionally, Ferreira et al. (2016) highlight a relevant retreat of the cliffed coast between Olhos de Àgua and Ancão, particularly in the decades of the 1970s to 1990s, induced by the construction of the Vilamoura jetties. The same authors also state that it is presently controlled by artificial nourishment actions.

Although a thorough review of the many Portuguese coastline (and shoreline) evolution studies is beyond the scope of this analysis, some results reported by other authors for areas prone to erosion are briefly summarized. However, caution is needed when comparing the results obtained in the scope of the present work with those presented in the literature, as they were obtained using different methodologies and using distinct time frames. For 1958–2007, SENER (2012) reported a maximum evolution rate of $-4.4\,\text{m year}^{-1}$ at Esmoriz, $-5.6\,\text{m year}^{-1}$ at Ovar, $+5.7\,\text{m year}^{-1}$ at S. Jacinto, $-5.8\,\text{m year}^{-1}$ near Costa Nova and -3.5 near Praia de Mira. Freire (1986) mentions a maximum of $-12.5\,\text{m year}^{-1}$ for the years 1958–1966 at Cova do Vapor/Costa da Caparica and Pinto et al. (2007) mention $-3.3\,\text{m year}^{-1}$ between 1999 and 2007. Ferreira and Matias (2013) acknowledge an important retreat of more than $3\,\text{m year}^{-1}$ in the western part of Culatra Island, while the eastern half exhibited accretion, between 1958 and 1976. Ferreira et al. (2016) also mention that Ria Formosas' system evolution is dominated by tidal inlet dynamics, which has directly and indirectly affected the entire system, including barriers' erosion/accretion. In general, at the areas where evolutions trends are reported by other authors, the magnitudes are similar to the ones obtained in this study.

Figure 8 summarizes results of long-term coastline change assessment at a country scale. Portuguese mainland coastal territory occupied by sandy beach–dune systems, which represents 40 % of the total extent, depict an overall erosional tendency of $-0.24 \pm 0.01\,\text{m year}^{-1}$. This trend results from high erosion values located at specific coastal stretches. In fact, only 46 % of all transects exhibit erosion, with an average trend of $-1.43\,\text{m year}^{-1}$. On the other

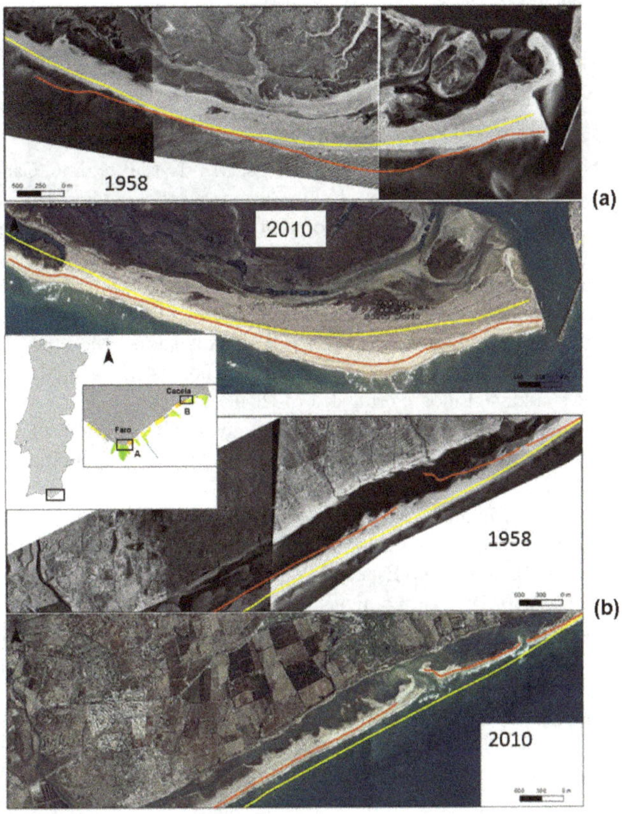

Figure 7. Detail of coastline change between 1958 (yellow) and 2010 (red) at (**a**) Faro-Olhão inlet and (**b**) Cacela inlet, both at the Ria Formosa barrier island system.

hand, 54 % of all analysed transects show values with either stable/within-uncertainty tendency or an accretion tendency ($+1.2\,\mathrm{m\,year^{-1}}$ at 34 %). Although beach erosion is dominant in average, this evolution is variable in signal and magnitude along and within each of the eight sediment cells (Fig. 8).

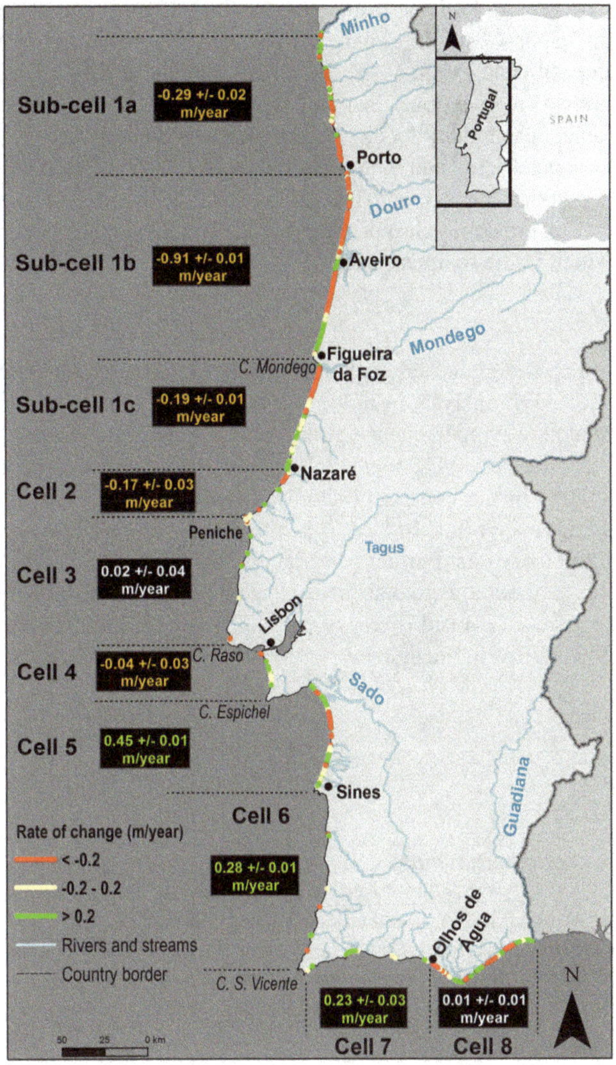

Figure 8. Rate-of-change map for the mainland Portuguese sandy beach–dune system coastline. Represented in red are sandy coastal stretches with $R < -0.2\,\mathrm{m\,year^{-1}}$, in green are sections in accretion ($R > +0.2\,\mathrm{m\,year^{-1}}$), and in light yellow are sections with $-0.2 < R < +0.2\,\mathrm{m\,year^{-1}}$. Boxed values refer to the mean coastline rate of change within each sediment cell (orange – erosion; green – accretion; white – within uncertainty measures).

5 Conclusions

This work characterizes coastline evolution of low-lying sandy coasts using a methodology that (1) is consistent from one coastal region to another and (2) is independent of short-term change, as it quantifies change using a single robust coastline indicator (foredune toe) adequate for applications at a national scale. Furthermore, a complete dataset derived with this methodology and applied to all sandy beach–dune systems in mainland Portugal is openly accessible at https://doi.pangaea.de/10.1594/PANGAEA.859136.

The dataset comprises two coastline sets (1958 and 2010); long-term (52 years) change rates estimated every 250 m; and minimum, maximum and mean coastline evolution rates for each considered sediment cell and for all cells. This dataset is the first systematic, national-scale and consistent long-term coastline evolution study of Portuguese mainland low-lying sandy coast. These data can be used in other studies and thus represent an important tool to support both coastal experts and end users.

Coastline evolution results showed an overall erosional tendency for the Portuguese mainland low-lying sandy coast, presenting a mean rate of $-0.24 \pm 0.01\,\mathrm{m\,year^{-1}}$. Despite this result, sandy beach–dune systems display a variable evolution, both in signal and intensity, along and within the eight coastal sediment cells of the studied coast. The most concerning beach erosion issues are to be found in cells 1 and 4, particularly on the coastal stretches of Espinho–Torreira,

Costa Nova–Praia de Mira, Cova da Gala–Leirosa and Cova do Vapor–Costa da Caparica. Cells standing out due to the intensity and extent of the erosional behaviour exhibit major human interventions, many of which originated and maintained a sediment deficit. In contrast, cells 5 and 6 have experienced less intervention and show stable or moderately accretional behaviour. Thus, sandy beach front areas where coastal erosion is chronic are associated with human-induced deficit in the sediment budget.

Long-term coastal evolution studies should also integrate short-term analysis in order to assess acceleration patterns related to climate change effects and also evaluate management measures. The methodology proposed here can also be applied in the context of short-term analysis (< 30 years), using smaller time frames. This integration will be addressed in future works. Further investigation is necessary to extend this methodology to other coastal geomorphological contexts (e.g. beach backed by rocky coasts), which will also be the objective of future studies.

Author contributions. C. Ponte Lira, A. N. Silva, R. Taborda and C. F. Andrade developed the methodology. C. Ponte Lira and A. N. Silva tested, validated and applied the methodology to all Portuguese mainland sand–dune systems. C. Ponte Lira wrote the manuscript with contributions from all co-authors.

Acknowledgements. The first author is supported by an FCT fellowship #SFRH/BPD/81800/2011. We acknowledge project CISML (*Criação e implementação de um sistema de monitorização no litoral abrangido pela área de jurisdição da Administração da Região Hidrográfica do Tejo*) (grant no. QREN – POVT: Operation POVT-12-0233-FCOES-000034, Portuguese Environmental Agengy – APA, I.P./Administração da Região Hidrográfica do Tejo) for contributing to making this study possible. The publication was supported by project FCT UID/GEO/50019/2013 – Instituto Dom Luiz. We thank Eng. A. Mota Lopes (APA) for the support in coastline extraction procedures and P. Brito and P. Terrinha (IPMA) for the opportunity to contribute to the Portuguese mainland coastline version submitted to the EMODnet database on the coastal behaviour area.

Edited by: H. Grobe

References

Abecassis, F., Castanho, J., and Matias, M.: Coastal regime. Carriage of material by sweel and currents. Model studies and in situ observations. influence of port structures. Coastal defence works, Breakwaters, Memória no. 362, LNEC, Lisbon, Portugal, 30 pp., 1970.

Andrade, C. and Freitas, M. C.: Coastal Zones, in: Climate Change in Portugal. Scenarios, Impacts and Adaptation Measures – SIAM project, edited by: Santos F. D., Forbes, K., and Moita, R., Gradiva, Liboa, Portugal, 173–2019, 2002.

Barceló, J. P.: Experimental study of the hydraulic behaviour of inclined groin systems, Memória no. 394, LNEC, Lisbon, Portugal, 20 pp., 1971.

Bettencourt, P. and Ângelo, C.: Faixa costeira Centro Oeste (Espinho – Nazaré): enquadramento geomorfológico evolução recente, Geonovas, no Especial 1 (A Geologia e o Ambiente), Lisboa, 7–30, 1992 (in Portuguese).

Boak, E. H. and Turner, I. L.: Coastline Definition and Detection: A Review, J. Coastal. Res., 214, 6880–703, doi:10.2112/03-0071.1, 2005.

Carapuço, M. M., Taborda, R., Silveira, T. M., Psuty, N. P., Andrade, C., and Freitas, M. C.: Coastal geoindicators: Towards the establishment of a common framework for sandy coastal environments, Earth-Sci. Rev., 154, 183–190, doi:10.1016/j.earscirev.2016.01.002, 2016.

Carter, R. W. G. and Woodroffe, C. D.: Coastal evolution: an introduction, in: Coastal Evolution: Late Quaternary Coastline Morphodynamics, edited by: Carter, R. W. G. and Woodroffe, C. D., Cambridge University Press, United Kingdom, 523 pp., 1994.

Castanho, J., Gomes, N., Carvalho, J., Vera-Cruz, D., Araújo, O., Teixeira, A., and Weinholtz, M.: Means of controlling littoral drift to protect beaches, dunes, estuaries and harbour entrances, Establishment of artificial beaches, Memória no 448, LNEC, 26 pp., 1974.

Catalão, J., Catita, C., Miranda, J., and Dias, J.: Photogrammetric analysis of the coastal erosion in the Algarve (Portugal), Géomorphologie Reli. Process. Environ., 2, 119–126, 2002.

Cenci, L., Disperati, L., Sousa, L. P., Phillips, M.m and Alves, F. L.: Geomatics for Integrated Coastal Zone Management: multitemporal coastline analysis and future regional perspective for the Portuguese Central Region, J. Coastal. Res., 65, 1349–1354, doi:10.2112/SI65-228.1, 2013.

Coyne, M. A., Fletcher, C. H. and Richmond, B. M.: Mapping coastal erosion hazards in Hawaii: observations and errors, J. Coastal. Res., SI 28, 171–184, 1999.

Del Río, L. and Gracia, F. J.: Error determination in the photogrammetric assessment of coastline changes, Nat. Hazards, 65, 2385–2397, doi:10.1007/s11069-012-0407-y, 2013.

Duarte Santos, F., Mota Lopes, A., Moniz, G., Ramos, L., and Taborda, R.: Gestão da Zona Costeira: O Desafio da Mudança – Relatório do Grupo do Litoral, Report delivered to the Portuguese Government in the Littoral Group framework, 242 pp.,

available at: http://www.apambiente.pt/_zdata/DESTAQUES/2015/GTL_RelatorioFinal_20150416.pdf (last access: December 2015), 2014a (in Portuguese).
Duarte Santos, F., Mota Lopes, A., Moniz, G, Ramos, L., and Taborda, R.: Gestão da Zona Costeira: O Desafio da Mudança – Sumário Executivo e Recomendações, Relatório do Grupo do Litoral, Report delivered to the Portuguese Government in the Littoral Group framework, 21 pp., available at: http://www.apambiente.pt/_zdata/DESTAQUES/2015/GTL_SumarioExecutivo_20141219.pdf (last access: December 2015), 2014b (in Portuguese).
EMODnet Database – European Marine Observation and Data Network, European Commission Directorate-General for Maritime Affairs and Fisheries (DG MARE), Marine Knowledge 2020 strategy, available at: http://www.emodnet.eu/ (last acess: May 2016), 2009.
European Commission: Living with coastal erosion in Europe – Sediment and space for sustainability, Luxembourg office for official publications of the European Commission, available at: http://www.curosion.org/reports-online/reports.html (last acess: September 2015), 2004.
EEA: Climate change, impacts and vulnerability in Europe 2012 – An indicator report. EEA Report, No 12/2012, Copenhagen, 300 pp., doi:10.2800/66071, 2012.
Ferreira, O.: Dune Erosion and Coastline Retreat between Aveiro and Cape Mondego (Portugal). Prediction of future evolution, in: Proceedings of the international coastal congress, edited by: Sterr, H., Hoststede, J., and Plag, H. P., Kiel, 187–200, 1992.
Ferreira, O. and Matias, A.: Portugal, in: Coastal Erosion and Protection in Europe, edited by: Williams, A. and Pranzini, E., Routledge, 457 pp., doi:10.4324/9780203128558, 2013.
Ferreira, O., Garcia, T., Matias, A., Taborda, R., and Dias, J. M. A.: An integrated method for the determination of set-back lines for coastal erosion hazards on sandy shores, Cont. Shelf Res., 26, 1030–1044, doi:10.1016/j.csr.2005.12.016, 2006.
Ferreira, O., Matias, A., and Pacheco, A.: The East Coast of Algarve: a Barrier Island Dominated Coast, Thalass, An Int. J. Mar. Sci., 1–11, doi:10.1007/s41208-016-0010-1, 2016.
Fletcher, C., Rooney, J., Barbee, M., Lim, S., and Richmond, B. M.: Mapping coastline change using digital ortophotogrammetry on Maui, Hawaii, J. Coastal. Res., SI 38, 106–124, 2003.
Fletcher, C. H., Romine, B. M., Genz, A. S., Barbee, M. M., Dyer, M., Anderson, T. R., Lim, S. C., Vitousek, S., Bochicchio, C., and Richmond, B. M.: National assessment of coastline change – Historical coastline change in the Hawaiian Islands, U.S. Geological Survey Open-File Report 2011-1051, 55 pp., available at: http://pubs.usgs.gov/of/2011/1051 (last acess: September 2015), 2012.
Freire, M. E. F.: A Planície Litoral entre a Trafaria e a Lagoa de Albufeira, MS Thesis, 204 pp., 1986 (in Portuguese).
IGeo: orthophotomaps: available at: http://www.igeo.pt (last access: May 2015), 2010.
Komar, P. D., Diaz-Mendez, G. M., and Marra, J. J.: Stability of the New River Spit, and the position of Oregon's beach-zone line, J. Coastal. Res., 17, 625–635, 2001.
Oliveira, S.: Evolução Recente da Linha de Costa no Troço Costeiro Forte Novo – Garrão (Algarve), Ms thesis, University of Lisbon, Portugal, 183 pp., available at: http://enggeoespacial.fc.ul.pt/ficheiros/teses/tese_sergio_oliveira.pdf (last acess: October 2015), 2005 (in Portuguese).
Phillips, M. R. and Jones, A. L.: Erosion and tourism infrastructure in the coastal zone: Problems, consequences and management, Tour. Manag., 27, 517–524, doi:10.1016/j.tourman.2005.10.019, 2006.
Pinto, C. and Teixeira, S. B.: Beach Rotation and Coastal Erosion in Armação de Pera Bay. The Example of Salgados – Galé Sector (Algarve – Portugal), International Conference on Coastal Conservation and Management (ICCCM) in the Atlantic and Mediterranean, Tavira, Portugal, 331–334, 2005.
Pinto, C. A., Taborda, R., and Andrade, C.: Evolução recente da linha de costa no troço Cova do Vapor – S. João da Caparica, in: Actas das Jornadas Portuguesas de Engenharia Costeira e Portuária, Lisboa, Portugal, 51–52, 2007 (in Portuguese).
Psuty, N. P., Duffy, M., Pace, J. F., Skidds, D. E., and Silveira, T. M.: Northeast Coastal and Barrier Network geomorphological monitoring protocol: part I – ocean coastline position, Natural Resource Report NPS/NCBN/NRR – 2010/185, National Park Service, Fort Collins, Colorado, 2010.
Rebêlo, L., Ferraz, M., Brito, P., and Terrinha, P.: Quantification of sediments accumulated in the NW sector of Tróia Peninsula (Portugal) between 1928 and 1995, J. Coast. Conserv., 16, 261–268, doi:10.1007/s11852-011-0171-2, 2011.
Rocchini, D. and Di Rita, A.: Relief effects on aerial photos geometric correction, Appl. Geogr., 252, 159–168, doi:10.1016/j.apgeog.2005.03.002, 2005.
SENER: Elaboração dos Estudos da Evolução e da Dinâmica Costeira e Estuarina, de Mobilidade e Navegabilidade na Laguna e de Reforço de Margens pela Recuperação de Diques e Motas com Vista à Prevenção de Riscos, Technical Report, 648 pp., available at: https://cld.pt/dl/download/177dcc91-ae9b-4a2d-b9f9-df1e39b0ca92/Estudo1_Relat1.pdf (last acess: May 2016), 2012 (in Portuguese).
Shoshany, M., Golik, A., Degani, A., Lavee, H., and Gvirtzman. G.: New evidence for sand transport direction along the coastline of Israel, J. Coastal. Res., 12, 311–325, 1996.
Silva, A., Lira, C., Andrade, C., Taborda, R., and Freitas, M. C.: Análise da evolução da linha de costa no litoral baixo arenoso nos últimos 50 anos. In: Criação e implementação de um sistema de monitorização no litoral abrangido pela área de jurisdição da Administração da Região Hidrográfica do Tejo, Technical Report. Entregável 1.2.2.2. a., available at: http://www.apambiente.pt/_zdata/Politicas/Agua/Ordenamento/SistemaMonitorizacaoLitoral/E_1.2.2.2.a_Linha20de20costa.pdf (last acess: June 2015), 2013 (in Portuguese).
Taveira Pinto, F., Pais Barbosa, J., and Veloso Gomes, F.: Coastline Evolution at Esmoriz-Furadouro Stretch (Portugal), J. Coastal. Res., SI 56, 673–677, 2009.
Thieler, E. R., Himmelstoss, E. A., Zichichi, J. L., and Ergul, A.: Digital Coastline Analysis System (DSAS) version 4.0 – An ArcGIS extension for calculating coastline change, U.S. Geological Survey Open-File Report 2008-1278, available at: http://woodshole.er.usgs.gov/project-pages/DSAS/version4/ (last acess: September 2015), 2009.
USGS – Coastal and Marine Geology Program: National Assessment of Coastline Change Project, available at: http://coastal.er.usgs.gov/coastline-change/ (last access: September 2015), 2004.

Veloso-Gomes, F., Costa, J., Rodrigues, A., Taveira-Pinto, F., Pais-Barbosa, J. and Neves, L.: Costa da Caparica Artificial Sand Nourishment and Coastal Dynamics, J. Coastal. Res., SI 56, 678–682, 2009.

Verhoeven, G., Doneus, M., Briese, C., and Vermeulen, F.: Mapping by matching: a computer vision-based approach to fast and accurate georeferencing of archaeological aerial photographs, J. Archaeol. Sci., 39, 2060–2070, doi:10.1016/j.jas.2012.02.022, 2012.

Wan, Y., Wang, D., Xiao, J., Lai, X., and Xu, J.: Automatic determination of seamlines for aerial image mosaicking based on vector roads alone, ISPRS J. Photogramm. Remote Sens., 76, 1–10, doi:10.1016/j.isprsjprs.2012.11.002, 2013.

2

Present-day surface deformation of the Alpine region inferred from geodetic techniques

Laura Sánchez[1], **Christof Völksen**[2], **Alexandr Sokolov**[1,2], **Herbert Arenz**[1], **and Florian Seitz**[1]

[1]Technische Universität München, Deutsches Geodätisches Forschungsinstitut (DGFI-TUM), Arcisstr. 21, 80333 Munich, Germany

[2]Bayerische Akademie der Wissenschaften, Erdmessung und Glaziologie, Alfons-Goppel-Str. 11, 80539 Munich, Germany

Correspondence: Laura Sánchez (lm.sanchez@tum.de)

Abstract. We provide a present-day surface-kinematics model for the Alpine region and surroundings based on a high-level data analysis of about 300 geodetic stations continuously operating over more than 12 years. This model includes a deformation model, a continuous surface-kinematic (velocity) field, and a strain field consistently assessed for the entire Alpine mountain belt. Special care is given to the use of the newest Global Navigation Satellite Systems (GNSS) processing standards to determine high-precision 3-D station coordinates. The coordinate solution refers to the reference frame IGb08, epoch 2010.0. The mean precision of the station positions at the reference epoch is ± 1.1 mm in N and E and ± 2.3 mm in height. The mean precision of the station velocities is $\pm 0.2\,\mathrm{mm\,a^{-1}}$ in N and E and $\pm 0.4\,\mathrm{mm\,a^{-1}}$ in height. The deformation model is derived from the point-wise station velocities using a geodetic least-squares collocation (LSC) approach with empirically determined covariance functions. According to our results, no significant horizontal deformation is detected in the Western Alps, while across the Southern and Eastern Alps the deformation vectors describe a progressive eastward rotation towards Pannonia. This kinematic pattern also makes evident an increasing magnitude of the deformation from $0.1\,\mathrm{mm\,a^{-1}}$ in the western part of Switzerland up to about $1.3\,\mathrm{mm\,a^{-1}}$ in the Austrian Alps. The largest shortening is observed along the southern front of the Eastern Alps (in the northern area of the Venetian-Friuli Basin) and in the northern part of the Apennine Peninsula, where rates reach 2 and $3\,\mathrm{mm\,a^{-1}}$, respectively. The average accuracy of the horizontal deformation model is $\pm 0.2\,\mathrm{mm\,a^{-1}}$. Regarding the vertical kinematics, our results clearly show an ongoing average uplift rate of $1.8\,\mathrm{mm\,a^{-1}}$ of the entire mountain chain, with the exception of the southern part of the Western Alps, where no significant uplift (less than $0.5\,\mathrm{mm\,a^{-1}}$) is detected. The fastest uplift rates (more than $2\,\mathrm{mm\,a^{-1}}$) occur in the central area of the Western Alps, in the Swiss Alps, and in the Southern Alps in the boundary region between Switzerland, Austria, and Italy. The general uplift observed across the Alpine mountain chain decreases towards the outer regions to stable values between 0.0 and $0.5\,\mathrm{mm\,a^{-1}}$ and, in some cases, to subsidence like in the Liguro-Provençal and Vienna basins, where vertical rates of -0.8 and $-0.3\,\mathrm{mm\,a^{-1}}$ are observed, respectively. In the surrounding region, three regional subsidence regimes are identified: the Rhône-Bresse Graben with $-0.8\,\mathrm{mm\,a^{-1}}$, the Rhine Graben with $-1.3\,\mathrm{mm\,a^{-1}}$, and the Venetian-Friuli Basin with $-1.5\,\mathrm{mm\,a^{-1}}$. The estimated uncertainty of our vertical motion model across the Alpine mountain belt is about $\pm 0.3\,\mathrm{mm\,a^{-1}}$. The strain field inferred from the deformation model shows two main contrasting strain regimes: (i) shortening across the south-eastern front of the Alps and the northern part of the Dinarides and (ii) extension in the Apennines. The pattern of the principal strain axes indicates that the compression directions are more or less perpendicular to the thrust belt fronts, reaching maximum values of $20 \times 10^{-9}\,\mathrm{a^{-1}}$ in the Venetian-Friuli and Po basins. Across the Alpine mountain belt, we observe a slight dilatation regime in the Western Alps, which smoothly changes to a contraction regime in western Austria and southern Germany, reaching maximum shortening values of $6 \times 10^{-9}\,\mathrm{a^{-1}}$ in north-eastern Austria. The numerical results of this study are available at https://doi.pangaea.de/10.1594/PANGAEA.886889.

1 Introduction

The Alpine orogeny is the result of the continental collision of the African and European plates, more precisely of the push of the African Plate through the Adriatic promontory (also called the Adria or Apulian microplate; Channell et al., 1979) against Central Europe. A wide range of multidisciplinary studies based on geologic, geophysical, and geodetic data have demonstrated that the Alpine orogeny has been active from the Jurassic until the present (e.g.; Dewey et al., 1973, 1989; Channell and Horváth, 1976; Dercourt et al., 1986; Le Pichon et al., 1988; Mueller and Kahle, 1993; Ustaszewski et al., 2008; Handy et al., 2010, 2015). The tectonic evolution and present geodynamic framework have been primarily deduced from seismotectonic syntheses, and it is clear that the Alpine area and, in a more extended sense, the Mediterranean represent a variety of lithosphere blocks of different age, thickness, and rheology, leading to complex kinematic processes, including subduction, back-arc spreading, rifting and thrust, reverse and strike-slip faulting. These processes are superimposed on those phenomena associated with any orogeny; i.e. uplift, deformation, erosion, metamorphism, and foredeep basin changes. During the last three decades, the Global Navigation Satellite Systems (GNSS), like GPS (Global Positioning System) and GLONASS (GLObalnaja NAwigazionnaja Sputnikowaja Sistema), became a fundamental tool to observe, model, and understand present-day kinematic processes, as they provide precise geodetic constraints with a high spatial resolution allowing the improvement of geodynamic models, especially at diffuse plate boundaries like the African–Eurasian boundary.

The use of GPS to detect tectonic surface deformations in Europe started already in the 1990s (e.g. Fejes et al., 1993; Kahle et al., 1994; Chéry et al., 1995; Kaniuth et al., 1995, 1999; van Mierlo et al., 1996; Calais et al., 1998, 2000; Ferhat et al., 1998; Calais, 1999; Grenerczy et al., 2000; Sue et al., 2000). Given the complexity of the deformation zone in the southern part of the continent, scientists concentrated on selected areas to deploy observing stations and to develop local deformation models. Recent studies can be classified following the geographic subdivision of the Alps into the Western, Central, Eastern, and Southern Alps (Fig. 1):

- The Western Alps in south-eastern France, with the boundary region France–Italy and south-western Switzerland; e.g. Tesauro et al. (2006), Delacou et al. (2008), Larroque et al. (2009), Nguyen et al. (2016), and Nocquet et al. (2016).
- The Central and Eastern Alps in Austria and the adjacent regions of Switzerland, Liechtenstein, Germany, Italy, and Slovenia; e.g. Haslinger et al. (2006), Weber et al. (2006), Gosar et al. (2007), Sue et al. (2007), Brückl et al. (2010), and Brockmann et al. (2012).
- The junction zone of the Eastern Alps, the Dinarides, and the Pannonian Basin; e.g. Bada et al. (2007), Bus et al. (2009), Caporali et al. (2009), and Möller et al. (2011).
- the Southern Alps in northern Italy (also considered as a boundary zone for the computation of deformation models in the Adriatic region); e.g. Grenerczy et al. (2005), Grenerczy and Kenyeres (2006), D'Agostino et al. (2005, 2011), Devoti et al. (2008, 2011), Serpelloni et al. (2005, 2013), Cuffaro et al. (2010), and Métois et al. (2015).

These studies describe in general deformation models of regional scope and cover isolated segments of the Alpine mountain chain and its forelands. The objective of this work is the consistent determination of a continuous horizontal and vertical surface-kinematic field of the Alpine area that provides an integral picture of the ongoing deformation processes in the entire region. A network of about 300 continuously operating GNSS (CO-GNSS) stations with observations collected over 12.4 years is used for the precise determination of station positions and velocities. Based on these results, a continuous kinematic field is derived using a geodetic least-squares collocation (LSC) approach with empirically determined covariance functions. Main results are a deformation model, a continuous surface-kinematic (velocity) field, and a strain field consistently assessed for the entire Alpine mountain belt. The core contribution of this work is the homogeneous analysis of a large number of freely available data from CO-GNSS stations located in the Alpine region, using unified processing standards and a common reference frame for the complete time span covered by the observations. Special care is given to the application of the newest GNSS processing standards like absolute corrections to the GNSS antenna phase centre variations, troposphere delay estimation based on numerical weather models, and atmospheric and oceanic tide loading effects. To ensure consistency, we also use reprocessed GNSS satellite orbits referring to the IGS08/IGb08 reference frame (Rebischung et al., 2012) and including the most recent models adopted by the International GNSS Service (IGS) for orbit determination, such as the albedo model of Rodríguez-Solano et al. (2012), the satellite antenna thrust effect, and the a priori solar radiation pressure model switch-off (Steigenberger et al., 2014). The IGS08/IGb08 reference frame corresponds to a subset of GNSS stations of the ITRF2008 (International Terrestrial Reference Frame 2008; Altamimi et al., 2011), which are used by the IGS as fiducial points for the computation of the satellite ephemeris and the Earth orientation parameters (EOPs).

In this context, after a brief description of the geological and tectonic framework of the Alpine region, the third section of this paper provides details about the geographical dis-

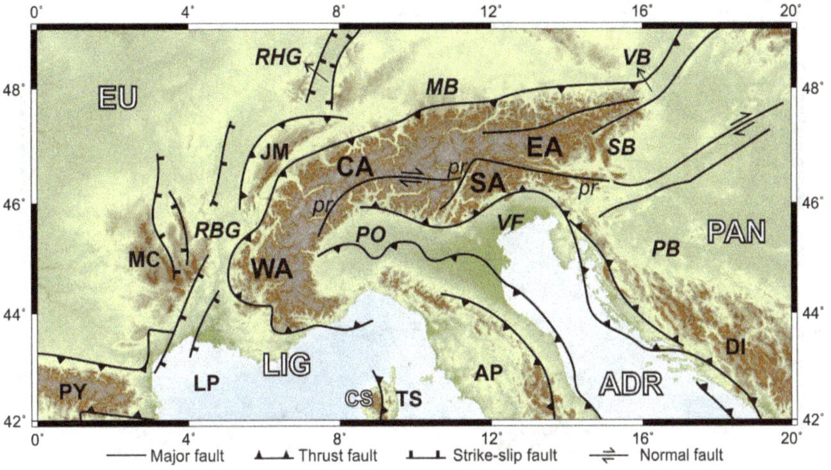

Figure 1. Simplified tectonic map of the Alps and surrounding areas showing the major tectonic features and faults, modified from Schmid et al. (2004) and Bousquet et al. (2012). Geographic subdivision of the Alps: WA – Western Alps, CA – Central Alps, EA – Eastern Alps, SA – Southern Alps. Tectonic plates/blocks: EU – Eurasia, ADR – Adria, LIG – Liguria, PAN – Pannonia, CS – Corso-Sardinia. Basins: LP – Liguro-Provençal, TS – Tyrrhenian, PO – Po, VF – Venetian-Friuli, MB – Molasse, PB – Pannonian, SB – Styrian, VB – Vienna. Graben: RBG – Rhône-Bresse, RHG – Rhine. Mountains and mountain belts: AP – Apennines, PY – Pyrenees, DI – Dinarides–Hellenides, JM – Jura, MC – Massif Central. Label *pr* identifies the Periadriatic Line. Topography and bathymetry after the ETOPO1 model (Amante and Eakins, 2009).

tribution of the GNSS stations, the availability of the GNSS observations, and the data sources. Afterwards, we present in Sect. 4 the GNSS data processing strategy, the geodetic datum realisation, the analysis of the position time series, and the computation of a cumulative (multi-year) solution for the estimation of precise station positions and velocities. Section 5 concentrates on the computation of the kinematic model: correlations between the station velocities and the tectonic settings are analysed, the modelling methodology is presented, and the vertical and horizontal deformation models are briefly described. Finally, the strain field computation is summarised in Sect. 6. The reliability of the results obtained in this study is validated by contrasting them with the deformation patterns and the displacement rates published in the existing literature. With this work, we are providing a present-day surface-kinematic model for the Alpine region based on a high-level GNSS data processing.

2 Geological and tectonic framework

We focus on the estimation of the present-day kinematics in the Alps and surrounding areas based on GNSS measurements gained during the last decade. Previous processes associated with the tectonic evolution and their effects are usually deduced from geophysical modelling, but they are beyond the scope of this work. The main tectonic blocks underlying the Alps and their forelands are the Eurasian Plate in the north and the west, the Adriatic and Ligurian microplates in the south, and the Pannonian fragment in the east (Fig. 1). The boundaries between these units are mainly inferred from (natural and controlled) seismic data and tomographic inversions; see details in Pfiffner et al. (1990), Waldhauser et al. (1998), Doglioni and Carminati (2002), Lippitsch et al. (2003), Kissling et al. (2006), and Brückl et al. (2007). The push of the African Plate against the Eurasian Plate causes an underthrusting motion beneath the Southern Alps as indicated, for instance, by the seismicity in the Friuli area. In addition, the Adriatic promontory rotates anticlockwise, indenting the Eastern Alps and Dinarides in the north (e.g. Nocquet and Calais, 2004; Handy et al., 2015) and subducting beneath the Hellenides in the south (Grenerczy et al., 2005; Bennett et al., 2008). The polarity of subduction changes from one area to the other: while the European lithosphere is subducted beneath the Alps, the African lithosphere is subducted beneath the Apennines and the Dinarides–Hellenides (Argnani, 2009). In this way, Alps, Apennines, and Dinarides–Hellenides are still active orogens, although at different rates (D'Agostino et al., 2005; Devoti et al., 2008; Cuffaro et al., 2010; Cheloni et al., 2014). Each subduction is associated with the usual vertical motions; i.e. subsidence in the foreland basin and uplift in the mountain belt. The continuous curved orogenic belt built by the Alps together with the Carpathians and the Dinarides–Hellenides bifurcates and encircles the Pannonian Basin system (Doglioni, 1987; Royden, 1993; Doglioni and Carminate, 2002; Horváth et al., 2006). South of the Alps are the Po and the Venetian-Friuli basins. As mentioned, these three basins – the Pannonian, Po, and Venetian-Friuli basins – are still active foreland basins; i.e. they continue subsiding and receiving sediments (e.g. Cuffaro et al., 2010; Serpelloni et al., 2006, 2013). The Po Basin is also part of the northern foreland basin of the Apennines (e.g. Laubscher, 1988;

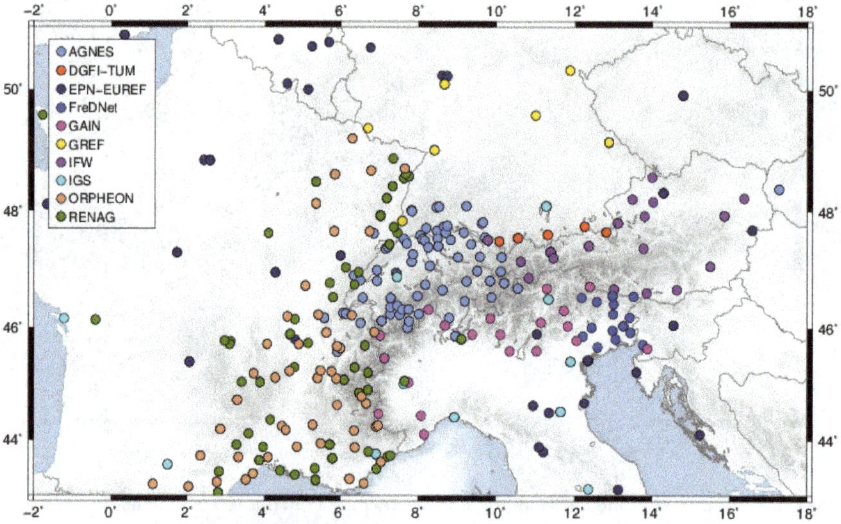

Figure 2. CO-GNSS stations processed for the estimation of the surface kinematics in the Alps. The stations of the contributing CO-GNSS arrays are colour coded. Topography and bathymetry after the ETOPO1 model (Amante and Eakins, 2009).

Argnani, 2009). In the west of the Apennines and south of the Western Alps, we find the Liguro-Provençal and the Tyrrhenian basins. They are separated by the Corso-Sardinia block (Serpelloni et al., 2006, 2013). The northern foreland basin of the Alps is the Molasse Basin (Ziegler, 1990; Kuhlemann and Kempf, 2002). It runs between the Jura Mountains and the Alps in the north-western part of Switzerland and continues along the Bavarian and Austrian Alps up to Vienna. In contrast to the Pannonian and Po basins, it is not active as a foreland basin; on the contrary, large parts of it are being uplifted and eroded (Becker, 1999). On the west, the Alps are boarded by the Rhône-Bresse Graben. In general, it can be stated that the deformation in the Alpine region is dominated by the anticlockwise motion of the Adria microplate, causing compression in the Eastern Alps, dextral shear in the Central Alps and a very slow deformation in the Western Alps.

3 Distribution of the CO-GNSS stations

The determination of the surface kinematics in the Alpine region is based on the observations collected at 306 CO-GNSS stations (Fig. 2). Stations located at distances up to some thousand kilometres from the Alps are included to improve the geometry of the network and to ensure a reliable geodetic datum realisation. Several sites are part of the global IGS network (Dow et al., 2009), the EUREF Permanent Network (EPN) (Bruyninx et al., 2012, 2013), and the Integriertes Geodätisches Referenznetz Deutschlands (GREF). EUREF stands for the Reference Frame Sub-Commission for Europe of the International Association of Geodesy (IAG). The CO-GNSS station distribution along the Alpine mountain belt is densified with more than 30 GAIN (Geodetic Alpine Integrated Network) stations. Five of these stations were installed and are maintained by DGFI-TUM (Seitz et al., 2014). The GAIN network was established in the frame of the ALPS-GPSQUAKENET project, which was funded by the European Union under the INTERREG IIIB Alpine Space Programme from 2004 to 2007 (Aoudia et al., 2007). Additional data from CO-GNSS sites in Austria were provided by the Space Research Institute (Institut für Weltraumforschung, Graz) of the Austrian Academy of Sciences (Haslinger and Strangle, 2006). Another valuable source of data for this study is the Friuli Regional Deformation Network (FreDNet) established in 2002 (Zuliani et al., 2002; Rossi et al., 2013). While FreDNet densifies the station distribution in the Eastern Alps, the Western Alps are very well covered by CO-GNSS sites of the RENAG (Réseau National GPS, France, Nocquet et al., 2016) and Orpheon networks. Orpheon is a geodetic network providing real time and augmentation services in France. The CO-GNSS network processed in this study does not include stations in Switzerland. However, to get a homogeneous data coverage in the Central Alps, we use IGb08-based station velocities of the Automated GNSS Network for Switzerland (AGNES) provided by Brockmann et al. (2012). Figure 2 shows the station distribution according to the data sources.

The GNSS data cover the time span from 1 January 2004 to 30 May 2016. However, the amount of the available GNSS observations varies over the years (Fig. 3). The period with the maximum number of active stations within the network is between 2012 and 2015, reaching nearly 180 CO-GNSS stations every day. The total number of the daily observation files processed in this study is about 0.57 million.

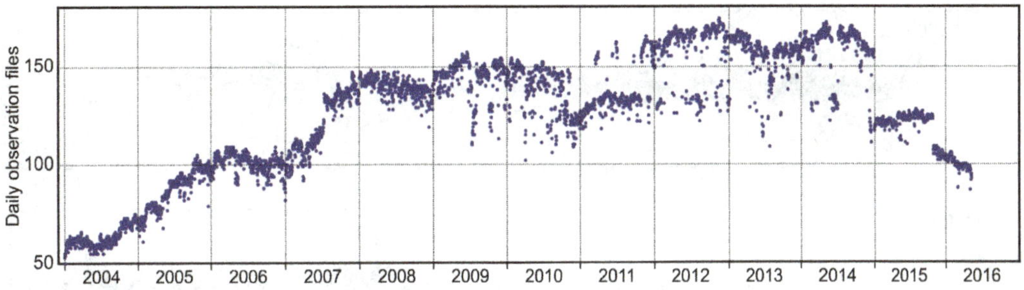

Figure 3. Amount of daily observation files processed in this study.

4 Analysis of GNSS data and determination of precise station positions and velocities

Effects of ongoing geodynamic processes may be inferred from crustal movements detected by the analysis of repeated (or continuous) geodetic observations. The basic idea is to determine (and compare) the geometry of the network formed by the geodetic stations at certain epochs. The changes in the network geometry over large time spans (e.g. years) are interpreted as a deformation of the network caused by geodynamic processes underlying the studied area. To ensure a high reliability in the determination of the network geometry, the geodetic measurements must be processed over the entire time span following strict standards and procedures. In this study, we therefore apply the same analysis strategy usually applied for the establishment of GNSS-based regional geodetic reference frames like SIRGAS (Sistema de Referencia Geocéntrico para las Américas, e.g. Sánchez et al., 2013; Sánchez and Drewes, 2016) or the EUREF Permanent Network (EPN) as a realisation of the European Terrestrial Reference System (ETRS); e.g. Völksen et al. (2009a, b). The data analysis is based on the conventions outlined by the International Earth Rotation and Reference Systems Service (IERS) for the determination of the ITRF (Petit and Luzum, 2010) and the GNSS-specific guidelines defined by the IGS for the precise processing of the global GNSS network (IGS, 2014).

In a first processing step, free weekly normal equations (NEQ) are computed. "Free" means in this context that the geometry of the GNSS network is fully consistent with the GNSS orbits, but the network's origin and orientation are not constrained yet. They will be constrained in an a posterior processing step, when the geodetic datum for the GNSS network is introduced. The free weekly NEQ covering the complete time span from January 2004 to May 2016 are combined to a multi-year solution with constant velocity for each station. This includes an iterative time series analysis to detect outliers and discontinuities in the weekly station positions. In the multi-year solution, the geodetic datum is realised with respect to the coordinates (positions and velocities) of selected reference stations. The station velocities represent the mean displacement per year of the CO-GNSS stations and they are the primary input data to model the surface kinematics. The following sections describe the key aspects of the GNSS data analysis performed in this study.

4.1 GNSS network weekly solutions

The processing of the daily GNSS observations is accomplished using the double-difference baseline approach and the least-squares adjustment implemented in the Bernese GNSS Software V5.2 (Dach et al., 2015). The main processing characteristics are as follows:

- GPS and GLONASS observations at a 30 s sampling rate are considered.
- The basic observation is the L3 ionosphere-free linear combination.
- The satellite orbits, satellite clock offsets and EOPs are fixed to the values generated by the IGS processing centre CODE (Center for Orbit Determination in Europe, Dach et al., 2017; Steigenberger et al., 2014; http://www.aiub.unibe.ch/download/REPRO_2013, last access: 20 August 2018). To ensure consistency, reprocessed (2004–2013) and final CODE (2014–1016) products based on the reference frame IGS08/IGb08 are used for the analysis.
- The antenna phase centre offsets and direction-dependent phase centre corrections for both satellite transmitting and ground receiving antennas are taken from the model igs08.atx (Schmid et al., 2016; ftp://www.igs.org/pub/station/general/pcv_archive/, last access: 20 August 2018).
- The tropospheric zenith delay is modelled using the global mapping function (GMF, Böhm et al., 2006). Here, the a priori zenith hydrostatic delay (ZPD) values are derived from gridded coefficients based on the global pressure temperature (GPT) model (Böhm et al., 2006) and are refined by computing zenith wet delays (ZWD) in a 1 h interval (also using GMF). In addition to model azimuthal asymmetries, horizontal gradient parameters are estimated using the model described by Chen and Herring (1997).

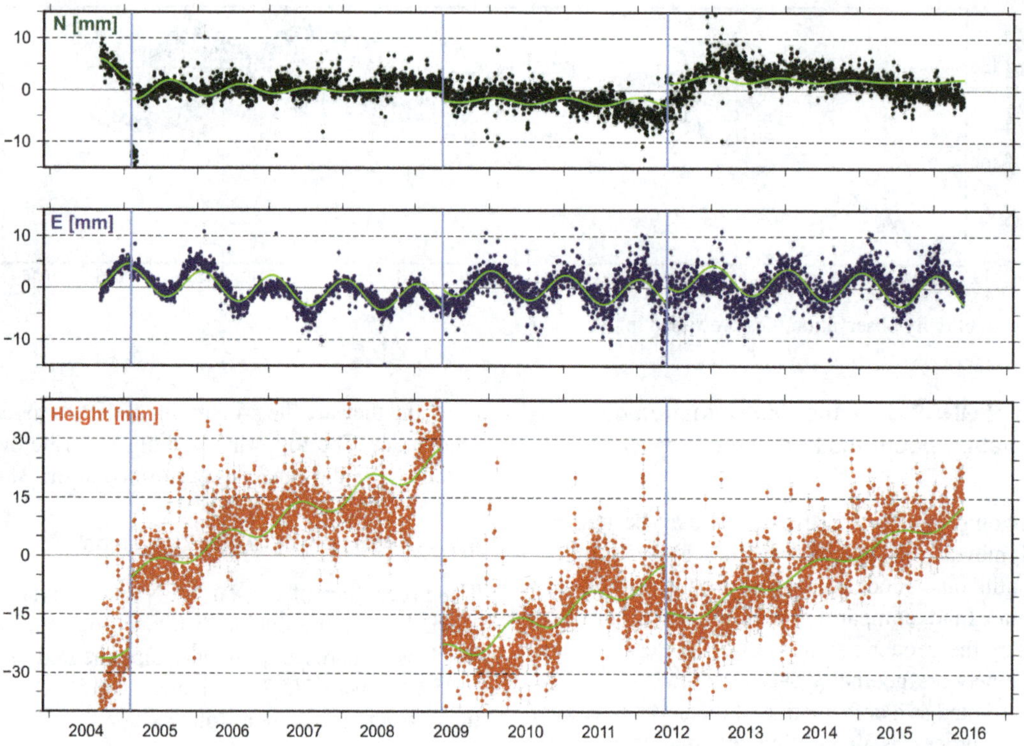

Figure 4. Position time series of the station BOLG located in Bologna, Italy. Discontinuities in 2005 and 2009 (indicated by blue vertical lines) are caused by the poor monument construction of the site. Freezing water in the outer tube incorporating an inner tube carrying the antenna raised the inner tube like a piston. In 2009 the inner tube was put in its original place (Bruni et al., 2014). The blue line in 2012 represents the discontinuity caused by seismic events that occurred on 20 May (M_W 6.0) and 29 May (M_W 5.8) 2012 in northern Italy (11.230° E, 44.890° N and 11.086° E, 44.851° N, respectively). Piecewise sinusoidal lines in light green represent a functional model approximating the seasonal motions detected at the station.

- Corrections for the solid Earth tide, permanent tide, and solid Earth pole tide are applied as described in Petit and Luzum (2010). The ocean tide loading is estimated with the FES2004 model (Letellier, 2004) and the atmospheric tide loading caused by the semidiurnal constituents S1 and S2 is estimated following the model of van Dam and Ray (2010). The coefficients for the ocean tide loading are provided by Machiel Simon Bos and Hans-Georg Scherneck at http://holt.oso.chalmers.se/loading/ (last access: 20 August 2018). The coefficients for the atmospheric tide loading are provided by Tonie van Dam at http://geophy.uni.lu/ggfc-atmosphere/tide-loading-calculator.html (last access: 20 August 2018).

- Non-tidal loadings such as atmospheric pressure, ocean bottom pressure, or surface hydrology are not reduced.

- The seven daily free normal equations corresponding to a week are combined to generate a free normal equation per week. In this combination, stations with large residuals in any daily normal equation (more than ±10 mm in the horizontal or more than ±15 mm in the vertical) are removed from the corresponding daily equation and the weekly combination is recomputed.

4.2 Determination of precise station positions and velocities

The determination of the station coordinates is performed within a multi-year solution. In a first step, all free weekly NEQ available from January 2004 to May 2016 are combined and solved using the programme ADDNEQ2 of the Bernese GNSS Software V5.2 (Dach et al., 2015). Based on this preliminary solution, in a second step, residual time series for each station are generated to identify outliers or discontinuities that may mislead the computation of the station velocities. In this study, we use as thresholds ±10 mm for the north and east components and ±20 mm for height (about 4-fold the mean RMS). Sporadically occurring outliers are removed from the respective weekly NEQ. If outliers reflect a discontinuity (i.e. they appear in successive NEQ), a new position is estimated for this station. Most of these discontinuities are caused by the replacement of antennas and receivers, either because of defects or to modernise the equipment. In some cases, the discontinuities are caused by local (poor construc-

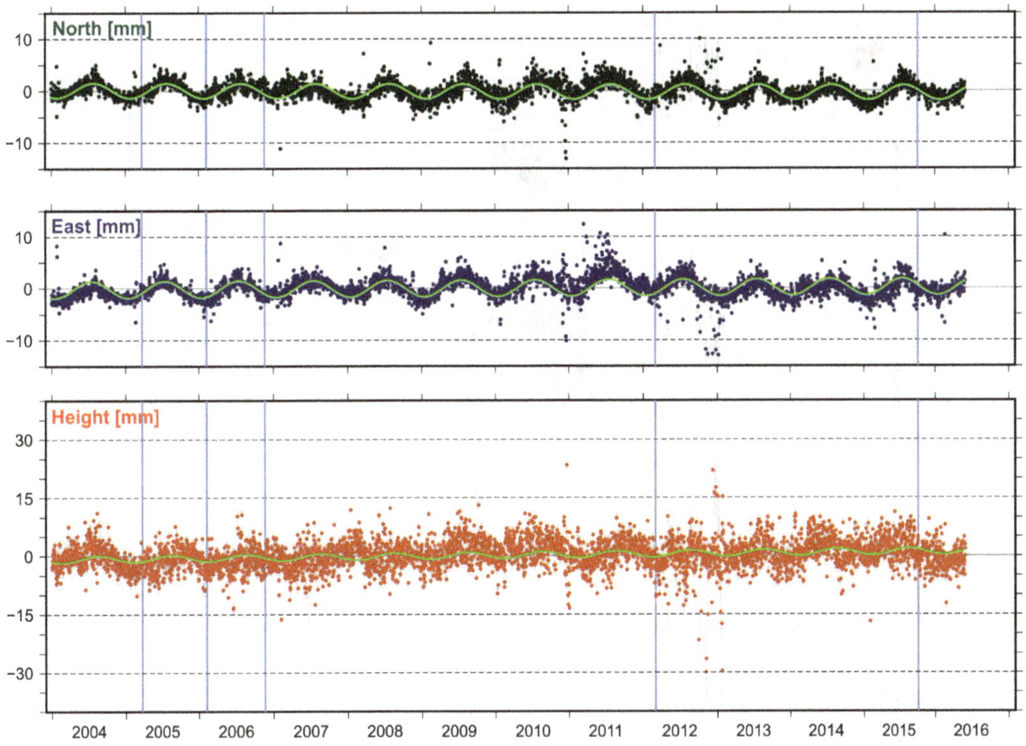

Figure 5. Position time series of the station KARL located in Karlsruhe, Germany. Annual signals are well observed in the north and east components. Equipment changes (indicated by the blue vertical lines) do not seem to affect the time series.

tion of the monument) or seismic effects and they may be followed by changes in the linear movement of the station; thus, new position and velocity values must be estimated for the affected stations (see an example in Fig. 4). Once outliers are removed and discontinuities are identified, a new multi-year solution is computed, new residual time series are generated, and the time series analysis is repeated to identify remaining outliers or discontinuities. This process is iteratively performed until no more outliers or discontinuities remain. The geodetic datum is realised by imposing two minimum constraint conditions with respect to the IGb08 coordinates of the reference stations. The no-net-translation (NNT) condition ensures that the origin $[X = Y = Z = 0]$ of our network coincides with the origin of the network built by the reference stations. The no-net-rotation (NNR) condition aligns the orientation of our network to the orientation defined by the network built by the reference stations. In this study, we selected six IGb08 sites (WTZR, ZIMM, GRAS, GRAZ, MEDI, LROC; Fig. 7) as reference stations (see Rebischung et al., 2012; ftp://igs-rf.ign.fr/pub/IGb08/IGb08.snx, last access: 20 August 2018). The main selection criteria is based on the homogenous distribution of the reference stations across the CO-GNSS network, rare incidents of equipment changes, and a nearly complete coverage of the time series from January 2004 to May 2016.

In general, the station position time series describe clear seasonal motions (Fig. 5), which are assumed to be a consequence of non-modelled geophysical loading or local effects superimposed with GNSS inherent systematic errors (e.g. Blewitt et al., 2001; Collilieux et al., 2010, 2012; Drewes et al., 2013; King et al., 2008; Ray et al., 2008). Therefore, to increase the reliability of the constant velocities required for modelling the surface kinematics, stations with time series shorter than 2 years are excluded from further computations. The station position time series of the entire network present an averaged RMS value of about ± 1 mm in the horizontal component and ± 2.5 mm in the height component (Fig. 6). The final multi-year solution for the CO-GNSS network covering the Alpine region refers to the IGb08, epoch 2010.0. While the averaged RMS precision for the station positions at the reference epoch is ± 1.1 mm in N and E and ± 2.3 mm in height, the averaged RMS precision for the station velocities is $\pm 0.2\,\mathrm{mm\,a^{-1}}$ in N and E and $\pm 0.4\,\mathrm{mm\,a^{-1}}$ in height. Figures 7 and 8 show the horizontal station velocities (after removing the Eurasian Plate motion) and the vertical station velocities, respectively.

4.3 Velocity solution

According to the results presented in Fig. 7, we observe in the Alps–Pannonian–Dinarides junction zone horizontal velocities in the range 1–4 $\mathrm{mm\,a^{-1}}$ relative to the stable part of the Eurasian Plate. These large relative velocities are controlled by the north-eastward motion of the Adria microplate

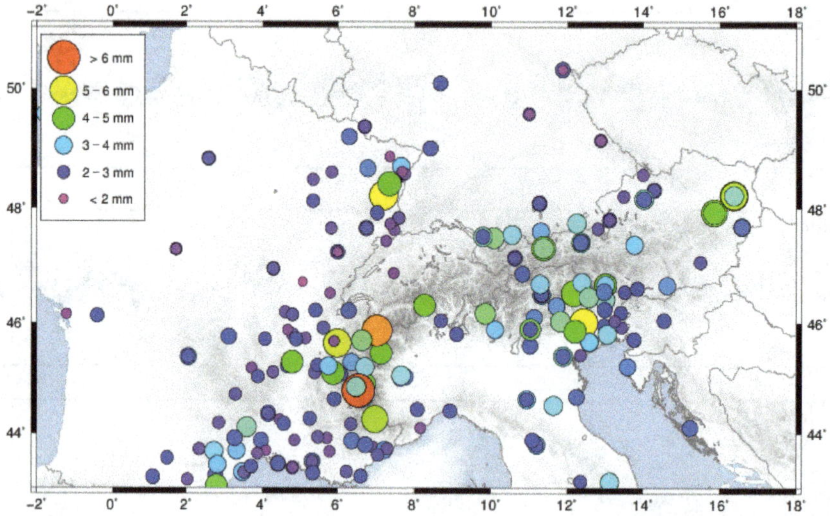

Figure 6. Mean RMS values of the station position time series for the period January 2004 to May 2016.

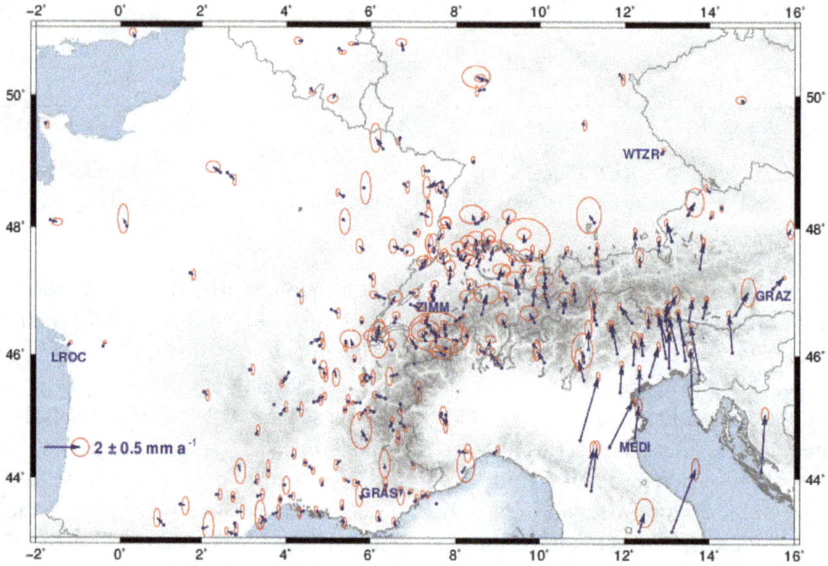

Figure 7. Horizontal station velocities relative to the Eurasian Plate with 95 % error ellipses. Labels identify the reference stations used for the geodetic datum definition: GRAS (Caussols, France), GRAZ (Graz, Austria), LROC (La Rochelle, France), MEDI (Medicina, Italy), WTZR (Bad Kötzting, Germany), and ZIMM (Zimmerwald, Switzerland). Topography and bathymetry after the ETOPO1 model (Amante and Eakins, 2009).

towards the Southern and Central Alps and the Dinarides. Based on our few CO-GNSS sites in the Apennines, we see apparently an extension of 2–4 mm a^{-1} across these mountains and a 0.5–2 mm a^{-1} shortening across the southern front of the Eastern Alps. These findings are in agreement with the conclusions presented by Devoti et al. (2017) in a recent study. They infer a crustal extension rate of about 3 mm a^{-1} across the Apennine belt and a compression of about 2 mm a^{-1} towards the Adriatic foreland (see Devoti et al., 2017, Fig. 9, profiles A–B and C–D). A progressive eastward rotation of the velocities towards the Pannonian Basin is also detected. The relative velocities of the stations located in the Western Alps and surrounding areas like the Po Basin, the Rhône-Bresse Graben, and southern France are very small (0.1–0.4 mm a^{-1}), allowing us to infer a very slow horizontal deformation in this region. Nocquet et al. (2016) call this a "virtually zero horizontal velocity". As expected, stations situated in the stable part of the Eurasian Plate show velocities around zero; some exceptions reflect mostly local effects.

While the motion of the Eurasian Plate dominates the horizontal velocities, the vertical velocities (Fig. 8) mirror in

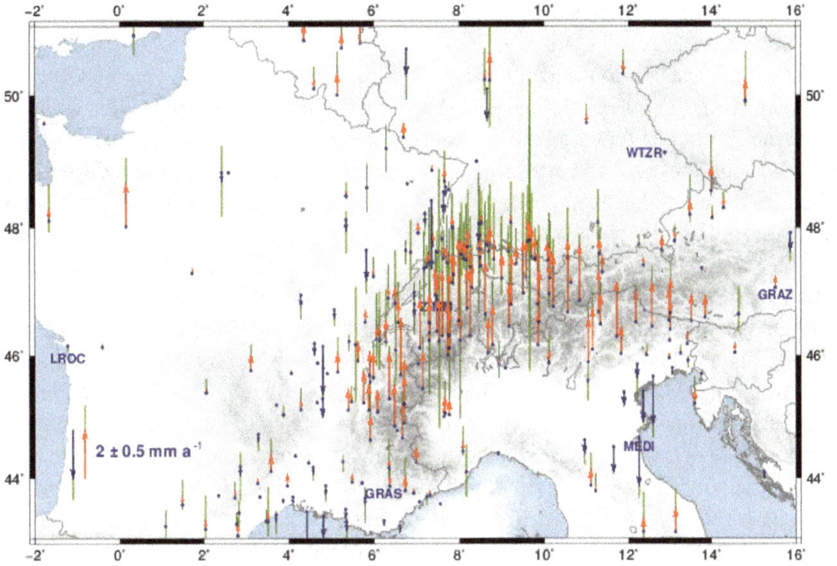

Figure 8. Vertical station velocities: red arrows indicate uplift, blue arrows represent subsidence, and green bars show 95 % confidence. Labels identify the reference stations used for the geodetic datum definition: GRAS (Caussols, France), GRAZ (Graz, Austria), LROC (La Rochelle, France), MEDI (Medicina, Italy), WTZR (Bad Kötzting, Germany), and ZIMM (Zimmerwald, Switzerland). Topography and bathymetry after the ETOPO1 model (Amante and Eakins, 2009).

general the orogenic processes in the Alpine region: uplift of the mountain chains and subsidence of the active foreland basins. A wide area of the Western Alps presents an uplift rate of approximately 2.2 mm a^{-1} decreasing westwards. In fact, stations located west of the Alpine foreland show zero vertical rates. The southern part of the Eastern Alps shows a slower vertical uplift rate of about 1.0–1.5 mm a^{-1}, which decreases towards the north. The CO-GNSS stations located in the northern margin of the Alps in Germany and Austria present uplift rates of less than 0.6 mm a^{-1} decreasing eastwards. Subsidence rates of about 1 mm a^{-1} are observed in the Rhine and Rhône-Bresse Grabens as well as in the Venetian-Friuli Basin. They increase southwards and reach up to 2 mm a^{-1} at the coasts in Marseille and Venice, respectively.

In general, it can be stated that the vertical velocities shown in Fig. 8 are a result of the combination of three main mechanisms: (i) uplift caused by the ongoing convergence of Eurasia and Africa; (ii) orogenic gravitational movement due to erosion, collapse, and associated isostatic compensation; and (iii) isostatic rebound caused by the melting of the Alpine ice shields since the last glacial maximum 22 000 years ago, i.e. glacial isostatic adjustment (GIA). The orogenic gravitational movement is much slower than the GIA; however, it is clear that the eroded rock masses are redistributed to the surroundings with lower elevation (delta valleys and sedimentary basins). This may cause a bedrock uplift and surface height lowering in the mountains as well as a bedrock subsidence and a regional surface height increase in the foreland basins. According to Persaud and Pfiffner (2004), Barletta et al. (2006), and Stocchi et al. (2005), the crustal uplift caused by GIA in the Alpine region varies from 0.06 to 0.20 mm a^{-1}. This assessment suggests that the first two mechanisms (plate convergence and orogenic gravitational movement) are the main contributors to the vertical rates obtained in this study.

5 Surface-kinematic modelling based on CO-GNSS network solutions

Based on the precise station velocities estimated previously, the objective of this section is to describe the surface-kinematic modelling performed within this study. We distinguish three main components: a deformation model, a continuous velocity field, and a strain field. A common trend motion is removed from the CO-GNSS station velocities to obtain the so-called residual velocities. The common trend motion represents a large-scale variation, while the residual velocities represent short-scale variations. The common trend motion in the horizontal component is usually well represented by the motion of the tectonic plate underlying the area of study, in this case the Eurasian Plate (see Sect. 5.1). The common trend motion in the vertical component may be assumed to be the average vertical movement of the area of study. In this work, we infer this mean vertical movement from the station velocities directly (see Sect. 5.2 and Appendix A). Once the trend motion is removed from the horizontal and vertical station velocities, the point-wise residual velocities are correlated with the existing tectonic structures (Fig. 1) and then they are interpolated to a regular grid. The interpolation of the residual velocities provides us with the deformation model (Sect. 5.2 and 5.3), which is the basis for the computation of the strain field (Sect. 6). The common trend motion (i.e. the Eurasian Plate motion and the average vertical motion) removed from the initial station velocities is restored to the deformation model to get a continuous velocity field (Sect. 5.4). Various geodetic and geophysical methods based on mathematical interpolation approaches or physical crust models may be applied to represent the continuous surface deformation from point-wise velocities. A frequently used interpolation method is the least-squares collocation (e.g. Moritz, 1973; Drewes, 1978); to include the physical properties of the deforming body, the finite element method is an appropriate tool (e.g. Meissl, 1981; Heidbach and Drewes, 2003). It has been demonstrated that for the sole representation of the Earth surface deformation, the results of both methods are nearly identical (e.g. Drewes and Heidbach, 2005). Therefore, we use here the least-squares collocation method.

5.1 Least-squares collocation (LSC) approach

The primary principle of LSC is the correlation of physical parameters (in this case, station velocities) between adjacent points. The observations are divided into a systematic linearised part (or trend) and two independent random parts: the signal and the observation error (or noise). The parameters describing the systematic component and the stochastically correlated signals are estimated by minimising the noise. The estimated parameters and signals together with a given spatial correlation allow the prediction (interpolation) of "fictitious measurements" at regions where no observations exist. In spatial applications, the correlation function is usually considered as a function dependent on the distance. The so-called correlation distance d should be in agreement with the spacing of the stations and the wavelength of the expected signal. If d is smaller than the mean station spacing, artefacts may appear in the interpolation. If d is too large, a smoothing similar to a filtering effect may occur. According to the station distribution in our network (Fig. 2), the initial maximum distance was set to 100 km for selecting the points of a collocation domain. If not enough points were found, the distance was enlarged until the necessary number of at least four points was available. The grid size for the interpolation was set to 25 km × 25 km. This grid size is chosen as appropriate since it corresponds to the mean station spacing in the most densely covered region within the network (i.e. the French and Swiss Alps). A larger grid size would average out the results and this filtering should be avoided. Appendix A provides a detailed description of the LSC formulation applied in this study.

According to the tectonic plate boundary model PB2002 (Bird, 2003), we assume that the Adria, Pannonia, and Liguria blocks constitute a deformation zone along the southern Eurasian Plate boundary. We do not consider small lithospheric blocks rotating with independent Euler vectors, but a continuous and viscous lithosphere deforming under certain kinematic boundary conditions as suggested, for instance, by Flesch et al. (2000); Vergnolle et al. (2007); or Copley (2008). However, as horizontal and vertical station motions are dominated by different mechanisms (see Sect. 4.3), the LSC procedure is applied separately: as a vector prediction for the horizontal component and as a scalar prediction for the height component. In the empirical procedure, the geocentric Cartesian station positions (X, Y, Z) and velocities (v_X, v_Y, v_Z) obtained within the multi-year solution (Sect. 4.2) are transformed together with their precision values into ellipsoidal coordinates (ϕ, λ, h) and $\left(v_\phi, v_\lambda, v_h\right)$. The Eurasian Plate motion is removed from horizontal velocities $\left(v_\phi, v_\lambda\right)$ of the stations. The corresponding rotation vector $\boldsymbol{\Omega}$ (260.74° ± 0.53° E, 55.14° ± 0.27° N, 0.2598 ± 0.0011° Ma^{-1}) is inferred from our CO-GNSS stations located on the stable part of the plate following the approach presented by Drewes (1982, 2009). Our values are very similar to the rotation vector derived from the ITRF2008 $\boldsymbol{\Omega}$ (261.15° ± 0.85° E, 55.23° ± 0.35° N, 0.2570 ± 0.0025° Ma^{-1}) (see Altamimi et al., 2012). The small differences are a consequence of the different station distribution (stations used for the estimation of $\boldsymbol{\Omega}$ in this study and in the ITRF2008 are not the same) and the different time span considered for the station velocity estimation (our study includes observations up to May 2016, while the ITRF2008 includes observations up to May 2009; Altamimi et al., 2011).

After removing the Eurasian Plate motion from the station horizontal velocities, the LSC interpolation is applied following the formulation described in Appendix A. In this procedure, outliers with discrepancies larger than 0.5 mm a^{-1} with respect to the velocities of the surrounding stations are considered to be affected by local effects and they are not included in the collocation.

5.2 Horizontal deformation model

The deformation model presented in Fig. 9 describes the surface kinematics of the Alpine region with respect to the Eurasian Plate. Two distinct kinematic patterns can clearly be seen: the Western Alps as well as the western and northern surrounding areas (Liguro-Provençal Basin, Massif Central mountains, Rhône-Bresse Graben, Jura Mountains, and Molasse Basin) show a small and statistically insignificant motion, indicating that these regions move north-eastward with the Eurasian Plate. This stationary pattern contrasts with the pronounced N- and NE-oriented movement predicted in the Southern and Eastern Alps as well as in the Apennine Peninsula, the Venetian-Friuli Basin, and the northern part of the Dinarides. It is also evident that the magnitude of deformation increases from 0.1 mm a^{-1} in the western part of Switzerland to the south and to the east, reaching 3.0 mm a^{-1} in the northern part of the Apennine Peninsula and about 1.5 mm a^{-1} in the Austrian Alps, respectively. The two kinematic patterns are apparently separated by the so-called Pe-

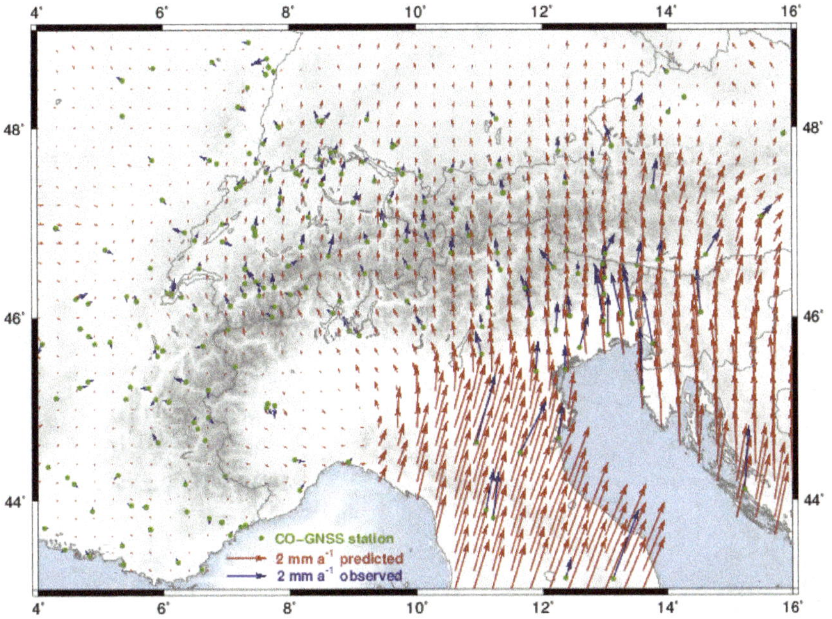

Figure 9. GNSS-inferred horizontal deformation model of the Alpine region. Red arrows represent the predicted surface deformation. Blue arrows represent the input residual velocities. Green dots show the CO-GNSS station distribution. Uncertainty ellipses are separately shown in Fig. 10. Topography and bathymetry after the ETOPO1 model (Amante and Eakins, 2009).

riadriatic Line, a fault system in EW direction along the boundary between the Southern Alps and the Adriatic region (around latitude 46° N, see Fig. 1). The magnitude of the deformation vectors across the westernmost part of the Po Basin and the south of the Western Alps is less than $0.2\,\mathrm{mm\,a^{-1}}$. The average residual motion with respect to the Western Alps is not larger than $0.1\,\mathrm{mm\,a^{-1}}$, indicating that presently there is no shortening in this region. These findings are confirmed by Nocquet et al. (2016), who suggest an upper bound of $0.3\,\mathrm{mm\,a^{-1}}$ (with 95 % confidence) for possible right-lateral strike slip motion in the western Alpine area.

Unlike the Western Alps, where the deformation vectors indicate a very small (insignificant) internal surface deformation, the Central Alps present an increasing north-oriented deformation from $0.2\,\mathrm{mm\,a^{-1}}$ at the longitude 6° E to $0.6\,\mathrm{mm\,a^{-1}}$ close to longitude 11° E. This deformation pattern continues over the Southern Alps up to longitude 13° E, where the deformation vectors describe a progressive eastward rotation towards the Pannonian segment, reaching magnitudes up to $1.3\,\mathrm{mm\,a^{-1}}$ and an orientation of about N20° E near longitude 16° E. These vectors also make evident a shortening of about $2\,\mathrm{mm\,a^{-1}}$ across the southern front of the Eastern Alps, in the northern area of the Venetian-Friuli Basin. This is in agreement with the results published by Devoti et al. (2017), Métois et al. (2015), and Cheloni et al. (2014). Although our CO-GNSS network presents a sparse station distribution in the northern Apennines and the Dinarides, our prediction supports the conclusions of D'Agostino et al. (2014) and Nocquet (2012) in those regions. In the former case, we identify a motion roughly oriented to the north-east with the largest magnitudes ($\sim 3\,\mathrm{mm\,a^{-1}}$) across the highest topographic features of the Apennines towards the Adriatic Sea. In the opposite direction, towards the Tyrrhenian Sea, the deformation magnitudes decrease to $1.5\,\mathrm{mm\,a^{-1}}$. According to Mantovani et al. (2015), this may be evidence of a faster motion of the outer Apennine belt with respect to the inner one. In the case of the Dinarides, a shortening of about $3.0\,\mathrm{mm\,a^{-1}}$ across the Adriatic coastline is found. This magnitude decreases towards the Pannonian Basin to $0.5\,\mathrm{mm\,a^{-1}}$. In the southern front of the Alps, the orientation of the deformation vectors presents a slightly progressive westward rotation from the area of Venice towards the Po Basin. These findings are quantitatively equivalent to the results presented by Möller et al. (2011) and Devoti et al. (2011, 2017). However, it should be kept in mind that, due to the low number of GNSS stations we processed in the Po Basin, our results are highly influenced by the linear gradient imposed by the LSC approach to the deformation model. To increase the reliability of our model in this particular zone, a major number of GNSS stations covering the Italian Alpine forelands should be considered. Figure 10 shows the uncertainty (95 % confidence) of the horizontal surface deformation model. The largest values (from ± 0.5 to $\pm 1.5\,\mathrm{mm\,a^{-1}}$) occur at those regions with poor station coverage (mainly the Apennines, Dinarides, and easternmost part of the Austrian Alps). In general, the mean accuracy of the model is $\pm 0.2\,\mathrm{mm\,a^{-1}}$.

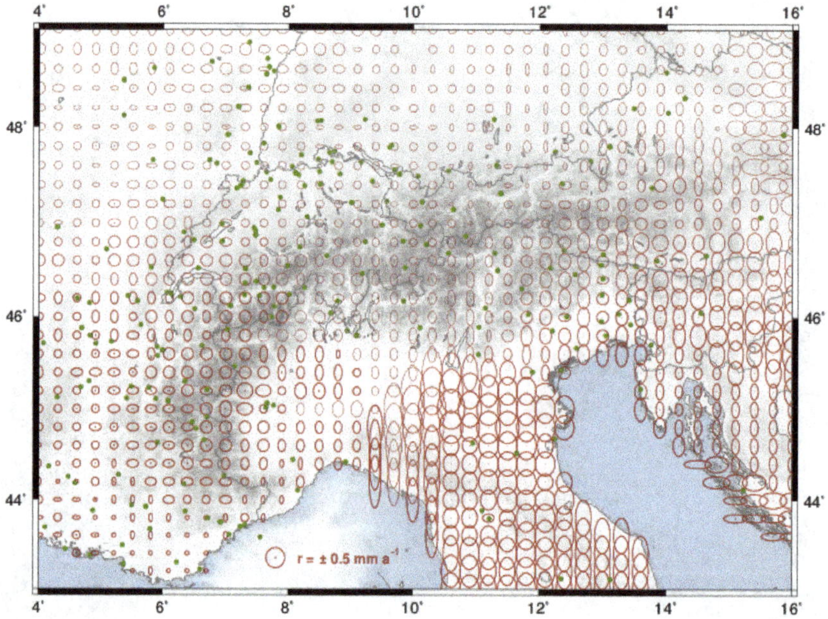

Figure 10. Estimated uncertainty (95 % confidence) of the GNSS-based horizontal deformation model of the Alpine region. Green dots show the CO-GNSS station distribution. Topography and bathymetry after the ETOPO1 model (Amante and Eakins, 2009).

5.3 Vertical deformation model

The present-day vertical motion of the Alps and their surrounding areas is displayed in Fig. 11. Our results clearly show an ongoing average uplift of $1.8\,\mathrm{mm\,a^{-1}}$ of the entire mountain chain, with the exception of the southern part of the Western Alps (6.5° E, 44° N), where no significant uplift (less than $0.5\,\mathrm{mm\,a^{-1}}$) is detected. The largest uplift rates (more than $2\,\mathrm{mm\,a^{-1}}$) occur in the central area of the Western Alps (7° E, 45° N), in the Swiss Alps (8° E, 46.5° N), and in the Southern Alps in the boundary region between Switzerland, Austria, and Italy (11° E, 46.6° N). The uplift kinematics we present here for the Western Alps are consistent with the vertical motion patterns identified by Nguyen et al. (2016) and Nocquet et al. (2016). Our vertical rates in the Swiss Alps also agree qualitatively with similar results inferred by Brockmann et al. (2012) using not only GNSS observations, but also levelling data.

In the north-west area, across the Rhône-Bresse and Rhine Grabens, a well-defined regional subsidence pattern with varying rates is observed. Apparently, the largest subsidence of around $-1.3\,\mathrm{mm\,a^{-1}}$ happens in the Rhine Graben (7° E, 48.5° N), while in the Rhône-Bresse Graben (5° E, 46–48° N) the largest rate reaches $-0.8\,\mathrm{mm\,a^{-1}}$. In the complete studied area, the largest subsidence rates are close to $-1.5\,\mathrm{mm\,a^{-1}}$ and are located west of the Venetian-Friuli Basin. We do not detect a significant subsidence in the western part of the Po Basin; this may be a consequence of the poor distribution of our CO-GNSS stations in that region. Actually, Devoti et al. (2017) inferred a mean subsidence rate of about $-0.8\,\mathrm{mm\,a^{-1}}$ in the Po Basin. The maximum ($-3\,\mathrm{mm\,a^{-1}}$) and minimum ($-0.5\,\mathrm{mm\,a^{-1}}$) magnitudes occur near Venice and in the eastern margin of the Western Alps, respectively (see Devoti et al., 2017, Fig. 6). Results of previous studies like Cuffaro et al. (2010) or Serpelloni et al. (2013) suggest subsidence rates up to $-11\,\mathrm{mm\,a^{-1}}$ near Modena, Italy (11° E, 44.6° N). However, we do not detect these extreme vales. This may be explained by the fact that we eliminate stations with strong local effects before the computation of the vertical-kinematics model, while the two studies mentioned base their conclusions on point-wise station velocities. It is well known that GNSS systematic errors and non-modelled local effects accumulate mainly in the estimation of the vertical coordinates. Therefore, we gave special care to identify and to remove those stations reflecting local effects (like water extraction, tunnel construction, and landslides) that may mislead the estimation of vertical motions mirroring mountain building or subsidence processes in the Alpine region.

The general uplift observed across the Alpine mountain chain decreases towards the outer regions to stable values between 0.0 and $0.5\,\mathrm{mm\,a^{-1}}$ and in some cases to subsidence like in the Liguro-Provençal (5° E, 43.6° N) and Vienna (15.5° E, 48.5°) basins, where vertical rates of -0.8 and $-0.3\,\mathrm{mm\,a^{-1}}$ are detected, respectively. Figure 12 displays the estimated uncertainty of the predicted uplift and subsidence rates. The largest values (from ± 0.6 to $\pm 0.8\,\mathrm{mm\,a^{-1}}$) occur in those areas with poor station coverage, mainly in the Apennine Peninsula and the Po Basin. The mean precision of the vertical motion model computed in this study for the Alpine mountain belt is about $\pm 0.3\,\mathrm{mm\,a^{-1}}$.

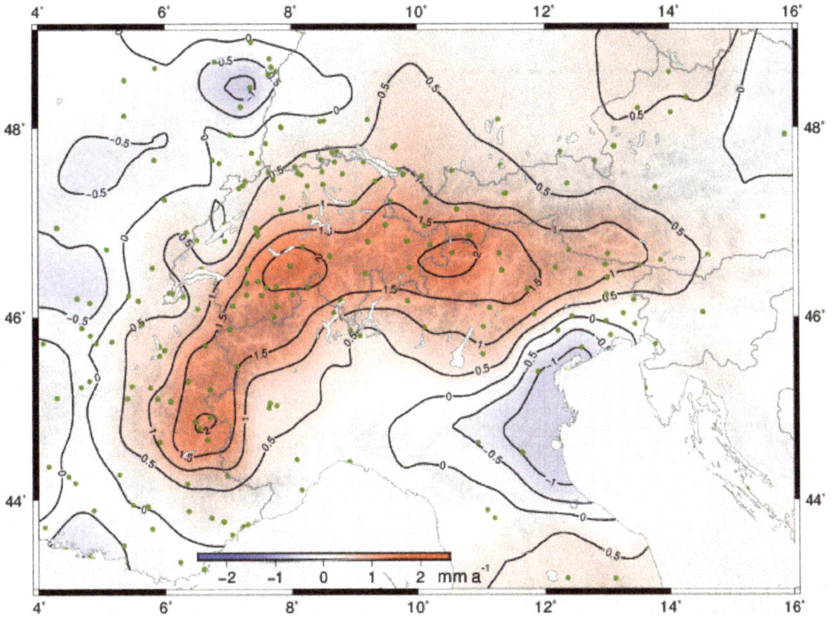

Figure 11. GNSS-inferred vertical deformation model of the Alpine region. Green dots show the CO-GNSS station distribution. Topography after the ETOPO1 model (Amante and Eakins, 2009).

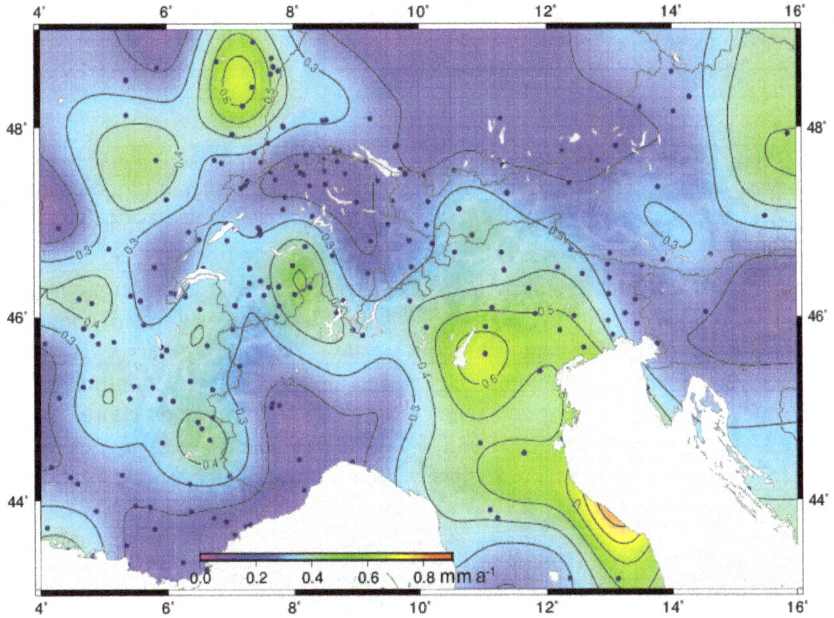

Figure 12. Estimated uncertainty (95 % confidence) of the GNSS-based vertical deformation model of the Alpine region. Dark blue dots show the CO-GNSS station distribution. Topography after the ETOPO1 model (Amante and Eakins, 2009).

Figure 13 shows two velocity profiles crossing the Alps along longitudes 6.5 and 13° E, respectively. In both cases, the vertical component (V_u) of the surface deformation presents a high correlation with the local topography. The profile in the Western Alps clearly shows the light subsidence ($-0.6\,\mathrm{mm\,a^{-1}}$) close the Liguro-Provençal Basin (near latitude 43° N); the large uplift rates (more than $1.5\,\mathrm{mm\,a^{-1}}$) in the border area between France and Switzerland (around latitude 45.5° N); and a decreasing uplift rate towards the north up to the Rhine Basin (latitude 48° N), where subsidence is observed. In the Eastern Alps, the gradient of the vertical velocity is stronger in the southern part than in northern area. South of latitude 46° N, the subsidence (up to $-1\,\mathrm{mm\,a^{-1}}$) detected in the Venetian-Friuli Basin can be well observed.

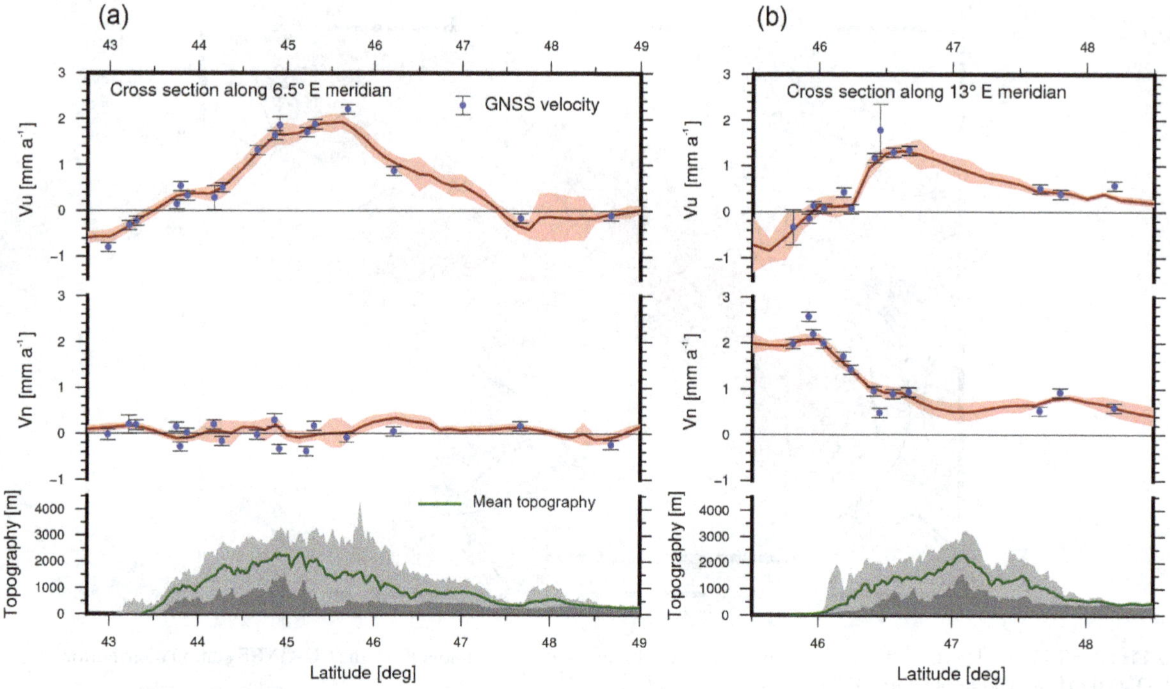

Figure 13. Velocity profiles in the Western Alps (cross section along longitude 6.5° E, **a**) and in the Eastern Alps (cross section along longitude 13° E, **b**). Blue dots with error bars represent the observed station velocities (with 68 % confidence). Red lines represent the north (V_n) and the vertical (V_u) components of the deformation model with their uncertainty (light red stripe). Green line shows the average topography in the profile swath, with light and dark grey shadows showing the maximum and minimum elevations, respectively.

North of latitude 46° N, we find uplift rates up to 1.2 mm a^{-1} in the border area between Italy and Austria (latitude 45° N). These rates diminish up to 0.3 mm a^{-1} in the border area between Austria and Germany (latitude 48° N). The north component (V_n) of the deformation model along longitude 6.5° suggests a very small (not significant) surface deformation in the Western Alps, while along longitude 13° E it captures the plate boundary region where the Adria microplate collides with the Eurasian Plate. The quite strong velocity gradient of nearly 2 mm a^{-1} over about 80 km (between latitudes 46 and 47° N) makes evident the NS compression occurring in the southern front of the Southern Alps.

5.4 Continuous velocity field

GNSS techniques are currently the primary tool for precise positioning. In applications of high reliability (accuracy at the 1 cm level or better), the GNSS data processing requires the reference station coordinates to be in the same reference frame and at the same observation epoch as the GNSS satellite orbits. In active seismic regions, strong earthquakes cause large changes in the station positions and velocities of the geodetic reference stations, disabling the use of previous coordinates (see for instance co-seismic and post-seismic effects of the El Maule earthquake in the Latin American reference frame presented by Sánchez and Drewes, 2016). To ensure the long-term stability of geodetic reference frames, the transformation of station positions between different epochs requires the computation of reliable continuous surface deformation (or velocity) models. Based on this kind of models, geodesists monitor the kinematics of reference frames, infer appropriate transformation parameters between pre-seismic and post-seismic (deformed) coordinates (specially in official matters like legal borders, cadastre, and land management), and interpolate surface motions arising from plate tectonics or crustal deformations in areas where no geodetic stations are established. Consequently, after computing the deformation model (Sect. 5.2), the Eurasian Plate motion removed from the CO-GNSS station velocities is restored to the deformation vectors to get a continuous velocity field referred to the ITRF. As shown in Fig. 14, the velocity field mirrors the present-day surface kinematics as a combination of deformation patterns and the Eurasian Plate tectonic motion, which is the dominant kinematic regime in Europe.

6 Strain field

The strain field represents the zones with compression, extension (or dilatation), and shear strain on the Earth's surface. In this study, the strain rates are computed from the deformation model (Fig. 9) as it provides a better (smoothed) surface coverage than the residual station velocities (Fig. 7). This strategy reduces random errors occurring at individual

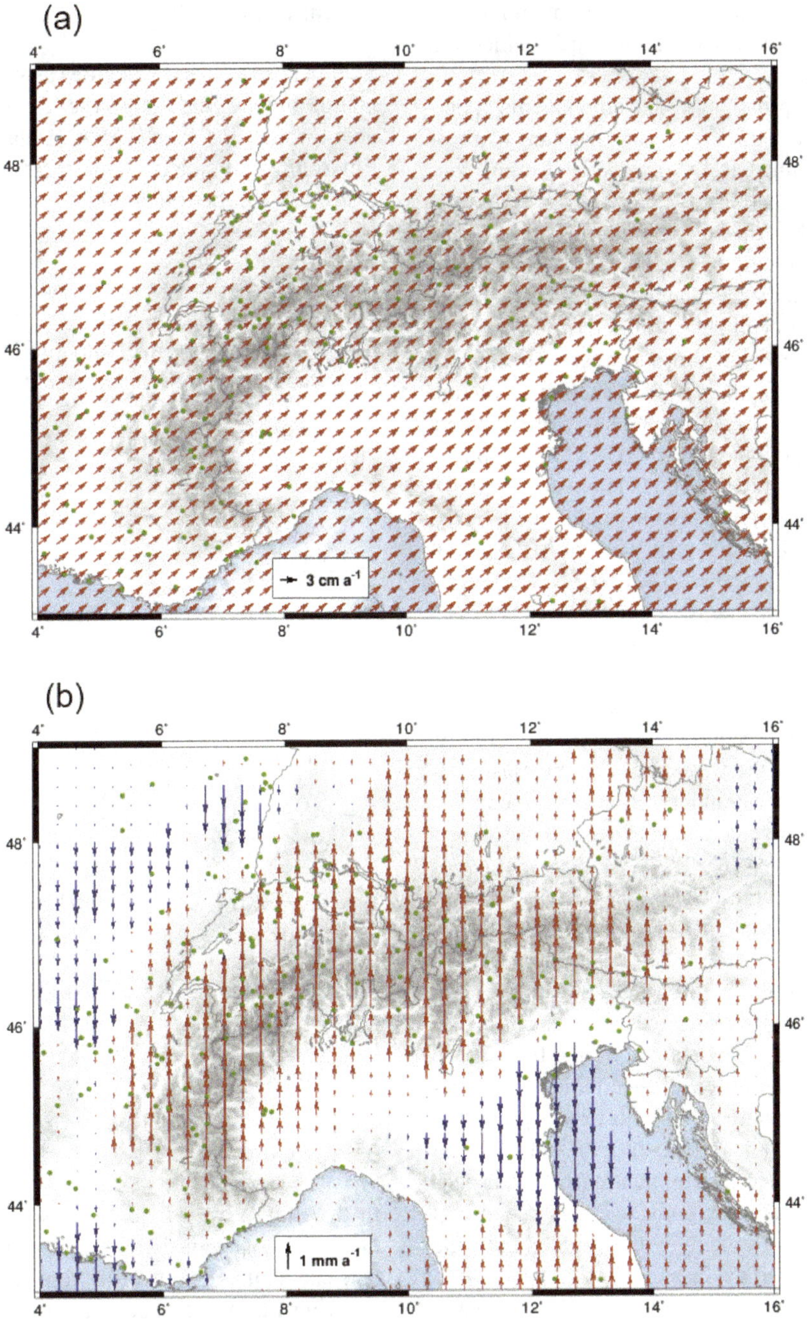

Figure 14. Continuous velocity field for the Alpine region referring to the ITRF2008 (**a**: horizontal velocity field, **b**: vertical velocity field). Topography and bathymetry after the ETOPO1 model (Amante and Eakins, 2009).

stations and enhances the major trends on strain patterns. Our computations are based on the infinitesimal strain theory, assuming that the deformation is much smaller (indeed, infinitesimally smaller) than any relevant dimension of the body, in this case the Alpine region. It means very small deformation changes (at the mm a^{-1} level) over large distances (at the km level) between grid points are presumed. In the empirical procedure, the deformation vectors computed at the grid points are used for the estimation of deformation gradients, and these gradients are used for the computation of the strain components following the equations described by Lambeck (1988, pp. 194–196). Even though the deformation field is available in three components, we concentrate only on the horizontal components. Thus, the obtained strain field is a horizontal projection of the surface deformation. The core contribution of a strain field is the possibility of distinguishing large-scale features (e.g. surface deformation and mountain building) from areas with large strain rates, which are often seismically active.

Figure 15 shows the principal strain components (compression and dilation) and the first strain invariant (sum of strain components) obtained in this study. Two main contrasting strain regimes are noticed: (i) shortening across the south-eastern front of the Alps and the northern part of the Dinarides and (ii) extension in the Apennines. The pattern of the strain principal axes indicates that the compression directions are more or less perpendicular to the thrust belt fronts (see Fig. 1). This compression presents maximum strain-rate values of about $20 \times 10^{-9}\,a^{-1}$ in the Venetian-Friuli and Po basins. These findings are in agreement with previous works published by Serpelloni et al. (2005, 2006), Caporali et al. (2009), Bennett et al. (2012), and Cheloni et al. (2014); in particular, the continuous compression regime we inferred along the Adriatic coastline fits very well with the strain model presented by Carafa and Bird (2016). Nevertheless, the lack of data in the eastern margin of the Adriatic Sea does not permit us to conclude if this regime continues with similar magnitudes to the southern Dinarides. As already stated by D'Agostino et al. (2014), dilatation is dominant in the north-western part of the Apennine Peninsula. Apparently, this dilatation continues to the west across the Po Basin, although at minor rates. However, poor station coverage in this area precludes any definitive conclusions from these observations.

Across the Alpine mountain belt, we observe a slight dilatation regime in the Western Alps, which smoothly changes to a contraction regime in western Austria and southern Germany, reaching maximum shortening values of $6 \times 10^{-9}\,a^{-1}$ in the north-eastern part of Austria (eastern margin of the Molasse Basin). The strain rates obtained for the French and Swiss Alps are in the same order of uncertainty of our results; however, they are in agreement with results presented in previous studies like Calais et al. (2002) and Delacou et al. (2008). The magnitude and orientation of the principal strain axes we observe in the boundary region between Switzerland, Austria, and Germany are very similar to the strain-rate patterns described by Tesauro et al. (2006).

The largest shear strain of about $20 \times 10^{-9}\,a^{-1}$ is detected in the south-eastern part of the Alps (Fig. 16). The sharp boundaries between the left-lateral and right-lateral shear zones and a quiet area reveal the triple junction between the Eurasian, Adriatic, and Pannonian plates (Möller et al., 2011). According to Brückl and Hammer (2014), the approximate location of the triple junction is 13.7° E, 46.6° N, which is well correlated with the strain model computed in this study. The Western Alps and surroundings do not reveal any specific pattern of the shear strain while the Dinarides show a compressional strain with a right-lateral shear of about 10 to $20 \times 10^{-9}\,a^{-1}$. The resulting principal and shear strain fields show that the main tectonic activity is ongoing on the plate boundaries.

7 Data availability

Station positions and velocities as well as velocity and deformation fields computed in the frame of this study are available through the PANGAEA (Data Publisher for Earth and Environmental Science) platform. The dataset collection is com-posed of the following.

- ALPS2017.SNX: multi-year solution for geocentric Cartesian station positions and velocities referring to the IGb08, epoch 2010.0, in SINEX, the Solution (Software/technique) INdependent EXchange format.
- ALPS2017_XYZ.CRD: geocentric Cartesian station positions.
- ALPS2017_XYZ.VEL: geocentric Cartesian station velocities.
- ALPS2017_NEH.CRD: ellipsoidal station positions.
- ALPS2017_NEH.VEL: ellipsoidal station velocities.
- ALPS2017_EPR.VEL: station velocities relative to the Eurasian Plate.
- ALPS2017_DEF_HZ.GRD: surface deformation model of the Alpine region.
- ALPS2017_DEF_VT.GRD: vertical deformation model of the Alpine region.
- ALPS2017_STR.GRD: strain field of the Alpine region.
- ALPS2017_VEL.GRD: continuous velocity field of the Alpine region.

8 Conclusions

In this study, we provide a present-day surface-kinematics model homogeneously assessed for the whole Alpine mountain chain. It is based on a high-level GNSS data analysis covering 12.4 years of continuous geodetic observations. Together with a coordinate solution for more than 300 CO-GNSS stations, our results comprise a deformation model, a continuous velocity field, and a strain field covering the area between 4–16° E and 43–50° N at a spatial interval of 25 km × 25 km. The average uncertainty of our model is $\pm 0.2\,mm\,a^{-1}$ in the horizontal component and $\pm 0.3\,mm\,a^{-1}$ in the height component. Although the station coverage is heterogeneous, our model describes coherent patterns of ongoing uplift processes and horizontal deformations with respect to the Eurasian Plate. The determination of the constant velocities, which form the basis for the deformation model and strain field, was particularly complex, because each little discontinuity had to be revealed, analysed, and taken into

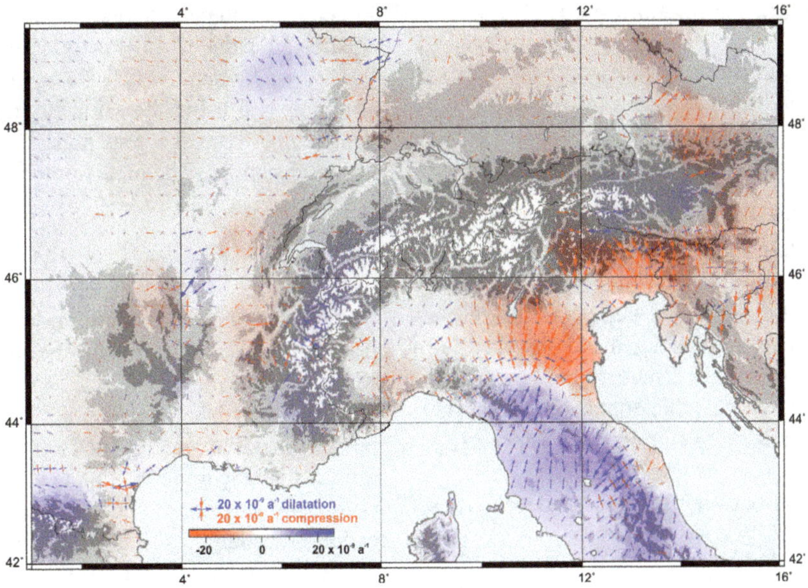

Figure 15. Strain distribution in the Alpine region. Blue shades represent dilatation; red shades represent compression. Topography after the ETOPO1 model (Amante and Eakins, 2009).

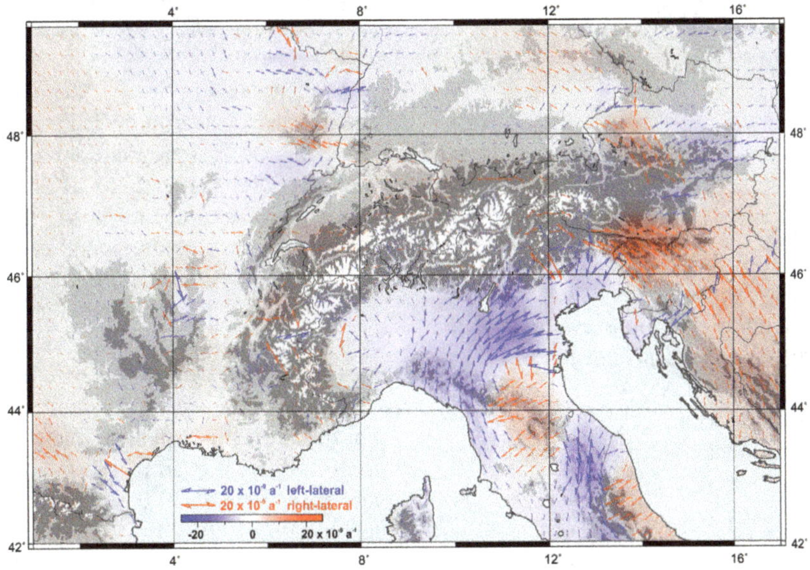

Figure 16. Shear strain distribution in the Alpine region. Blue and red shades represent the left-lateral and right-lateral shear, respectively. For illustration purposes, the left-lateral shear is considered as negative and the right-lateral shear as positive. Topography after the ETOPO1 model (Amante and Eakins, 2009).

account in determining the constant velocity of each site. Because the geodetic network consists of many permanently operating GNSS stations, which have to fulfil different purposes, the analysis was difficult and required a lot of manual intervention. Some sites fulfil geodetic services requiring frequent updates in the set-up, which in turn cause inconsistencies in the time series that had to be recovered. Not all of these changes were well documented. The GNSS data processed in this study are unique as they cover the entire Alpine range and the associated foreland. The comparison of our derived deformation patterns with the results of other studies, whose research areas were mostly limited to smaller regions of the Alps, has proven the high reliability of our model. With this data set, a deformation model of the Alps derived from a uniform analysis is now available.

Appendix A: Least-squares collocation equations

The basic least-squares collocation formula is given by (e.g. Moritz, 1973; Drewes, 1978)

$$\boldsymbol{v}_{\text{pred}} = \mathbf{C}_{\text{new}}^T (\mathbf{C}_{\text{obs}} + \mathbf{C}_{\text{nn}})^{-1} \boldsymbol{v}_{\text{obs}}. \tag{A1}$$

$\boldsymbol{v}_{\text{obs}}$ contains the velocities observed at the geodetic stations (Figs. 7 and 8). $\boldsymbol{v}_{\text{pred}}$ contains the velocities to be predicted at the continuous grid of 25 km × 25 km (Figs. 9 and 11). $\mathbf{C}_{\text{obs}}$ is the correlation matrix between observed velocities. $\mathbf{C}_{\text{new}}$ is the correlation matrix between predicted and observed velocities. $\mathbf{C}_{\text{nn}}$ is the noise covariance matrix; i.e. it contains the uncertainty of the station velocities obtained within the multi-year solution (error ellipses and error bars in Figs. 7 and 8).

The empiric correlation between the observed velocities may be written as

$$\mathbf{C}_{\text{obs}}(d_{ik}) = E\{v_i \cdot v_k\}, \tag{A2}$$

where E is the expectation; i, k represent the geodetic stations; and $(d_{i,k})$ is the distance between the station i and k. The C_{obs} values are classified in Δ_j class intervals with $j = 1\ldots10$. The cross-covariance $\mathbf{C}_{\text{obs}}(\Delta_j)$ and auto-covariance $\mathbf{C}_{\text{obs}}(d=0) = C_0$ are determined for each interval Δ_j:

$$\mathbf{C}_{\text{obs}}(\Delta_j) = \frac{1}{n_j}\sum_{i<k}^{j} v_i \cdot v_k \quad ;$$

$$\mathbf{C}_{\text{obs}}(d=0) = C_0 = \frac{1}{n}\sum_{i=1}^{n} v_i^2, \tag{A3}$$

where n stands for the number of stations available at the domain defined by d, while n_j represents the number of stations available at each class interval Δ_j. After estimating the discrete empirical covariances with Eq. (A3), they are approximated by a continuous function $C(d_{ik})$: in this case, the exponential function

$$C(d_{ik}) = a\, e^{-b \cdot d_{ik}}. \tag{A4}$$

The function parameters a and b are estimated by a least-squares adjustment. $\mathbf{C}_{\text{obs}}$ is symmetrical and its main diagonal $(i = k)$ contains the values C_0. The elements of $\mathbf{C}_{\text{new}}$ are computed using the same Eq. (A4) as a function of the distance between the grid node to be interpolated and the geodetic stations. The precision σ_{new}^2 of the predicted values is assessed with (Moritz, 1973)

$$\sigma_{\text{new}}^2 = C_0 - \mathbf{C}_{\text{new}}^T (\mathbf{C}_{\text{obs}} + \mathbf{C}_{\text{nn}})^{-1} \mathbf{C}_{\text{new}}. \tag{A5}$$

To fulfil the statistical condition of stationarity, the correlation function Eq. (A4) has to be identical for all points located in the d region; i.e. the exponential function must be homogeneous over the entire domain of known and predicted points. A consequence of the isotropy and stationarity conditions is that any trend (e.g. plate tectonics) in the given set of point parameters has to be removed before starting the collocation procedure. As horizontal and vertical station motions are dominated by different mechanisms (see Sect. 4.3), it is necessary to apply the LSC procedure separately. For the horizontal model, the north-east movement of the Eurasian Plate defines the trend motion to be removed from the station velocities. For the vertical model, the trend motion is given by the mean value of the vertical station velocities available at each d region.

As mentioned in Sect. 5, the deformation model is computed at a regular grid of 25 km spacing interval. This grid size is chosen as appropriate since it corresponds to the mean station spacing (about 20 to 30 km) in the most densely covered region within the network (the French and Swiss Alps, see Fig. 2). For the horizontal component of the deformation model, the minimum correlation distance d is defined to be 100 km. This radius provides an overlap between adjacent grid points and allows a certain degree of smoothing. For areas with poor station coverage, the correlation distance is extended until at least four stations are available for the LSC. For the vertical component, a larger correlation distance (up to 300 km) is needed as the number of rejected outliers is higher for the vertical station velocities than for the horizontal ones.

Author contributions. LS contributed to the GNSS data analysis strategy, computed the deformation model and the strain field, interpreted the station velocities with respect to the tectonic features, and wrote the paper. CV collected and compiled the GNSS observation data, defined the GNSS data analysis strategy, and took part in and supervised the GNSS data processing. AS performed the GNSS data analysis and contributed to the strain field. HA installed and maintains the GNSS stations operated by DGFI-TUM in the German Alps. FS initiated and coordinated the work. All authors contributed to the scientific discussion of the results, the editing, and revision of the paper.

Competing interests. The authors declare that they have no conflict of interest.

Acknowledgements. This study has been possible thanks to the support of different organisations providing us with the observational data of their CO-GNSS stations. We thank the International GNSS Service (IGS); the Reference Frame Sub-Commission for Europe of the International Association of Geodesy (EUREF); the Federal Agency for Cartography and Geodesy of Germany (BKG); the Space Research Institute of the Austrian Academy of Sciences; the Geodetic Alpine Integrated Network (GAIN); the Centro Ricerche Sismologiche (CRS) of the Istituto Nazionale di Oceanografia e di Geofisica Sperimentale (OGS); the Réseau National GPS, France; and the Orpheon network. We are also grateful to our colleague Elmar Brockmann from the Swiss Federal Office of Topography (swisstopo) for providing us with the coordinate solution of the AGNES (Automated GNSS Network for Switzerland) network. Jeff Freymueller and Roberto Devoti are gratefully thanked for their careful and constructive reviews that have greatly improved the quality of the paper. Figures in this paper were compiled using the GMT (Generic Mapping Tools) software package (Wessel et al., 2013).

Edited by: David Carlson

References

Altamimi, Z., Collilieux, X., and Métivier, L.: ITRF2008: an improved solution of the international terrestrial reference frame, J. Geodesy, 85, 457–473, https://doi.org/10.1007/s00190-011-0444-4, 2011.

Altamimi, Z., Métivier, L., and Collilieux, X.: ITRF2008 plate motion model, J. Geophys. Res., 117, B07402, https://doi.org/10.1029/2011JB008930, 2012.

Amante, C. and Eakins, B. W.: ETOPO1 Global Relief Model converted to PanMap layer format. NOAA-National Geophysical Data Center, PANGAEA, https://doi.org/10.1594/PANGAEA.769615, 2009.

Aoudia, A., Amodio, A., Troisi, C., Manzino, A., Luchetta, A., Völksen, C., Drewes, H., Zivcic, M., Van der Voerd, J., Fosson, J. P., Carraro, C., Zampedri, G., Walpersdorf, A., Laffi, R., Fioroni, M., Barzaghi, R., Sabadini, R., Pasetti, L., and Sguero, D.: ALPS GPSQUAKENET. Alpine Integrated GPS Network: Real-Time Monitoring and Master Model for Continental Deformation and Earthquake Hazard, Intergared III B Project – Alpine Space, available at: http://www.alpine-space.org/2000-2006/uploads/media/ALPS-GPS_QUAKENET_Project_booklet.pdf (last access: 20 August 2018), 2007.

Argnani, A.: Plate Tectonics and the Boundary between Alps and Apennines, Ital. J. Geosci., 128, 317–330, https://doi.org/10.3301/IJG.2009.128.2.317, 2009.

Bada, G., Horváth, F., Dövényi, P., Szafián, P., Windhoffer, G. and Cloetingh, S.: Present-day stress field and tectonic inversion in the Pannonian basin, Glob. Planet. Change, 58, 165–180, https://doi.org/10.1016/j.gloplacha.2007.01.007, 2007.

Barletta, V. R., Ferrari, C., Diolaiuti, G., Carnielli, T., Sabadini, R., and Smiraglia, C.: Glacier shrinkage and modeled uplift of the Alps, Geophys. Res. Lett., 33, L14307, https://doi.org/10.1029/2006GL026490, 2006.

Becker, A.: In situ stress data from the Jura Mountains – new results and interpretation, Terra Nova, 11, 9–15, https://doi.org/10.1046/j.1365-3121.1999.00215.x, 1999.

Bennett, R. A., Hreinsdóttir, S., Buble, G., Bašić, T., Bašić, Ž, Marjanović, M., Casale, G., Gendaszek, A., and Cowan, D.: Eocene to present subduction of southern Adria mantle lithosphere beneath the Dinarides, Geology, 36, 3–6, https://doi.org/10.1130/G24136A.1, 2008.

Bennett, R. A., Serpelloni, E., Hreinsdóttir, S., Brandon, M. T., Buble, G., Basic, T., Casale, G., Cavaliere, A., Anzidei, M., Marjonovic, M., Minelli, G., Molli, G., and Montanari, A.: Syn-convergent extension observed using the RETREAT GPS network, northern Apennines, Italy, J. Geophys. Res., 117, B04408, https://doi.org/10.1029/2011JB008744, 2012.

Bird, P.: An updated digital model for plate boundaries, Geochem. Geophys. Geosyst., 4, 1027, https://doi.org/10.1029/2001GC000252, 2003.

Blewitt, G., Lavallée, D., Clarke, P., and Nurudinov, K.: A new global model of Earth deformation: Seasonal cycle detected, Science, 294, 2342–2345, https://doi.org/10.1126/science.1065328, 2001.

Böhm, J., Niell, A., Tregoning, P., and Schuh, H.: Global mapping function (GMF): a new empirical mapping function based on numerical weather model data, Geophys. Res. Lett., 33, L07304, https://doi.org/10.1029/2005GL025546, 2006.

Bousquet, R., Schmid, S. M., Zeilinger, G., Oberhänsli, R., Rosenberg, C., Molli, G., Robert, C., Wiederkehr, M., and Rossi, Ph.: Tectonic framework of the Alps, CCGM/CGMW, available at: http://www.geodynalps.org (last access: 20 August 2018), 2012.

Brockmann, E., Ineichen, D., Marti, U., Schaer, S., Schlatter, A., and Villiger, A.: Determination of Tectonic Movements in the Swiss Alps using GNSS and Levelling, in: Geodesy for Planet Earth, edited by: Kenyon S., Pacino M., Marti, U., IAG Symposia, 136, 689–695, https://doi.org/10.1007/978-3-642-20338-1_85, 2012.

Brückl, E. and Hammer, C.: Eduard Suess' conception of the Alpine orogeny related to geophysical data and models, Austrian J. Earth Sc., 107, 94–114, 2014.

Brückl, E., Bleibinhaus, F., Gosar, A., Grad, M., Guterch, A., Hrubcová, P., Keller, R., Majdański, M., Šumanovac, F., Tiira, T., Yliniemi, J., Hegedüs, E., and Thybo, H.: Crustal structure due to collisional and escape tectonics in the Eastern Alps region based on profiles Alp01 and Alp02 from the

ALP 2002 seismic experiment, J. Geophys. Res., 112, B06308, https://doi.org/10.1029/2006JB004687, 2007.

Brückl, E., Behm, M., Decker, K., Grad, M., Guterch, A., Keller, G., and Thybo, H.: Crustal structure and active tectonics in the eastern Alps, Tectonics, 29, TC2011, https://doi.org/10.1029/2009TC002491, 2010.

Bruni, S., Zerbini, S., Raicich, F., Errico, M., and Santi, E.: Detecting discontinuities in GNSS coordinate time series with STARS: case study, the Bologna and Medicina GPS sites, J. Geodesy, 88, 1203–1214, https://doi.org/10.1007/s00190-014-0754-4, 2014.

Bruyninx, C., Habrich, H., Söhne, W., Kenyeres, A., Stangl, G., and Völksen, C.: Enhancement of the EUREF permanent network services and products, IAG Symposia, 138, 27–34, https://doi.org/10.1007/978-3-642-20338-1_4, 2012.

Bruyninx, C., Habrich, H., Kenyeres, A., Söhne, W., Stangl, G., and Völksen, C.: EUREF Permanent Network. International GNSS Service 2012 Technical Report, edited by: Dach, R. and Jean, Y., Astronomical Institute University of Bern, 101–107, https://doi.org/10.7892/boris.80303, 2013.

Bus, Z., Grenerczy, G., Tóth, L., and Mónus, P.: Active crustal deformation in two seismogenic zones of the Pannonian region – GPS versus seismological observations, Tectonophysics, 474, 343–352, https://doi.org/10.1016/j.tecto.2009.02.045, 2009.

Calais, E.: Continuous GPS measurements across the Western Alps, 1998–1998, Geophys. J. Int., 38, 221–230, https://doi.org/10.1046/j.1365-246x.1999.00862.x, 1999.

Calais, E., Barlier, F., Bayer, R., Boucher, C., Chéry, J., Cotton, F., Gariel, J. C., Jouanne, F., Martinod, J., and Vigny, C.: A permanent GPS network for monitoring crustal deformation in the Western Alps, Ann. Geophys. 6, Suppl. 1, C393, 1998.

Calais, E., Galisson, L., Stéphan, J.-F., Delteil, J., Deverchère, J., Larroque, C., Mercier de Lépinay, B., Popoff, M., and Sosson, M.: Crustal strain in the Southern Alps, 1948–1998, Tectonophysics, 319, 1–17, https://doi.org/10.1016/S0040-1951(00)00029-9, 2000.

Calais, E., Nocquet, J. M., Jouanne, F., and Tardy, M.: Current strain regime in the Western Alps from continuous Global Positioning System measurements, 1996–2001, Geology, 30, 651–654, https://doi.org/10.1130/0091-7613(2002)030<0651:CSRITW>2.0.CO;2, 2002.

Caporali, A., Aichhorn, C., Barlik, M., Becker, M., Fejes, I., Gerhatova, L., Ghitau, D., Grenerczy, G., Hefty, J., Krauss, S., Medak, D., Milev, G., Mojzes, M., Mulic, M., Nardo, A., Pesec, P., Rus, T., Simek, J., Sledzinski, J., Solaric, M., Stangl, G., Stopar, B., Vespe, F., and Virag, G.: Surface kinematics in the Alpine-Carpathian-Dinaric and Balkan region inferred from a new multi-network GPS combination solution, Tectonophysics, 474, 295–321, https://doi.org/10.1016/j.tecto.2009.04.035, 2009.

Carafa, M. M. C. and Bird, P.: Improving deformation models by discounting transient signals in geodetic data: 2. Geodetic data, stress directions, and long term strain rates in Italy, J. Geophys. Res.-Sol. Ea., 121, 5557–5575, https://doi.org/10.1002/2016JB013038, 2016.

Channell, J. E. T. and Horváth, F.: The African/Adriatic promontory as a paleogeographical premise for Alpine orogeny and plate movements in the Carpathó-Balkan region, Tectonophysics, 35, 71–101, https://doi.org/10.1016/0040-1951(76)90030-5, 1976.

Channell, J. E. T., D'Argenio, B., and Horvárth, F.: Adria, the African promontory, in mesozoic Mediterranean palaeogeography, Earth-Sci. Rev., 15, 213–292, https://doi.org/10.1016/0012-8252(79)90083-7, 1979.

Cheloni, D., D'Agostino, N., and Selvaggi, G.: Interseismic coupling, seismic potential and earthquake recurrence on the southern front of the Eastern Alps (NE Italy), J. Geophys. Res.-Sol. Ea., 119, 4448–4468, https://doi.org/10.1002/2014JB010954, 2014.

Chen, G. and Herring, T. A.: Effects of atmospheric azimuthal asymmetry on the analysis of space geodetic data, J. Geophys. Res., 102, 20489–20502, https://doi.org/10.1029/97JB01739, 1997.

Chéry, J., Vigny, C., Meyer, B., Ferhat, G., Anzidei, M., Bayer, R., Boloh, L., Briole, P., Deschamps, A., Feigl, K., Gamond, J. F., Geiger, A., Jouanne, F., Kasser, M., Le Pape, M., Martinod, J., Ménard, G., Ruegg, J. C., Scheubel, J. M., and Walsh, J. J.: Global Positioning System network monitors the Western Alps, EOS Trans. Am. Geophys. Union August, 27, 489–489, https://doi.org/10.1029/95EO00299, 1995.

Collilieux, X., Altamimi, Z., Coulot, D., van Dam, T., and Ray, J.: Impact of loading effects on determination of the International Terrestrial Reference Frame, Adv. Space Res., 45, 144–154, https://doi.org/10.1016/j.asr.2009.08.024, 2010.

Collilieux, X., van Dam, T., Ray, J., Coulot, D., Métivier, L., and Altamimi, Z.: Strategies to mitigate aliasing of loading signals while estimating GPS frame parameters, J. Geodyesy, 86, 1–14, https://doi.org/10.1007/s00190-011-0487-6, 2012.

Copley, A.: Kinematics and dynamics of the southeastern margin of the Tibetan Plateau, Geophys. J. Int., 174, 1081–1100, https://doi.org/10.1111/j.1365-246X.2008.03853.x, 2008.

Cuffaro, M., Riguzzi, F., Scrocca, D., Antonioli, F., Carminati, E., Livani, M., and Doglioni, C.: On the geodynamics of the northern Adriatic plate, Rend. Fis. Acc. Lincei, 21, S253–S279, https://doi.org/10.1007/s12210-010-0098-9, 2010.

Dach, R., Lutz, S., Walser, P., and Fridez, P. (Eds.): Bernese GNSS Software Version 5.2. University of Bern, 2015.

Dach, R., Schaer, S., Arnold, D., Prange, L., Sidorov, D., Sušnik, A., Villiger, A., and Jäggi, A.: CODE final product series for the IGS. Published by Astronomical Institute, University of Bern, available at: http://www.aiub.unibe.ch/download/CODE (last access: 20 August 2018), https://doi.org/10.7892/boris.75876.2, 2017.

D'Agostino, N., Cheloni, D., Mantenuto, S., Selvaggi, G., Michelini, A., and Zuliani, D.: Interseismic strain accumulation in the eastern Southern Alps (NE Italy) and deformation at the eastern boundary of the Adria block observed by CGPS measurements, Geophys. Res. Lett., 32, L19306, https://doi.org/10.1029/2005GL024266, 2005.

D'Agostino, N., Mantenuto, S., D'Anastasio, E., Giuliani, R., Mattone, M., Calcaterra, S., Gambino, P., and Bonci, L.: Evidence for localized active extension in the central Apennines (Italy) from global positioning system observations, Geology, 39, 291–294, https://doi.org/10.1130/G31796.1, 2011.

D'Agostino, N., England, P., Hunstad, I., and Selvaggi, G.: Gravitational potential energy and active deformation in the Apennines, Earth Planet. Sc. Lett., 397, 121–132, https://doi.org/10.1016/j.epsl.2014.04.013, 2014.

Delacou, B., Sue, Ch., Nocquet, J. M., Champagnac, J. D., Allanic, C., and Burkhard, M.: Quantification of strain rate in the Western Alps using geodesy: comparisons with seismotectonics, Swiss J. Geosci., 101, 377–385, https://doi.org/10.1007/s00015-008-1271-3, 2008.

Dercourt, J., Zonenshain, L. P., Ricou, L.-E., Kazmin, V. G., Le Pichon, X., Knipper, A. L., Grandjacquet, C., Sbortshikov, I. M., Geyssant, J., Lepvrier, C., Pechersky, D. H., Boulin, J., Sibuet, J.-C., Savostin, L. A., Sorokhtin, O., Westphal, M., Bazhenov, M. L., Lauer, J. P., and Biju-Duval, B.: Geological evolution of the tethys belt from the atlantic to the pamirs since the LIAS, Tectonophysics, 123, 241–315, https://doi.org/10.1016/0040-1951(86)90199-X, 1986.

Devoti, R., Riguzzi, F., Cuffaro, M., and Doglioni, C.: New GPS constraints on the kinematics of the Apennines subduction, Earth Planet. Sc. Lett., 273, 163–174, https://doi.org/10.1016/j.epsl.2008.06.031, 2008.

Devoti, R., Esposito, A., Pietrantonio, G., Pisani, A. R., and Riguzzi, F.: Evidence of large scale deformation patterns from GPS data in the Italian subduction boundary, Earth Planet. Sc. Lett., 311, 230–241, https://doi.org/10.1016/j.epsl.2011.09.034, 2011.

Devoti, R., D'Agostino, N., Serpelloni, E., Pietrantonio, G., Riguzzi, F., Avallone, A., Cavaliere, A., Cheloni, D., Cecere, G., D'Ambrosio, C., Falco, L., Selvaggi, G., Métois, M., Esposito, A., Sepe, V., Galvani, A., and Anzidei, M.: A combined velocity field of the Mediterranean Region, Ann. Geophys., 60, S0215, https://doi.org/10.4401/ag-7059, 2017.

Dewey, J. F., Pitman III, W. C., Ryan, W. B. F., and Bonnin, J.: Plate Tectonics and the Evolution of the Alpine System, Geol. Soc. Am. Bull., 84, 3137–3180, https://doi.org/10.1130/0016-7606(1973)84<3137:PTATEO>2.0.CO;2, 1973.

Dewey, J. F., Helman, M. L., Knott, S. D., Turco, E., and Hutton, D. H. W.: Kinematics of the western Mediterranean, Geological Society, London, Special Publications, 45, 265–283, https://doi.org/10.1144/GSL.SP.1989.045.01.15, 1989.

Doglioni, C.: Tectonics of the Dolomites (southern alps, northern Italy), J. Struct. Geol., 9, 181–193, https://doi.org/10.1016/0191-8141(87)90024-1, 1987.

Doglioni, C. and Carminati, E.: The effects of four subductions in NE-Italy, Mem. Sci. Geol., 54, 1–4, 2002.

Dow, J. M., Neilan, R. E., and Rizos, C.: The International GNSS Service in a hanging landscape of Global Navigation Satellite Systems, J. Geodesy, 83, 191–198, https://doi.org/10.1007/s00190-008-0300-3, 2009.

Drewes, H.: Experiences with least squares collocation as applied to interpolation of geodetic and geophysical quantities, Proceedings 12th Symposium on Mathematical Geophysics, Caracas, Venezuela, 1978.

Drewes, H.: A geodetic approach for the recovery of global kinematic plate parameters, Bull. Geod., 56, 70–79, https://doi.org/10.1007/BF02525609, 1982.

Drewes, H.: The Actual Plate Kinematic and Crustal Deformation Model APKIM2005 as Basis for a Non-Rotating ITRF, in: Geodetic Reference Frames, edited by: Drewes, H., IAG Symposia, 134, 95–99, https://doi.org/10.1007/978-3-642-00860-3_15, 2009.

Drewes, H. and Heidbach, O.: Deformation of the South American crust estimated from finite element and collocation methods, in: A Window on the Future of Geodesy, edited by: Sansò, F., IAG Symposia, 128, 544–549, https://doi.org/10.1007/3-540-27432-4_92, 2005.

Drewes, H., Angermann, D., and Seitz, M.: Alternative definitions of the terrestrial reference system and its realization in reference frames, in: Reference Frames for Applications in Geosciences, edited by: Altamimi, Z. and Collilieux, X., IAG Symposia, 138, 39–44, https://doi.org/10.1007/978-3-642-32998-2_7, 2013.

Fejes, I., Barlik, M., Busics, I., Packelski, W., Rogowsky, J., Sledzinsky, J., and Zielinsky, J. B.: The Central Europe Regional Geodynamics Project, paper presented at 2nd International Seminar on "GPS in Central Europe", Hun. Acad. Sci. Penc., Hungary, 27–29 April 1993.

Ferhat, G., Feigl, K. L., Ritz, J. F., and Souriau, A.: Geodetic measurement of tectonic deformation in the Southern Alps and Provence, France, 1947–1994, Earth Planet. Sc. Lett., 159, 35–46, https://doi.org/10.1016/S0012-821X(98)00065-X, 1998.

Flesch, L. M., Holt, W. E., Haines, A. J., and Bingming Shen-Tu, B.: Dynamics of the Pacific-North American Plate Boundary in the Western United States, Science, 287, 834–836, https://doi.org/10.1126/science.287.5454.834, 2000.

Gosar, A., Šebela, S., Košťák, B., and Stemberk, J.: Micro-deformation monitoring of active tectonic structures in W Slovenia, Acta Geodyn. Geomater., 4, 87–98, 2007.

Grenerczy, G. and Kenyeres, A.: Crustal deformation between Adria and the European platform from space geodesy, in: the Adria microplate: GPS geodesy, tectonics and hazards, edited by: Pinter, N., Grenerczy, G., Weber, J., Stein, S., and Medak, D., NATO Sci. S. IV Ear. En., 61, 321–334, 2006.

Grenerczy, G., Kenyeres, A., and Fejes, I.: Present crustal movements and strain distribution in Central Europe inferred from GPS measurements, J. Geophys. Res., 105, 21835–21846, https://doi.org/10.1029/2000JB900127, 2000.

Grenerczy, G., Sella, G., Stein, S., and Kenyeres, A.: Tectonic implications of the GPS velocity field in the northern Adriatic region, Geophys. Res. Lett., 32, L16311, https://doi.org/10.1029/2005GL022947, 2005.

Handy, M. R., Schmid, S. M., Bousquet, R., Kissling, E., and Bernoulli, D.: Reconciling plate-tectonic reconstructions with the geological-geophysical record of spreading and subduction in the Alps, Earth.-Sci. Rev., 102, 121–158, 2010.

Handy, M. R., Ustaszewski, K., and Kissling, E.: Reconstructing the Alps–Carpathians–Dinarides as a key to understanding switches in subduction polarity, slab gaps and surface motion, Int. J. Earth Sci., 104, 1–26, https://doi.org/10.1007/s00531-014-1060-3, 2015.

Haslinger, C. and Stangle, G.: The First Austrian Velocity Field derived from GPS, in: Report on the Symposium of the IAG Sub-commission for Europe (EUREF) held in Riga, Latvia, 14–17 June 2006, available at: http://www.euref.eu/symposia/2006Riga/Symposium2006-Riga.html (last access: 20 August 2018), 2006.

Haslinger, C., Kraus, S., and Stangl, G.: The Intra-Plate Velocities of GPS Permanent Stations of the Eastern Alps, Vermessung & Geoinformation, 2, 66–72, 2006.

Heidbach, O. and Drewes, H.: 3-D finite element model of major tectonic processes in the Eastern Mediterranean, Geol. Soc. Spec. Publ., 212, 261–274, https://doi.org/10.1144/GSL.SP.2003.212.01.17, 2003.

Horváth, F., Bada, G., Szafián, P., Tari, G., Ádám, A., and Cloetingh, S.: Formation and deformation of the Pannonian Basin: Constraints from observational data, Geological Society Mem., 32, 191–206, https://doi.org/10.1144/GSL.MEM.2006.032.01.11, 2006.

IGS: Specifications for the 2nd data reprocessing campaign of the International GNSS Service, available at: http://acc.igs.org/reprocess2.html (20 August 2018), 2014.

Kahle, H. G., Müller, M. V., Mueller, St., Veis, G., Billiris, H., Paradissis, D., Drewes, H., Kaniuth, K., Stuber, K., Tremel, H., Zerbini, S., Pezzoli, L., and Corrado, G.: The convergent African/Eurasian plate boundary and associated crustal movements: GPS investigations in the Calabrian and West Hellenic arcs, Proc. 8th Int. Symp. on Recent Crustal Movements, Kobe/Japan, 333–342, 1994.

Kaniuth, K., Drewes, H., Tremel, H., Stuber, K., Zerbini, S., Pezzoli, L., and Corrado, G.: Crustal deformations in the Calabrian Arc area from five years of GPS measurements, Proc. of ISTANBUL-94, 1st Turkish Internat. Symp. on Deformations, Istanbul, 781–790, 1995.

Kaniuth, K., Drewes, H., Stuber, K., Tremel, H., Kahle, H. G., Peter, Y., Zerbini, S., Tonti, G., and Fagard, H.: Crustal deformations in the central Mediterranean derived from the WHAT A CAT GPS project, in: Proceedings of the 13th Working Meeting on European VLBI for Geodesy and Astrometry, Viechtach, 12–13 February, edited by: Schlüter, W. and Hase, H., 192–197, 1999.

King, M. A., Watson, C. S., Penna, N. T., and Clarke, P. J.: Subdaily signals in GPS observations and their effect at semi-annual and annual periods, Geophys. Res. Lett., 35, L03302, https://doi.org/10.1029/2007GL032252, 2008.

Kissling, E., Schmid, S. M., Lippitsch, R., Ansorge, J., and Fügenschuh, B.: Lithosphere structure and tectonic evolution of the Alpine arc: new evidence from high-resolution teleseismic tomography, Geological Society, London, Memoirs, 32, 129–145, https://doi.org/10.1144/GSL.MEM.2006.032.01.08, 2006.

Kuhlemann, J. and Kempf, O.: Post-Eocene evolution of the North Alpine Foreland Basin and its response to Alpine tectonics, Sediment. Geol., 152, 45–78, https://doi.org/10.1016/S0037-0738(01)00285-8, 2002.

Lambeck, K.: Geophysical Geodesy: The slow deformations of the Earth, Oxford University Press, New York, 1988.

Larroque, C., Delouis, B., Godel, B., and Nocquet, J. M.: Active deformation at the southwestern Alps-Ligurian basin junction (France-Italy boundary): Evidence for recent change from compression to extension in the Argentera massif, Tectonophysics, 467, 22–34, https://doi.org/10.1016/j.tecto.2008.12.013 , 2009.

Laubscher, H. P.: The arcs of the Western Alps and the Northern Apennines: an updated view, Tectonophysics, 146, 67–78, https://doi.org/10.1016/0040-1951(88)90082-0, 1988.

Le Pichon, X., Bergerat, F., and Roulet, M. J.: Plate kinematics and tectonics leading to the Alpine belt formation, A new analysis, in: Processes in Continental Lithospheric Deformation, edited by: Clark Jr., S. P., Clark, Burchfiel, B., and Suppe, J., Geological Society of America, special paper 2018, 111–131, 1988.

Letellier, T.: Etude des ondes de marée sur les plateux continentaux. Thèse doctorale, Université de Toulouse III, Ecole Doctorale des Sciences de l'Univers, de l'Environnement et de l'Espace, 237 pp., 2004.

Lippitsch, R., Kissling, E., and Ansorge, J.: Upper mantle structure beneath the Alpine orogen from high-resolution teleseismic tomography, J. Geophys. Res., 108, 2376, https://doi.org/10.1029/2002JB002016, 2003.

Mantovani, E., Viti, M., Cenni, N., Babbucci, D., and Tamburelli, C.: Present Velocity Field in the Italian Region by GPS Data: Geodynamic/Tectonic Implications, Int. J. Geosci., 6, 1285–1316, https://doi.org/10.4236/ijg.2015.612103, 2015.

Meissl, P.: The use of finite elements in physical geodesy, Rep. Dep. Geod. Sci. and Surv., The Ohio State Univ., Columbus, No. 313, 1981.

Métois, M., D'Agostino, N., Avallone, A., Chamot-Rooke, N., Rabaute, A., Duni, L., Kuka, N., Koci, R., and Georgiev, I.: Insights on continental collisional processes from GPS data: Dynamics of the peri-Adriatic belts, J. Geophys. Res.-Sol. Ea., 120, 8701–8719, https://doi.org/10.1002/2015JB012023, 2015.

Möller, G., Brückl, E., and Weber, R.: Active tectonic deformation at the transition from the European and Pannonian domain monitored by a local GNSS network, Vermessung & Geoinformation, 2, 138–148, 2011.

Moritz, H.: Least squares collocation, Dt. Geod. Komm., Nr. A 75, Munich, ISBN 3 7696 8162 2, 1973.

Mueller, S. and Kahle, H.-G: Crust-Mantle Evolution, Structure and Dynamics of the Mediterranean-Alpine Region, in: Contributions of Space Geodesy to Geodynamics: Crustal Dynamics, edited by: Smith, D. E. and Turcotte, D. L., American Geophysical Union, Washington, DC, https://doi.org/10.1029/GD023p0249, 1993.

Nguyen, H. N., Vernant, P., Mazzotti, S., Khazaradze, G., and Asensio, E.: 3-D GPS velocity field and its implications on the present-day post-orogenic deformation of the Western Alps and Pyrenees, Solid Earth, 7, 1349–1363, https://doi.org/10.5194/se-7-1349-2016, 2016.

Nocquet, J. M.: Present-day kinematics of the Mediterranean: A comprehensive overview of GPS results, Tectonophysics, 579, 220–242, https://doi.org/10.1016/j.tecto.2012.03.037, 2012.

Nocquet, J. M. and Calais, E.: Geodetic measurements of crustal deformation in the Western Mediterranean and Europe, Pure Appl. Geophys., 161, 661–681, https://doi.org/10.1007/s00024-003-2468-z, 2004.

Nocquet, J. M., Sue, C., Walpersdorf, A., Tran, T., Lenôtre, N., Vernant, P., Cushing, M., Jouanne, F., Masson, F., Baize, S., Chéry, J., and van der Beek, V. A.: Present-day uplift of the western Alps, Nature, Sci. Rep.-UK, 6, 28404, https://doi.org/10.1038/srep28404, 2016.

Persaud, M. and Pfiffner, O. A.: Active Deformation in the Eastern Swiss Alps: Post-Glacial Faults, Seismicity and Surface Uplift, Tectonophysics, 385, 59–84, https://doi.org/10.1016/j.tecto.2004.04.020, 2004.

Petit, G. and Luzum, B. (Eds.): IERS Conventions 2010. IERS Technical Note 36, Verlag des Bundesamtes für Kartographie und Geodäsie, Frankfurt a.M., 2010.

Pfiffner, O. A., Frei, W., Valasek, P., Stäuble, M., Levato, L., DuBois, L., Schmid, S. M., and Smithson, S. B.: Crustal shortening in the Alpine Orogen: Results from deep seismic reflection profiling in the eastern Swiss Alps, Line NFP 20-east, Tectonics, 9, 1327–1355, https://doi.org/10.1029/TC009i006p01327, 1990.

Ray, J., Altamimi, Z., Collilieux, X., and van Dam, T. M.: Anomalous harmonics in the spectra of GPS position estimates, GPS Solutions, 12, 55–64, https://doi.org/10.1007/s10291-007-0067-7, 2008.

Rebischung, P., Griffiths, J., Ray, J., Schmid, R., Collilieux, X., and Garayt, B.: IGS08: the IGS realization of ITRF2008, GPS Solutions, 16, 483–494, https://doi.org/10.1007/s10291-011-0248-2, 2012.

Rodríguez-Solano, C. J., Hugentobler, U., Steigenberger, P., and Lutz, S.: Impact of Earth radiation pressure on GPS position estimates, J. Geodesy, 86, 309–317, https://doi.org/10.1007/s00190-011-0517-4, 2012.

Rossi, G., Fabris, P., and Zuliani, D.: Slow transients recorded by the cGPS network FreDNet at the northern Adria microplate boundary (NE-Italy), Gephys. Res. Abstr., EGU General Assembly, Vienna, Austria, 15, 8417, 2013.

Royden, L. H.: Evolution of retreating subduction boundaries formed during continental collision, Tectonics, 12, 629–638, https://doi.org/10.1029/92TC02641, 1993.

Sánchez, L. and Drewes, H.: Crustal deformation and surface kinematics after the 2010 earthquakes in Latin America, J. Geodyn., 102, 1–23, https://doi.org/10.1016/j.jog.2016.06.005, 2016.

Sánchez, L., Seemüller, W., Drewes, H., Mateo, L., González, G., Silva, A., Pampillón, J., Martinez, W., Cioce, V., Cisneros, D., and Cimbaro, S.: Long-term stability of the SIRGAS reference frame and episodic station movements caused by the seismic activity in the SIRGAS region, IAG Symposia, 138, 153–161, https://doi.org/10.1007/978-3-642-32998-2_24, 2013.

Schmid, R., Dach, R., Collilieux, X., Jäggi, A., Schmitz, M., and Dilssner, F.: Absolute IGS antenna phase center model igs08.atx: status and potential improvements, J. Geodesy, 90, 343–364, https://doi.org/10.1007/s00190-015-0876-3, 2016.

Schmid, S. M., Fügenschuh, B., Kissling, E., and Schuster, R.: Tectonic map and overall architecture of the Alpine orogeny, Eclogae Geol. Helv., 97, 93–117, https://doi.org/10.1007/s00015-004-1113-x, 2004.

Seitz, F., Arenz, H., and Leismüller, F.: Position measurements at five permanent GPS stations in the Bavarian Alps as part of the Geodetic Alpine Integrated Network (GAIN) of the ALPS-GPSQUAKENET project, Deutsches Geodätisches Forschungsinstitut, Munich, https://doi.org/10.1594/PANGAEA.834193, 2014.

Serpelloni, E., Anzidei, M., Baldi, P., Casula, G., and Galvani, A.: Crustal velocity and strain-rate fields in Italy and surrounding regions: New results from the analysis of permanent and non-permanent GPS networks, Geophys. J. Int., 161, 861–880, https://doi.org/10.1111/j.1365-246X.2005.02618.x, 2005.

Serpelloni, E., Adziei, M., Baldi, P., Casula, G., and Galvani, A.: GPS measurement of active strains across the Apennines, Ann. Geophys., Supplement to Vol. 49, https://doi.org/10.4401/ag-5756, 2006.

Serpelloni, E., Faccenna, C., Spada, G., Dong, D., and Williams, S. D.: Vertical GPS ground motion rates in the Euro-Mediterranean region: New evidence of velocity gradients at different spatial scales along the Nubia-Eurasia plate boundary, J. Geophys. Res.-Sol. Ea., 118, 6003–6024, https://doi.org/10.1002/2013JB010102, 2013.

Steigenberger, P., Lutz, S., Dach, R., Schaer, S., and Jäggi, A.: CODE repro2 product series for the IGS. Published by Astronomical Institute, University of Bern, available at: http://www.aiub.unibe.ch/download/REPRO_2013 (last access: 20 August 2018), https://doi.org/10.7892/boris.75680, 2014.

Stocchi, P., Spada, G., and Cianetti, S.: Isostatic rebound following the Alpine deglaciation: impact on the sea level variations and vertical movements in the Mediterranean region, Geophys. J. Int., 162, 137–147, https://doi.org/10.1111/j.1365-246X.2005.02653.x, 2005.

Sue, C., Martinod, J., Tricart, P., Thouvenot, F., Gamond, J. F., Frechet, J., Marinier, D., Glot, J. P., and Grasso, J. R.: Active deformation in the inner western Alps inferred from comparison between 1972-classical and 1996-GPS geodetic surveys, Tectonophysics, 320, 17–29, https://doi.org/10.1016/S0040-1951(00)00024-X, 2000.

Sue, Ch., Delacou, B., Champagnac, J. D., Allanic, C., and Burkhard, M.: Aseismic deformation in the Alps: GPS vs. seismic strain quantification, Terra Nova, 19, 182–188, 2007.

Tesauro, M., Hollenstein, Ch., Egli, R., Geiger, A., and Kahle, H.-G.: Analysis of central western Europe deformation using GPS and seismic data, J. Geodyn., 42, 194–209, https://doi.org/10.1016/j.jog.2006.08.001, 2006.

Ustaszewski, K., Schmid, S. M., Fügenschuh, B., Tischler, M., Kissling, E., and Spakman, W.: A map-view restoration of the Alpine-Carpathian-Dinaridic system for the Early Miocene, Swiss J. Geosci., 101, 273–294, https://doi.org/10.1007/s00015-008-1288-7, 2008.

van Dam, T. and Ray, R.: S1 and S2 atmospheric tide loading effects for geodetic applications, available at: http://geophy.uni.lu/ggfc-atmosphere/tide-loading-calculator.html (last access: 20 August 2018), 2010.

van Mierlo, J., Oppen, S., and Vogel, M.: Monitoring of recent crustal movements in the Eastern Alps with the Global Positioning System (GPS), Tectonophysics, 275, 273–283, https://doi.org/10.1016/S0040-1951(97)00032-2, 1996.

Vergnolle, M., Calais, E., and Dong, L.: Dynamics of continental deformation in Asia, J. Geophys. Res., 112, B11403, https://doi.org/10.1029/2006JB004807, 2007.

Völksen, C., Walpersdorf, A., Aoudia, A., Barzaghi, R., Borghi, A., and Cannizzaro, L.: The ALPS GPSQUAKENET project – A permanent GPS network in the Alps, in: Bollettino di Geodesia e Science Affini, Journal of Istituto Geografico Militare, Vol. LXVIII, 1, 1–17, 2009a.

Völksen, C., Árnadóttir, T., Geirsson, H., and Valsson, G.: Present day geodynamics in Iceland monitored by a permanent network of continuous GPS stations, J. Geodyn., 48, 279–283, https://doi.org/10.1016/j.jog.2009.09.033, 2009b.

Waldhauser, F., Kissling, E., Ansorge, J., and Mueller, St.: Three dimensional interface modelling with two-dimensional seismic data: the Alpine crust-mantle boundary, Geophys. J. Int., 135, 264–278, https://doi.org/10.1046/j.1365-246X.1998.00647.x, 1998.

Weber, J., Vrabec, M., Stopar, B., Pavlovčič Prešeren, P., and Dixon, T.: The PIVO-2003 experiment: a GPS study of Istria peninsula and Adria microplate motion, and active tectonics in Slovenia, in: The Adria Microplate: GPS Geodesy, Tectonics, and Hazards, edited by: Pinter, N., Gyula, G., Weber, J., Stein, S., and Medak, D., 305–320, Springer, 2006.

Wessel, P., Smith, W. H. F., Scharroo, R., Luis, J. F., and Wobbe, F.: Generic Mapping Tools: Improved version released, EOS Trans. AGU, 94, 409–410, 2013.
Ziegler, P. A.: Geological Atlas of Western and Central Europe Shell, Int. Petrol. Maatschap. Geol. Soc., London, 239 pp., 1990.
Zuliani, D., Battaglia, M., Murray, M. H., Michellini, A., Burgmann, R., and Marson, I.: FREDNET: A continuous GPS geodetic network monitoring crustal deformation in NE Italy, AGU Fall Meeting, Poster presentation, 2002.

3

Multibeam bathymetry and CTD measurements in two fjord systems in southeastern Greenland

Kristian Kjellerup Kjeldsen[1,2,3], Reimer Wilhelm Weinrebe[4], Jørgen Bendtsen[5,6], Anders Anker Bjørk[1,7,8], and Kurt Henrik Kjær[1]

[1]Centre for GeoGenetics, Natural History Museum, University of Copenhagen, 1350 Copenhagen K, Denmark
[2]Department of Earth Sciences, University of Ottawa, Ottawa, K1N 6N5, Canada
[3]DTU Space – National Space Institute, Technical University of Denmark, Department of Geodesy, 2800 Kgs. Lyngby, Denmark
[4]GEOMAR, Helmholtz-Zentrum für Ozeanforschung, 24148 Kiel, Germany
[5]Arctic Research Centre, Aarhus University, 8000 Aarhus, Denmark
[6]ClimateLab, Symbion Science Park, Fruebjergvej 3, 2100 Copenhagen Ø, Denmark
[7]Department of Earth System Science, University of California, Irvine, CA 92697, USA
[8]NASA Jet Propulsion Laboratory, Pasadena, CA 91109, USA

Correspondence to: Kristian Kjellerup Kjeldsen (kkjeldsen@snm.ku.dk)

Abstract. We present bathymetry and hydrological observations collected in the summer of 2014 from two fjord systems in southeastern Greenland with a multibeam sonar system. Our results provide a detailed bathymetric map of the fjord complex around the island of Skjoldungen in Skjoldungen Fjord and the outer part of Timmiarmiut Fjord and show far greater depths compared to the International Bathymetric Chart of the Arctic Ocean. The hydrography collected shows different properties in the fjords with the bottom water masses below 240 m in Timmiarmiut Fjord being 1–2 °C warmer than in the two fjords around Skjoldungen, but data also illustrate the influence of sills on the exchange of deeper water masses within fjords. Moreover, evidence of subglacial discharge in Timmiarmiut Fjord, which is consistent with satellite observations of ice mélange set into motion, adds to our increasing understanding of the distribution of subglacial meltwater. Data are available through the PANGAEA website at https://doi.pangaea.de/10.1594/PANGAEA.860627.

1 Introduction

During the past decades the Greenland ice sheet has experienced a considerable increase in mass loss associated with the speeding up of glaciers and enhanced melting (Khan et al., 2015). An anomalous inflow of subtropical waters driven by atmospheric changes, multidecadal natural ocean variability, and a long-term increase in the North Atlantic's upper ocean heat content since the 1950s has, in conjunction with increased meltwater runoff, led to enhanced submarine melting. This is believed to have triggered the retreat of Greenland's outlet glaciers, partly responsible for the increased ice loss (Straneo and Heimbach, 2013).

In order to better understand oceanic, cryospheric, and geological processes and their interaction, detailed knowledge of the morphology of a given survey area is essential. The International Bathymetry Chart of the Arctic Ocean (IBCAO; Jakobsson et al., 2012) provides an overview of the circum-Arctic bathymetry. However, in many of the Greenland fjord systems there is little, if any, bathymetric information, and therefore the resulting gridded bathymetry may have large errors. Narrow fjords, bays, or islands only slightly wider than the IBCAO resolution (500 m) are difficult to preserve, yet this is at least partly aided by including land topography, which in Greenland is based on a 2000 × 2000 m resolution digital elevation model (DEM; Jakobsson et al., 2012). As a result, many glaciers appear grounded near or at sea level, whereas in reality the glaciers are grounded in deeper waters (Andresen et al., 2014; Rignot et al., 2016).

The publication of different datasets is continuously improving the bathymetry around Greenland (e.g., Arndt et al., 2015; Fenty et al., 2016; Rignot et al., 2015, 2016; Schumann et al., 2012; Williams et al., 2017), even if it is only based on a suite of single-point observations (Andresen et al., 2014) or inversion of gravity data to obtain bathymetry (Porter et al., 2014). Whereas the topography of the onshore area is more easily determined from air and satellite imagery and altimetry (e.g., Korsgaard et al., 2016; Willis et al., 2015), the relief of the submarine parts is hidden by the water column and can only be determined with hydroacoustic methods or aerial gravimetry. Thus, extensive bathymetric mapping is necessary to create a full image of the seafloor morphology.

During the summer of 2014 a multibeam echo sounder was temporarily installed on SS *ACTIV* (Fig. 1) and used to map the seafloor in two fjord systems in southeastern Greenland. To obtain knowledge of the water column, sound velocity profiles were determined at 11 positions using a CTD sensor (CTD: conductivity, temperature, depth). The data presented here supplement data acquired during the 2016 field campaign of Oceans Melting Greenland (Fenty et al., 2016) and as a consequence enabled better use of the ship's resources during the 2016 campaign.

Figure 1. The three-masted wooden schooner SS *ACTIV*.

Study site

The study area consists of two fjord systems in southeastern Greenland, namely Timmiarmiut Fjord and the fjord systems around Skjoldungen (Fig. 2b). Timmiarmiut Fjord (Fig. 2c) is ca. 57 km long and reaches the Timmiarmiut Glacier, a large marine outlet glacier from the ice sheet. Along the fjord there are different branches and islands of variable sizes, in particular towards the north where the fjord system reaches Heimdal Glacier, another large marine outlet glacier from the ice sheet. Both glaciers are fast-moving outlets of the Greenland ice sheet (Rignot and Mouginot, 2012), with Timmiarmiut Glacier being the 15th fastest in Greenland. The fjord system also drains a large number of smaller glaciers and ice caps, both land- and marine-terminating, that are either isolated or in contact with the ice sheet. At the fjord mouth, depths extent down to ca. 620 m according to IBCAO (Jakobsson et al., 2012). These depths are based on single-beam echo soundings retrieved from the Olex database (http://www.olex.no) as the median values on a 0.12×0.12 arcmin grid. Subsequently, these have been filtered onto a 2×2 km grid using the median z value within each grid cell before being resampled to a 500×500 m grid and evaluated (Jakobsson et al., 2012). The fjord system surrounding Skjoldungen (Fig. 2d) is comprised of a northern and a southern sound, Nørre Skjoldungesund (NSsund; ca. 49 km) and Søndre Skjoldungesund (SSsund; ca. 52 km), respectively, and a narrow, shallow section (ca. 10 km) towards the northwest that links the two sounds. The fjord system drains both local and ice sheet glaciers and ice caps with the vast majority being land-terminating, while the large ice-sheet-based Thrym Glacier enters the fjord directly in the very northwestern part of the fjord system at the head of NSsund. Around Skjoldungen the official nautical chart supplied by the Danish Geodata Agency is comprised of low-density point depth measurements ranging between 179 and 548 m in NSund and between 46 and 702 m in SSsund, while for both NSsund and SSsund IBCAO suggests depths down to ca. 20 m. The Fjord system around the island of Skjoldungen is annually visited by cruise ships, thus adding to the importance of mapping the fjords.

Outside the fjord systems, oceanographic conditions are dominated by the East Greenland Coastal Current (EGCC) above the shelf and the East Greenland Current (EGC) located at the shelf break (Sutherland and Pickart, 2008). These drive polar surface water originating from the Arctic Ocean and Atlantic water originating from the Irminger Sea.

2 Data and methods

The data in this study were collected during July 2014 and are comprised of a multibeam dataset and 11 CTD profiles. Below follows a description of the system, sensors and components, and the deployment. Figure 3 shows the deployment of the system onboard SS *ACTIV*. The software used for data acquisition was ELAC Nautik Hydrostar 3.5.3, while post-processing was accomplished using the software packages ELAC Nautik HDPpost and CARIS HIPS as well as the open-source software packages MB-System (Caress and Chayes, 1996) and GMT (Wessel and Smith, 1991).

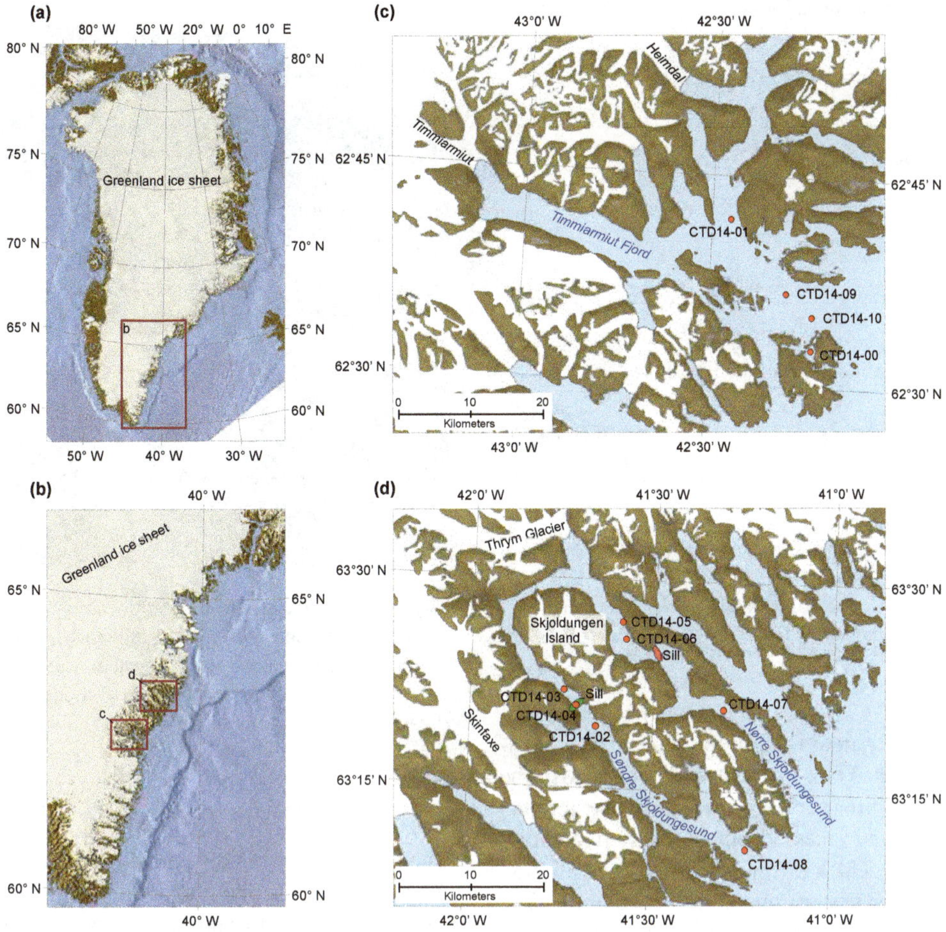

Figure 2. Maps of Greenland **(a)** and the southeast coast **(b)** with red boxes show location of the study sites. Map of the Timmiarmiut Fjord system **(c)** and the fjord complex around Skjoldungen **(b)** with Nørre Skjoldungesund (NSsund) towards the north and Søndre Skjoldungesund (SSsund) towards the south **(d)**. Note the location of two sills found in NSsund and SSsund. Also shown in **(c–d)** are the locations of the CTD stations.

2.1 Seabeam 1050

Multibeam systems are echo sounders that map the seafloor along a wide swath beneath the vessel. Due to their complexity, such systems are generally permanently installed in the hull of survey vessels. Onboard SS *ACTIV* we temporarily installed a portable ELAC Nautik Seabeam 1050 multibeam system provided by the GEOMAR Helmholtz-Centre for Ocean Research Kiel.

The Seabeam 1050 multibeam echo sounder collects bathymetry data with beams as narrow as $1.5° \times 1.5°$ and a swath width of up to 153°. The 50 kHz signal achieves a depth range of more than 2000 m. Different to other multibeam systems, which transmit a full swath of acoustic energy and use directed reception (beam forming), the Seabeam 1050 applies beam forming at transmission and reception, thus achieving a very high side lobe suppression of 36 dB with very low error rates. The disadvantage of this technique is a slightly lower ping rate. The system is completely stabilized for roll of the ship if an adequate motion sensor is installed. The Seabeam 1050 employs two transducer arrays, port and starboard, both capable of transmitting and receiving. Their acoustic planes are tilted 38° to the vertical. The arrays are normally installed fixed to the ship's hull, but on SS *ACTIV* they were mounted using a steel pole over the port side of the vessel. An overview of the technical specifications of the Seabeam 1050 is provided in Table 1.

2.2 Navigation, heading, and motion sensor F180R

The F180R inertial attitude and positioning system from Coda Octopus is an instrument designed to make precise measurements of the vessel attitude (including heading), dynamics, and geographical position for use in any marine hydrographic survey application. It is a multisensor system consisting of two survey-grade GPS receivers and an inertial measurement unit (IMU) made up of three solid-state gyros and three inertial-grade accelerometers. The IMU of the

Table 1. Overview of the technical specifications of the Seabeam 1050.

Frequency:	50 kHz
Maximum depth:	3000 m
Beam width:	1.5° × 1.5°
Pulse length:	0.15–10 ms
Side lobe suppression:	36 dB (transmission and reception)
Maximum swath width:	153°
Maximum number of beams:	126
Beam spacing:	equiangular

Table 2. Performance specifications of the F180R system.

	According to specifications	Achieved during *ACTIV* survey
Heading accuracy	0.05°	0.46°
Attitude accuracy	0.025°	0.045°
Position accuracy	0.5 m (without DGPS DGPS or RTK)	0.49 m
Heave accuracy	5 % of heave amplitude, minimum 5 cm	Not determined

F180R is built into a separate waterproof pod in order to allow the IMU to be located close to the transducer heads. The F180R integrates the information provided by the attitude and position sensors and takes advantage of their complementary attributes to yield a position and attitude solution more stable than either system operating in isolation. The performance specifications of the F180R system are provided in Table 2.

The F180R system performance depends on a variety of external factors, e.g., poor GPS satellite constellation and multipaths of the GPS signals that cause errors and reduced accuracy of attitude and heading determinations.

2.3 CTD sensor Sea & Sun CTD48M

The CTD48M (Sea & Sun Marine Tech) memory probe is a small (1.2 kg) microprocessor-controlled multi-parameter titanium (except screws) probe for precise marine measurements. The accuracy of the conductivity and temperature sensor is specified as $\pm 0.003\,\mathrm{mS\,m^{-1}}$ and $\pm 0.002\,°\mathrm{C}$, respectively. Considering the instrumental drift we estimate the temperature and salinity error to be about $\pm 0.05\,°\mathrm{C}$ and ± 0.05, respectively. This is well below the temperature and salinity range in the water column reported below. The CTD48M data acquisition rate was set to record data every 1 dbar during both downcast and upcast. CTD profiles were obtained manually using rope and a diving lead. Practical salinity (S_P) was calculated by the Sea & Sun Marine Tech software and potential temperature was calculated from IOC et al. (2010). Only downcast profiles were considered in the analysis.

Figure 3. Deployment of the multibeam system onboard SS *ACTIV* using a clamp and a U-shaped clip **(a-b)**. **(c)** The system during the surveying mode and **(d)** the system in the safe position used for transit. The multibeam system and computers were in installed in the wheelhouse **(e)**.

2.4 System installation onboard *ACTIV*

The two multibeam transducers were mounted to a 6 m pole, which was attached to the hull outboard at the port side midship using a clamp and a U-shaped clip (Fig. 3a, b). The transducer head was secured by two Dyneema ropes attached to the vessel (Fig. 3c). During transit and in ice-covered areas, the entire pole could be hoisted up into a safe position (Fig. 3d). The construction was sufficiently stable and easy to manage throughout the entire cruise; however, the hydrodynamic pressure of the water against the pole during the surveys caused a slight bending of the pole. In addition, a collision with an ice floe on 22 July might have also shifted the transducers slightly out of position. Both events likely had an influence on the accuracy of the multibeam data. The multibeam system and the computer for data acquisition were installed in the wheelhouse (Fig. 3e).

The IMU of the motion sensor F180R was installed on the deck close to the transducer pole. The two GPS antennas were mounted at a distance of 2.003 m apart on a wooden bar, which was temporarily installed on top of the wheelhouse. The motion sensor was set up to determine the motion of the vessel at the positions of the multibeam transducers (remote lever arm calculation). The reference point for the multibeam depth measurements was defined to be a point exactly above the center point of the transducers at water level.

2.5 Post-setup calibration and limitations

Once the temporary installation was completed, several components of the system had to be calibrated. The motion sensor F180R runs a self-calibration scheme, which requires considerable movement of the vessel and a continuously good GPS signal. Initially, the self-calibration failed. After remounting the GPS antennas at a position near the stern of the vessel to reduce the multipath of the GPS signals caused by the masts and rig, a new self-calibration procedure of the F180R was performed. At this position a stable operation of the motion sensor was achieved; however, the accuracy of the attitude and heading determinations missed the attainable specifications (Table 2).

The multibeam system requires calibration to determine the angular offsets of the transducers from the horizontal (roll offset) and the vertical (pitch offset) axis, respectively. To perform a roll calibration, a profile of multibeam data over an essentially flat seafloor is recorded on the same line in both directions. When both lines are processed separately, a potential roll offset will lead to different inclinations of the seafloor in the recorded data and the angular offset can be determined. To determine a possible pitch offset, a structure on the seafloor is mapped on two lines in opposite directions along the same profile. A possible pitch offset leads to different positions of the structure in separately processed lines. As no information on the seafloor depth was available beforehand, the calibrations had to be run over "unknown" seafloor. It turned out that the area was not well suited for a roll calibration; however, an estimate of the roll offset of 0.6° could be determined from the recordings and was used during the following days. A significant vertical offset of the transducers had not been detected, so the pitch bias was set to 0°.

The slight bending of the pole likely influenced the accuracy of the multibeam data. We assume that the offset in angle from the vertical was within a few degrees, and as a 4° offset at a water depth of 1000 m increases the depth by 2.4 m and also shifts the horizontal position of the measured point by 70 m, we anticipate that the uncertainty related to the bending and collision was within a few meters. However, as the system was only temporally installed and the seafloor "unknown", we were unable to test how our setup was affected.

3 Results and discussion

3.1 Multibeam surveys

The multibeam system was installed onboard *ACTIV* on 15 July 2014 when the vessel was anchored in the fjord south of the old Timmiarmiut weather station. Multibeam surveys were carried whenever weather and ice conditions as well as the scientific program allowed until the system was dismounted on the evening of 28 July 2014. The total track length of the multibeam surveys was 402 km recorded in 54.86 h and based on a total of 74 502 pings.

3.1.1 Timmiarmiut area survey

Multibeam surveys in the Timmiarmiut area were carried out on 16, 17, 18, and 28 July. The days of 16 and 17 July were used to calibrate the setup and reconfigure the original installation. The survey in the Timmiarmiut Fjord area was continued on 18 and 28 July; however, due to numerous ice floes and icebergs only a part of the fjord was mapped (Fig. 4a). The survey shows that at least the outer part of the fjord consists of depths down to ca. 1060 m. An enlargement of the mouth of Timmiarmiut Fjord is provided in Fig. 5a. At a cross section at the fjord mouth, the deepest part is found along a ca. 2 km wide section in the middle where depths range between 740 and 830 m. North of this midsection, depths increase from ca. 780 to ca. 95 m over a distance of ca. 1.2 km. South of the midsection, depths increase from 780 to ca. 190 m over a ca. 2 km stretch intervened by a 1 km wide and ca. 150 m deep trough that extends down to ca. 550 m, which further up fjord increases to ca. 450 m. The general configuration of a wide, deep cross section with great depths is characteristic of the outer well-surveyed area, which covers ca. 15 km from the mouth and up fjord, but it also seems to resemble the branch towards the north covered by the red line in Fig. 4a despite the poorer multibeam coverage. In order to compare to existing data, we extract data from the deepest part of the fjord in the along-fjord direction (Fig. 4b, black and red lines), which ranges between ca. 800 and 880 m. Figure 4b shows data extracted from IBCAO (Jakobsson et al., 2012) along the three center lines: dashed red, solid red, and black. Our new data show that along the surveyed center line, the fjord is on average 325 m deeper where data overlap. Note that along the inner part of the fjord (dashed red line), IBCAO contains values above sea level. Following our data collection in 2014, the NASA-led Oceans Melting Greenland (OMG) mission (Fenty et al., 2016) acquired data in Timmiarmiut Fjord during their 2016 cruise. With knowledge of our data, OMG was able to collect data to supplement our mapping effort, thus saving time and enabling better use of the ship's resources.

3.1.2 Skjoldungen Fjord survey

In the Skjoldungen Fjord complex around Skjoldungen, the multibeam system was deployed on 20, 21, 22, 23, 24, and 26 July. Nearly the entire fjord area was surveyed, leaving only a small gap of 5 km in the innermost part of the fjord system and small unmapped stripes close to the shorelines. Whenever possible, the fjord was mapped along three lines: one in the center, one closer to the northern shore, and one closer to the southern shore. The innermost stretch of the fjord area could not be mapped as the slopes of the surrounding mountains were too steep for the motion sensor to main-

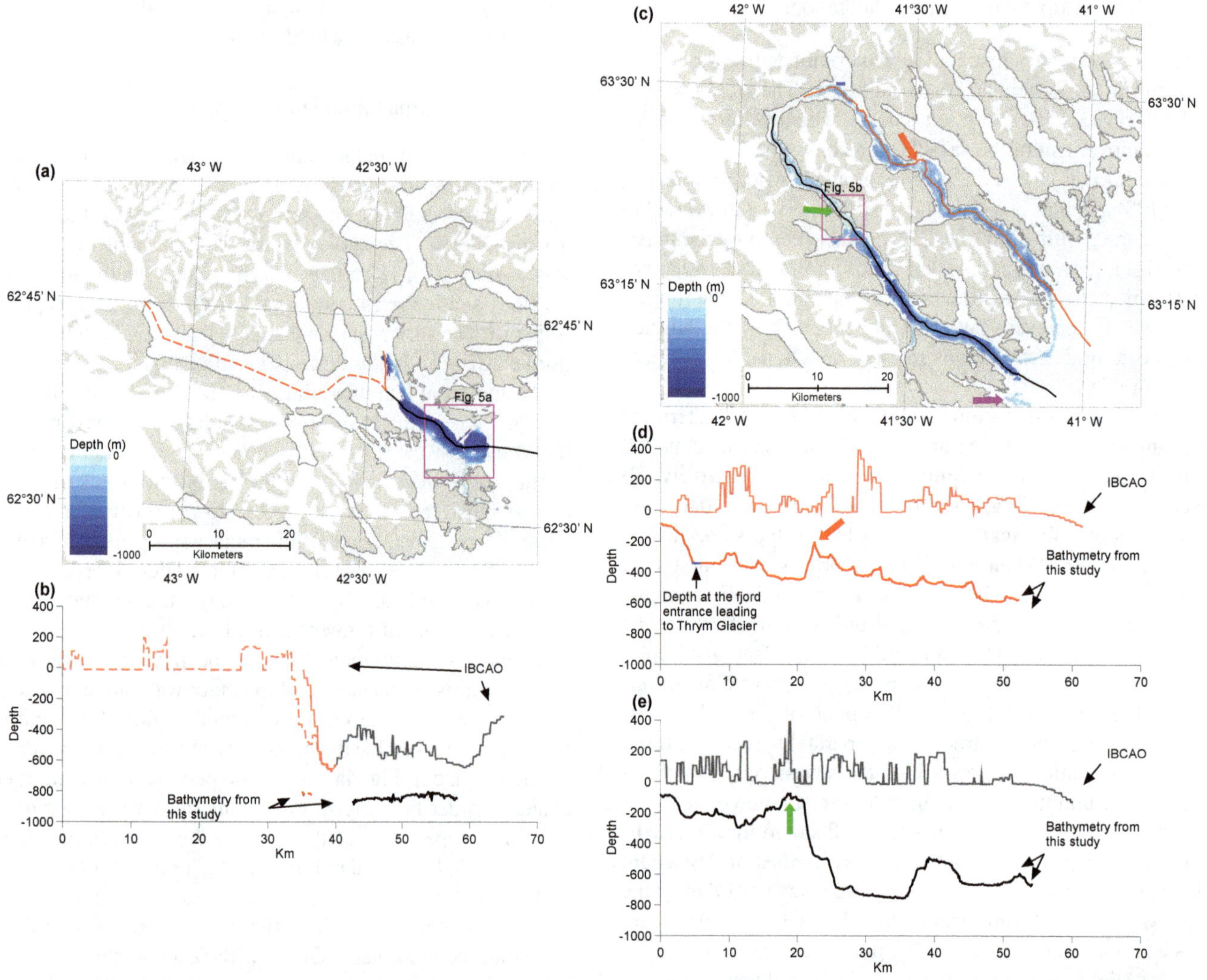

Figure 4. Surveyed bathymetry in the Timmiarmiut Fjord system **(a)**. The three colored lines follow the deepest part along the fjords and are illustrated in panel **(b)**, representing the outer well-surveyed area (black line), the branch towards the north leading to Heimdal Glacier at the head, and towards Timmiarmiut Glacier (dashed red line). The surveyed bathymetry around Skjoldungen is illustrated in panel **(c)**, while panel **(d)** shows the longitudinal profiles of Søndre Skjoldungesund (black lines) and Nørre Skjoldungesund (red lines). Green, red, and magenta arrows in panels **(c)** and **(d)** indicate the prominent sill in SSsund, the local high in NSsund, and the offshore shoal, respectively. For the profiles in panels **(b)** and **(d)**, bathymetry obtained from IBCAO is included.

tain its GPS connection; thus the multibeam system lacked navigation, heading, and motion information. Besides this small gap, a comprehensive map of the entire Skjoldungen Fjord area can now be presented for the first time (Fig. 4c). Figure 4d shows the bathymetry from our new data and IBCAO-extracted data at the center lines of the two fjords. Note that the IBCAO dataset here contains values above sea level.

Generally, the fjord width around Skjoldungen varies between 1.1 and 3.1 km except for a narrow stretch along SSsund, where the fjord width narrows down to ca. 600 m, and along a small stretch along the innermost unmapped part, where the width narrows down to ca. 580 m.

The bathymetry map of the Skjoldungen Fjord complex displays water depths up to 800 m (Fig. 4c and d). The deepest part is a stretch of about 10 km in SSsund. Generally, the southern fjord features an outer deep part showing water depths between 500 and 800 m and a shallow inner part with depths never reaching more than 300 m. Between the outer and inner parts along the narrowing of the fjord, there is a prominent sill with depths only extending down to ca. 77 m (green arrow, Fig. 4c, d; an enlargement is provided in Fig. 5b). Interestingly, in the field we observed multiple lateral moraines on land on either valley side dipping concave down towards the sill or near it. On digital elevation models and orthophotos from 1981 (Korsgaard et al., 2016), the

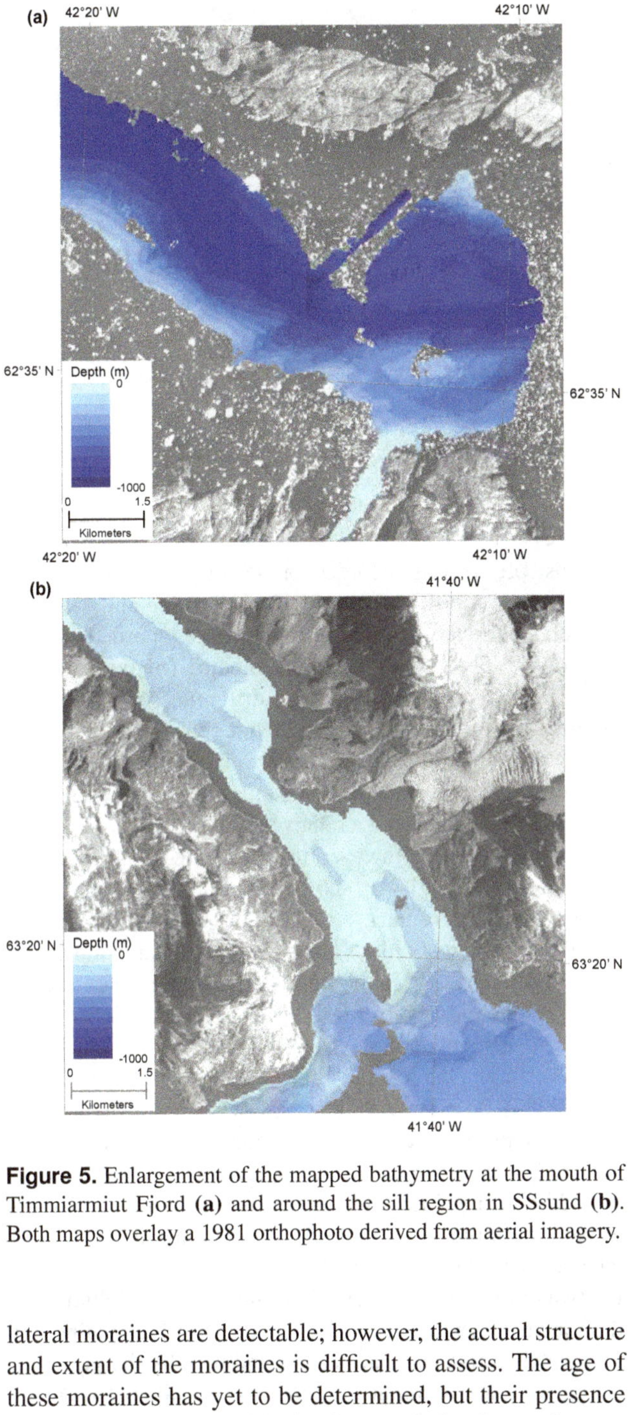

Figure 5. Enlargement of the mapped bathymetry at the mouth of Timmiarmiut Fjord **(a)** and around the sill region in SSsund **(b)**. Both maps overlay a 1981 orthophoto derived from aerial imagery.

lateral moraines are detectable; however, the actual structure and extent of the moraines is difficult to assess. The age of these moraines has yet to be determined, but their presence suggests that the sill and bedrock outcrop that forces the narrowing of the fjord acted as a pinning point for the receding glacier during deglaciation of the fjord.

The bathymetry of the NSsund is remarkably different from the bathymetry of the SSsund. NSsund shows a more gradual increase of water depths from 85 m in the very inner part to ca. 590 m at the entrance with a few undulations along the fjord and a local high of ca. 210 m (red arrow, Fig. 4c, d). However, in the vicinity of this local high point, only part of the area was surveyed and therefore a deeper path in the non-surveyed area of the bend could be possible, thus making our data a minimum estimate of the limiting depth of NSsund. At the fjord entrance leading to Thrym Glacier the water depth in NSsund is ca. 340 m (blue bar, Fig. 4c, d).

In the non-surveyed area in the northwestern part of the fjord complex where SSsund and NSsund are connected, water depth is likely limited and perhaps the deepest points in either end reflected the maximum depth. However, a moraine complex marking the Little Ice Age maximum extent of a local glacier originating from Skjoldungen extends halfway into the fjord and likely causes the fjord to be fairly shallow.

Around the island of Skjoldungen, IBCAO (Jakobsson et al., 2012) displays only a limited number of grid cells with negative values. Along the longitudinal profile overlapping data from this study and IBCAO range between 15 and 390 m with an increase in depth out on the shelf (Fig. 4d). The values above sea level in the IBCAO dataset, both around Skjoldungen and in Timmiarmiut Fjord, are in part due to the 500×500 m resolution of the data not fully being capable of resolving narrow fjords, but more likely data not being available or incorporated into the final dataset.

Offshore of the Skjoldungen fjord complex, we further surveyed a small shoal (magenta arrow, Fig. 4b). At the position 63°07.5′ N, 41°11.0′ W a very shallow area was mapped; the seafloor at this position showed water depths as shallow as 13 m. Surveying along two additional profiles, the extent of this feature could be outlined; unfortunately, an iceberg was stranded right on the shallowest position, so this spot could not be mapped. This submarine feature has dimensions of ca. 800×400 m and a height of ca. 200 m.

3.1.3 Constraints and limitations of the multibeam survey

The part of the area mapped during the SS *ACTIV* cruise in 2014 has never been surveyed before using a multibeam system. Furthermore, for great parts of the region no sounding information was available at all. The conditions for multibeam surveys in this area, particularly on this cruise, were difficult and of course this has affected the quality of the data.

Multibeam transducers are usually permanently installed in the hull of a vessel. For precise measurements of the water depth, the transducers have to be mounted such that their three axes are aligned horizontally, vertically, and in the direction of the center line of the vessel. Minimal deviations are determined by a calibration scheme following the installation. On SS *ACTIV* the transducers were installed temporarily at the lower end of a pole 6 m in length, which was then mounted over the side of the vessel. Of course this cannot be as stable as an installation flush in the hull. The pole is prone to vibrations, and the hydrostatic pressure on the pole while deployed in the water and when moving led to slight bending of the pole. Floating ice in the water also collided with the pole, moving it slightly out of position. A proper survey would have required at least a roll calibration each time the

Table 3. Overview of the 11 CTD profiles.

Station no.	Date (dd-mm-yyyy)	Fjord system	Latitude (degrees)	Longitude (degrees)	Depth of the CTD profile (m)	Water depth (m)
CTD14-00	16-07-2014	TF	62.52633	−42.18650	14	56
CTD14-01	18-07-2014	TF	62.69767	−42.44867	707	795
CTD14-02	21-07-2014	SSsund	63.32350	−41.65467	420	454
CTD14-03	21-07-2014	SSsund	63.36550	−41.74217	135	161
CTD14-04	21-07-2014	SSsund	63.34800	−41.70750	75	59–90*
CTD14-05	22-07-2014	NSsund	63.44933	−41.58950	13	13
CTD14-06	23-07-2014	NSsund	63.42883	−41.57967	412	420
CTD14-07	23-07-2014	NSsund	63.34717	−41.30383	323	407
CTD14-08	24-07-2014	SSsund	63.18167	−41.23200	238	238
CTD14-09	28-07-2014	TF	62.61100	−42.28933	673	826
CTD14-10	28-07-2014	TF	62.58450	−42.21767	289	400

* Water depth registered by the ship while obtaining the CTD profile.

pole was deployed into the water; however, the area was not well suited for a roll calibration, which should be run over absolutely planar seafloor. As an ad hoc solution, transit lines run in opposite directions roughly along the same line were used throughout the survey to check and correct the roll bias values.

The motion sensor F180R used during the survey needs a continuously good connection to GPS satellites, and this was very difficult to maintain inside the fjords. Thus, the achieved heading accuracies and motion information often failed to achieve the intended specifications, leading to decreased quality of the multibeam data. Other sources of error in the data are missing tidal information, frequent changes in the water sound velocity due to the influence of meltwater, and abrupt course changes of the vessel due to ice floes.

However, in spite of the aforementioned problems the results of the multibeam surveys clearly demonstrate that it is actually possible using a ship such as SS *ACTIV* and a temporary installation to achieve bathymetric maps of satisfactory quality, even under difficult conditions in remote areas.

3.2 CTD measurements

During the SS *ACTIV* cruise, the CTD was deployed 11 times (Table 3). Locations of the profiles are displayed in Fig. 2c and d. Two of the 11 profiles were obtained in shallow waters while the ship was anchored and only the top 13–14 m of the water column was sampled. Three CTD profiles were obtained from Timmiarmiut Fjord (TF), four from SSsund, and two from NSsund. Note that the vessel drifted during the recording of CTD14-03, causing the water depth to vary.

Figure 6 illustrates the temperature and salinity profiles and the corresponding temperature–salinity (T–S) plots from the nine deeper profiles, divided into those from Timmiarmiut Fjord (a, d), SSsund (b, e), and NSsund (c, f).

In Timmiarmiut Fjord (Fig. 6a and d) three profiles were recorded reaching depths between 285 and 700 m. The innermost profile was recorded on 18 July, roughly midway between the fjord mouth and Timmiarmiut and Heimdal glaciers, while the two other profiles were recorded 10 days later on 28 July closer to the fjord mouth. The three profiles show certain similarities but also differences, with the two outermost profiles being almost identical throughout the profiles. For all three profiles the top consists of a pronounced cold layer between ca. 120 and 30 m with temperatures less than 0 °C and underlying a relatively low-saline ($S < 32$) and warm surface layer. However, the transition to the layers below differs as CTD14-01 shows a gradual warming and increasing salinity from ca. 110 to ca. 175 m, while at the fjord mouth the same conditions are already reached at a depth of ca. 125 m. At ca. 240 m the three profiles again start to deviate, with the midway profile generally being cooler by more than 1 °C at some depths.

Interestingly, the top ca. 40 m of the water column in profile CTD14-01 shows a clear difference compared to the two other profiles near the fjord mouth. Here temperatures are cooler and more saline compared to the fjord mouth. This is a strong indication of mixing with subglacial water and therefore subglacial discharge from a tidewater outlet glacier in the fjord (Mortensen et al., 2013). Subglacial meltwater discharge causes a buoyant ascending plume near the glacier where meltwater is mixed with high-saline bottom water (Bendtsen et al., 2015), and thereby the salinity in the upper water column may increase significantly (Kjeldsen et al., 2014). Thus, we speculate that the high salinities observed in the layer between ca. 50 and 10 m at station CTD14-01 can be explained as subglacial water from a tidewater outlet glacier. This is further supported by satellite observation from the period before the CTD profile was measured. MODIS imagery from 10 to 22 July 2014 (Fig. 7) shows that a large pool of ice mélange near the calving front of Heimdal Glacier north of the CTD station 14-01 is set into motion and directed away from the terminus by 12 July. By 17 July a large area of the inner fjord is covered by ice reach-

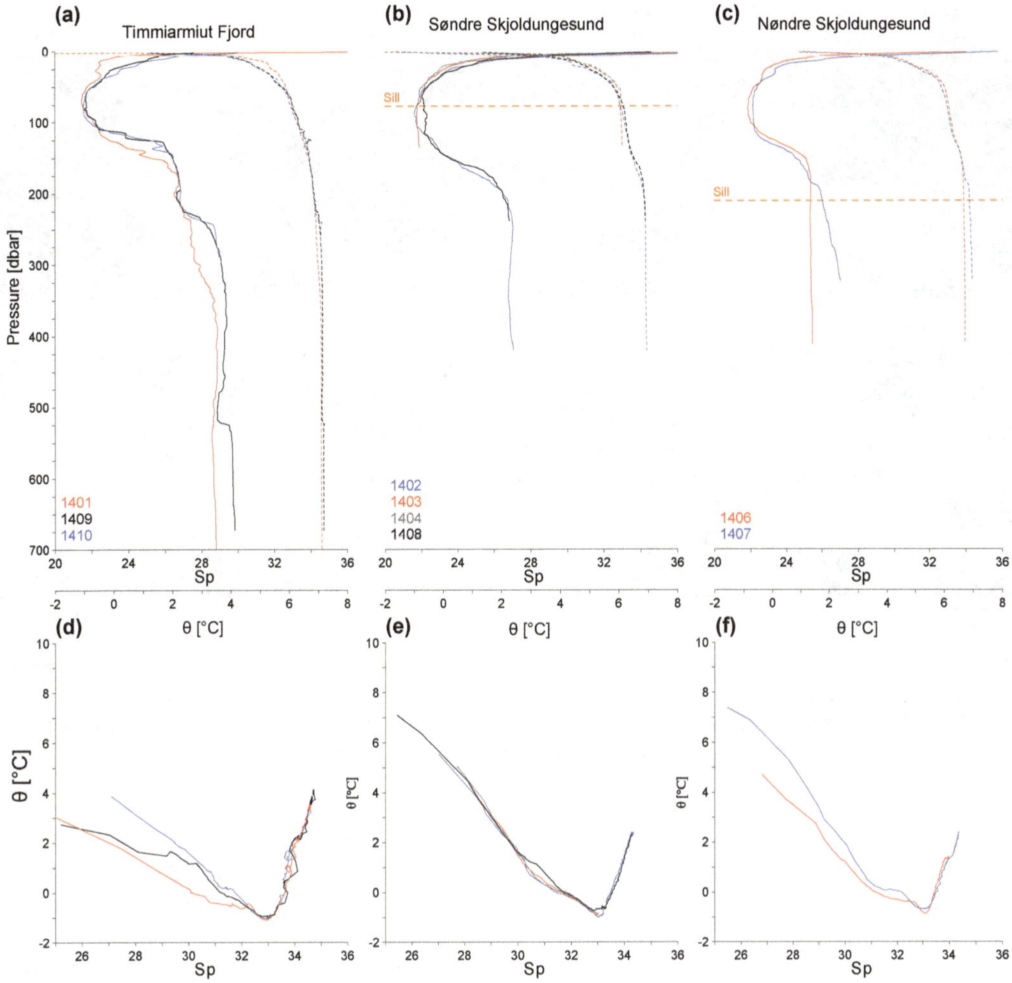

Figure 6. Temperature and salinity profiles and the corresponding *T–S* diagrams for Timmiarmiut Fjord **(a, d)**, Søndre Skjoldungesund **(b, e)**, and Nørre Skjoldungesund **(c, f)**. Approximate sills depths retrieved from the bathymetry data are illustrated in **(b, c)**.

ing midway between Heimdal Glacier and the CTD station. Thus, the high salinities between 50 and 10 m of depth can be explained by subglacial water originating from the Heimdal tidewater outlet glacier, and the relatively low temperatures in the upper 50 m can be explained by heat loss due to the melting of the icebergs. However, the unsteady nature of these fjord systems can limit conclusions based on a few observations.

In SSsund four profiles (Fig. 6b and e) were recorded on 21 and 24 July, two seawards of the sill (CTD14-02 and CTD14-08; Fig. 4c, d, green arrow), one on the sill (CTD14-04), and one landwards of the sill (CTD14-03). The deepest reached 420 m and was recorded mid-fjord seawards of the sill, CTD14-02. Generally, the four profiles show the same properties down through the water column, with a fresh warm surface layer overlaying a layer of polar water that extends down to 200 m. Further below at greater depths, the water column appears homogenous with a constant temperature and salinity, though this is based only on a single profile that reaches 238 m (CTD14-08) and another that reaches 420 m (CTD14-02).

Interestingly, below ca. 110 m, profiles landwards and seawards of the sill begin to deviate from each other, with CTD14-03 (landwards) being ca. 0.6 °C cooler and 0.4 units more fresh. Possibly, this reflects a deeper-lying local water mass occupying the basin between the sill and the northwestern part of the fjord complex, as the inner part connecting SSsund and NSsund is likely shallow due the presence of a moraine complex, thus limiting the exchange of deeper water masses. This water mass would, however, likely be affected by water masses that flow just over the sill, possibly during winter and spring as observed in Godthåbsfjorden in western Greenland (Mortensen et al., 2011).

In NSsund (Fig. 6c and f) two deeper and one shallow profile were recorded. The shallow profile (CTD14-05) is omitted as it reassembles the nearby deeper profile, CTD14-06. The latter is recorded landwards of the sill (Fig. 4c, d, red arrow), while CTD14-07 was recorded midway between

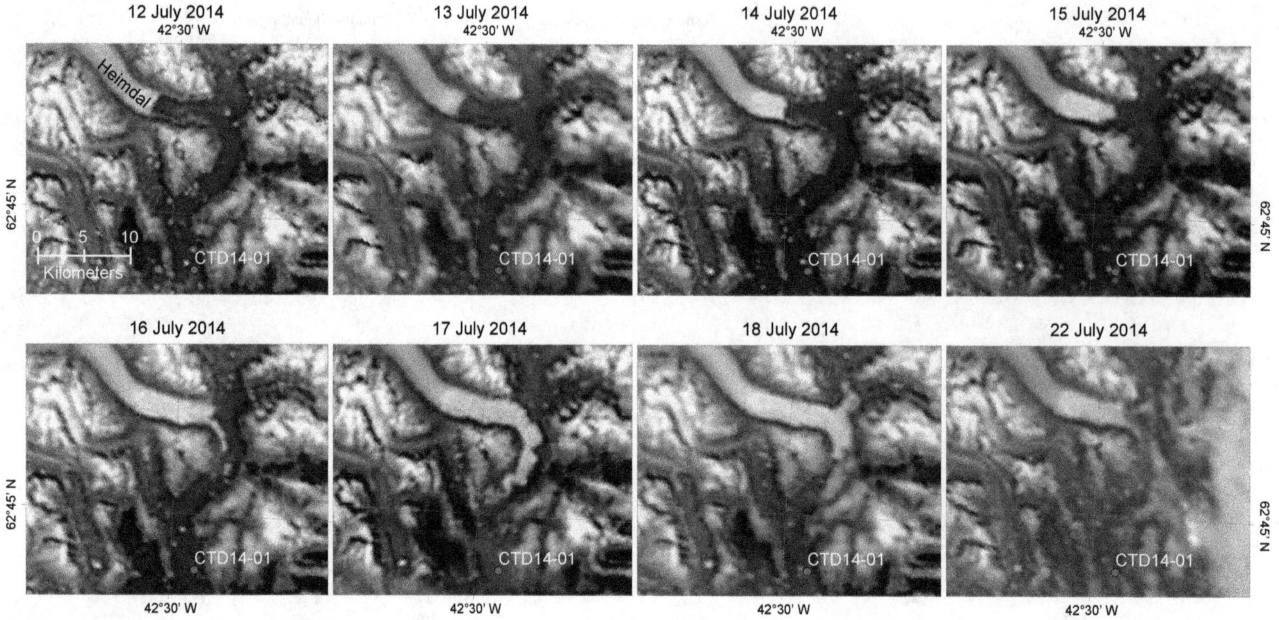

Figure 7. MODIS imagery of the inner part of the branch of Timmiarmiut Fjord that leads towards Heimdal at the head. The imagery illustrates the distribution of ice mélange set into motion on 13 July near the calving front of Heimdal Glacier due to the release of subglacial meltwater. During the following days the ice mélange expands and is directed southwards. In our CTD measurements on 18 July from CTD14-01, we detect this as a cooling and an increase in the salinity of the layer between 10 and 50 m of depth.

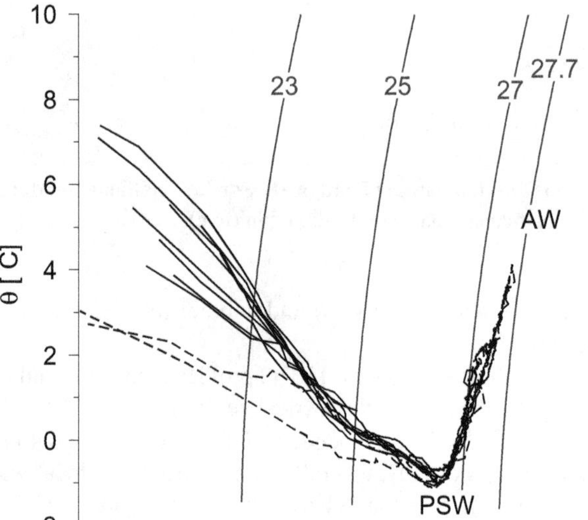

Figure 8. Potential temperature (θ) and salinity (S_P) diagram of all profiles from the three fjords. The two stations in Timmiarmiut Fjord near the Heimdal tidewater outlet glacier (CTD14-01 and 14-09) are shown with dashed lines. Temperature–salinity characteristics of two water masses, i.e., Atlantic water (AW) and polar surface water (PSW), are indicated and lines of constant density (σ_θ, $kg\,m^{-3}$) are contoured.

the sill and the fjord mouth. In the uppermost part, the profiles look alike with CTD14-06 showing a slightly stronger stratification near 140 m, while at depth the deviation between the two profiles starts to become more obvious. Below ca. 185 m CTD14-07 gradually becomes warmer and more saline through to the deepest measurement at 323 m. In contrast, below ca. 185 m CTD-1406 remains unchanged with depth. This difference reflects limited water exchange in the deeper parts of the fjord, and thus supports the presence of a sill in accordance with the recorded bathymetry.

Relatively cold water masses characterize the three fjords in the upper ca. 120 m and, except for the presence of subglacial water in the innermost profile in Timmiarmiut Fjord, which has a local origin, the upper water masses share the same temperature and salinity characteristics. The deeper part of the fjords show some differences; the bottom water masses below 240 m in Timmiarmiut Fjord are 1–2 °C warmer than in the other two fjords. This indicates a stronger influence from mixing with warmer water masses in Timmiarmiut Fjord. Also, it may partly reflect the absence of sills such that Atlantic water flows into the fjord unimpeded.

The regional oceanographic conditions outside the fjord systems are dominated by the East Greenland Coastal Current (EGCC) above the shelf and the East Greenland Current (EGC) located at the shelf break (Sutherland and Pickart, 2008). The shelf is ca. 200–300 m deep outside the fjords and ca. 40–50 km wide. Upper water masses on the shelf are characterized by polar surface water ($\theta < 0\,°C$, $\sigma_\theta < 27\,kg\,m^{-3}$;

Sutherland and Pickart, 2008) originating from the Arctic Ocean and influenced by melted ice and runoff. The Atlantic water is located on the shelf break and originates from the Irminger Sea and from recirculated water from the North Atlantic Current. Bottom water masses inside the fjords are relatively cold and from the temperature–salinity diagram (Fig. 8) they are located at the mixing line between Atlantic water (AW; $4.5 < \theta < 6.5$ °C, $34.8 < S_P < 35.0$) and PSW. Water masses below 120 m are relatively cold and below 4 °C, which indicates a limited exchange between the deeper part of the fjords and the warm Atlantic water mass. However, the relatively warm bottom water masses in Timmiarmiut Fjord compared with the two other fjords further north could indicate a larger influence from warm Atlantic water here. This could possibly be due to the relatively narrow width of the shelf outside the fjord, but also, at least partly, the absences of sills.

4 Data availability

Multibeam bathymetry and hydrographic data are available through the PANGAEA. Additional data used here include the International Bathymet-ric Chart of the Arctic Ocean (IBCAO) (Jakobsson et al., 2012), a digital elevation model and orthophotos derived from stereophotogrammetric imagery recorded in 1981 (Korsgaard et al., 2016), and Moderate Resolu-tion Imaging Spectroradiometer (MODIS) archive satellite imagery.

5 Conclusions

In this study we have presented bathymetric data obtained during a cruise in 2014 using a multibeam system temporarily installed on SS *ACTIV*. The data collected provide new insights into fjords with limited information of the bathymetry, but they also supplement subsequent large-scale data collection from the NASA OMG mission. The need for new data is evident when comparing to existing data, such as IBCAO or single-point measurements from nautical maps. These new efforts provide better spatial resolution and, importantly, also a better description of depths and the seafloor morphology. For instance, comparison to IBCAO shows not only greater depths (Fig. 4b), but also that the 2000 × 2000 m land DEM that is implemented in IBCAO and its final 500 m of spatial resolution is insufficient to properly distinguish between land and fjord in these narrow fjord settings. It is also likely that the single-point measurements from the nautical charts, for instance around Skjoldungen, were not included in IBCAO, leading to considerable discrepancy. However, with the single major objective of IBCAO being to provide a portrayal of the Arctic Ocean seafloor (Jakobsson et al., 2012), it is understandable that not all fjord complexes are as well resolved as offshore bathymetry based on extensive multibeam mapping. However, large-scale efforts, such as the NASA OMG mission measuring bathymetry (Fenty et al., 2016) and the synthetic bathymetry datasets (Williams et al., 2017), are beginning to shed new light on the configurations of fjords around Greenland.

The CTD measurements collected during the cruise provide information about the properties of the water column in relation to sound velocity and the bathymetrical mapping, but they also allow insight into the distribution of water masses in the different fjord settings. Here we find considerable difference between the water masses located in the three fjord settings, influenced not only by the local bathymetry, such that sills in SSsund and NSsund hinder the exchange of deeper, warmer water masses that might affect the marine-terminating Thrym Glacier at the head of the fjord and the variability in the influence of Atlantic water flowing on the shelf outside the fjords. The latter causes deeper water in Timmiarmiut Fjord to be warmer relative to the fjords around Skjoldungen further north and possibly also reflects the absence of sills. Moreover, high salinities are observed in the layer between ca. 50 and 10 m in the inner part of Timmiarmiut Fjord, suggesting the release of subglacier meltwater from Heimdal Glacier up fjord of the CTD station, which is consistent with satellite imagery that shows a large pool of ice mélange set into motion 5 days prior to our CTD measurements.

The observations presented in this study complement large-scale efforts to obtain knowledge about the bathymetry and hydrography in the fjords around Greenland, but they also add to our understanding of how subglacial meltwater is distributed in fjord systems and the impact on fjord circulation (Bendtsen et al., 2015; Kjeldsen et al., 2014). This is important for understanding how a future warming ocean will affect the stability of marine-terminating outlet glaciers.

Acknowledgements. We thank the GEOMAR Helmholtz-Centre for Ocean Research Kiel for graciously lending us their Seabeam 1050 multibeam system, Wärtsilä ELAC Nautik for lending us a transducer mounting frame, and Teledyne CARIS for providing the software CARIS HIPS. Moreover, we thank the crew onboard SS *ACTIV* for their help collecting the data during the field campaign. Kristian Kjellerup Kjeldsen acknowledges support from the Danish Council Research for Independent Research (grant no. DFF – 4090-00151). Anders Anker Bjørk was supported by the Danish Council for Independent Research (grant no. DFF – 610800469) and by the Inge Lehmann Scholarship from the Royal Danish Academy of Science and Letters.

Edited by: Michael E. Contadakis

References

Andresen, C. S., Kjeldsen, K. K., Harden, B., Nørgaard-Pedersen, N., and Kjær, K. H.: Outlet glacier dynamics and bathymetry at Upernavik, Geol. Surv. Den. Greenl., 31, 79–82, 2014.

Arndt, J. E., Jokat, W., Dorschel, B., Myklebust, R., Dowdeswell, J. A., and Evans, J.: A new bathymetry of the Northeast Greenland continental shelf: Constraints on glacial and other processes, Geochem. Geophy. Geosy., 16, 3733–3753, https://doi.org/10.1002/2015GC005931, 2015.

Bendtsen, J., Mortensen, J., Lennert, K., and Rysgaard, S.: Heat sources for glacial ice melt in a west Greenland tidewater outlet glacier fjord: The role of subglacial freshwater discharge, Geophys. Res. Lett., 42, 4089–4095, https://doi.org/10.1002/2015GL063846, 2015.

Caress, D. W. and Chayes, D. N.: Improved processing of Hydrosweep DS multibeam data on the R/V *Maurice Ewing*, Mar. Geophys. Res., 18, 631–650, https://doi.org/10.1007/BF00313878, 1996.

Fenty, I., Willis, J. K., Khazendar, A., Dinardo, S., Forsberg, R., Fukumori, I., Holland, D., Jakobsson, M., Moller, D., Morison, J., Münchow, A., Rignot, E., Schodlok, M., Thompson, A. F., Tinto, K., Rutherford, M., and Trenholm, N.: Oceans Melting Greenland: Early Results from NASA's Ocean-Ice Mission in Greenland, Oceanography, 29, 72–83, https://doi.org/10.5670/oceanog.2016.100, 2016.

IOC, SCOR and IAPSO: The international thermodynamic equation of seawater – 2010: Calculation and use of thermodynamic properties, Intergovernmental Oceanographic Commision, UNESCO, Manuals and Guides, 56, 1–196, 2010.

Jakobsson, M., Mayer, L., Coakley, B. J., Dowdeswell, J. A., Forbes, S., Fridman, B., Hodnesdal, H., Noormets, R., Pedersen, R., Rebesco, M., Schenke, H. W., Zarayskaya, Y., Accettella, D., Armstrong, A., Anderson, R. M., Bienhoff, P., Camerlenghi, A., Church, I., Edwards, M., Gardner, J. V., Hall, J. K., Hell, B., Hestvik, O., Kristoffersen, Y., Marcussen, C., Mohammad, R., Mosher, D., Nghiem, S. V., Pedrosa, M. T., Travaglini, P. G., and Weatherall, P.: The International Bathymetric Chart of the Arctic Ocean (IBCAO) Version 3.0, Geophys. Res. Lett., 39, L12609, https://doi.org/10.1029/2012GL052219, 2012.

Khan, S. A., Aschwanden, A., Bjørk, A. A., Wahr, J., Kjeldsen, K. K., and Kjær, K. H.: Greenland ice sheet mass balance: a review, Rep. Prog. Phys., 78, 046801, https://doi.org/10.1088/0034-4885/78/4/046801, 2015.

Kjeldsen, K. K., Mortensen, J., Bendtsen, J., Petersen, D., Lennert, K., and Rysgaard, S.: Ice-dammed lake drainage cools and raises surface salinities in a tidewater outlet glacier fjord, west Greenland, J. Geophys. Res.-Earth, 119, 1310–1321, https://doi.org/10.1002/2013JF003034, 2014.

Korsgaard, N. J., Nuth, C., Khan, S. A., Kjeldsen, K. K., Bjørk, A. A., Schomacker, A., and Kjær, K. H.: Digital elevation model and orthophotographs of Greenland based on aerial photographs from 1978–1987, Scientific Data, 3, 160032, https://doi.org/10.1038/sdata.2016.32, 2016.

Mortensen, J., Lennert, K., Bendtsen, J., and Rysgaard, S.: Heat sources for glacial melt in a sub-Arctic fjord (Godthåbsfjord) in contact with the Greenland Ice Sheet, J. Geophys. Res., 116, C01013, https://doi.org/10.1029/2010JC006528, 2011.

Mortensen, J., Bendtsen, J., Motyka, R. J., Lennert, K., Truffer, M., Fahnestock, M., and Rysgaard, S.: On the seasonal freshwater stratifi cation in the proximity of fast-flowing tidewater outlet glaciers in a sub-Arctic sill fjord, J. Geophys. Res.-Oceans, 118, 1382–1395, https://doi.org/10.1002/jgrc.20134, 2013.

Porter, D. F., Tinto, K. J., Boghosian, A., Cochran, J. R., Bell, R. E., Manizade, S. S., and Sonntag, J. G.: Bathymetric control of tidewater glacier mass loss in northwest Greenland, Earth Planet. Sc. Lett., 401, 40–46, https://doi.org/10.1016/j.epsl.2014.05.058, 2014.

Rignot, E. and Mouginot, J.: Ice flow in Greenland for the International Polar Year 2008–2009, Geophys. Res. Lett., 39, L11501, https://doi.org/10.1029/2012GL051634, 2012.

Rignot, E., Fenty, I., Xu, Y., Cai, C., and Kemp, C.: Undercutting of marine-terminating glaciers in West Greenland, Geophys. Res. Lett., 42, 5909–5917, https://doi.org/10.1002/2015GL064236, 2015.

Rignot, E., Fenty, I., Xu, Y., Cai, C., Velicogna, I., Cofaigh, C., Dowdeswell, J. A., Weinrebe, W., Catania, G., and Duncan, D.: Bathymetry data reveal glaciers vulnerable to ice-ocean interaction in Uummannaq and Vaigat glacial fjords , west Greenland, Geophys. Res. Lett., 43, 2667–2674, https://doi.org/10.1002/2016GL067832, 2016.

Schumann, K., Völker, D., and Weinrebe, W. R.: Acoustic mapping of the Ilulissat Ice Fjord mouth, West Greenland, Quaternary Sci. Rev., 40, 78–88, https://doi.org/10.1016/j.quascirev.2012.02.016, 2012.

Straneo, F. and Heimbach, P.: North Atlantic warming and the retreat of Greenland's outlet glaciers, Nature, 504, 36–43, https://doi.org/10.1038/nature12854, 2013.

Sutherland, D. A. and Pickart, R. S.: Progress in Oceanography The East Greenland Coastal Current: Structure, variability, and forcing, Prog. Oceanogr., 78, 58–77, https://doi.org/10.1016/j.pocean.2007.09.006, 2008.

Wessel, P. and Smith, W. H. F.: Free software helps map and display data, Eos T. Am. Geophys. Un., 72, 441–446, 1991.

Williams, C. N., Cornford, S. L., Jordan, T. M., Dowdeswell, J. A., Siegert, M. J., Clark, C. D., Swift, D. A., Sole, A., Fenty, I., and Bamber, J. L.: Generating synthetic fjord bathymetry for coastal Greenland, The Cryosphere, 11, 363–380, https://doi.org/10.5194/tc-11-363-2017, 2017.

Willis, M. J., Herried, B. G., Bevis, M. G., and Bell, R. E.: Recharge of a subglacial lake by surface meltwater in northeast Greenland, Nature, 518, 223–227, https://doi.org/10.1038/nature14116, 2015.

4

Global and regional phosphorus budgets in agricultural systems and their implications for phosphorus-use efficiency

Fei Lun[1], Junguo Liu[2], Philippe Ciais[3], Thomas Nesme[4], Jinfeng Chang[3], Rong Wang[3], Daniel Goll[3], Jordi Sardans[5], Josep Peñuelas[6], and Michael Obersteiner[7]

[1]College of Resources and Environmental Sciences, China Agricultural University, Beijing 100193, China

[2]School of Environmental Science and Engineering, South University of Science and Technology of China, Shenzhen 518055, China

[3]IPSL – LSCE, CEA CNRS UVSQ, Centre d'Etudes Orme des Merisiers, 91191 Gif-sur-Yvette, France

[4]Bordeaux Sciences Agro, Université de Bordeaux, UMR 1391 ISPA, CS 40201, 33175 Gradignan CEDEX, France

[5]CREAF, Cerdanyola del Vallès 08193, Catalonia, Spain

[6]CSIC, Global Ecology Unit CREAF-CSIC-UAB, Cerdanyola del Vallès 08193, Catalonia, Spain

[7]International Institute for Applied Systems Analysis, 2361 Laxenburg, Austria

Correspondence: Junguo Liu (liujg@sustc.edu.cn, junguo.liu@gmail.com)

Abstract. The application of phosphorus (P) fertilizer to agricultural soils increased by 3.2 % annually from 2002 to 2010. We quantified in detail the P inputs and outputs of cropland and pasture and the P fluxes through human and livestock consumers of agricultural products on global, regional, and national scales from 2002 to 2010. Globally, half of the total P inputs into agricultural systems accumulated in agricultural soils during this period, with the rest lost to bodies of water through complex flows. Global P accumulation in agricultural soil increased from 2002 to 2010 despite decreases in 2008 and 2009, and the P accumulation occurred primarily in cropland. Despite the global increase in soil P, 32 % of the world's cropland and 43 % of the pasture had soil P deficits. Increasing soil P deficits were found for African cropland vs. increasing P accumulation in eastern Asia. European and North American pasture had a soil P deficit because the continuous removal of biomass P by grazing exceeded P inputs. International trade played a significant role in P redistribution among countries through the flows of P in fertilizer and food among countries. Based on country-scale budgets and trends we propose policy options to potentially mitigate regional P imbalances in agricultural soils, particularly by optimizing the use of phosphate fertilizer and the recycling of waste P. The trend of the increasing consumption of livestock products will require more P inputs to the agricultural system, implying a low P-use efficiency and aggravating P-stock scarcity in the future. The global and regional phosphorus budgets and their PUEs in agricultural systems are publicly available at https://doi.pangaea.de/10.1594/PANGAEA.875296.

1 Introduction

Population increases and dietary changes require higher food production, which increases global demand for fertilizers (Grote et al., 2005; Foley et al., 2011). Phosphorus (P) is an essential element for all organisms, and a lack of P limits growth. Fertilizer P enhances agricultural production, but P is also fixed in soils and can accumulate. In countries with high fertilizer use, much P is lost to leaching and run-off, leading to the eutrophication of both inland and coastal waters (Carpenter et al., 1998; MacDonald et al., 2011).

To supply the growing need for P in fertilizer, the mining of phosphate rock has quadrupled in the past half century, increasing from 46 Mt in 1961 to 198 Mt in 2011 (Scholz et al., 2013). Despite some short-term fluctuations in the price of phosphate rock, the global production of fertilizer P has been steadily increasing at a rate of 3 to 4 % annually during the half century before 2011 and is projected to increase by 50 to 100 % by 2050 (Cordell et al., 2009, 2012). Extractable phosphate rock is a non-renewable resource, and significant depletion of the resource is projected by the end of this century if the current intensive use continues, possibly leading to resource shortages (Cordell et al., 2009; van Vuuren et al., 2010; Peñuelas et al., 2013).

The mining of P and its application as fertilizer in cultivated land is a major anthropogenic perturbation of the natural biogeochemical P cycle (Carpenter and Bennett, 2011; Elser and Bennett, 2011; Steffen et al., 2015). The negative impacts of this perturbation on the natural environment depend on how much P is lost from regions with intensive fertilizer use (Smil, 2000; Bennett et al., 2001).

P application differs significantly between countries and crop types (Grote et al., 2005), and previous researchers have attempted to estimate the P flows in agricultural systems in Europe (Ott & Rechberger, 2012), the United States (Suh & Yee, 2011), China (Ma et al., 2011), France (Senthilkumar et al., 2012), Australia (Cordell et al., 2013), and the world (Smil, 2000; Liu et al., 2008; MacDonald et al., 2011; Schipanski & Bennett, 2012). International trade and regional agricultural policies affect P budgets by increasing or decreasing the gap between P inputs and P outputs in agricultural land (Grote et al., 2005). Previous research mainly focused on cropland, and P fluxes in pasture and livestock production systems have also received more and more attention recently, especially due to diet change. The differences in methodologies, system boundaries, and data sources have made it difficult to assess the differences in the phosphorus-use efficiencies (PUEs) among agricultural sectors and to extrapolate regional findings to the global scale.

To mitigate these problems, we (1) compiled a detailed and harmonized dataset of P fluxes in agriculture for countries around the world, including detailed analysis of input and output fluxes for cropland, managed grassland (hereafter, pasture), livestock, and human consumers of agricultural products; (2) characterized P budgets and P-use efficiencies in those different subsystems; and (3) examined how the international trade of phosphate fertilizer and agricultural commodities influences regional P fluxes. We performed this analysis on the scale of countries, regions, and the world; wherever possible, we distinguished different crop types. The study period was from 2002 to 2010, allowing us to study temporal trends.

2 Materials and methods

In this study, we obtained data for 224 countries (Table S1 in the Supplement). We defined the agriculture system as cropland and pasture ecosystems plus human and livestock consumers of agricultural production and of other products containing P (Fig. 1). External P inputs to the agriculture system came from mined phosphate rock and atmospheric deposition. Several processes cause P losses from the system into the external environment (here, defined as non-agricultural land and bodies of water). Figure 1 presents the fluxes of P into and out of the agriculture system on a global scale, including internal fluxes between ecosystems and consumers. We quantified these fluxes in the present study based on a mass-balance approach (Cordell et al., 2012). We defined the phosphorus-use efficiency (PUE) of the agricultural system and of its subsystems as the ratio of the total P harvested in economic outputs (e.g. crops, meat, milk, and eggs) to the total P input. International trade in fertilizer and food is discussed separately in Sect. 2.3. The data sources and an overview of the mass-balance equations are presented in the rest of this section; details and equations are presented in the Supplement.

2.1 P flows into and out of the agricultural system

Inputs into the agricultural system, which are within the grey box in Fig. 1, are from mined phosphate rocks and atmospheric deposition. We did not include P from the in situ weathering of soil particles because the rate of this process is insignificant compared with the magnitude of other inputs (Liu et al., 2008). Outputs included P emission into the atmosphere from fires and P loss to uncultivated land or bodies of water.

2.1.1 P inputs

Data on agricultural inputs of phosphate P in fertilizers were collected from the International Fertilizer Industry Association (http://www.fertilizer.org) and divided between cropland and pasture uses based on information from FAO (2002) and the FAOSTAT database (http://www.fao.org/faostat/en/#data). A small fraction (8 %) of P from mined phosphate rock is used to produce animal feed additives. Apart from fertilizer and animal feed additives, the rest of the mined P is used to produce detergents and other products directly consumed by humans (Ringeval et al., 2014). Atmospheric

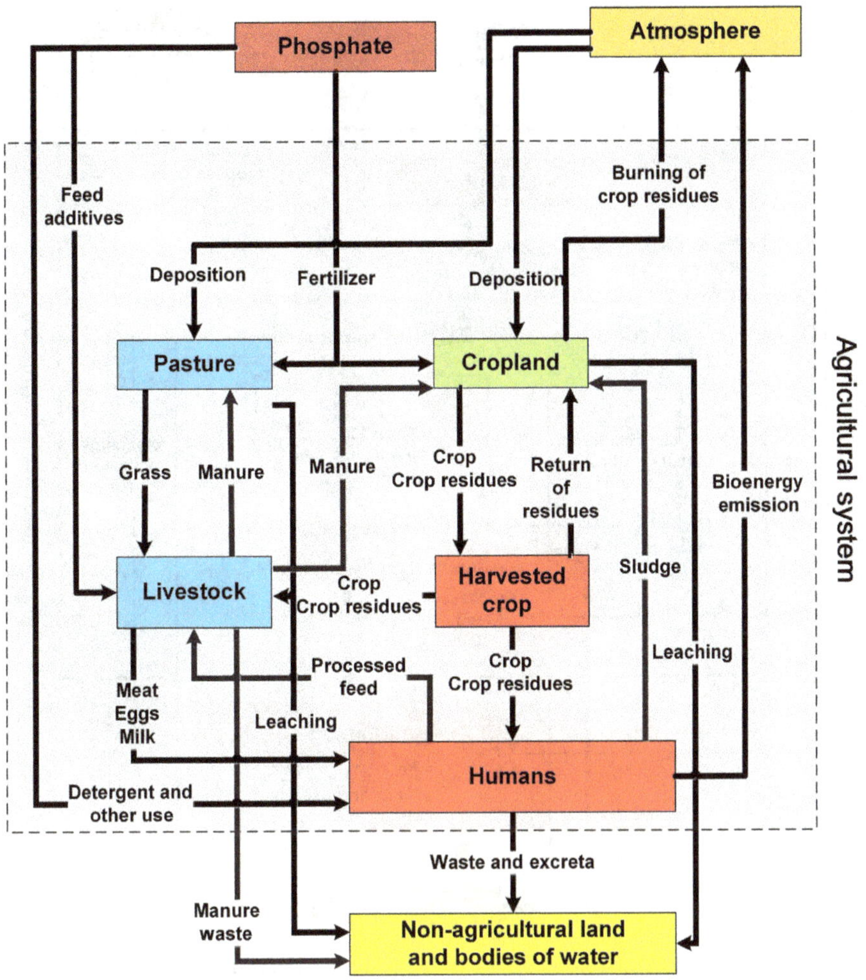

Figure 1. Scheme of the P pools and fluxes used to diagnose global P budgets for the agricultural sector. The agricultural sector (or system) in the grey box includes cropland and pasture soils, livestock, human consumers of livestock and crop products, and users of phosphate-derived products. National and regional P budgets are calculated using the same scheme, but including in addition exports and imports of P embedded in traded crop and livestock products and fertilizers.

P deposition in cropland and pasture areas was calculated separately in each country using gridded global P-deposition maps obtained using the LMDz-INCA aerosol chemistry transport model of Wang et al. (2014, 2015) and agricultural land-use maps. Details are provided in the Supplement (Table S2).

2.1.2 P outputs

P emissions from agricultural fires were obtained from the gridded dataset of Wang et al. (2015) and cover the burning of crop residues in the field, by households, and for the production of bioenergy from crop biomass. Leaching from cropland and pasture soils was assumed to be a constant fraction (12.5 %) of P inputs for each agricultural land-use type (Bouwman et al., 2013). P outputs from non-recycled livestock and human manure were calculated based on the mass balance. Note that erosion-induced losses of P are important in many agricultural regions (Quinton et al., 2010), but were not considered in this study because we lack data on the redeposition of P in eroded soil material from agricultural soils. In future studies, it will be important to quantify this source of P, particularly in agricultural areas that receive large annual inputs of sediment (e.g. in river floodplains and sites on steep terrain that experience significant erosion farther up the slope followed by deposition).

2.2 P flows within the agricultural system

2.2.1 P in harvested crop biomass and crop residues

The flux of P in harvested crop biomass was estimated from yield data (FAOSTAT) using crop-specific P concentrations after grouping 178 different crops into 13 crop types (COMIFER, 2007; USDA-NRCS, 2009; Waller, 2010;

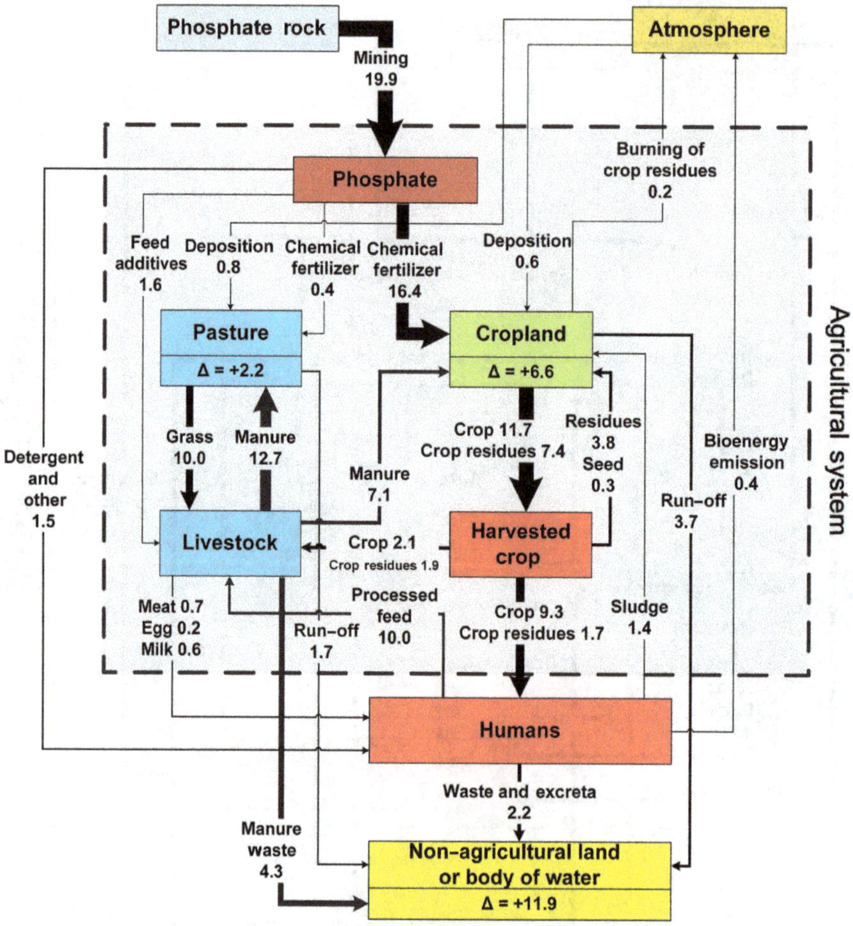

Figure 2. Annual P flows in the global agriculture system from 2002 to 2010. Values are Tg P yr^{-1}. The notation Δ denotes the average change of P in pasture and cropland soils, respectively. By convention, a positive value means accumulation. Note that livestock and human changes of P are assumed to be zero.

Table S2). P in harvested crop biomass was partitioned into crops (for human and livestock consumption) and crop residues (Fig. 1). We estimated the P fluxes of crop residues from FAOSTAT data and from Liu et al. (2008) to account for residue that is recycled in the field (50 %), transformed into livestock feed (25 %), and burned or used by other human activities (25 %).

2.2.2 P in grazed biomass

The P removed from pasture by livestock grazing was estimated by combining forage grass consumption data with the P concentrations in grass biomass (Antikainen et al., 2005; COMIFER, 2007; USDA-NRCS, 2009; Waller, 2010). Gridded data on grass biomass consumption by livestock were obtained by combining the global livestock production systems dataset of Herrero et al. (2013) with pasture net primary productivity simulated by the ORCHIDEE-GM global pasture model (Chang et al., 2013, 2015). We chose the ORCHIDEE-GM model for this analysis because it is able to separate the intake of grazed vs. cut forage grass.

2.2.3 P in animal feed products

Animal feed products used as complementary diet ("feed additives") represent direct inputs to the livestock subsystem (Fig. 1). This flux was deduced from the mass balance of the known input and output fluxes for the livestock P pool, but did not account for long-term changes in P storage in that pool. See the Supplement for more details.

2.2.4 P embedded in livestock products

This flux of P leaving the livestock subsystem and entering the human subsystem (Fig. 1) through the harvesting of products was calculated by multiplying the FAOSTAT production data for meat, eggs, and milk by the product-specific P concentrations reported by Grote et al. (2005).

2.2.5 P in livestock manure

We calculated the manure P production based on FAOSTAT data about N in livestock manure and P : N values for each types of livestock manure (MWPS-18, 1985; OECD Secretariat, 1991; Levington Agriculture, 1997; Sheldrick et al., 2003; ASAE, 2005; see Table S3). Once produced, manure P is either applied to cropland, left in the pasture, or lost to the environment as waste (Fig. 1), following the same partitioning as that for N in the manure from FAOSTAT.

2.2.6 P in human sewage sludge

We assumed that the P output from humans equaled the inputs from non-fertilizer P-ore products and the consumed crop and livestock products (Fig. 1) and used this to calculate the total P production in human excreta. P in human sewage sludge was estimated using population data and values of per capita production of P in excreta (Smil, 2000; Cordell et al., 2009). Following the method of Liu et al. (2008), we assumed that 30 % of the excreta P from urban populations and 70 % of P from rural populations were returned to cropland, either directly or after the treatment of sewage sludge, with the remaining P assumed to be lost to the environment (e.g. in landfills or bodies of water).

2.3 P flows from international trade

We compiled the flows of P in international trade both from the P embodied in crops and livestock products and in P embodied in fertilizers exchanged between countries. For agricultural commodities, we used FAOSTAT data that provided a matrix of commodities exchanged between countries and converted this data into P fluxes using commodity-specific P content data. For P fertilizers, we used the International Fertilizer Industry Association trade statistics. By convention, a positive trade balance for a country means that it is a net P importer. In addition, P fluxes associated with the international trade of fertilizers, food, feed, and fibre commodities can be associated with local cropland PUE and pasture PUE. We defined the dependency on chemical fertilizer imports (F_{fer}) as the ratio of the P in imported chemical fertilizers ($P_{\text{fer-imp}}$) to the P in all chemical fertilizers consumed by a country ($P_{\text{fer-con}}$). Similarly, we defined the dependency on food imports (F_{food}) as the ratio of P in food imports ($P_{\text{food-imp}}$) to the P in all food consumed by a country. Furthermore, we defined F_{total} as the ratio of the total P imported (food and fertilizers) to the total P consumed as fertilizers and food in a country. The equations for these calculations are presented in Sects. 2 to 6 of the Supplement.

2.4 Annual P budgets of cropland and pasture soils

Annual changes in P stocks in cropland and pasture soils (ΔP) were estimated as the difference between inputs and outputs (i.e. the budget); $\Delta P > 0$ indicates net P accumulation in the soil, $\Delta P < 0$ indicates a net deficit, and $\Delta P = 0$ represents no net change. ΔP calculated in this manner does not reflect the legacy effects from previous management and fertilization practices (Ringeval et al., 2014), but it is a useful metric to identify regions with a P surplus or deficit at any point in time and to compare countries.

Annual soil ΔP values were calculated as the differences between annual inputs and outputs. Details and the equations are presented in Sect. 2 of the Supplement.

2.5 Cumulative P budgets of cropland and pasture soils

Following the method of Sattari et al. (2012), we separated the P inputs to soils (except inputs in seeds) into two pools: (1) a stable P pool, which represents P that is unavailable to plants on an annual basis, such as the P absorbed onto iron and aluminum oxides (20 % of total P inputs, including fertilizers, manure, sludge, and deposition); and (2) a labile P pool that is assumed to be available for plant uptake (80 % of total P inputs). P can be exchanged between the two pools. If inputs of labile P are larger than P removal in crop biomass, we assumed that the surplus labile P gets transferred into the stable P pool at the end of the year. In the opposite case, in which inputs of labile P are lower than P removal, plants can take up P from the stable pool (Sattari et al., 2012). This approach assumes that the P loss by run-off and leaching into bodies of water is from the labile P pool only and that P stored in seeds does not belong to either the stable pool or the labile pool. This approach is simplistic, as more research will be required to allow for more realistic modelling of these two pools and the flows they are involved in.

2.6 Phosphorus-use efficiency

We defined PUE as the ratio of P in the harvested economic outputs to P in the inputs for the entire agricultural system (the grey area in Fig. 1) or for a given subsystem. PUE indicates how much of the input P is transferred into value-added products. If PUE > 1, the input of P is insufficient to sustain the output (harvested P), suggesting a net reduction of the system's P reservoir. For cropland PUE, we defined P in harvested crops as the economic P output of the crops and the sum of phosphate fertilizer, livestock manure, human sewage sludge, and P from atmospheric deposition as the P input. For pasture PUE, harvested P refers to the P consumed by grazing animals and the sum of phosphate fertilizer, livestock manure going to the pasture, and P from atmospheric deposition as the total inputs. For the livestock subsystem, the harvested P output represents the P in livestock products (meat, eggs, and milk), whereas the inputs represent the input into livestock. We also defined the PUE of human food ($\varepsilon_{\text{food}}$) as the ratio of the P content in human excreta to the total P input in human food; this represents an inconsistency with our previous definitions, since human excreta currently have no economic value. The equations for all the PUE terms are provided in Sect. 5 of the Supplement.

2.7 Uncertainty estimates

Uncertainties in each flux originate both from the material flux data and from data on the P concentration in each material considered by our analysis, including crop products, crop residues, livestock, meat, eggs, milk, livestock, and human excreta. Many of the global statistical datasets used in our analysis are not replicated, and no alternative dataset is available for establishing a range of uncertainty values for the different P fluxes. National datasets have usually not been formally analysed to determine their uncertainty, and many of the sources of uncertainty are difficult to trace (e.g. clerical errors, differences between countries in product definitions). Thus, we have only addressed the effect of uncertainties in the P concentration by means of Monte Carlo simulations (3000 iterations) using the range of P concentrations reported in the literature (Table S5).

3 Results

3.1 Global agricultural P flows and their trends

3.1.1 Global P fluxes in and out of the agricultural system

Figure 2 summarizes the annual average of global P flows for the period from 2002 to 2010. P from phosphate fertilizers was the largest single input flux, representing 93 % of the 21.3 $Tg P yr^{-1}$ of global input, and most of it (82.4 %) goes to cropland and pasture. Outputs from the agriculture system amounted to 12.5 $Tg P yr^{-1}$, which combines outputs from leaching and run-off into bodies of water (5.4), non-recycled manure waste (4.3), and sewage (2.2), bioenergy (0.4), and burned crop residues (0.2). The global annual P balance of agricultural systems was therefore positive during the entire study period, with 8.8 $Tg P yr^{-1}$ accumulating in soil, of which 6.6 $Tg P yr^{-1}$ accumulated in cropland and 2.2 $Tg P yr^{-1}$ in pasture. On average, 41 % of the P input accumulated in soils from 2002 to 2010.

3.1.2 Temporal trends

Figure 3 shows the trends for the four largest P fluxes in the agriculture system, illustrating that chemical fertilizer inputs, P loss to the environment, and P harvested in crops presented significantly increasing trends during this period. The application of phosphate fertilizer increased at an average annual rate of 3.2 % from 2002 to 2010 despite a decrease in 2008 that reflected reduced fertilizer application at a time when the price of phosphate fertilizers increased (Cordell et al., 2009, 2012). The trend for P in harvested crop biomass was also a steady increase, but at a lower annual rate (2.4 %) and with no decrease in 2008, probably because of the availability of

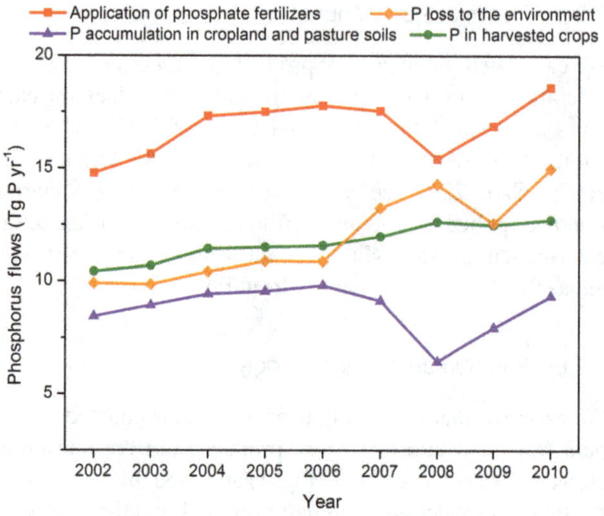

Figure 3. Time series in the four largest global annual P flows within, in, and out of the agriculture system from 2002 to 2010.

P that accumulated in the soil from previous years (as described in Sect. 2.5). Overall, P in agricultural soils increased by 1.3 % annually, whereas P losses to the environment increased faster (6.4 % yr^{-1}) than fertilizer inputs.

3.1.3 Global P fluxes in cropland

Cropland received the largest fraction (82 %) of phosphate fertilizer, 29 % of the manure produced by livestock, and all of the recycled human sewage sludge (Fig. 2). Atmospheric deposition contributed an additional 0.6 $Tg P yr^{-1}$ of inputs to croplands. Harvesting of cropland removed 11.7 $Tg P yr^{-1}$, which can be divided into crop products used for human nutrition (9.3 $Tg P yr^{-1}$, including 5.3 for food, 2.7 for processing, 0.4 for waste, and 0.9 for other use) and for livestock feed (2.1 $Tg P yr^{-1}$), with a small pool in seeds returned to the cropland (0.3 $Tg P yr^{-1}$). On average, 50 % of the P contained in crop residues was recycled to cropland during the study period, with 0.2 $Tg P yr^{-1}$ lost to the atmosphere from the burning of crop residues. The remaining 3.6 $Tg P yr^{-1}$ contained in harvested crop residues is removed from cropland and redistributed to livestock and humans. Globally, 3.7 $Tg P yr^{-1}$ was lost from cropland soils through leaching and run-off. The sum of all these fluxes results in an annual soil P accumulation of 6.6 $Tg P yr^{-1}$ (Fig. 2).

The global cropland PUE averaged 0.46, with a maximum of 0.51 in 2008 and a minimum of 0.44 in 2006. The annual cropland P accumulation ratio (cropland soil P accumulation / total P input to cropland) was 23 %, which is lower than the accumulation ratio of 48 % found for the overall agriculture system. In countries where labile P inputs were lower than P removal in crops, the soil's labile P pool was depleted by 1.9 $Tg P yr^{-1}$ through the harvesting of crop biomass. In countries where labile P inputs are higher than P removal by

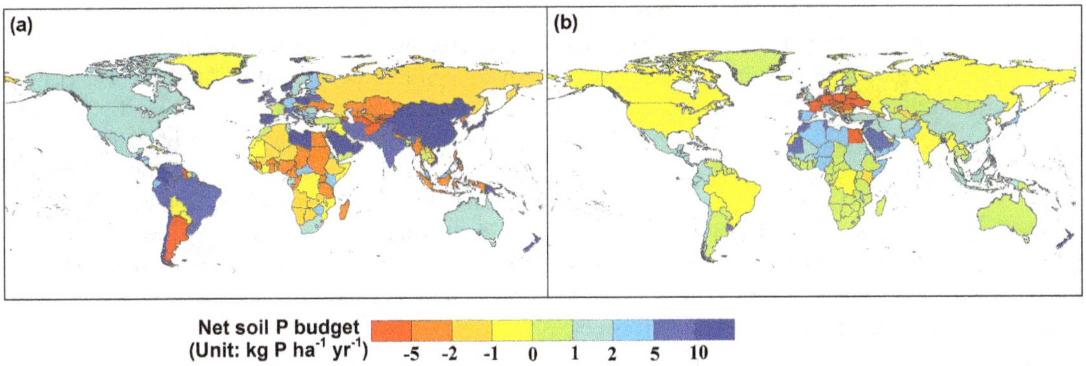

Figure 4. Map of global net soil P budgets (positive values, increase; negative values, decrease) for **(a)** cropland and **(b)** pasture.

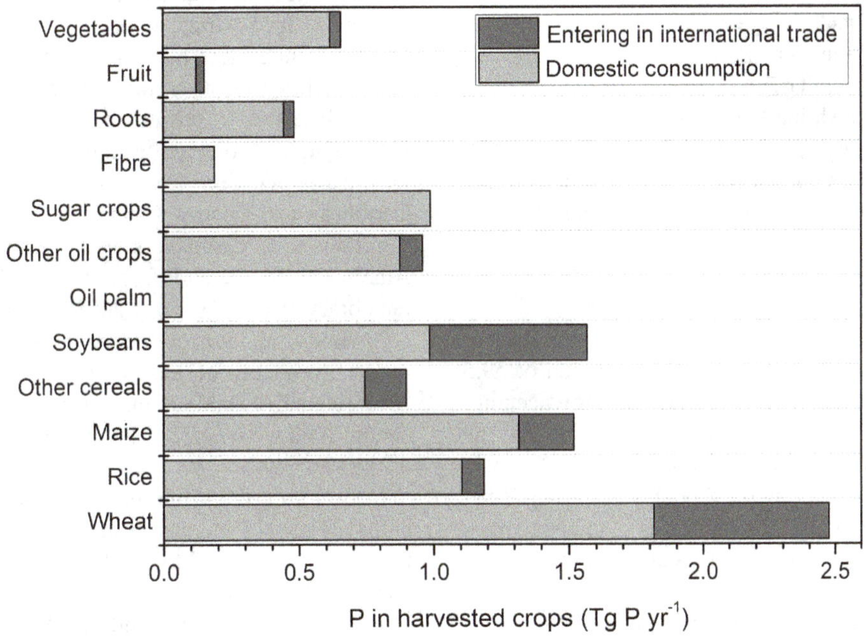

Figure 5. P flows embedded in different crop products, including the fraction of these flows entering into international trade circuits vs. being used for domestic consumption, for the year 2010.

crops, the accumulation of soil labile P was 6.0 Tg P yr^{-1}. Thus, there is an asymmetry between these two groups of countries, with accumulation being larger than depletion on a global scale. In addition, the global stable P pool in cropland increased by an average of 5.6 Tg P yr^{-1} from 2002 to 2010.

3.1.4 Global P fluxes in pasture

Different from croplands, most P inputs to pasture were from livestock manure (12.7 Tg P yr^{-1}), with small additional contributions from atmospheric deposition (0.8 Tg P yr^{-1}) and phosphate fertilizers (0.4 Tg P yr^{-1}). The primary production of pasture incorporates 10.0 Tg P yr^{-1} of P into grass biomass that is digested by animals, and the leaching and run-off loss averages 1.7 Tg P yr^{-1}. From all these fluxes, we estimated a global pasture PUE of 0.72 and a net accumulation of 2.2 Tg P yr^{-1} in the soil. In the countries where grass P removal exceeded the labile P inputs, the labile soil P pool was depleted by 1.4 Tg P yr^{-1}. In the countries where the labile P input exceeded grass P removal, an average of 5.3 Tg P yr^{-1} was transferred from the labile to the stable soil P pool from 2002 to 2010.

3.1.5 Global P fluxes in livestock

The annual P input to livestock was 25.6 Tg P yr^{-1}, with most of the contributions from grazed grass (10.0 Tg P yr^{-1}) and processed feed (10.0 Tg P yr^{-1}). The economic P output in the form of livestock products averaged 1.5 Tg P yr^{-1}, which gives a PUE of 0.06. Averages of 29 and 56 % of the P produced in livestock manure were recycled into cropland and pasture, respectively; the rest of this manure (4.3 $\mathrm{Tg\,P\,yr^{-1}}$) was lost to the environment.

3.1.6 Global P fluxes in human use

Humans receive an annual input of 14.0 $\mathrm{Tg\,P\,yr^{-1}}$ from harvested crop products, livestock products, and the use of detergents and other products manufactured from phosphate rock. Although P inputs as food (crop food and livestock products) amounted to 6.8 $\mathrm{Tg\,P\,yr^{-1}}$, humans only absorbed 3.0 $\mathrm{Tg\,P\,yr^{-1}}$ (44 %); the remainder was either wasted before consumption (e.g. in food processing) or transferred back to livestock as processed feed. Thus, only 14.3 % of the total P inputs into the agriculture system end up as food being actually consumed by humans. P lost to the environment by human use amounts to 2.6 $\mathrm{Tg\,P\,yr^{-1}}$, which is divided among 2.2 $\mathrm{Tg\,P\,yr^{-1}}$ lost through inefficient processing and excreta and 0.4 $\mathrm{Tg\,P\,yr^{-1}}$ through bioenergy-related emissions. The fate of non-recycled P in human waste was not separated between bodies of water (untreated sewage) and landfills.

3.2 Regional P budgets

Globally, both cropland and pasture presented soil P accumulation from 2002 to 2010, with an accumulation of 59.6 and 19.4 Tg P, respectively. For croplands, the net P accumulation in the stable P pools amounted to 52.7 Tg P, and the remaining 6.9 Tg P accumulated in soil labile pools. For pasture, the accumulation in the stable P pool was 25.0 Tg P, but 5.6 Tg P was transferred from the stable P pool to be incorporated by grass in regions where P inputs are lower than grass P uptake.

Those global numbers mask large regional differences (Table 1 and Fig. 4) and there were also differences between cropland and grassland. About 32 % of the global cropland area (in 75 countries) had annual soil P deficits from 2002 to 2007, with a net cropland soil P accumulation of 6.20–7.66 $\mathrm{Tg\,P\,yr^{-1}}$. This fraction increased to 50 % in 2008 and 2009 but the net cropland soil P accumulation decreased to 4.38 $\mathrm{Tg\,P\,yr^{-1}}$ in 2008 and 5.39 $\mathrm{Tg\,P\,yr^{-1}}$ at the time of the global financial crisis as a result of high P prices and the resulting reduction in fertilizer application (Cordell et al., 2009, 2012). However, the fraction of cropland soil P deficits returned close to the decadal mean value in 2010, with a net soil P accumulation of 7.30 $\mathrm{Tg\,P\,yr^{-1}}$. On average, 48 % of cropland P uptake was supplied by stable P that accumulated in previous years according to the equations in Sect. 3 of the Supplement. Including the United States, France, Russia, Argentina, and Paraguay, 89 countries had labile P inputs into cropland that were lower than crop P removal from 2002 to 2010. However, if we consider stable P inputs, cropland soil still presented a net soil P surplus in the United States during the same period. Compared with cropland, a slightly larger proportion of the total global pasture area had a net annual soil P deficit from 2002 to 2010, mostly in Europe and North America. The deficit proportion of grassland was only about 38 % in 2002 and 2003, with an annual net soil P accumulation of 2.26 $\mathrm{Tg\,P\,yr^{-1}}$; however, it increased to 43 % during the period of 2004–2010 and the annual pasture soil P accumulation was about 2.10 $\mathrm{Tg\,P\,yr^{-1}}$, with the smallest of 2.00 $\mathrm{Tg\,P\,yr^{-1}}$ in 2009. However, only 48 countries had labile P inputs into pasture that were lower than the P removal in grass.

3.2.1 Regional cropland budgets

Examining Fig. 4a reveals that cropland in all African countries experienced an annual soil P deficit, especially in western and central Africa, with soil P loss rates per unit area ranging from 2.5 $\mathrm{kg\,P\,ha^{-1}\,yr^{-1}}$ in 2002 to 2.7 $\mathrm{kg\,P\,ha^{-1}\,yr^{-1}}$ in 2010. In contrast, cropland in eastern Asia accumulated 23.4 $\mathrm{kg\,P\,ha^{-1}\,yr^{-1}}$ during the period from 2002 to 2010, a cumulative storage equivalent to more than 4 years of P fertilizer application. Cropland in Oceania, Europe, and the Caribbean and Central America also annually accumulated P in their soils. Cropland soils in North America and South America accumulated P from 2002 to 2007, but experienced temporary P deficits from 2008 to 2010. Yet despite this, crop yields did not decrease from 2008 to 2010 in those two regions, probably because of the re-mobilization of P that accumulated in stable pools. Cropland soils in western and central Asia were nearly balanced, with a mean areal flux of 0.2 $\mathrm{kg\,P\,ha^{-1}\,yr^{-1}}$.

Considering the different countries (Fig. 4a), the largest cumulative soil P increase was found in China (34.6 Tg P) for the 9 years from 2002 to 2010, followed by India (11.4 Tg P) and Brazil (3.6 Tg P). Pakistan (1.8 Tg P), the United States (1.8 Tg P), and New Zealand (1.8 Tg P) also had net soil P accumulation, yet of a smaller magnitude. These six countries accounted for 77 % of the global accumulation of P in countries where cropland had a positive soil P balance. Furthermore, a large amount of P accumulated in the soil labile P pools of cropland in China and India, at about 20.0 and 4.5 Tg P, respectively; however, in the United States, about 6.0 Tg P accumulated in the cropland stable P pool from 2002 to 2010; thus, 4.2 Tg P was absorbed from the previous cropland soil P. In contrast, most African countries experienced persistent cropland soil P deficits from 2002 to 2010. This was especially true in Nigeria, which had a cumulative deficit of 1.7 Tg P (Fig. 4a). We also found cumulative soil P deficits in Russia, the Ukraine, and Kazakhstan, but with a smaller magnitude (1.1, 0.9, and 0.7 Tg P, respectively) for the 9 years. Comparing the rates of change of crop soil P per unit area, New Zealand had the fastest rate of increase (> 100 $\mathrm{kg\,P\,ha^{-1}\,yr^{-1}}$), whereas Argentina had the fastest rate of decrease (–7.9 $\mathrm{kg\,P\,ha^{-1}\,yr^{-1}}$). In terms of the difference between inputs and outputs, loss rates in Argentina were about 5 times input rates.

Table 1. Regional annual agricultural P budgets and P-use efficiency (PUE).

Subsystem	World	Eastern and southern Africa	Northern Africa	Western and central Africa	Eastern Asia	Southern and south-eastern Asia	Western and central Asia	Oceania	Europe	North America	Caribbean and Central America	South America
Agricultural land P budget ($Tg\,P\,yr^{-1}$)[a]												
Cropland	6.6	−0.1	−0.1	−0.2	4.1	1.4	0.0	0.3	0.7	0.3	0.0	0.3
Pasture	2.2	0.1	0.5	0.3	0.5	0.2	0.2	0.4	−0.3	−0.1	0.1	0.3
Agricultural land P budget per unit area ($kg\,P\,ha^{-1}\,yr^{-1}$)[a]												
Cropland	4.7	−1.0	−1.5	−2.7	23.4	4.1	0.2	5.2	2.8	1.5	3.8	2.3
Pasture	0.4	0.2	1.6	0.9	1.0	0.4	0.6	0.8	−0.4	−0.1	3.4	0.4
Food consumption ($Tg\,P\,yr^{-1}$)												
Crops	5.3	0.2	0.3	0.2	1.4	1.3	0.3	0.0	0.8	0.4	0.0	0.3
Meat	0.71	0.01	0.01	0.01	0.29	0.05	0.01	0.01	0.17	0.10	0.00	0.05
Eggs	0.16	0.00	0.00	0.00	0.07	0.02	0.01	0.00	0.03	0.02	0.00	0.01
Milk	0.60	0.01	0.03	0.00	0.04	0.13	0.04	0.01	0.18	0.09	0.01	0.05
PUE												
Cropland	0.46	0.80	0.84	1.51	0.27	0.43	0.64	0.31	0.54	0.57	0.53	0.63
Pasture	0.72	0.77	0.61	0.46	0.58	0.80	0.61	0.42	1.25	0.98	0.37	0.75
Livestock	0.06	0.02	0.02	0.01	0.08	0.05	0.04	0.04	0.09	0.08	0.03	0.03
Food	0.45	0.60	0.50	0.64	0.40	0.64	0.44	0.26	0.28	0.32	0.68	0.42
International trade of P in commodities ($Tg\,P\,yr^{-1}$)[b]												
Crops	–	0.02	0.12	0.03	0.35	0.03	0.12	−0.07	−0.02	−0.41	0.03	−0.24
Meat	–	0.000	0.001	0.000	0.013	0.000	0.002	−0.004	−0.002	−0.005	0.001	−0.010
Eggs	–	0.0001	0.0000	0.0000	0.0001	−0.0004	0.0001	0.0000	0.0000	−0.0002	0.0000	−0.0001
Milk	–	0.000	0.003	0.001	0.004	0.005	0.004	−0.015	−0.012	0.003	0.001	−0.001
Fertilizer	–	0.05	−0.67	0.03	0.00	1.45	−0.05	0.21	−0.37	−1.00	0.06	1.10

[a] The positive values represent a soil P surplus, whereas negative values represent a soil P deficit. [b] The positive values represent net P importers, whereas negative values represent net P exporters.

3.2.2 Regional pasture budgets

We found mainly net losses of P in pasture soils (Fig. 4b), most likely because of the net removal of P through animal grazing, followed by the export of manure P to enrich cropland soils. Pasture soil P loss rates per unit area in Europe averaged $0.4\,kg\,P\,ha^{-1}\,yr^{-1}$ and reached high values in countries (Denmark, Luxembourg, Germany, and Belgium) with intensive livestock production systems (Chang et al., 2015) and large grass consumption by livestock, with loss rates $> 10\,kg\,P\,ha^{-1}\,yr^{-1}$. North American pastures had a smaller average loss rate of about $0.1\,kg\,P\,ha^{-1}\,yr^{-1}$. The United States, India, and Russia had the largest cumulative P deficits at 2.1, 1.5, and 0.7 Tg P, respectively, from 2002 to 2010. In contrast, pasture in the Caribbean and Central America had greater P inputs than P removals. Consequently, these regions had the largest soil P accumulation rates. Pasture in northern and eastern Africa also had net soil P accumulation. For instance, Mauritania, Tunisia, and Morocco had net soil P accumulation rates of 9.8, 9.4, and $5.5\,kg\,P\,ha^{-1}\,yr^{-1}$, respectively. The reason for this excess is not clear, but one possibility is that these countries apply P fertilizer to some of their pasture.

3.3 Phosphorus-use efficiencies in different regions

Table 1 gives the values of PUE for cropland, pasture, livestock, and food (human use) in the world's different regions. Globally, 116 countries have cropland PUE values above the global mean value of 0.46, mostly in Africa, and these countries account for 64 % of the global cropland area. In addition, 16 % of the countries had a PUE of around 0.6 (0.55 to 0.65). In particular, African countries had the highest overall cropland PUE (≥ 0.80) because of their low P input. On the other hand, eastern Asia and Oceania have cropland PUE below the global average. Conversely, pasture had high PUE in Europe (1.25) and North America (0.98) but low values in Africa (≤ 0.77) and particularly low values in the Caribbean and Central America (0.37). P removal from pasture exceeded P inputs in Europe, resulting in pasture PUE > 1, largely because of P inputs from feed given to animals.

The livestock subsystem generally had a low PUE (< 0.1), with the highest values in Europe, North America, and eastern Asia (Table 1). Regarding human food PUE, our data indicate that only 25 to 40 % of the P in food products in eastern Asia, Oceania, Europe, and North America is actually consumed by humans (Table 1). The resulting low PUE of human use in these regions results from both large P inputs and high food waste. Eastern and southern Africa, western and central Africa, southern and southeastern Asia, and the Caribbean and Central America had the highest PUE for human use, with more than 60 % of P in food being consumed by humans. Globally, most of the P consumed by humans (78 %) originates from crops, and the fraction of P from livestock differs among regions; it ranges from 35 % of the total human food P consumption in Oceania, Europe, and North America to 10 % in less developed regions (Africa and the Caribbean and Central America) and to 4 % in western and central Africa.

3.4 P flows through international trade

Approximately 2.1 $\mathrm{Tg\,P\,yr^{-1}}$ entered into international trade in 2010, amounting to about 17 % of the total harvested crop P (Fig. 5). The remainder (10.6 $\mathrm{Tg\,P\,yr^{-1}}$) is consumed domestically. Differences in crop types as a result of their specific P content (Table S1) strongly determine the magnitude of the traded P fluxes. For example, 37 % of the P in soybeans and 27 % of the P in wheat produced each year were traded internationally in 2010. Also significant fractions of the P in maize, other cereals, and fruit were traded internationally, but almost all of the P in sugar crops and fibre were consumed or processed in the countries where they were grown.

Considering the P fluxes in phosphate fertilizers and food products, we examined how international trade influences regional P budgets and redistributes P between regions. We found that southern and southeastern Asia have the largest net P imports (Table 1), with imports of phosphate fertilizer amounting to 1.4 $\mathrm{Tg\,P\,yr^{-1}}$ and P exports as food products being much smaller, mainly to China and South Korea. South America is the second-largest exporter of P in food, but imports 56 % of its P fertilizer. North America is a large exporter of P in both crop products and fertilizer, yet it also imports P-rich milk products. Most European countries imported nearly all their phosphate fertilizers, but Europe as a whole is a net exporter because of large exports (0.9 $\mathrm{Tg\,P\,yr^{-1}}$) from Russia (Fig. 6). Western European countries were the main exporters of P-rich livestock products. Some northern African countries (especially Morocco and Tunisia, which have the largest mines of P-rich ores) exported a total of 0.7 $\mathrm{Tg\,P\,yr^{-1}}$ in fertilizer. The remaining regions (eastern and southern Africa, northern Africa, and the Caribbean and Central America) imported P in both food and fertilizer, although much less than other regions (Table 1).

Figure 6 illustrates the disparities among countries with respect to the role of international trade in crops, livestock, and fertilizer for the main exporters and importers. Based on data for all 224 countries, a country can be categorized into one of the following four groups (Fig. 7).

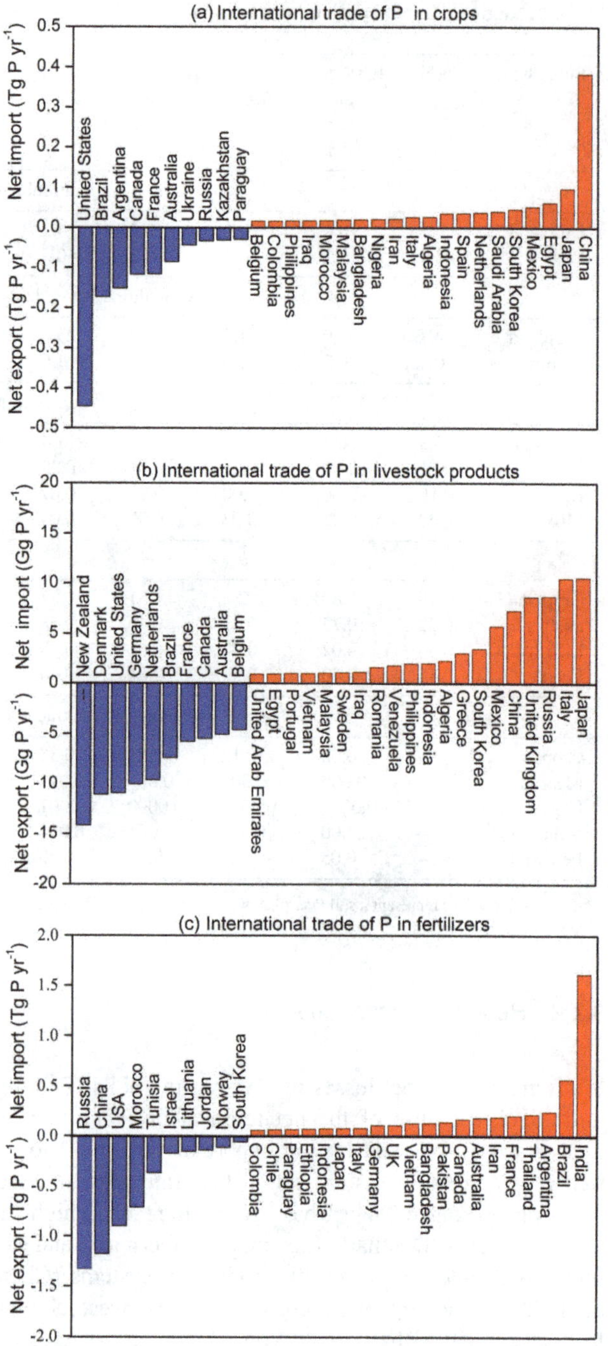

Figure 6. Annual P flows embedded in traded crop products **(a)**, livestock products **(b)**, and fertilizers **(c)** in 2010. By convention, a positive flow is P received (imported) by a country.

Food and fertilizer P exporters. P storage in these countries has been decreasing due to their international exports of both fertilizer and food. Examples include the United States and Russia.

Food P importers and fertilizer P exporters. This group mainly comprises countries that export phosphate fertilizers and import food to meet domestic consumption. Examples includes Tunisia, Morocco, and China.

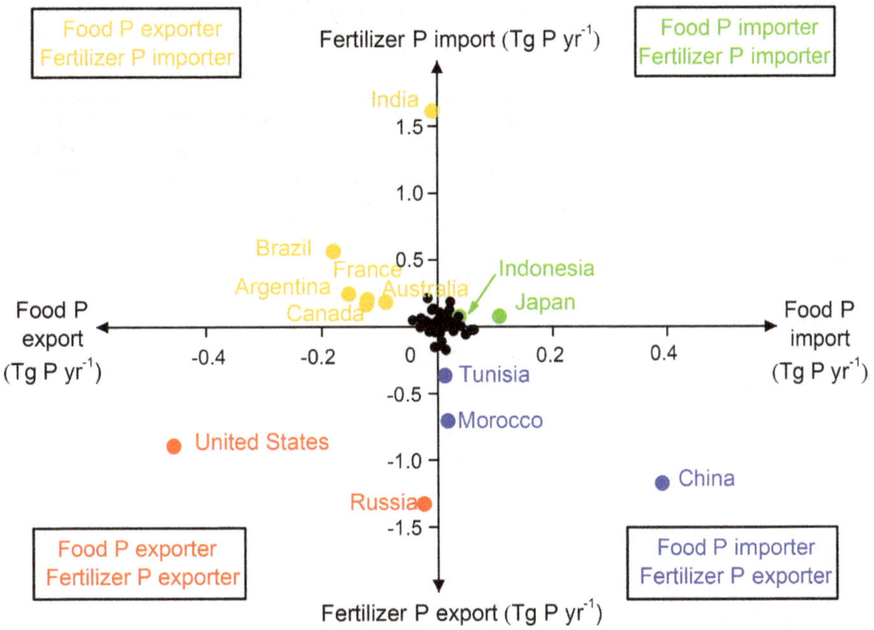

Figure 7. Groupings of the countries based on whether they import or export P through their international trade in food and fertilizer.

Food P exporters and fertilizer P importers. These countries have high food and livestock production, but this depends strongly on phosphate fertilizer imported from other countries. Examples include Brazil, Argentina, Canada, France, Australia, and India.

Food and fertilizer P importers. These countries depend on imports for both food and fertilizers; they are thus vulnerable to economic shocks that result from changing food prices. Examples include Japan and Indonesia.

International trade affects the global P cycle by physically moving the P contained in traded crops, livestock products, and phosphate fertilizers (Grote et al., 2005). Imports of P fertilizers accounted for 55 and 79 %, respectively, of the total application of P fertilizer for countries that are food P exporters and fertilizer P importers or food and fertilizer P importers. The P trade in food followed a similar trend. Countries that are food P importers and fertilizer P exporters or food and fertilizer P importers depended more on food imports than countries that are food and fertilizer P exporters or food P exporters and fertilizer P importers. International trade also increased the connections among countries (Table 2). For example, although the United States and China are clearly major P fertilizer exporters, they also import fertilizer from each other; 2.6 % of the P fertilizer applied in the United States originated in China, and 3.6 % of the phosphate fertilizer applied in China originated in the United States. In addition, 11.4 % of the phosphate fertilizer consumption in the United States originated from Russia, Morocco, Tunisia, and other countries. About 1.5 % of Chinese domestic P consumption originates from the United States, which is higher than the fraction of domestic P consumption in the United States from China. Countries with small or no reserves of P-containing minerals imported large amounts of phosphate fertilizer; for example, imports accounted for 61 and 46 % of total P consumed in France and Brazil (food P exporters and fertilizer P importers) and 76 % of total P consumed in Japan.

3.5 Uncertainties in soil P changes result from uncertain P concentrations

We estimated the net cropland soil P balance in 2000 by means of Monte Carlo simulations, as described in Sect. 2.7. We found a net accumulation of $5.8 \pm 0.6\,\mathrm{Tg\,P\,yr^{-1}}$. More detailed calculations suggest that uncertainty in the crop P concentrations contributed $\pm 0.2\,\mathrm{Tg\,P\,yr^{-1}}$ of the uncertainty in the net cropland soil P balance; this is because of the dominance of calculations using cereals, which have low uncertainty due to the narrow range of reported P concentrations (Antikainen et al., 2005; COMIFER, 2007; USDA-NRCS, 2009; Waller, 2010). Uncertainty in P concentrations in crop residues contributed an additional $\pm 0.2\,\mathrm{Tg\,P\,yr^{-1}}$ to the total uncertainty, and uncertainty in P concentrations in the livestock manure applied to cropland added $\pm 0.4\,\mathrm{Tg\,P\,yr^{-1}}$. In addition, the uncertainty in the pasture soil P balance attributed to uncertainty in the P concentrations in grass biomass and manure was $\pm 1.3\,\mathrm{Tg\,P\,yr^{-1}}$. This relative uncertainty is higher than that for the cropland soil P balance and the results from the large range of grass P concentrations found in our review of the available data. See Table S5 for more details.

Table 2. Proportions of total consumption and total international trade accounted for by P in fertilizer and food imports and exports.

Group	Proportion (%)			
	P fertilizer imports as a proportion of total consumption	P fertilizer exports as a proportion of the total international P fertilizer trade	P in food imports as a proportion of total consumption	P in food exports as a proportion of the total international P in the food trade
Group level				
Food and fertilizer exporter	22	43	7	31
Food importer and fertilizer exporter	5	48	22	5
Food exporter and fertilizer importer	55	5	5	48
Food and fertilizer importer	79	4	28	15
Country level				
United States (food and fertilizer exporter)	13	18	6	26
China (food importer and fertilizer exporter)	2	20	14	2
France (food exporter and fertilizer importer)	52	0	19	8
Brazil (food exporter and fertilizer importer)	44	1	4	10
Japan (food and fertilizer importer)	40	0	60	0

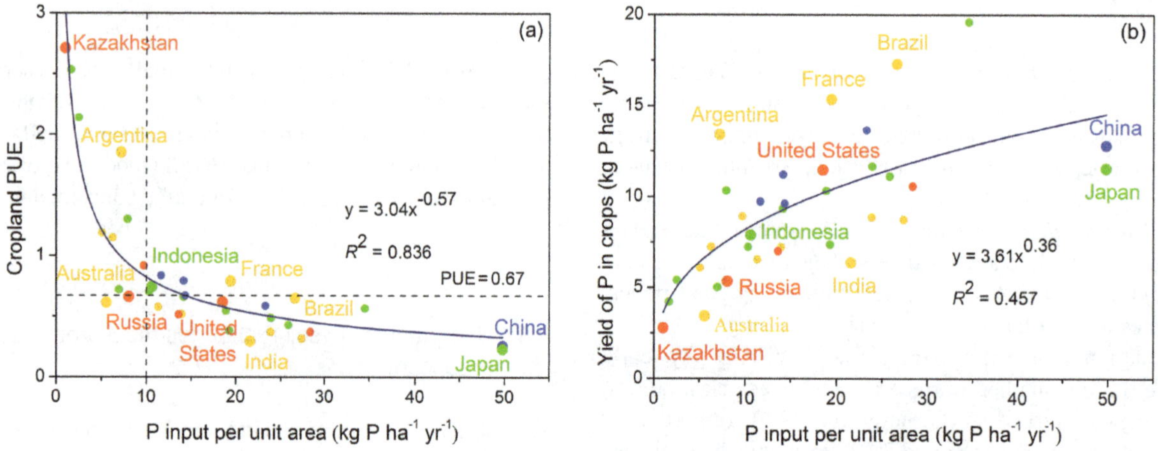

Figure 8. The relationships between P input per unit area of cropland and **(a)** phosphorus-use efficiency (PUE) The horizontal line at PUE = 0.67 represents the global average. **(b)** P in harvested crops for the 35 largest crop producers representing 90 % of global crops. The equations give the fit to the data represented by black curves.

4 Discussion

4.1 Cropland PUE and P in harvested crops as a function of cropland P inputs

Figure 8a shows the relationship between the cropland PUE and cropland P inputs for 35 countries that are large crop producers. PUE decreased exponentially with increasing input; that is, P was used most efficiently at low application rates. PUE decreased rapidly as P inputs increased to $10\,\text{kg}\,\text{P}\,\text{ha}^{-1}\,\text{yr}^{-1}$ and then decreased more slowly. High PUE values were associated with countries that had a low P input and a soil P deficit. This suggests that there is a trade-off between the efficient use of P in cropland and the avoidance of soil P deficits that limit crop yields (Obersteiner et al., 2013). Figure 8a also indicates that cropland soils have a net soil P deficit if their inputs are lower than $10\,\text{kg}\,\text{P}\,\text{ha}^{-1}\,\text{yr}^{-1}$, which is a threshold value that corresponds to PUE = 0.67. Argentina, South Africa, Indonesia, Mexico, and Paraguay are below this threshold (Fig. 9).

P in harvested crops increased exponentially with increasing P inputs, but the response slowed at high P inputs (Fig. 8b). The P in harvested crops in countries with cropland PUE > 0.67 (except Argentina) is only half of that in countries with high P in the harvested crops, such as the United States and China. P in the harvested crops was very low in Australia due to low cropland P input, which was less

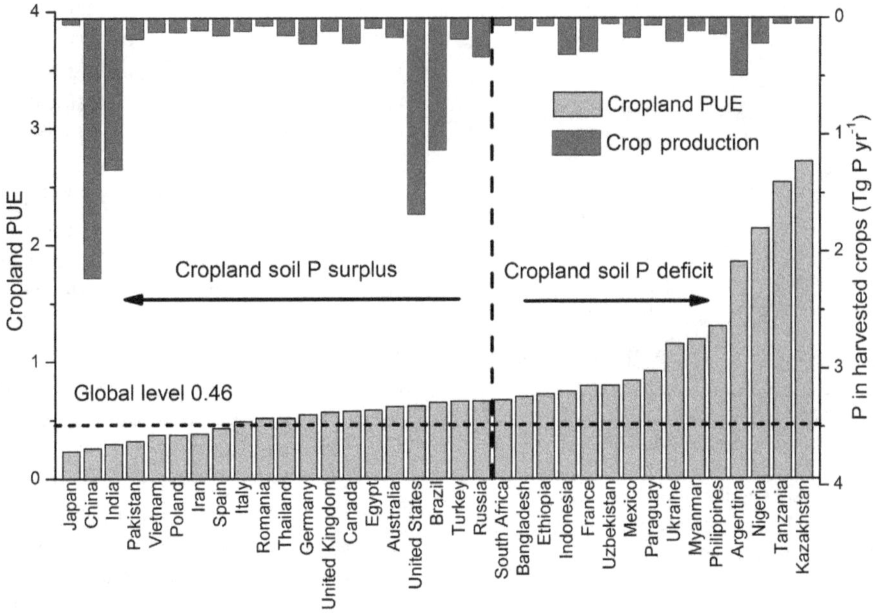

Figure 9. Phosphorus-use efficiency (PUE) and P in harvested crops for the 35 large countries shown in Fig. 8. Cropland soil P surplus or deficit is separated by the vertical dashed line.

than 25 % of the inputs in the United States and China. P already present in the soil may be sufficient to sustain high crop yields for some time without additional inputs in some countries (e.g. France) that formerly had large P fertilization rates, despite currently having a negative annual P balance. Comparing Fig. 8a and b suggests that total cropland P inputs of 20 to 25 $\mathrm{kg\,P\,ha^{-1}\,yr^{-1}}$ may be a good compromise that will achieve high yields while creating a near-equilibrium soil P balance. Both excessive P inputs (e.g. China and Japan) and low PUE (e.g. India) can lead to high P accumulation in cropland soil, leading to high losses into the environment.

The data in Fig. 8 indicate that different countries face different challenges for P resource management, implying a need for country-specific policy options and solutions. Countries like Kazakhstan and Argentina may have to increase P inputs to their cropland in order to prevent long-term depletion of soil P, which could be realized by increasing the application of phosphate fertilizer or reducing losses to leaching and erosion. Countries like France that are currently experiencing a net negative soil P balance (Fig. 9) following a period of sustained accumulation (Senthilkumar et al., 2012; van Dijk et al., 2016) may need to progressively adjust fertilizer inputs in coming years to balance inputs with removals and avoid the risk of a long-term soil fertility decline due to inadequate levels of P. In contrast, countries such as Japan and China are rapidly accumulating P in cropland soils due to high and sustained P inputs and will urgently need to consider how to improve their cropland PUE. This could be initiated by identifying crop types that are being over-fertilized and regions with excessive application of phosphate fertilizer; they can then consider a range of options such as precision agriculture (i.e. applying only as much P as the crop requires). We estimate that if Chinese cropland PUE could be increased to the global average of 0.46 (Fig. 9), China would save 3.8 $\mathrm{Tg\,P\,yr^{-1}}$ of phosphate fertilizer, which is equivalent to 60 % of its phosphate fertilizer consumption in 2010. Last, in countries like India where crop P harvests are lower than average despite high average P inputs and positive soil ΔP, improvements in agricultural management (such as the use of precision fertilization) appear necessary. We did not have access to subnational data for this study, but it is likely that in a country as large as India, excessive or insufficient P may occur in different regions, for different crop types, or for different region–crop-type combinations.

4.2 Pasture P budget, livestock consumption, and international trade

Figure 10 shows that the soil P balance is negatively related to the flux of P in livestock products per unit area of pasture. Several western European countries (Germany, the Netherlands, Denmark, and Belgium) achieve high P yields in livestock products (defined by the amount of P in livestock products per unit area of pasture), and all of these countries export livestock products. In these countries, only a small fraction of livestock manure is recycled to pasture, so there is currently a soil P deficit; in the long term, this may result in a loss of soil fertility. Therefore, these countries should increase P fertilization in pasture or import forage or feed to supply the P required to sustain high livestock production. New Zealand, Australia, and Canada are also large exporters of P in live-

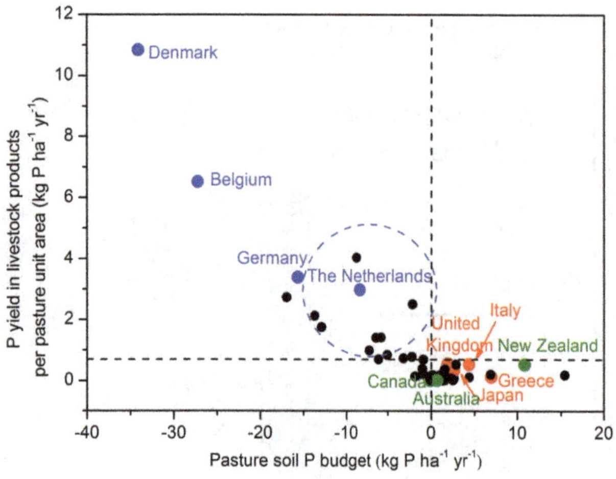

Figure 10. The relationship between the P yield of livestock products, defined by the amount of P in livestock products per unit area of pasture and the P balance of pasture soils.

stock products. However, given their low-input production systems and large areas of pasture (Fig. 4b), P removals per unit area through grazing are much lower than in western Europe, and the soil P balance of pasture ranges from slightly negative to slightly positive.

4.3 Livestock and human food PUE and trends in P consumption

The increasing consumption of livestock products by humans is an essential factor that is responsible for increasing P mining and increasing P inputs to agricultural systems (Metson et al., 2012; van Dijk et al., 2016). Where socioeconomic development is improving the income of residents, especially in Africa and the Caribbean and Central American region, residents are consuming more P from livestock products (Fig. S3). Unfortunately, the livestock PUE in countries in these two regions is much smaller (0.01 to 0.03) than the global average of 0.06 (Table 1), indicating that only a small proportion of livestock P inputs is used by humans. This may be because countries in these regions are primarily importers of livestock products. Therefore, animal husbandry has important implications for global P security and special attention will be required to improve livestock PUE (Wu et al., 2014). If livestock PUE reaches the global level of 0.06 in these two regions, both regions could more than double their livestock production by about $0.16\,\mathrm{Tg\,P\,yr^{-1}}$.

In addition, the management of manure differs greatly among regions due to different livestock production systems. The yield of livestock products is very low in African countries, resulting in low livestock PUE. Almost all livestock manure is applied to cropland for which this resource is an important P input. In contrast, with the application of phosphate fertilizer to pasture in Europe and eastern Asia, only a small fraction of livestock manure is recycled for pasture (36 and 17 %, respectively); a larger fraction of the manure is applied to cropland in eastern Asia (40 %) and Europe (60 %). Consequently, improving the manure utilization efficiency and applying more livestock manure to pasture will be important strategies in eastern Asia and Europe (Wu et al., 2014).

As shown in Sect. 3.3, only 45 % of the P that enters the food production subsystem was absorbed by humans; thus, large amounts of food (and the P it contains) are wasted, although some parts of the waste were consumed by livestock. Despite this recycling, $2.2\,\mathrm{Tg\,P\,yr^{-1}}$ flowed into the environment as waste either before or after food consumption, and only 14.3 % of the total P inputs to the agriculture system ended up in food consumed by humans. In eastern Asia, Oceania, Europe, and North America, the PUE of human food was very low, reflecting the high proportion of livestock products in the diet and a high degree of waste. Therefore, decreasing food waste before consumption, recycling P in food waste, and better treatment of organic waste could significantly decrease the amount of P required to support humans (Metson et al., 2012; van Dijk et al., 2016). In eastern Asia, Oceania, Europe, and North America, fully absorbing the 45 % of the P that enters food produced for humans could reduce agricultural inputs of P by $0.7\,\mathrm{Tg\,P\,yr^{-1}}$ globally. Thus, decreasing food waste and improving the PUE of human food represent key challenges that must be solved to achieve sustainable P management.

Population increases and dietary changes are requiring higher P inputs in cultivated land and the increased mining of P ores (Grote et al., 2005; Foley et al., 2011). From 2002 to 2010, this mining increased by 33 % in our estimate, during a period when the global population and per capita food P consumption increased by 10 and 5 %, respectively. In 2010, humans consumed 8.0 and 3.8 % more P in livestock products and crops, respectively. Since livestock PUE was much lower than cropland PUE, the consumption of more livestock products resulted in lower external P inputs in food that flowed into the human subsystem; this proportion decreased from 36 % in 2002 to 31 % in 2010. Therefore, consuming more livestock products will require increasing P inputs. Thus, human dietary shifts may have been responsible for half of the increase of P-ore mining.

4.4 International trade and global P flows

International trade also increased the connections among countries. Whether international trade is good or bad for humans and the environment in terms of its impact on the management of P resources is a complex question. International trade can increase cropland P deficits if countries that export large amounts of P in crop and livestock products do not counteract these exports by increasing inputs of phosphate fertilizer to soils. For example, Argentina exported lots of food to other countries (about $0.15\,\mathrm{Tg\,P\,yr^{-1}}$) and has developed a serious cropland soil P deficit of $0.38\,\mathrm{Tg\,P\,yr^{-1}}$

($10.3\,kg\,P\,ha^{-1}\,yr^{-1}$). Massive P imports through trade can result in an excess supply of P to cropland soils as manure (Schipanski and Bennett, 2012), with potentially significant negative environmental effects. On the one hand, trade can hamper the proper recycling of P resources from waste and manure to agricultural soils through local food webs (Schipanski and Bennett, 2012). On the other hand, trade may contribute to more efficient use of P resources if traded products flow from countries with lower PUE to countries with higher PUE, as is generally observed for water resources (Dalin et al., 2014). This confirms that more integrated studies are required to fully assess the effects of trade on P resource recycling, efficiency, and conservation. Our study identified world regions and countries with lower PUE and others with high PUE and regions and countries with a net loss of P in soils and others with a net gain. This provides valuable information to policymakers on how to improve the trade relationships for a global optimization of PUE and therefore global food security.

4.5 Comparison with previous studies

Previous studies have estimated P flows in agriculture on a global scale (Smil, 2000; Sheldrick et al., 2003; Liu et al., 2008; Cordell et al., 2009; Bouwman et al., 2009, 2013; Potter et al., 2010; MacDonald et al., 2011). However, to the best of our knowledge, the present analysis provides the first consistent multi-year overview of the P flows in agriculture. In addition, it provides national and regional P budgets, calculates agricultural PUE, and quantifies P fluxes in international trade based on a combination of datasets for cropland and pasture inputs (fertilizers, manure, atmospheric deposition, and recycling of crop residues) and outputs (crop harvests, residue removal, and P loss by burning and leaching or surface run-off into bodies of water). For data from 2000, our results are consistent with the abovementioned studies for most P flows (Table 3). For data from 2000, our results are generally consistent with those in the previous studies for cropland soil P inputs, harvested crop P, cropland soil P lost by erosion or surface run-off into bodies of water, pasture soil P inputs, and harvested grass P (Table 3). However, methods, data sources, and system boundaries differed among the studies, making an accurate comparison difficult. Our estimate of a net accumulation of $5.8 \pm 0.6\,Tg\,P\,yr^{-1}$ is in line with the reported net accumulation in soils, which ranged between 0 and $11.5\,Tg\,P\,yr^{-1}$ (Smil, 2000; Bennett et al., 2001; Bouwman et al., 2009; MacDonald et al., 2011), but disagrees with the estimate of Liu et al. (2008), who calculated a net loss of $9.6\,Tg\,P\,yr^{-1}$. The difference from the present results can be explained by accounting for large P losses ($19.3\,Tg\,P\,yr^{-1}$) due to soil erosion caused by land-use change and overgrazing. The quantification of erosional losses of P from arable land is prone to high uncertainties due to the unknown amount of redeposited soil material, and other studies have reported much lower losses (e.g. $2.5\,Tg\,P\,yr^{-1}$; Quinton et al., 2010).

The main cropland P fluxes estimated in our study agreed with previous results, except for the production and recycling of crop residues (Table 3). Smil (2000) and Liu et al. (2008) used harvest index data (defined as the ratio of total aboveground biomass to crop residues) for estimating the P in crop residues, whereas we estimated P in crop residues by combining data from Liu et al. (2008) and FAO. MacDonald et al. (2011) estimated that 29 % of the global cropland area was subject to soil P deficits in 2000, which is similar to our estimate (32 %) based on data from 2002 to 2010. In addition, our estimate of $22.3\,Tg\,P\,yr^{-1}$ in animal manure for the livestock subsystem in 2000 is within the reported range of 17.1 to $24.3\,Tg\,P\,yr^{-1}$ from Potter et al. (2010). We defined global cropland PUE as the ratio of P in harvested crops to total P inputs without accounting for the recycling of crop residues. Under this definition, global PUE was estimated to be 0.43 by Liu et al. (2008) and 0.40 by Smil (2000), both of which are comparable to our estimate of 0.46 from 2002 to 2010. Since we applied the same methods across the globe to calculate agricultural P fluxes, we were able to compare the P fluxes and budgets for different regions and countries on a consistent basis. This information is of critical importance for the development of more appropriate agricultural policy and to support the development of technological and other solutions for different types of countries, which better integrate cultivated ecosystems, livestock production, and the human food supply.

4.6 Limitations and novelty of our study

Due to limited data sources for some parameters, our study and most previous studies focused on P in livestock products and manure as the outputs of the livestock system and did not consider the fate of P in non-edible livestock products (e.g. bones, blood, leather products). Xu et al. (2005) pointed out that from 12 to 23 and 72 % of P were contained in livestock meat and bones, respectively. If these percentages are applied to our data, this gives an annual flux of $2.5\,Tg\,P\,yr^{-1}$ in the bones of slaughtered animals. Although most livestock bones are currently wasted or landfilled, some countries have begun to use them as fertilizers, protein sources, and condiments (Wu and Ma, 2005; Li, 2008). In addition, as we focused on the annual P budgets for livestock and human beings, we did not account for P accumulation in humans. From 2002 to 2010, the global population increased by 635×10^6 persons. If we assume that a typical adult body contains 600 g of P, then about 0.38 Tg more P would have accumulated in humans. Therefore, the annual human P accumulation would be $0.04\,Tg\,P\,yr^{-1}$, accounting for only 0.3 % of the P inputs into humans.

Despite the abovementioned limitations in our study, we were able to achieve some interesting and novel results. First, we have provided a detailed and harmonized summary of

Table 3. Comparison of the present results for P flows and budgets in 2000 with the results of other studies at a global level ($Tg\,P\,yr^{-1}$).

Global P flux		Previous studies	Our study	Reasons for differences
Cropland	Fertilizer input	14–15[1–3]	13.7	—
	Animal manure to cropland	6–8[2,3]	6.7 ± 0.4	Method
	Human sewage sludge to cropland	1.5[1,3]	1.3	Method
	Crop production	8.2–12.3[1–5]	10.2 ± 0.4	Boundary/data
	Crops (human food)	3.5[3]	4.8 ± 0.2	Method/data
	Crops (animal feed)	2.6[3]	1.9 ± 0.1	Data
	Crop residues	3.75–4.5[1–2]	6.7 ± 0.2	Method/data
	Recycling of residues	1–2.2[1–3]	3.5 ± 0.1	Method/data
	Leaching and run-off from cropland	4[6]	3.2	Method
Pasture	Livestock manure	17.1–24.3[5,7,8]	22.3 ± 1.3	Method/data
	Manure wasted (released into the environment)	2–8[1–3]	4.1 ± 0.2	Method/data
	Grass	6–12.1[3,4]	8.9 ± 1.3	Method/data
	Animal feed additives	0.9[3]	1.4	Data
	Leaching and run-off from pasture	1.0[5]	1.6	Method
Humans	Excreta	3–3.3[1,3]	2.8	Method

Sources: [1] Liu et al. (2008), [2] Smil (2000), [3] Cordell et al. (2009), [4] MacDonald et al. (2011), [5] Bouwman et al. (2009), [6] Bouwman et al. (2013), [7] Sheldrick et al. (2003), [8] Potter et al. (2010).

the P fluxes as inputs and outputs for the agricultural system and the internal P flows within the agricultural system on national, regional, and global scales. In addition, we have characterized the P budgets and P-use efficiencies in the subsystems of the overall agricultural system and discussed their influences and impacts. Finally, we have discussed how changes in population, diets, and food consumption have influenced the global mining of P ore and how international trade has influenced P fluxes. These insights will support the development of policies to use P more sustainably at national, regional, and global levels.

5 Conclusions

The estimation of global and regional phosphorus budgets in agricultural systems and their PUE is a major effort by the anthropogenic nutrient cycle research community that requires lots of work. We quantified in detail the P inputs and outputs of cropland and pasture and the P fluxes through human and livestock consumers of agricultural products on global, regional, and national scales from 2002 to 2010. The results from this analysis confirmed that P from phosphate fertilizers is the largest single input flux into the agricultural system, while one-half of this input was lost to freshwaters and one-third accumulated in soils. Chemical fertilizer inputs, P losses to the environment, and P harvested in crop biomass all present increasing trends. The positive global P balance in cropland and pasture soils is the sum of regional excess and deficits, with differences in the sign of the balance between cropland and pastures as well. Compared with croplands, a slightly larger proportion of the global pasture area had a net annual soil P deficit. The hot spots of cropland P budgets are on the one hand eastern Asian countries with excessive inputs and sub-Saharan African countries with deficits. Hot spots of pasture P budgets are European and North American grazing lands showing significant soil P deficits. There are great differences in PUE for croplands, pastures, livestock products, and food products on global, regional, and national scales. Livestock products generally had the lowest PUE. We showed that cropland PUE decreased exponentially with increasing input implying that P is used more efficiently at low application rates. In parallel, P in harvested crop biomass increased with P inputs following a saturating exponential relationship. International trade plays a significant role in the P redistribution among countries when considering P embedded in imported livestock and food products consumed in each country and the trade of fertilizers. Nearly one-fifth of total harvested crop P entered into international trade during the period 2002–2010. Population increases and dietary changes are requiring higher P inputs in cultivated land and the increased production of P fertilizers from minerals. Human dietary shifts may have been responsible for half of the increase in P mining during the period 2002–2010. Regional P imbalances in agricultural soils can be mitigated both by optimizing phosphate fertilizer application and recycling P.

Competing interests. The authors declare that they have no conflict of interest.

Acknowledgements. This study was supported by the National Natural Science Foundation of China (41625001, 41571022), the Beijing Natural Science Foundation (grant 8151002), the Southern University of Science and Technology (grant no. G01296001), and a Synergy Grant (ERC-2013-SyG-610028 IMBALANCE-P) from the European Research Council.

Edited by: Attila Demény

References

Antikainen, R., Lemola, R., Nousiainen, J. I., Sokka, L., Esala, M., Huhtanen, P., and Rekolainen, S.: Stocks and flows of nitrogen and phosphorus in the Finnish food production and consumption system, Agr. Ecosyst. Environ., 107, 287–305, 2005.

ASAE: Manure Production and Characteristics. D384.2, American Society of Agricultural Engineers, St. Joseph, MI, USA, 2005.

Bennett, E. M., Carpenter, S. R., and Caraco, N. F.: Human impact on erodable phosphorus and eutrophication: a global perspective: increasing accumulation of phosphorus in soil threatens rivers, lakes, and coastal oceans with eutrophication, BioScience, 51, 227–234, 2001.

Bouwman, A. F., Beusen, A. H., and Billen, G.: Human alteration of the global nitrogen and phosphorus soil balances for the period 1970–2050, Global Biogeochem. Cy., 23, https://doi.org/10.1029/2009GB003576, 2009.

Bouwman, L., Goldewijk, K. K., Van Der Hoek, K. W., Beusen, A. H., Van Vuuren, D. P., Willems, J., Rufino, M. C., and Stehfest, E.: Exploring global changes in nitrogen and phosphorus cycles in agriculture induced by livestock production over the 1900–2050 period, P. Natl. Acad. Sci. USA, 110, 20882–20887, 2013.

Carpenter, S. R. and Bennett, E. M.: Reconsideration of the planetary boundary for phosphorus, Environ. Res. Lett., 6, 014009, https://doi.org/10.1088/1748-9326/6/1/014009, 2011.

Carpenter, S. R., Caraco, N. F., Correll, D. L., Howarth, R. W., Sharpley, A. N., and Smith, V. H.: Nonpoint pollution of surface waters with phosphorus and nitrogen, Ecol. Appl., 8, 559–568, 1998.

Chang, J., Viovy, N., Vuichard, N., Ciais, P., Campioli, M., Klumpp, K., Matin, R., Leip, A., and Soussana, J. F.: Modeled changes in potential grassland productivity and in ruminant livestock density in Europe over 1961–2010, PLOS ONE, 10, e0127554, https://doi.org/10.1371/journal.pone.0127554, 2015.

Chang, J. F., Viovy, N., Vuichard, N., Ciais, P., Wang, T., Cozic, A., Lardy, R., Graux, A.-I., Klumpp, K., Martin, R., and Soussana, J.-F.: Incorporating grassland management in ORCHIDEE: model description and evaluation at 11 eddy-covariance sites in Europe, Geosci. Model Dev., 6, 2165–2181, https://doi.org/10.5194/gmd-6-2165-2013, 2013.

COMIFER: Teneur en P, K et Mg des organes végétaux récoltés pour les cultures de plein champ et les principaux fourrages, Comité Français d'Étude et de Développement de la Fertilisation Raisonneé, Paris, 2007 (in French).

Cordell, D., Drangert, J., and White, S.: The story of phosphorus: global food security and food for thought, Global Environ. Chang., 19, 292–305, 2009.

Cordell, D., Neset, T. S., and Prior, T.: The phosphorus mass balance: identifying "hotspots" in the food system as a roadmap to phosphorus security, Curr. Opin. Biotech., 23, 839–845, 2012.

Cordell, D., Jackson, M., and White, S.: Phosphorus flows through the Australian food system: identifying intervention points as a roadmap to phosphorus security, Environ. Sci. Policy, 29, 87–102, 2013.

Dalin, C., Hanasaki, N., Qiu, H., Mauzerall, D. L., and Rodriguez-Iturbe, I.: Water resources transfers through Chinese interprovincial and foreign food trade, P. Natl. Acad. Sci. USA, 111, 9774–9779, 2014.

Elser, J. and Bennett, E.: Phosphorus cycle: a broken biogeochemical cycle, Nature, 478, 29–31, 2011.

FAO: Fertilizer Use by Crop (5th Edition), Food and Agriculture Organization of the United Nations, Rome, 2012.

Foley, J. A., Ramankutty, N., Brauman, K. A., Cassidy, E. S., Gerber, J. S., Johnston, M., Muller, N. D., O'Connell, C., Ray, D. K., West, P. C., Balzer, C., Bennett, E. M., Carpenter, S. R., Hill, J., Monfreda, C., Polasky, S., Rockström, J., Sheehan, J., Siebert, S., Tilman, D., and Balzer, C.: Solutions for a cultivated planet, Nature, 478, 337–342, 2011.

Grote, U., Craswell, E., and Vlek, P.: Nutrient flows in international trade: ecology and policy issues, Environ. Sci. Policy, 8, 439–451, 2005.

Herrero, M., Havlík, P., Valin, H., Notenbaert, A., Rufino, M. C., Thornton, P. K., Blümmelb, M., Weissc, F., Grace, D., and Obersteiner, M.: Biomass use, production, feed efficiencies, and greenhouse gas emissions from global livestock systems, P. Natl. Acad. Sci. USA, 110, 20888–20893, 2013.

Levington Agriculture: A Report for the European Fertiliser Manufacturers' Association. Levington Agriculture Ltd., Ipswich, UK, 1997.

Li, J.: The development and prospective of nutrition in livestock bones, Meat Ind., 322, 41–44, 2008 (in Chinese).

Liu, Y., Villalba, G., Ayres, R. U., and Schroder, H.: Global phosphorus flows and environmental impacts from a consumption perspective, J. Ind. Ecol., 12, 229–247, 2008.

Ma, W., Ma, L., Li, J., Wang, F., Sisák, I., and Zhang, F.: Phosphorus flows and use efficiencies in production and consumption of wheat, rice, and maize in China, Chemosphere, 84, 814–821, 2011.

MacDonald, G. K., Bennett, E. M., Potter, P. A., and Ramankutty, N.: Agronomic phosphorus imbalances across the world's croplands, P. Natl. Acad. Sci. USA, 108, 3086–3091, 2011.

Metson, G. S., Bennett, E. M., and Elser, J. J.: The role of diet in phosphorus demand, Environ. Res. Lett., 7, 044043, https://doi.org/10.1088/1748-9326/7/4/044043, 2012.

MWPS-18: Livestock Waste Facilities Handbook, Midwest Plan Service, University of Missouri, Ames, IA, USA, 1985.

Obersteiner, M., Peñuelas, J., Ciais, P., Van Der Velde, M., and Janssens, I. A.: The phosphorus trilemma, Nat. Geosci., 6, 897–898, 2013.

OECD: Secretariat National Soil Surface Nutrient Balances, 1985 to 1995. Explanatory Notes. Coefficients to Convert Livestock Numbers Into Manure Nitrogen Quantities From National Sources, Organisation for Economic Cooperation and Development, Paris, 1991.

Ott, C. and Rechberger, H.: The European phosphorus balance, Resour. Conserv. Recy., 60, 159–172, 2012.

Penuelas, J., Poulter, B., Sardans, J., Ciais, P., Van Der Velde, M., Bopp, L., Boucher, O., Godderis, Y., Hinsinger, P., Llusia, J., Nardin, E., Vicca, S., Obersteiner, M., and Nardin, E.: Human-induced nitrogen–phosphorus imbalances alter natural and managed ecosystems across the globe, Nat. Commun., 4, 2934, https://doi.org/10.1088/1748-9326/7/4/044043, 2013.

Potter, P., Ramankutty, N., Bennett, E. M., and Donner, S. D.: Characterizing the spatial patterns of global fertilizer application and manure production, Earth Interact., 14, 1–22, 2010.

Quinton, J. N., Govers, G., Van Oost, K., and Bardgett, R. D.: The impact of agricultural soil erosion on biogeochemical cycling, Nat. Geosci., 3, 311–314, 2010.

Ringeval, B., Nowak, B., Nesme, T., Delmas, M., and Pellerin, S.: Contribution of anthropogenic phosphorus to agricultural soil fertility and food production, Global Biogeochem. Cy., 28, 743–756, 2014.

Sattari, S. Z., Bouwman, A. F., Giller, K. E., and van Ittersum, M. K.: Residual soil phosphorus as the missing piece in the global phosphorus crisis puzzle, P. Natl. Acad. Sci. USA, 109, 6348–6353, 2012.

Schipanski, M. E. and Bennett, E. M.: The influence of agricultural trade and livestock production on the global phosphorus cycle, Ecosystems, 15, 256–268, 2012.

Scholz, R. W., Ulrich, A. E., Eilittä, M., and Roy, A.: Sustainable use of phosphorus: a finite resource, Sci. Total Environ., 461, 799–803, 2013.

Senthilkumar, K., Nesme, T., Mollier, A., and Pellerin, S.: Regional-scale phosphorus flows and budgets within France: the importance of agricultural production systems, Nutr. Cycl. Agroecosys., 92, 145–159, 2012.

Sheldrick, W., Syers, J. K., and Lingard, J.: Contribution of livestock excreta to nutrient balances, Nutr. Cycl. Agroecosys., 66, 119–131, 2003.

Smil, V.: Phosphorus in the environment: natural flows and human interferences, Annu. Rev. Mater. Sci., 25, 53–88, 2000.

Steffen, W., Richardson, K., Rockström, J., Cornell, S. E., Fetzer, I., Bennett, E. M., Biggs, R., Carpenter, S. R., de Vries, W., de Wit, C. A., Folke, C., Gerten, D., Heinke, J., Mace, G. M., Persson, L. M., Ramanathan, V., Reyers, B., and Sörlin, S.: Planetary boundaries: guiding human development on a changing planet, Science, 347, 1259855, https://doi.org/10.1126/science.1259855, 2015.

Suh, S. and Yee, S.: Phosphorus use-efficiency of agriculture and food system in the US, Chemosphere, 84, 806–813, 2011.

USDA-NRCS: Crop Nutrient Tool: Nutrient Content of Crops. United States Department of Agriculture, Natural Resource Conservation Service, Washington, 2009.

van Dijk, K. C., Lesschen, J. P., and Oenema, O.: Phosphorus flows and balances of the European Union Member States, Sci. Total Environ., 542, 1078–1093, 2016.

Van Vuuren, D. P., Bouwman, A. F., and Beusen, A. H. W.: Phosphorus demand for the 1970–2100 period: a scenario analysis of resource depletion, Global Environ. Chang., 20, 428–439, 2010.

Waller, J. C.: Byproducts and unusual feedstuffs, Feedstuffs, 9, 18–22, 2010.

Wang, R., Tao, S., Balkanski, Y., Ciais, P., Boucher, O., Liu, J., Piao, S., Shen, H., Vuoloc, M. R., Valarie, M., Chen, H., Chen, Y., Cozic, A., Huang, Y., Li, B., Li, W., Shen, G., Wang, B., and Zhang, Y.: Exposure to ambient black carbon derived from a unique inventory and high-resolution model, P. Natl. Acad. Sci. USA, 111, 2459–2463, 2014.

Wang, R., Balkanski, Y., Boucher, O., Ciais, P., Peñuelas, J., and Tao, S.: Significant contribution of combustion-related emissions to the atmospheric phosphorus budget, Nat. Geosci., 8, 48–54, 2015.

Wu, H., Yuan, Z., Zhang, Y., Gao, L., and Liu, S.: Life-cycle phosphorus use efficiency of the farming system in Anhui Province, Central China, Resour. Conserv. Recy., 83, 1–14, 2014.

Wu, L. and Ma, M.: The comprehensive utilization of animals' bone in China, Modern Food Sci. Tech., 83, 38–46, 2005 (in Chinese).

Xu, J., Liu, X., Wang, F., Zhang, F., Ma, W., and Ma, L.: Phosphorus balance and environmental effect of animal production in China, Acta Ecol. Sinica, 11, 2911–2918, 2005 (in Chinese).

5

Thickness of marine Holocene sediment in the Gulf of Trieste (northern Adriatic Sea)

Ana Trobec[1], Martina Busetti[2], Fabrizio Zgur[2], Luca Baradello[2], Alberto Babich[2,3], Andrea Cova[2], Emiliano Gordini[2], Roberto Romeo[2], Isabella Tomini[2], Sašo Poglajen[4], Paolo Diviacco[2], and Marko Vrabec[1]

[1]University of Ljubljana, Faculty of Natural Sciences and Engineering, Department of Geology, Aškerčeva cesta 12, 1000 Ljubljana, Slovenia

[2]OGS – Istituto Nazionale di Oceanografia e di Geofisica Sperimentale, Borgo Grotta Gigante 42/c, 34010 Sgonico, Trieste, Italy

[3]Università di Trieste, Dipartimento di Matematica e Geoscienze, Via Weiss 2, 34127 Trieste, Italy

[4]Harpha Sea d.o.o., Čevljarska ulica 8, 6000 Koper, Slovenia

Correspondence: Ana Trobec (ana.trobec@geo.ntf.uni-lj.si)

Abstract. We use various geophysical datasets (multibeam and singlebeam echosounder data, sub-bottom profiling Chirp and sonar data and very high-resolution boomer seismic data) along with published sedimentological data and depth data from nautical charts in order to create models of the depth of the seafloor and the base of Holocene marine sediment in the Gulf of Trieste. The two models are later used in order to calculate the thickness of marine Holocene sediment which has been depositing on the Late Pleistocene alluvial plain since the Holocene transgression in the Italian and Slovenian parts of the gulf. Thicker Holocene marine sedimentary sequences averaging at around 5 m are characteristic for the southeastern part of the gulf. In other parts of the gulf the Holocene marine sedimentary cover is very thin or even absent, except in close proximity to the shoreline and fluvial sediment sources, in the area of the Trezza Grande paleodelta and above topographic depressions of the Late Pleistocene base. The presented datasets available from the OGS SNAP data repository (http://doi.org/cpz2) represent a valuable reference for a wide variety of research disciplines dealing with the dynamic Earth system in the Gulf of Trieste and could be used as a valuable tool for designing sampling and geophysical campaigns in the studied area.

1 Introduction

The Gulf of Trieste is a shallow gulf with an average depth at around 20 m located in the northernmost part of the Adriatic Sea where it is surrounded by Italian, Slovenian and Croatian coasts (Fig. 1). After the Last Glacial Maximum with the onset of the Holocene transgression, marine sediment started depositing approximately 10 000 years ago in this area (Marocco, 1991; Ogorelec et al., 1991, 1997; Lambeck et al., 2004; Covelli et al., 2006; Ogrinc et al., 2007; Trincardi et al., 2011b).

Early research of Holocene marine sediment in the Gulf of Trieste was mainly limited to sedimentological and geochemical investigations of data acquired with a relatively small number of cores and wells located onshore and offshore the gulf in the 80s and 90s (Ogorelec et al., 1981, 1991, 1997; Marocco et al., 1984; Marocco, 1989, 1991; Faganeli et al., 1991; Gordini et al., 2002; Covelli et al., 2006; Ogrinc et al., 2007) and investigations for the Italian Marine Geological Map 1 : 250 000 of Venice (Trincardi et al., 2011a, b). The development of cost-effective geophysical methods in recent years resulted in a number of geophysical surveys undertaken in the last decade that were focused on high-resolution geophysical investigation of the seafloor and the sub-seafloor geological structure of the gulf (Gordini et al., 2003, 2004, 2006; Gordini, 2007, 2009; Romeo,

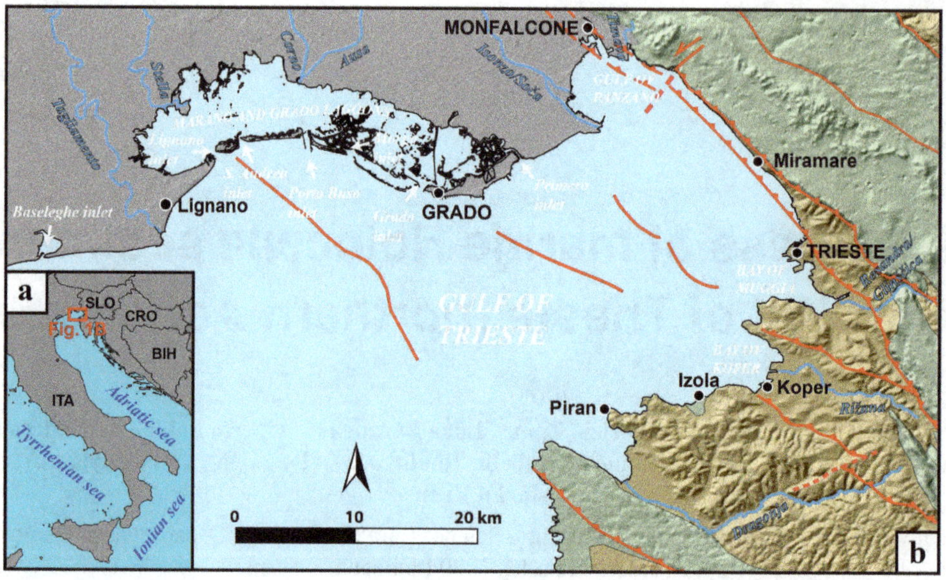

Figure 1. Location of the study area; **(a)** regional map; **(b)** a simplified geological map of the area (following Biolchi et al., 2016; Busetti et al., 2010a; Carulli et al., 2006; Carulli, 2011; EMODnet, 2018; Jurkovšek et al., 2016; Placer et al., 2010; Placer, 2015; Pleničar et al., 1969). Mesozoic carbonates are marked with green, Eocene flysch with orange and Quaternary sediment with grey.

2009; Trincardi et al., 2011a, b, 2014; Kolega and Poklar, 2012; Slavec, 2012; Zampa et al., 2015; Zecchin et al., 2015; Trobec, 2015; Trobec et al., 2016, 2017). Some of these extensive datasets along with additional unpublished data spanning over a major part of the gulf allowed us to assess the thickness of the Holocene marine sediment in the Gulf of Trieste.

This work aims to present the first comprehensive model of the distribution and thickness of Holocene marine sediment in the Slovenian and Italian parts of the Gulf of Trieste. The model is derived from geophysical, core/well, and nautical chart data and in parts is complemented by the published Holocene marine thicknesses from Trincardi et al. (2011b).

Geological setting of the study area

A simplified geological overview of the surroundings of the Gulf of Trieste is shown in Fig. 1b and consists of Quaternary alluvial sediment (grey) of the Friuli plain in the north, Cretaceous–Paleogene carbonates (green) of the Classical Karst in the northeast and Paleogene marls and sandstones of the flysch (orange) of the Istria peninsula and Trieste coastline in the southern and eastern parts of the gulf. A similar geological sequence can be observed in geophysical data offshore from the gulf area where the carbonate platform is followed by a flysch succession which is overlain by a few hundred metres of Quaternary sediment deposited during the transgressive–regressive cycles (Busetti et al., 2010a, b; Cimolino et al., 2010; Vrabec et al., 2014). The youngest sedimentary sequence in the Gulf of Trieste is represented by Holocene marine sediment which has been depositing for the last 10 000 years since the onset of the Holocene transgression following the Younger Dryas (Ogorelec et al., 1981, 1997; Lambeck et al., 2004; Covelli et al., 2006; Trincardi et al., 2011b; Zecchin et al., 2015). Many authors suggest that the base on which marine sediment has been depositing represents relict continental-paralic sedimentary environments predating the Holocene sea transgression in the Gulf of Trieste (Ogorelec et al., 1981, 1991, 1997; Marocco, 1989; Lambeck et al., 2004; Covelli et al., 2006; Trincardi et al., 2011b; Slavec, 2012; Zecchin et al., 2015; Trobec, 2015; Trobec et al., 2017). The sediments of the Holocene marine transgression provided a thin and discontinuous deposit occurring on an erosive surface of the Late Pleistocene sediments due to subaerial exposure (Trincardi et al., 2011a).

2 Data used and modelling of the different surfaces

Numerous geophysical surveys were conducted between 2000 and 2015 in the Gulf of Trieste by the Istituto Nazionale di Oceanografia e di Geofisica Sperimentale (OGS) and by Harpha Sea d.o.o. in cooperation with the Department of Geology of the University of Ljubljana (Figs. 4 and 5, Tables 2 and 3). Most of the data and results from these surveys were so far published only in internal reports and/or theses and are here presented to a wider audience for the first time.

Geophysical data were correlated and calibrated with core/well data previously published in other scientific publications (Table 1 and Figs. 2 and 3; Ogorelec et al., 1981, 1991, 1997; Marocco et al., 1984; Marocco, 1989; Gordini et al., 2002; Covelli et al., 2006; Romeo, 2009; Trincardi et al., 2011b; Zecchin et al., 2015). The cores/wells usually doc-

Table 1. Previously published wells and cores used in this study with reported ^{14}C ages. The > symbol indicates that the core contains only Holocene marine sediment. The "/" symbol indicates that data were not reported or investigated in the original publication.

Core/ well	Water depth (m)	Holocene thickness (m)	Dated material	Core depth of dated material (m)	Reported conventional ^{14}C age (years BP)	Reported calibrated ^{14}C age (years BP)	Reference	Label colour
VE04-06	15.6	> 3.27	peat	2.90–2.91	8530 ± 40	9477–9546	Trincardi et al. (2011a)	green
VE04-11	24.5	1.63	peat	1.67–1.68	8610 ± 50	9516–9689	Trincardi et al. (2011a)	green
VE04-14	21.7	1.61	freshwater gastropod shell	3.48	9630 ± 40	10 784–11 034	Trincardi et al. (2011a)	green
VE04-15	21.9	2.48	peat	2.67–2.68	8750 ± 55	9549–9920	Trincardi et al. (2011a)	green
VE04-18	22	1.12	*Cerastoderma glaucum* shell	0.97–0.98	8860 ± 40	9265–9507	Trincardi et al. (2011a)	green
VE04-23	12.5	0.05	/	/	/	/	Trincardi et al. (2011a)	green
VE05-06	21	0	peat	0.82–0.83	48210 ± 45	9021–9302	Trincardi et al. (2011a)	green
S12	0	6.55	/	/	/	/	Marocco et al. (1984)	blue
S9	−1.7	2.45	beach shells	1.05	2300 ± 105	/	Marocco (1989)	blue
			Cerastoderma glaucum shell	2.60	1400 ± 290			
			organic material	5.76	3660 ± 290			
			Cerastoderma glaucum shell	10.05	5540 ± 225			
			peat	22.0	20 200 ± 720			
GT2	13.5	> 5.43	/	/	/	/	Gordini et al. (2002), Zecchin et al. (2015)	orange
GT3	20	0.83	peat	0.86–0.88	18630 ± 60	21 650–22 610	Gordini et al. (2002), Zecchin et al. (2015)	orange
GT4	14	1.32	/	/	/	/	Zecchin et al. (2015)	orange
S19	−2.5	17.5	organic material	2.2	300 ± 110	/	Marocco (1991), Gordini et al. (2002),	orange
			mollusc shells	15.2	2040 ± 70		Zecchin et al. (2015)	
GT1	23	1.55	sedimentary organic carbon	0.12	4020 ± 30	/	Covelli et al. (2006)	purple
			sedimentary organic carbon	0.31	4170 ± 55			
			sedimentary organic carbon	0.72	7320 ± 40			
			sedimentary organic carbon	1.60	9030 ± 70			
			sedimentary organic carbon	2.00	9140 ± 40			
			Viviparus contectus shell	2.38	9610 ± 40			
GT2	15	1.75	sedimentary organic carbon	0.15	3370 ± 40	/	Covelli et al. (2006)	purple
			sedimentary organic carbon	0.55	3820 ± 50			
			sedimentary organic carbon	1.55	5860 ± 40			
			sedimentary organic carbon	1.95	9380 ± 40			
			sedimentary organic carbon	2.75	9040 ± 50			
			sedimentary organic carbon	2.95	8270 ± 50			
			sedimentary organic carbon	3.15	8770 ± 40			
GT3	25	0.85	sedimentary organic carbon	0.64	4560 ± 35	/	Covelli et al. (2006)	purple
			Cerastoderma glaucum shell	0.87	8810 ± 40			
			sedimentary organic carbon	1.20	9160 ± 50			
M1	4.9	1.9	/	/	/	/	Romeo (2009)	yellow
M2	5.15	1.4	/	/	/	/	Romeo (2009)	yellow
M3	5.4	14.5	/	/	/	/	Romeo (2009)	yellow
M4	5.2	7.9	/	/	/	/	Romeo (2009)	yellow
AIII-7	13	14.2	/	/	/	/	Ogorelec et al. (1997)	red
V1-95	12	19	/	/	/	/	Ogorelec et al. (1997)	red
MK6	7	21	/	/	/	/	Ogorelec et al. (1991, 1997)	red
V3	4.5	25	/	/	/	/	Ogorelec et al. (1991, 1997)	red
V6	0	26.5	peat	26.5	9160 ± 120	/	Ogorelec et al. (1981)	red

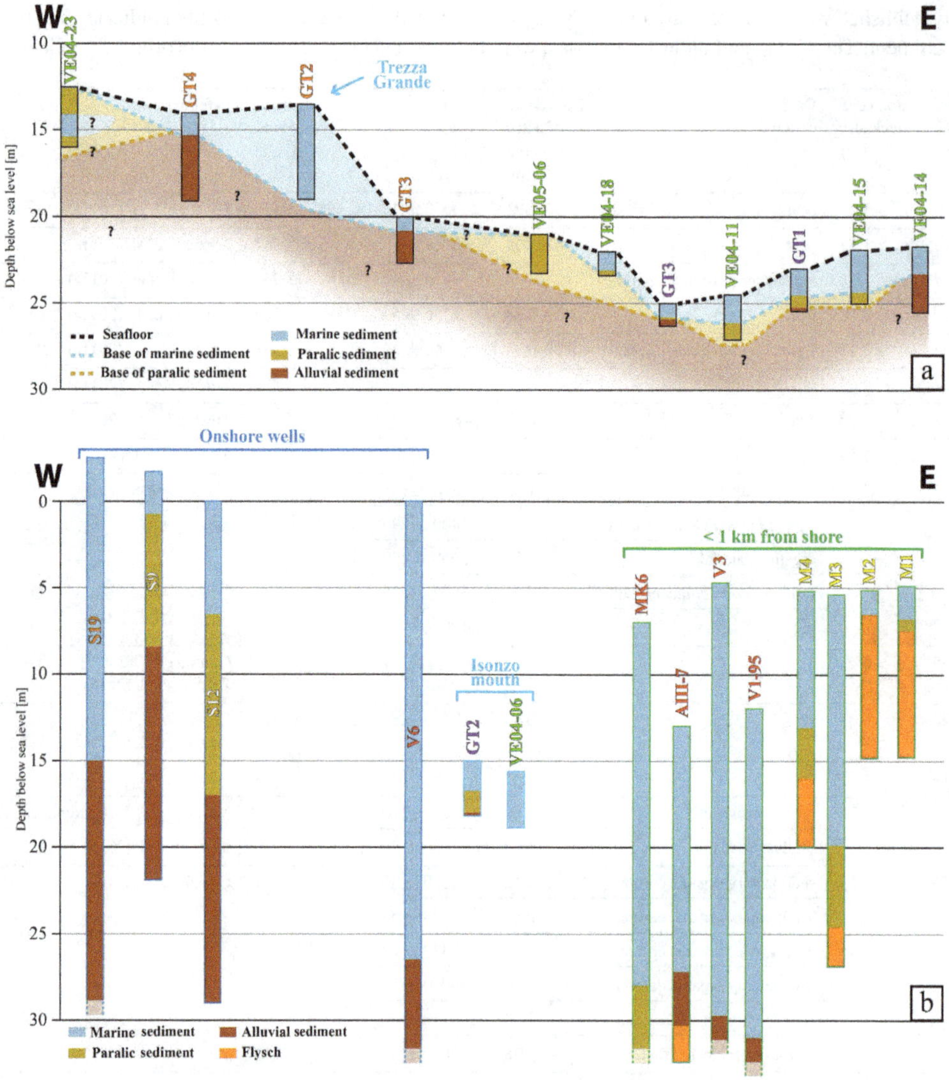

Figure 2. Simplified stratigraphy of all the cores and wells used in this study (Table 1; for details see Ogorelec et al., 1981, 1991, 1997; Marocco et al., 1984; Marocco, 1989; Gordini et al., 2002; Covelli et al., 2006; Romeo, 2009; Trincardi et al., 2011b; Zecchin et al., 2015). Note that the distance between the cores/wells is not in scale, **(a)** a simplified W–E stratigraphic profile of the cores located in the central part of the Gulf of Trieste; **(b)** wells and cores used in the study that were not suitable for the stratigraphic profile due to their location or proximity to the shore or fluvial sedimentary sources (see Figs. 4 and 5 for location). Wells outlined in dark blue are located onshore. Cores outlined in bright blue are located in close proximity of the present-day Isonzo river mouth. Cores outlined in green are located less than 1 km from the shoreline. Dashed bottoms of wells indicate that their whole stratigraphy is not illustrated in the figure and that the reader should refer to the original publications for further information.

ument a transition from an alluvial to paralic and later marine sedimentary environment. However, in some cores the marine or paralic sedimentary sequence is absent or the marine and paralic sediments are deposited directly on the bedrock (Figs. 2, 3 and 8d).

The calibration of the geophysical data with the stratigraphy of core/well data available for the Gulf of Trieste (Table 1) indicates that the Holocene marine sediment is represented on high-resolution acoustic sonar and seismic boomer profiles as a characteristic transparent acoustic/seismic facies which shows absent or very faint low-amplitude internal reflections (e.g. Fig. 3). On geophysical profiles Holocene paralic and/or Late Pleistocene continental sedimentary units are separated from the Holocene marine sediment by a prominent middle- to high-amplitude reflection and furthermore demonstrate significantly dissimilar acoustic facies containing several reflections of varying reflection geometries, amplitudes and continuity (Romeo, 2009; Slavec, 2011, 2012; Trobec, 2015; Trobec et al., 2017).

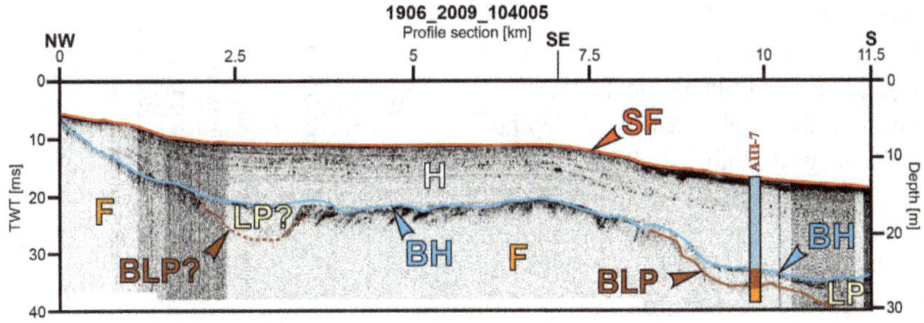

Figure 3. An example of an acoustic profile, acquired along the northern coast of the Istria Peninsula, which is correlated with the core AIII-7 from Ogorelec et al. (1997) (SF: seafloor, H: Holocene marine sediment, BH: reflector that marks the base of the Holocene sediment, LP: Late Pleistocene sediment, BLP: base of the Late Pleistocene sediment, F: flysch, TWT: two-way travel time). For the location of the profile, see Fig. 5.

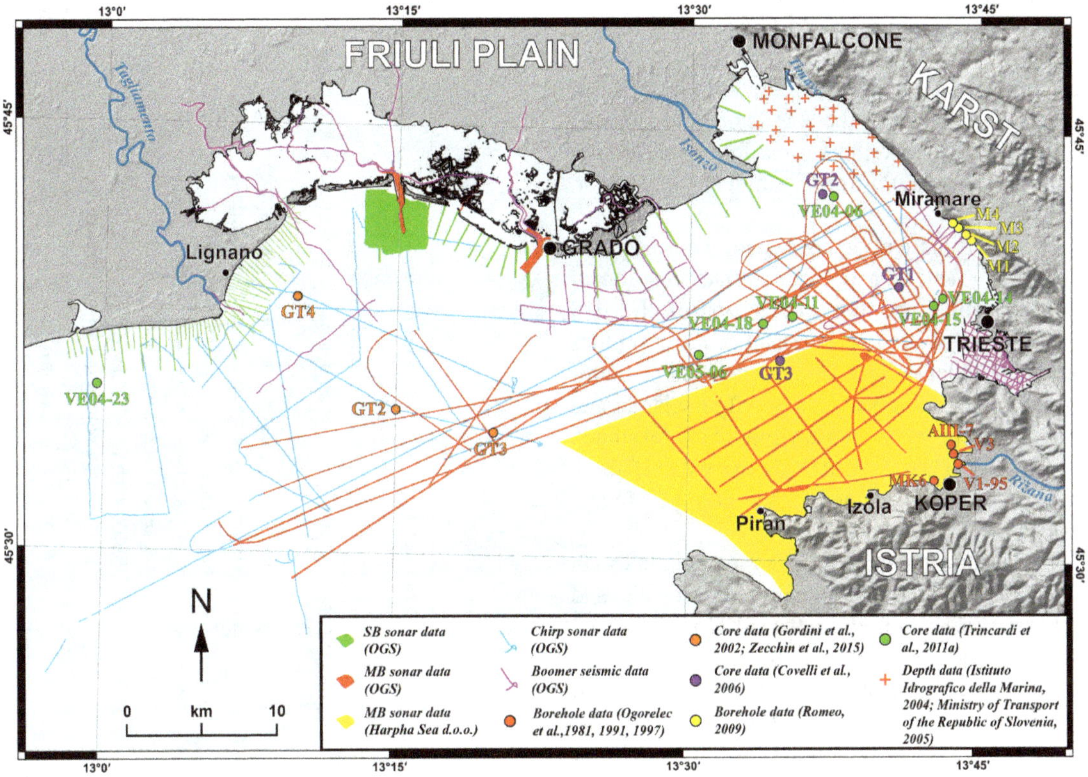

Figure 4. Locations of the datasets used for the bathymetric model of the Gulf of Trieste.

Reported conventional radiocarbon ages from published core/well data (Table 1) show that Holocene marine sediment started depositing on the alluvial sedimentary sequence approximately 9150 years BP in the Gulf of Trieste. The Holocene marine sequence started depositing later where it overlies paralic sequences for which conventional radiocarbon ages younger than approximately 8800 years BP are reported (Table 1). The reported radiocarbon ages in Table 1 clearly show that the alluvial- or paralic-Holocene marine boundary in the Gulf of Trieste is diachronous.

Geophysical datasets along with previously published core/well data allowed us to create a model of the distribution and thickness of Holocene marine sediment in the Italian and Slovenian parts of the gulf. The different datasets and methods used to create each model are described in detail in the subsequent subsections.

2.1 The bathymetric model

In order to create the bathymetric model of the Gulf of Trieste we used multibeam and singlebeam sonar data complemented with high-resolution single-channel seismic data (boomer and Chirp profiles) (Fig. 4, Tables 2 and 3). In addition, water depths of 28 points in the Gulf of Panzano were determined from the Slovenian nautical chart of the Gulf of Trieste (Ministry of Transport of the Republic of Slovenia, 2005) and the Da Punta Tagliamento a Pula nautical chart (Istituto Idrografico della Marina, 2004). We also used depths of published cores/wells in the gulf (Table 1, Fig. 2). The used datasets are depicted in Fig. 4. In order to constrain the model, we used the coastline contour for which we assumed a seafloor depth value equal to 0 m above sea level.

Multibeam sonar data included data acquired during the reflection seismic surveys in the central and Slovenian parts of the gulf in the years 2009 and 2013 (Table 2). In addition, a multibeam-based bathymetric model of Slovenian territorial waters (Slavec, 2012) acquired by Harpha Sea d.o.o. was also used. Smaller areas surveyed by OGS by multibeam sonar also include three areas in front of the GradoMarano lagoon: the Porto Buso inlet, Morgo inlet and Grado inlet (Figs. 1 and 4, Table 2).

Singlebeam sonar data were acquired along the coast of the Tagliamento delta between the Baseleghe and Lignano inlets, from the Morgo inlet to the Marina Julia beach and in a dense grid in front of the Porto Buso inlet (Figs. 1 and 4, Table 2).

Single-channel seismic and acoustic data were acquired in various surveys (Table 2 and Fig. 4). Interpretation of the geophysical profiles was done by means of the IHS Kingdom® software. We used seismic and acoustic profiles to determine the seafloor depth where singlebeam or multibeam bathymetric data were not acquired. The seismic and acoustic profiles used to create the bathymetric model stretch over a large part of the western part of the gulf, offshore Grado, southeast of Miramare and in the Bay of Muggia (Figs. 1 and 4).

The multibeam and singlebeam sonar data coupled with the interpreted depths of the seafloor from acoustic and seismic profiles were imported in the SKUA-GOCAD™ Paradigm software package in order to perform quality control and consequent adjustments to the dataset. In order to convert the seismic and acoustic data from the time to the depth domain, $1514\,\mathrm{m\,s^{-1}}$ was assumed as the velocity of sound in the water column. This value represented the average value of sound velocity profiler measurements acquired together with the datasets used in this study. After the conversion gridding of the bathymetric model was done with the discrete smooth interpolation (DSI) method (Mallet, 1992, 1997) in the SKUA-GOCAD™ software package. The coordinates of the pointsets building the model were clipped to the coastline extent and later exported as a column-based text file and a geotiff grid with a $50\,\mathrm{m} \times 50\,\mathrm{m}$ cell size. The resulting model contained some areas with negative depth values that resulted from lack of geophysical data in the near proximity of the coast. In addition, the morphologically very dynamic northern part of the research area starting from the Isonzo mouth towards the west also contained negative depth values due to transient sedimentary bodies elevated above or just below the sea level that were present at the time of

Table 2. Multibeam and singlebeam datasets used for the bathymetric model (previously published only in internal reports/theses).

Area	Year	Institute	Type	Instrument	Quantity	Reference
Porto Buso inlet	2006	OGS	Multibeam	Reson Seabat 8125	$1.6\,\mathrm{km^2}$	Cova (2008)
Grado inlet	2007	OGS	Multibeam	SWATHplus-Wide swath Bathymetry and Side scan sonar system Sea	$1.7\,\mathrm{km^2}$	Gordini (2008a)
Morgo inlet	2007	OGS	Multibeam	SWATHplus-Wide swath Bathymetry and Side scan sonar system Sea	$0.3\,\mathrm{km^2}$	Gordini (2008a)
Central part of the gulf	2009	OGS	Multibeam	Reson SeaBat 8111 hull mounted	$49\,\mathrm{km^2}$	Zgur et al. (2010)
Slovenian and central part of the gulf	2013	OGS	Multibeam	Reson SeaBat 8111 hull mounted	$20.6\,\mathrm{km^2}$	Zgur et al. (2013)
Slovenian part of the gulf	2006–2009	Harpha Sea d.o.o.	Multibeam	Reson SeaBat 8125 and Elac HydroStar 4300	$205\,\mathrm{km^2}$	Poglajen (2012), Slavec (2012)
Porto Buso–Monfalcone	2007	OGS	Singlebeam	Garmin echosounder	62 km	Gordini (2008b)
Porto Buso	2007	OGS	Singlebeam	Garmin echosounder	250 km	Gordini (2008c)
Tagliamento Delta	2002–2003	OGS	Singlebeam	Lowrance LCX-18C echosounder	120 km	Gordini et al. (2006)

Table 3. Seismic and acoustic data used for the model of the base of the Holocene marine sediment (previously published only in internal reports/theses).

Area	Year	Institute	Type	Instrument	Quantity	References
Lignano	2000	OGS	Boomer	Boomer CEA PULSAR 2002 and UWAK 05	61.7 km	Della Vedova et al. (2002)
Lignano	2002–2003	OGS	Chirp	Datasonic Sub-bottom Profiler CHIRP CAP-6600	153.1 km	Gordini (2009), Zecchin et al. (2015)
Grado and Grado-Marano lagoon	2008	OGS	Boomer	Boomer CEA PULSAR 2002 and UWAK 05	60.9 km	Baradello et al. (2009)
Miramare	2003	OGS	Boomer	Boomer CEA PULSAR 2002 and UWAK 05	16.2 km	Romeo (2009)
Gulf of Trieste	2005	OGS	Chirp	Benthos CHIRP II Hull mounted 16 transducers	179.2 km	Busetti et al. (2005)
Gulf of Trieste	2009	OGS	Chirp	Benthos CHIRP II Hull mounted 16 transducers	516.4 km	Zgur et al. (2010)
Gulf of Trieste	2013	OGS	Chirp	Benthos CHIRP II Hull mounted 16 transducers	160.3 km	Zgur et al. (2013)
Muggia	2013	OGS	Boomer	Boomer CEA PULSAR 2002 and UWAK 05	76.5 km	Baradello et al. (2013)
Piran–Koper	2013	OGS	Boomer	Boomer CEA PULSAR 2002 and UWAK 05	26.1 km	Luca Baradello, personal communication, 2017
Slovenian waters	2009–2015	Harpha Sea d.o.o.	Parametric sub-bottom sonar	Innomar Parametric Sediment Echo Sounder SES-2000 Compact	560.8 km	Žerjal et al. (2015), Sašo Poglajen, personal communication, 2017
Strunjan bay	2013–2014	Harpha Sea d.o.o.	Parametric sub-bottom sonar	Innomar Parametric Sediment Echo Sounder SES-2000 Compact	53.3 km	Trobec (2015)

the various surveys but were not delineated by the coastline shapefile used to clip the dataset. In order to still represent these dynamic shallow areas while eliminating artificial negative depths from our dataset, we replaced all the negative values in the geotiff grid and the column-based text file with a depth value of 0.2 m.

2.2 Model of the base of the Holocene marine sediment

Seismic (boomer) and acoustic data (Chirp and parametric sub-bottom sonar) were used in order to create a model of the base of the Holocene marine sediment in the Gulf of Trieste (Fig. 5 and Table 3). In general, the used profiles span throughout the Gulf of Trieste but are also located in the inner parts of the Grado Lagoon and Bay of Muggia, and in the Bay of Koper (Fig. 5). The base of the marine Holocene sediment was determined from reflection characteristics visible from profiles that were interpreted by means of the IHS Kingdom® software and calibrated with the cores listed in Table 1. In general, the base of the Holocene marine sediment on geophysical profiles can be recognized as a prominent reflector underlying the acoustically transparent marine sediment (Figs. 3 and 8; Slavec, 2011, 2012; Trobec, 2015; Trobec et al., 2017). Additionally, depths of the base of Holocene marine sediment determined from offshore and onshore well and core data from Italy and Slovenia were included in the model and are shown in Table 1 and Figs. 2 and 5.

This dataset was imported, quality controlled and adjusted in the SKUA-GOCAD™ software package. A velocity of $1530\,\mathrm{m\,s^{-1}}$ was assumed for the velocity of sound travelling through the Holocene marine sediment and used for the conversion of seismic and acoustic data from the time to depth domain. Afterwards, gridding of the model of the base of the Holocene was done in the SKUA-GOCAD™ software package with the DSI method. Even though the sound velocity of the sediment overlying the flysch in the Gulf of Trieste was determined at $1610\,\mathrm{m\,s^{-1}}$ (Masoli et al., 2015), we decided for a lower value because Masoli et al. (2015) considered a Late Pleistocene–Holocene sedimentary sequence a few tens of metres thick where compaction most probably already affected the velocity value. Considering (a) that the typical sound velocity of a marine water saturated sediment is approximately $1500\,\mathrm{m\,s^{-1}}$ (Anderson and Hampton, 1980; Yuan et al., 1992), (b) that the average value of sound velocity profiler measurements acquired together with the datasets used in this study is $1514\,\mathrm{m\,s^{-1}}$, and (c) that the Holocene marine sediment thickness in our study is an order of magnitude smaller than the sediment considered in Masoli et al. (2015) leads us to believe that our chosen velocity is reasonable for our conversion. The coordinates of the pointsets building the model were clipped to the coastline extent and later exported as a column-based text file and a geotiff grid with a $50\,\mathrm{m} \times 50\,\mathrm{m}$ cell size.

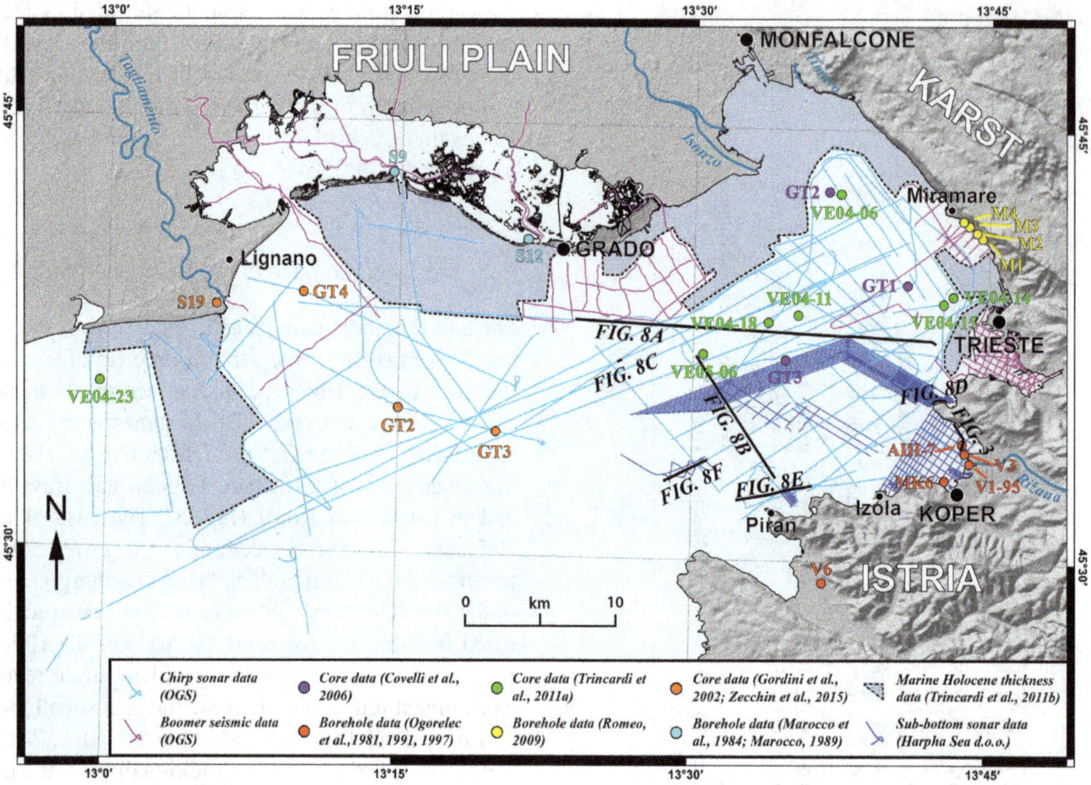

Figure 5. Locations of the datasets used for the model of the base of the Holocene. Black lines indicate locations of acoustic profiles shown in Figs. 3 and 8. Note that the black dashed lines mark areas where published data from Trincardi et al. (2011b) were used for the model of the thickness of Holocene marine sediment.

2.3 Modelling the thickness of Holocene marine sediment

A model of the thickness of Holocene marine sediment in the Gulf of Trieste was created with the SKUA-GOCAD™ software. In order to create the model, the difference between the bathymetric model and the model of the base of the Holocene was calculated. Data used to create the model of the base of the Holocene spanned over a smaller area compared to the bathymetric model (Figs. 4, 5, 6 and 7). In areas where data coverage was insufficient or even absent (grey areas bounded by black dashed lines in Fig. 5), we used data from the Carta degli spessori dei sistemi di stazionamento alto (HST) published by Trincardi et al. (2011b), for which we assumed $1530\,\mathrm{m\,s^{-1}}$ as the velocity of sound travelling through the sediment. We excluded the areas inside the Grado-Marano lagoon from the model due to dubious interpretation of geophysical profiles (signal reverberation) resulting from shallow seafloor depth and because these Holocene sediments were deposited in a lagoonal rather than marine sedimentary environment. The coordinates of the pointsets building the model were clipped to the coastline extent and later exported as a column-based text file and a geotiff grid with a 50 m × 50 m cell size. The model contained negative thickness values in areas of erosion (e.g. Fig. 8e) and in the western part of the research area, where Holocene marine sediment is very thin or even absent and is therefore very difficult to model. In this area thicker sedimentary sequences can seldom be found only as infill of Late Pleistocene channels (Fig. 8a and c). When modelling such geometries, the base of the Holocene surface tends to rise above the seafloor outside the channel as a convex more or less pronounced bulge. This in turn results in artificial negative thickness values. In order to honour the data, we replaced all the negative values in the geotiff grid and the column-based text file with a thickness value of zero.

3 Results

3.1 Bathymetry of the Gulf of Trieste

The bathymetry of the seabed of the Gulf of Trieste has a generally smooth morphology. Depth values in our model vary between 0.2 and 32.6 m below sea level with a mean of 16.2 m and standard deviation of 6.2 (Fig. 6). In the southeastern and eastern parts of the gulf depth exceeds the 20 m isobath approximately 4 km away from the shore at most, while in the northern and northwestern parts the distance needed is approximately twice as much (but can reach up to 20 km).

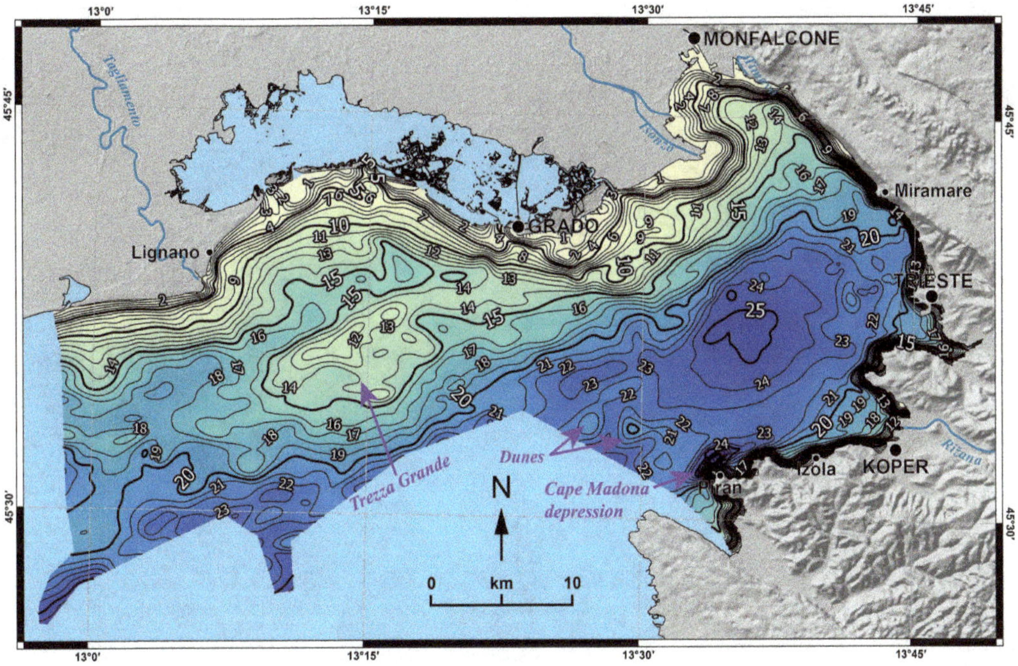

Figure 6. Map of the bathymetry of the Gulf of Trieste (in metres) with prominent morphological features indicated with purple arrows.

The Trezza Grande paleodelta (Gordini et al., 2002; Zecchin et al., 2015) and the dune-shaped features NW of Piran (Slavec, 2012) are topographically higher areas visible in Fig. 6. The deepest feature on the bathymetric model is the elongated Cape Madona depression stretching in a SW–NE direction north of Piran (Fig. 6), which also represents the deepest point of the Gulf of Trieste, with a depth of 38 m (Ministry of Transport of the Republic of Slovenia, 2005; Slavec, 2012). A deeper area where depths exceed 25 m is located north of Piran and west of the Bay of Muggia.

3.2 Base of the Holocene marine sediment

Contrary to the smoothness of the bathymetric model, the model of the base of the Holocene marine sediment presents rougher morphologies (Fig. 7), which are evident also from Figs. 3 and 8. The model of the base of the Holocene is located between 6.6 and 31.0 m below sea level with a mean of 19.8 m and a standard deviation of 5.7. In general, the model can be divided into two parts, the shallower northwestern part and the deeper southeastern part (Fig. 7). This difference in depth is also evident from geophysical profiles (Fig. 8a and b) and core data (Fig. 2).

The northwestern part is shallower than 20 m and in general rises as we approach the coastline. Local topographical highs are located in the southern part of the Trezza Grande area (Fig. 7). In places the model exhibits channel-like features that are a few metres deeper than the surrounding topography, but their exact course is difficult to map due to insufficient profile coverage.

The base of Holocene marine sediment in the southeastern part of the gulf forms a basin that is in general deeper than 25 m below sea level (Fig. 7) and is also visible in geophysical profiles and recognizable in core data (Figs. 8 and 2). Shallower depths can only be observed in the Bay of Muggia and in the immediate vicinity of the coastline where it rises rapidly in a very short distance (Fig. 7). A slightly deeper section oriented in a WNW–ESE direction approximately 6 km north of Piran is evident and corresponds to the meander belt of the Paleorižana (Slavec, 2012; Trobec et al., 2017).

3.3 Thickness of Holocene marine sediment

The modelled thickness of the Holocene marine sediment in the Gulf of Trieste ranges between 0 and 24.0 m with a mean thickness of 3.2 m and a standard deviation of 2.8. In general, the inner parts of the gulf are covered by very thin drapes of Holocene marine sediment that can also be partially absent (Figs. 9, 8a and b). Towards the shore, the Holocene sedimentary sequence gradually thickens and appears as a coastal sedimentary wedge that can exceed thicknesses of 10 m near the coastline and in the internal parts of bays.

The map of the thickness of the Holocene marine sediment in the Gulf of Trieste can be divided into two parts, the central–western part with thinner marine sequences and the southeastern part of the gulf with thicker Holocene marine sediments (Fig. 9). This prominent difference is clearly evident from geophysical profiles spanning through a larger part of the gulf (Fig. 8a and b) and from a simplified W–E stratigraphic profile from the published cores (Fig. 2).

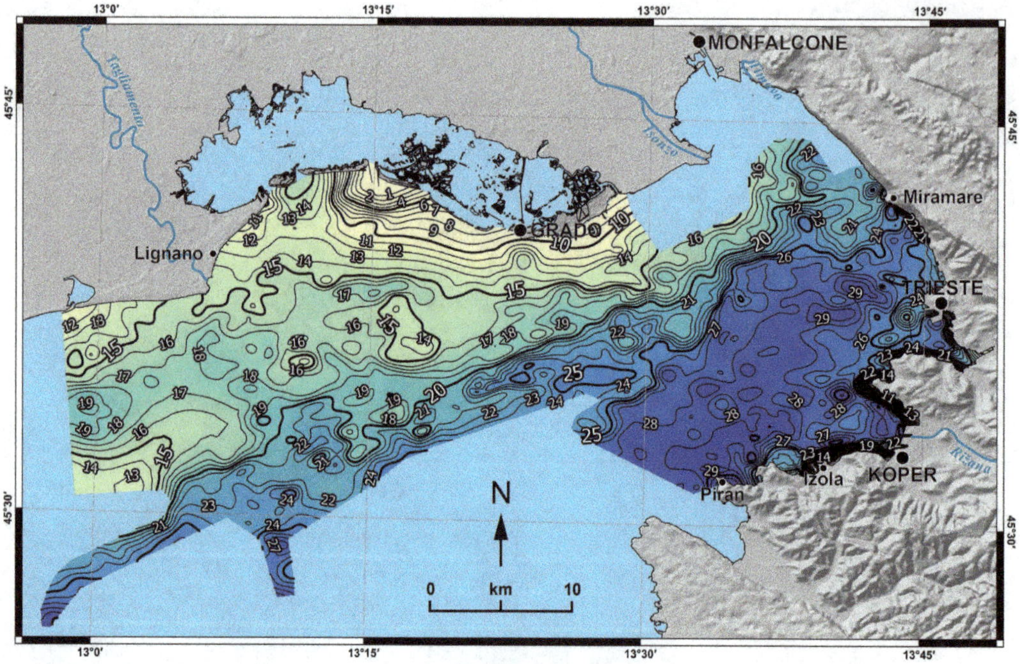

Figure 7. Map of the depth of the base of the Holocene marine sediment in metres.

The central–western part of the gulf is characterized by a very thin Holocene sedimentary cover which rarely exceeds a thickness of 3 m (Figs. 9, 8a and b). Where the thin sediments are not well resolved or even unresolvable with the high-resolution geophysical methods used in this study, we assume they are absent or only a few centimetres thick. Thicker Holocene sedimentary sequences can be observed only in the vicinity of the Tagliamento delta (Figs. 9, 3 and 4 from Zecchin et al., 2015), the Trezza Grande paleodelta (Figs. 2 and 9; Figs. 4 and 5 from Zecchin et al., 2015) and in the central part of the bay where thicker sedimentary sequences fill pre-Holocene paleochannels (Figs. 9, 7, 8a, b and c).

The southeastern part of the gulf is characterized by a thicker sedimentary cover which averages at about 5 m (Figs. 9, 2, 8a and b). Further examination reveals that in some places in close proximity to the shoreline the Holocene sedimentary cover often thins out or is unresolvable in geophysical profiles (for example near Debeli rtič, Fig. 8d). This effect is a consequence of basement rock (predominantly flysch) outcropping and/or subcropping near the seafloor. In the Cape Madona depression in front of Piran (Fig. 6) the marine Holocene sedimentary sequence is not recognizable due to erosion which can also be identified on geophysical profiles down to the Late Pleistocene sediment (Fig. 8e). In addition to thick marine Holocene sedimentary deposits in the internal parts of bays (Figs. 9 and 2), thicker sedimentary sequences can be observed over buried paleochannels (Figs. 9, 8a, b and c; Slavec, 2012; Trobec, 2015, Trobec et al., 2017) and in the area of the dune-shaped features NW of Piran (Figs. 6, 8f and 9).

4 Model quality assessment

While our work was done with great care and attention to detail in order to produce accurate results, we would like to emphasize that the model is primarily intended to illustrate the general trend of the thickness of Holocene marine sediment in the Gulf of Trieste and cannot account for strong local thickness variations. This is a result of several factors that are described in the following section.

As mentioned in Sects. 3.2 and 3.3, the surface of the base of the Holocene marine sediment can be very undulated due to the pre-existing paleotopography and/or due to subcropping or outcropping of the basement rock on the seafloor near the eastern and southern coastlines (Fig. 8d). Combined with areas where density of acoustic and seismic profiles is not adequate, this undulation can lead to artefacts that are later outstretched throughout the model of the base of the Holocene and consequentially affect the values of the model of the thickness. Discrepancies of the map with the natural state are also a result of the gridding algorithm used (Mallet, 1992, 1997). Due to the inherent two-way travel time attribute of the acquired data in the case of geophysical profiles, major influencing factors on the models are the chosen velocity for the conversion to the time domain and the local sound velocity variations in the Holocene sediment which are unknown. As there are no measurements of speed of sound travelling solely through the Holocene sediment of the Gulf of Trieste, we used a conservative approximation of the speed of sound travelling through the sedimentary column, which could lead to systematic underestimation of the

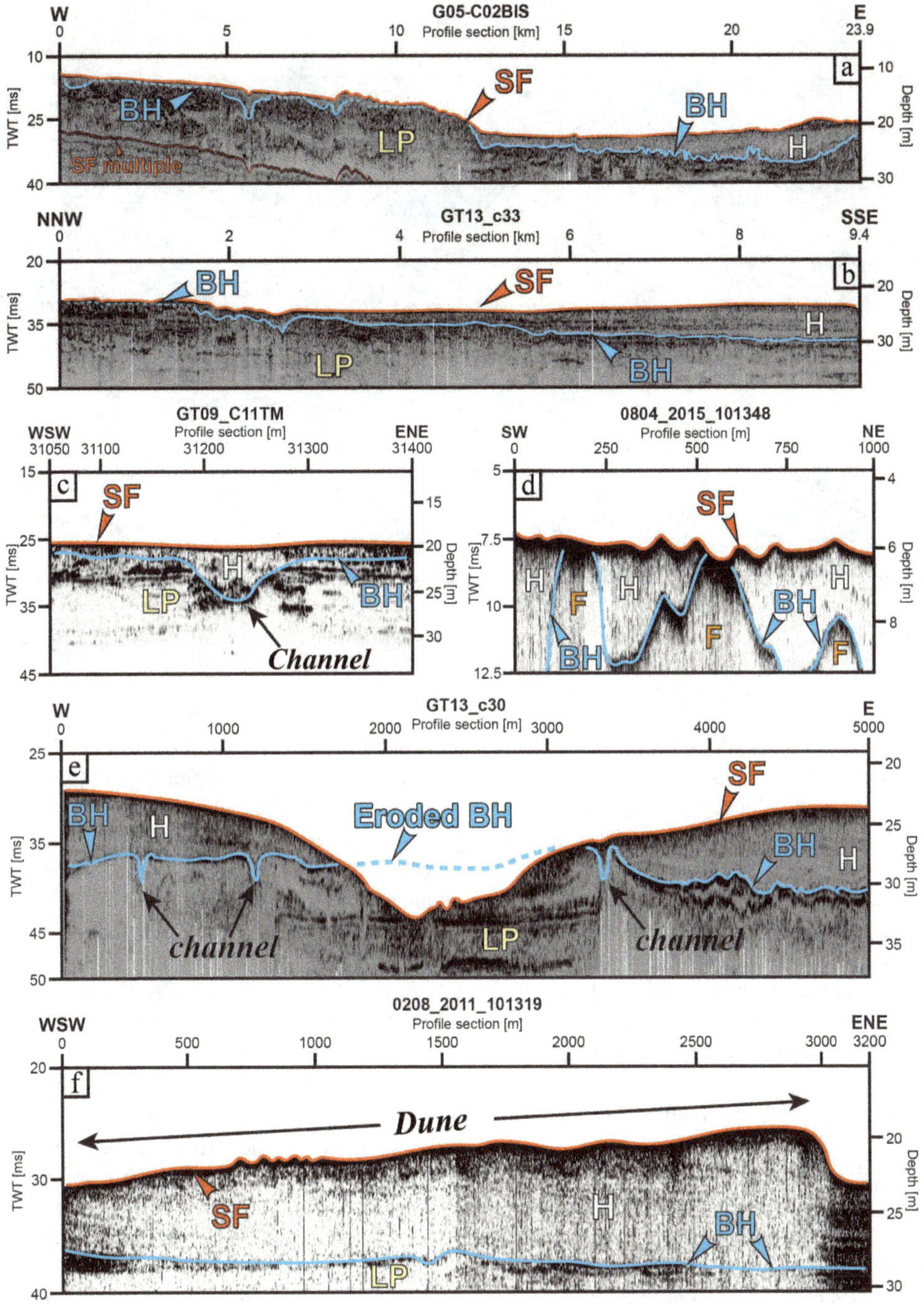

Figure 8. Examples of sub-bottom and Chirp sonar profiles (for explanation of the symbols, refer to the caption of Fig. 3); **(a–b)** regional Chirp profiles show that the Late Pleistocene paleotopography forms a basin in the eastern part which serves as a depocentre for the Holocene marine sediment (dashed blue line – base of the Holocene sediment where the sedimentary sequence is very thin or even absent or eroded); **(a)** a W–E oriented profile showing a difference in marine Holocene thickness in the western and eastern parts of the gulf. Note that thicker Holocene marine sequences in the western part of the gulf are filling channel-like features; **(b)** a NNW–SSE oriented profile showing the same difference in Holocene marine thickness; **(c–f)** variations of the thickness of marine Holocene sediment; **(c)** a profile from the central part of the gulf, where a thin Holocene marine sediment cover is characteristic, except where it fills pre-existing channels; **(d)** variation of Holocene sediment thickness controlled by the pre-existing Flysch paleotopography in the area of Debeli rtič; **(e)** Holocene sediment erosion in the seafloor depression in front of Piran (Fig. 6); note also Holocene filling of pre-existing channels; **(f)** thicker Holocene sedimentary sequences visible on a profile crossing the area of dunes in front of Piran (Fig. 6). For locations of the profiles, see Fig. 5.

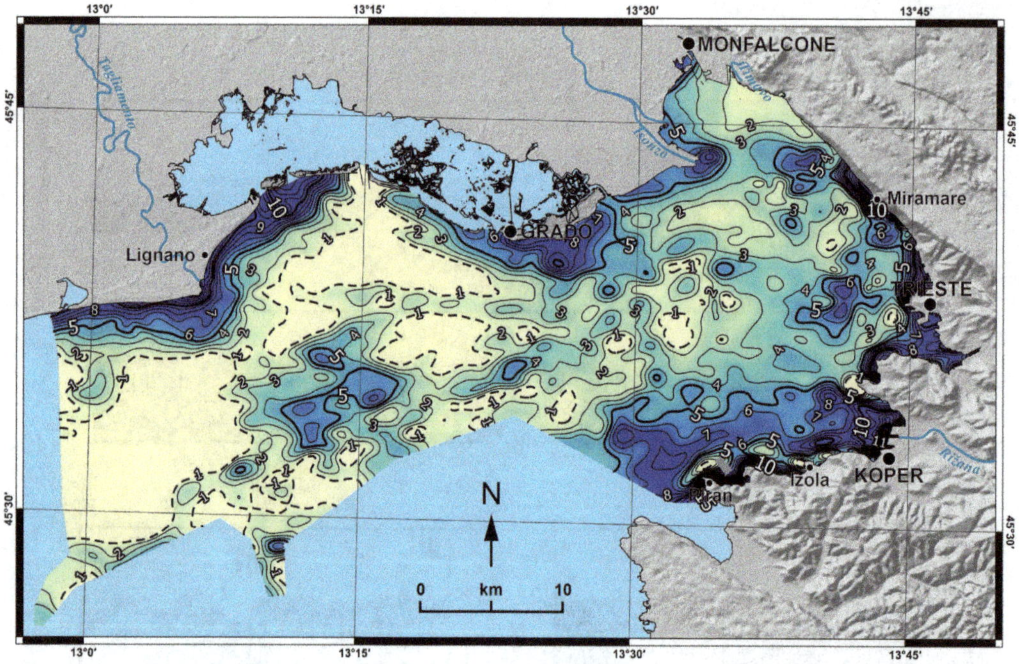

Figure 9. Map of the thickness of the Holocene marine sediment in metres. Dashed black lines mark areas where Holocene marine sediment is thinner than 1 m or even absent.

thickness throughout the whole study area. Due to a lack of sound velocity measurements we also assumed a model without velocity variations. Finally, due to scarce cores/wells, the presence of marine sediment in the Holocene sedimentary sequence was assessed mostly by following visual criteria when investigating geophysical profiles. We considered the acoustically transparent Holocene seismic facies to represent sediment primarily deposited in a marine environment. In order to avoid including predominantly paralic Holocene sediment in the model, we excluded the Grado-Marano lagoon area from the model of the thickness of Holocene marine sediment.

The first opportunity to test the quality of our model was presented by two 1.5 m long cores composed entirely of Holocene marine sediment (Mautner et al., 2018; Tomašových et al., 2017) that were published after the creation of our dataset. The cores M28 and M53 were collected in the Bay of Panzano and NW of Piran. In both locations the modelled thicknesses of the Holocene marine sequence exceed 1.5 m, which demonstrates the quality of our model and the potential of our dataset for planning of future core collecting campaigns.

5 Data availability

An important moment of the data lifecycle is that when observations are shared with other team members or domain experts in general. This introduces the possibility of replicating the experiment, of re-using the data in other contexts, and of providing data with additional value. OGS is deeply involved in the curation and dissemination of its data assets and has in this perspective developed a web-based data system called SNAP (Diviacco et al., 2015; Diviacco and Busato, 2013) which allows the contained data to be findable, accessible, interoperable and reusable (FAIR). The system is based on fully compliant OGC O&M and SensorML metadata and offers previewing facilities that allow remote data access and download once the end user has checked whether they are what he/she is looking for. DOIs are assigned to the datasets residing in the system so that they can be directly accessed resolving the DOIs or simply following the corresponding URLs. In the case of the models used in this study, they are accessible at the following URL: http://doi.org/cpz2. This points to a landing page which has been built following the standards set in the UNESCO Ocean data Publication Cookbook (Leadbetter et al., 2013), and offers further metadata details and an interactive preview of the dataset.

The data described in this study (Trobec and Busetti, 2017) are available as column-based text files containing data points with X, Y and Z/thickness coordinates in the UTM 33 North coordinate system (datum: WGS 84). In order to facilitate their use (especially with GIS software) these datasets are also available as 50×50 m elevation grids clipped to the extent of the coastline in the widely compatible geotiff format along with georeferencing information (the *.proj and *.tfw files). Depth contours of the models are also available in the shapefile format.

6 Conclusions

Together with the published datasets (http://doi.org/cpz2), this paper represents the second assessment of the thickness of Holocene marine sediment in a major part of the Gulf of Trieste, employing new datasets from a wide range of different geophysical methods and published sedimentological data from wells and cores.

Geophysical profiles along with core/well data show a basin in the Late Pleistocene topography in the southeastern part of the gulf that served as a depocentre for the significantly thicker Holocene marine sedimentary sequences compared to the rest of the gulf. While the exact cause of the basin formation exceeds the scope of this paper, we suspect it represents an interplay between different tectonic, sedimentologic, climatic and oceanographic factors that need to be further examined in order to better understand the distribution of Holocene marine sediment in the Gulf of Trieste. Additionally, thicker Holocene marine sequences can be observed above paleochannels, in the Trezza Grande area and in the internal parts of bays and in close proximity to the shoreline, where the thickness is probably governed by increased amounts of available sediment transported by rivers and surface run-off from rocky coasts of the southeastern part and due to increased amounts of fluvially transported sediment from the alluvial plains of the northern and northwestern coasts of the gulf (Covelli et al., 2004; Ogorelec et al., 1991; Mandac Soczka and Faganeli, 2015). In the central part of the gulf the Holocene marine sequences are very thin or even absent (unresolvable on geophysical profiles).

Our work provides a solid reference for a wide variety of disciplines involved in future studies of Holocene sediment in the area, especially regarding sampling sites and/or survey selection. Furthermore, the thickness model provides important implications for further sedimentological, geomorphological, paleoenvironmental, neotectonic and oceanographic studies of the Gulf of Trieste.

Author contributions. AT interpreted acoustic data from the Slovenian part of the gulf, modelled the surfaces and thickness, prepared the figures and wrote the manuscript. MB conceived and managed the study, interpreted the geophysical data from the Italian part of the gulf and contributed in writing the manuscript. FZ, LB, AC, EG, RR and IT acquired the multibeam, singlebeam and sonar data in various campaigns in the Italian part of the gulf. AB, EG and RR interpreted various geophysical datasets in the Italian part. LB, EG, IT and FZ processed the geophysical data from the Italian part. SP acquired and processed the multibeam dataset and the sub-bottom sonar profiles from the Slovenian part of the gulf. PD managed data and metadata handling on the online repository. MV interpreted acoustic data from the Slovenian part of the gulf.

Competing interests. The authors declare that they have no conflict of interest.

Acknowledgements. Figures in the paper were produced using Copernicus digital elevation model data and information funded by the European Union – EU-DEM layers. A substantial part of this work was conducted during an Erasmus+ internship of the first author (SMP-2014-2015-3307) at the Istituto Nazionale di Oceanografia e di Geofisica Sperimentale (OGS). Additionally, we would like to acknowledge the Slovenian Research Agency for the Young Researcher grant of the first author (1000-15-0510-6/15). We would also like to acknowledge IHS Kingdom for their Educational grant which allowed us to use the Kingdom software.

The data used in this work were acquired during several projects. The authors acknowledge project "Mapping of the regional geothermal resource and guidelines for the use" supported by the Geological Service of the Autonomous Region of Friuli Venezia Giulia from 2004 to 2006, Italian FISR project VECTOR from 2006 to 2010, OGS project "Gulf of Trieste" in 2009, service activity "Survey in the SIN Site of Trieste, the Muggia Bay" for the Multiproject srl of Gorizia (Italy) in 2013, project "SLOMARTEC" of OGS, the Department of Geology (University of Ljubljana) and Harpha Sea d.o.o. in 2013, project "Application of sonar in research of active tectonics and paleoseismology in low-strain environments" (L1-5452) supported by the Slovenian Research Agency and Harpha Sea d.o.o. from 2013 to 2016 and the "Preliminary survey for submarine cable interconnection between Italy and Slovenia" service for Next Geosolutions Europe SpA in 2015.

We would like to acknowledge the reviewer comments suggested by Thomas M. Cronin and Cristina Bernardes. Their input along with the comments provided by Adam Tomašových allowed us to further improve our manuscript.

Edited by: David Carlson

References

Anderson, A. L. and Hampton, L. D.: Acoustics of gas-bearing sediments I. Background, J. Acoust. Soc. Am., 67, 1865–1889, https://doi.org/10.1121/1.384453, 1980.

Baradello, L., Busetti, M., and Visnovich, G.: Progetto VECTOR: Rilievo sismico ad alta risoluzione nella Laguna di Grado e Marano, Rep. 8443, Rep. 2009/89 GDL 39 GEBA, Istituto Nazionale di Oceanografia e di Geofisica Sperimentale – OGS, Sgonico, Italy, 16 pp., 2009.

Baradello, L., Busetti, M., Nieto, D., Romeo, R., and Visnovic, G.: Survey VHR Sismico Boomer del Sito di Interesse Nazionale (SIN) di Trieste, Rep. OGS 2013/14 – IRI 2, Istituto Nazionale di Oceanografia e di Geofisica Sperimentale – OGS, Sgonico, Italy, 29 pp. and 33 att., 2013.

Biolchi, S., Furlani, S., Covelli, S., Busetti, M., and Cucchi, F.: Morphoneotectonics and lithology of the eastern sector of the Gulf of Trieste (NE Italy), J. Maps, 12, 936–946, https://doi.org/10.1080/17445647.2015.1099572, 2016.

Busetti, M., Cova, A., Marchi, M., Romeo, R., Volpi, V., and Zgur, F.: Realizzazione della carta geologico-tecnica della risorsa geotermica regionale, Rep. 80/2005 – RIMA 9, 47 pp. and 5 att., 2005.

Busetti, M., Volpi, V., Nicolich, R., Barison, E., Romeo, R., Baradello, L., Brancatelli, G., Giustiniani, M., Marchi, M., Zanolla, C., Wardell, N., Nieto, D., and Ramella, R.: Dinaric tectonic features in the Gulf of Trieste (Northern Adriatic Sea), Bollettino di Geofisica Teorica ed Applicata, 51, 117–128, 2010a.

Busetti, M., Volpi, V., Barison, E., Giustiniani, M., Marchi, M., Ramella, R., Wardell, N., and Zanolla, C.: Meso-Cenozoic seismic stratigraphy and the tectonic setting of the Gulf of Trieste (northern Adriatic), GeoActa, SP 3, 1–14, 2010b.

Carulli, G. B.: Structural model of the Trieste Gulf: A proposal, J. Geodynam., 51, 156–165, https://doi.org/10.1016/j.jog.2010.05.004, 2011.

Carulli, G. B., Podda, F., Venturini, C., Zanferrari, A., Cucchi, F., Monegato, G., Nicolich, R., Paiero, G., Piano, C., Slejko, D., Tunis, G., Zanolla, C., Bartolini, C., Masetti, D., Ponton, M., Stefani, C., Zini, L., Burelli, G., and Marchesini, A.: Carta Geologica del Friuli Venezia Giulia, Scale: 1 : 150000, Regione Autonoma Friuli Venezia Giulia, Direzione Centrale Ambiente e Lavori Pubblici, Servizio Geologico, Trieste, Italy, 2006.

Cimolino, A., Della Vedova, B., Nicolich, R., Barison, E., and Brancatelli, G.: New evidence of the outer Dinaric deformation front in the Grado area (NE-Italy), Rend. Fis. Acc. Lincei, 21, 167–179, https://doi.org/10.1007/s12210-010-0096-y, 2010.

Cova, A.: Progetto VECTOR: Acquisizione dati multibeam nella bocca lagunare di Porto Buso (Laguna di Grado e Marano), Rep. 2008/40 – RIMA 4 ADAM, Istituto Nazionale di Oceanografia e di Geofisica Sperimentale – OGS, Sgonico, Italy, 20 pp., 2008.

Covelli, S., Piani, R., Faganeli, J., and Brambati, A.: Circulation and suspended matter distribution in a microtidal deltaic system: The Isonzo river mouth (northern Adriatic Sea), J. Coastal Res., 41, 130–140, 2004.

Covelli, S., Fontolan, G., Faganeli, J., and Ogrinc, N.: Anthropogenic markers in the Holocene stratigraphic sequence of the Gulf of Trieste (northern Adriatic Sea), Mar. Geol., 230, 29–51, https://doi.org/10.1016/j.margeo.2006.03.013, 2006.

Della Vedova, B., Brancolini, G., Bongiovanni, S., Brusadin, A., Busetti, M., Codermatz, R., Giustiniani, M., Nicolich, R., Paganini, P., and Paoli G.: Studio preliminare degli acquiferi profondi della Pianura friulana, Agreement with the Autonomous Region Friuli Venezia Giulia dd. 09/12/1999, Rep. 7170, Final Report, 54 pp. and 8 thematic maps, 2002.

Diviacco, P. and Busato, A.: The Geo-Seas Seismic data viewer: a Tool to facilitate the control of data access, Bollettino di Geofisica Teorica e Applicata, 54, 257–270, https://doi.org/10.4430/bgta0078, 2013.

Diviacco, P., Wardell, N., Forlin, E., Sauli, C., Burca, M., Busato, A., Centonze, J., and Pelos, C.: Data rescue to extend the value of vintage seismic data: The OGS-SNAP experience, GeoResJ, 6, 44–52, https://doi.org/10.1016/j.grj.2015.01.006, 2015.

EMODnet: Map Viewer, Sea-floor Geology (Tectonic lines), available at http://www.emodnet-geology.eu/map-viewer/, last access: 12 June 2018.

Faganeli, J., Planinc, R., Pezdič, J., Smodiš, B., Stegnar, P., and Ogorelec, B.: Marine geology of the Gulf of Trieste (northern Adriatic): Geochemical aspects, Mar. Geol., 99, 93–108, https://doi.org/10.1016/0025-3227(91)90085-I, 1991.

Gordini, E.: Evolutionary trends in the beaches of the Tagliamento river delta, Bollettino di Geofisica Teorica e Applicata, 48, 287–304, 2007.

Gordini, E.: Progetto VECTOR Rilievo mutlibeam delle bocche laguanri di Grado e Morgo (Laguna di Grado e Marano) realizzato con sistema interferometrico swathplus wide swath bathymetry & side scan sonar system, Rep. 119/2008 – RIMA 24 GEA, Istituto Nazionale di Oceanografia e di Geofisica Sperimentale – OGS, Sgonico, Italy, 38 pp., 2008a.

Gordini, E.: Progetto VECTOR: Rilievo batimetrico singlebeam dell'Area marina-litorale della Regione Friuli Venezia Giulia, Report 118/2008 RIMA 22 GEA, Istituto Nazionale di Oceanografia e di Geofisica Sperimentale – OGS, Sgonico, Italy, 53 pp., 2008b.

Gordini, E.: Progetto VECTOR: Rilievo batimetrico singlebeam dell' ebb-tidal delta di Porto Buso, Report 117/2008 RIMA 22 GEA, Istituto Nazionale di Oceanografia e di Geofisica Sperimentale – OGS, Sgonico, Italy, 37 pp., 2008c.

Gordini, E.: Integrazione di metodologie geofisiche, geomorfologiche, sedimentologiche e geochimice, per la definizione della genesi e dell'età degli affioramenti rocciosi presenti sul fondale marino dell' Adriatico settentrionale, PhD thesis, University of Trieste, Trieste, Italy, 363 pp., 2009.

Gordini, E., Marocco, R., and Vio, E.: Stratigrafia del sottosuolo della "Trezza Grande" (Golfo di Trieste, Adriatico Settentrionale), Gortania, 24, 31–63, 2002.

Gordini, E., Caressa, S., and Marocco, R.: Nuova carta morfosedimentologica del Golfo di Trieste (da Punta Tagliamento alla Foce dell'Isonzo), Gortania, 25, 5–29, 2003.

Gordini, E., Marocco, R., Tunis, G., and Ramella, R.: I depositi cementati del Golfo di Trieste (Adriatico Settentrionale): Distribuzione areale, caratteri geomorfologici e indagini acustiche ad alta risoluzione, Il Quaternario, 17, 555–563, 2004.

Gordini, E., Marocco, R., and Ramella, R.: Dinamica morfologica del litorale del delta del fiume Tagliamento (Adriatico settentrionale) in relazione ai possibili interventi di ripascimento, Il Quaternario, 19, 45–65, 2006.

Istituto Idrografico della Marina: Carta nautica Da Punta Tagliamento a Pula 1:100 000, 2004.

Jurkovšek, B., Biolchi, S., Furlani, S., Kolar-Jurkovšek, T., Zini, L., Jež, J., Tunis, G., Bavec, M., and Cucchi, F.: Geology of the Classical Karst Region (SW Slovenia–NE Italy), J. Maps, 12, 352–362, https://doi.org/10.1080/17445647.2016.1215941, 2016.

Kolega, N. and Poklar, M.: Morphological analysis of the Slovenian coast with data from lidar and sonar ranging, Acta Geogr. Slov., 52, 121–140, https://doi.org/10.3986/AGS52105, 2012.

Lambeck, K., Antonioli, F., Purcell, A., and Silenzi, S.: Sea-level change along the Italian coast for the past 10,000 yr, Quaternary Sci. Rev., 23, 1567–1598, https://doi.org/10.1016/j.quascirev.2004.02.009, 2004.

Leadbetter, A., Raymond, L., Chandler, C., Pikula, L., Pissierssens, P., and Urban, E.: Ocean Data Publication Cookbook, Executive Summary, Intergovernmental Oceanographic Commission, Unesco, Paris, IOC Manuals and Guides 64, 2013.

Mallet, J. L.: Discrete smooth interpolation in geometric modelling, Comput. Des., 24, 178–191, https://doi.org/10.1016/0010-4485(92)90054-E, 1992.

Mallet, J. L.: Discrete modeling for natural objects, Math. Geol., 29, 199–219, https://doi.org/10.1007/BF02769628, 1997.

Mandac Soczka, R. and Faganeli, J.: Deposition of riverine suspended solids in a shallow bay (Bay of Koper, Gulf of Trieste, northern Adriatic Sea), J. Soils Sediments, 15, 2433–2442, https://doi.org/10.1007/s11368-015-1146-y, 2015.

Marocco, R.: Evoluzione quaternaria della Laguna di Marano (Friuli-Venezia Giulia), Il Quaternario, 2, 125–137, 1989.

Marocco, R.: Evoluzione tardopleistocenica-olocenica del delta del F. Tagliamento e delle Lagune di Marano e Grado (Golfo di Trieste), Il Quaternario, 1b, 223–232, 1991.

Marocco, R., Pugliese, N., and Stolfa, D.: Some remarks on the origin and evolution of the Grado lagoon (Northern Adriatic Sea), Bollettino di Oceanologia Teorica ed Applicata, 2, 11–17, 1984.

Masoli, C. A., Petronio, L., Gordini, E., Deponte, M., Cotterle, D., Romeo, R., Böhm, G., Barbagallo, A., Belletti, R., Maffione, S., and Meneghini, F.: Marine geophysical and geological investigations in support to the construction of new harbour infrastructures: the Trieste Marine Terminal extension, in Atti del 34 convegno nazionale del Gruppo Nazionale di Geofisica della Terra Solida (GNGTS), tema 3: Geofisica Applicata, edited by: Slejko, D., Riggio, A., Rebez, A., Caielli, G., Cassiani, G., Florio, G., Mazzucchelli, P., Nicolich, R., and Sambuelli, L., Istituto Nazionale di Oceanografia e di Geofisica Sperimentale, Trieste, 63–70, 2015.

Mautner, A. K., Gallmetzer, I., Haselmair, A., Schnedl, S. M., Tomašových, A., and Zuschin, M.: Holocene ecosystem shifts and human-induced loss of Arca and Ostrea shell beds in the north-eastern Adriatic Sea, Mar. Pollut. Bull., 126, 19–30, https://doi.org/10.1016/j.marpolbul.2017.10.084, 2018.

Ministry of Transport of the Republic of Slovenia: Male karte Tržaški zaliv Merilo 1 : 100 000, 2005.

Ogorelec, B., Mišič, M., Šercelj, A., Cimerman, F., Faganeli, J., and Stegnar, P.: Sediment sečoveljske soline, Geologija, 24, 180–216, 1981.

Ogorelec, B., Mišič, M., and Faganeli, J.: Marine geology of the Gulf of Trieste (northern Adriatic): Sedimentological aspects, Mar. Geol., 99, 79–92, https://doi.org/10.1016/0025-3227(91)90084-H, 1991.

Ogorelec, B., Faganeli, J., Mišič, M., and Čermelj, B.: Reconstruction of paleoenvironment in the Bay of Koper (Gulf of Trieste, Northern Adriatic), Annales, 11, 187–200, 1997.

Ogrinc, N., Faganeli, J., Ogorelec, B., and Čermelj, B.: The origin of organic matter in Holocene sediments in the Bay of Koper (Gulf of Trieste, northern Adriatic Sea), Geologija, 50, 179–187, https://doi.org/10.5474/geologija.2007.014, 2007.

Placer, L.: Simplified structural map of Kras Kras (Slovene), Carso (Italian) = Geographical unit, Geologija, 58, 89–93, https://doi.org/10.5474/geologija.2015.008, 2015.

Placer, L., Vrabec, M., and Celarc, B.: The bases for understanding of the NW Dinarides and Istria Peninsula tectonics, Geologija, 53, 55–86, https://doi.org/10.5474/geologija.2010.005, 2010.

Pleničar, M., Polšak, A., and Šikić, D.: Osnovna geološka karta SFRJ. / L 33–88, Trst, 1969.

Poglajen, S.: 3D-karta morskega dna slovenskega morja in njena uporaba v geoprostorski podpori pomorstvu Slovenske vojske, in: Geoprostorska podpora obrambnemu sistemu Republike Slovenije, edited by: Balas, J., Kokalj, A., and Kovič, B., Ministrstvo za obrambo Republike Slovenije, Direktorat za obrambne zadeve, Sektor za načrtovanje, Inštitu za antropološke in prostorske študije ZRC SAZU, Geodetski inštitut Slovenije, Ljubljana, Slovenia, 93–106, 2012.

Romeo, R.: Studio geofisico integrato ad alta risoluzione dei depositi marini e della struttura del substrato della Riviera di Miramare (Golfo di Trieste), PhD thesis, University of Trieste, Trieste, Italy, 175 pp., 2009.

Slavec, P.: Spreminjanje višine morske gladine v kvartarju, Geografski vestnik, 83, 43–56, 2011.

Slavec, P.: Analiza morfologije morskega dna slovenskega morja, MS thesis, University of Ljubljana, Ljubljana, Slovenia, 58 pp., 2012.

Tomašových, A., Gallmetzer, I., Haselmair, A., Kaufman, D. S., Vidović, J., and Zuschin, M.: Stratigraphic unmixing reveals repeated hypoxia events over the past 500 yr in the northern Adriatic Sea, Geology, 45, 363–366, https://doi.org/10.1130/G38676.1, 2017.

Trincardi, F., Correggiari, A., Cattaneo, A., Remia, A., Taviani, M., Angeletti, L., Foglini, F., and Campiani, E.: Carta Geologica dei mari italiani alla scala 1:250.000 foglio NL 33-7 Venezia, Istituto Superiore per la Protezione e la Ricerca Ambientale, Servizio Geologico d' Italia, Roma, Italy, 2011a.

Trincardi, F., Argnani, A., Correggiari, A., Foglini, F., Rovere, M., Angeletti, L., Asioli, A., Campiani, E., Cattaneo, A., Gallerani, A., Piva, A., Remia, A., Ridente, D., and Taviani, M.: Note illustrative della Carta Geologica dei mari italiani alla scala 1:250.000 foglio NL 33-7 Venezia, 151, Istituto Superiore per la Protezione e la Ricerca Ambientale, Servizio Geologico d' Italia, Roma, Italy, 2011b.

Trincardi, F., Campiani, E., Correggiari, A., Foglini, F., Maselli, V. and Remia, A.: Bathymetry of the Adriatic Sea: The legacy of the last eustatic cycle and the impact of modern sediment dispersal, J. Maps, 10, 151–158, https://doi.org/10.1080/17445647.2013.864844, 2014.

Trobec, A.: Raziskave zgradbe sedimentnega morskega dna v Strunjanskem zalivu s podpovršinskim sonarjem, MS thesis, University of Ljubljana, Ljubljana, Slovenia, 66 pp., 2015.

Trobec, A. and Busetti, M.: Models of the bathymetry, of the base and of the thickness of Holocene marine sediment in the Gulf of Trieste (Northern Adriatic Sea) [Data set], OGS SNAP System, available at: https://doi.org/10.6092/6ad9b1e6-c977-cec9-8a2d-db10c7f90adc (last access: 12 June 2018), 2017.

Trobec, A., Šmuc, A., Poglajen, S., and Vrabec, M.: Burried MIS 5 abrasion platforms in the Bay of Koper (Gulf of Trieste, Northern Adriatic) confirm long-term subsidence of the Northern Adriatic region, EGU General Assembly, Vienna, Austria, 17–22 April 2016, EPSC2016-11963, 2016.

Trobec, A., Šmuc, A., Poglajen, S., and Vrabec, M.: Submerged and buried Pleistocene river channels in the Gulf of Trieste (Northern Adriatic Sea): Geomorphic, stratigraphic and tectonic inferences, Geomorphology, 286, 110–120, https://doi.org/10.1016/j.geomorph.2017.03.012, 2017.

Vrabec, M., Busetti, M., Zgur, F., Facchin, L., Pelos, C., Romeo, R., Sormani, L., Slavec, P., Tomini, I., Visnovic, G., and Žerjal, A.: Refleksijske seizmične raziskave v slovenskem morju SLOMARTEC 2013, in: Raziskave s področja geodezije in geofizike 2013, edited by: Kuhar, M., Čop, R., Gosar, A., Kobold, M., Kralj, P., Malačič, V., Rakovec, J., Skok, G., Stopar, B., and Vreča, P., University of Ljubljana, Faculty of Civil and Geodetic Engineering, Ljubljana, 97–101, 2014.

Yuan, F., Bennell, J. D. and Davis, A. M.: Acoustic and physical characteristics of gassy sediments in the western Irish Sea,

Cont. Shelf Res., 12, 1121–1134, https://doi.org/10.1016/0278-4343(92)90073-S, 1992.

Zampa, L. S., Busetti, M., Furlani, S., Baradello, L., and Romeo, R.: Evidence of neo-tectonic tilting in the Gulf of Trieste, in: Proceedings: GeoSUB – Underwater geology, Trieste, 13–14 October 2015, edited by: Furlani, S., Antonioli, F., Anzidei, M., Busetti, M., Ferranti, L., Mastronuzzi, G., and Orrù, P., 72–73, 2015.

Zecchin, M., Gordini, E., and Ramella, R.: Recognition of a drowned delta in the northern Adriatic Sea, Italy: Stratigraphic characteristics and its significance in the frame of the early Holocene sea-level rise, The Holocene, 25, 1027–1038, https://doi.org/10.1177/0959683615575358, 2015.

Zgur, F., Codiglia, R., Busetti, M., Petronio, L., De Vittor, R., Loreto, M. F., Pelos, C., Pettenati, F., and Romeo, R.: Studio dell'evoluzione geologica e tettonica del Golfo di Trieste (Adriatico Settentrionale), Rep. OGS 2010/24 RIMA 6 ADEST dd. 18-03-2010, Istituto Nazionale di Oceanografia e di Geofisica Sperimentale – OGS, Sgonico, Italy, 47 pp., 2010.

Zgur, F., Facchin, L., Busetti, M., Vrabec, M., Slavec, P., Romeo, R., Tomini, I., Sormani, L., Visnovic, P., and Pelos C.: SLO-MARTEC: Correlation of tectonic structures occurring onland in the Istria Peninsula and in the Gulf of Trieste (Northern Adriatic), and investigation of their neotectonic activity, Cruise Report OGS Explora, 18–21 March 2013, Rep. 2013/64 – IRI 9 ADEST, Istituto Nazionale di Oceanografia e di Geofisica Sperimentale – OGS, Sgonico, Italy, 47 pp., 2013.

Žerjal, A., Poglajen, S., Rant, I., Soczka Mandac, R., Lovrič, E., and Mencin, E.: Preliminary survey for submarine cable interconnection between Italy and Slovenia, Final Report (Version 1.0), Harpha Sea d.o.o., Koper, Slovenia, 33 pp., 2015.

6

A 156 kyr smoothed history of the atmospheric greenhouse gases CO_2, CH_4, and N_2O and their radiative forcing

Peter Köhler[1], Christoph Nehrbass-Ahles[2], Jochen Schmitt[2], Thomas F. Stocker[2], and Hubertus Fischer[2]

[1]Alfred-Wegener-Institut Helmholtz-Zentrum für Polar- und Meeresforschung (AWI), P.O. Box 12 01 61, 27515 Bremerhaven, Germany

[2]Climate and Environmental Physics, Physics Institute, and Oeschger Centre for Climate Change Research, University of Bern, Bern, Switzerland

Correspondence to: Peter Köhler (peter.koehler@awi.de)

Abstract. Continuous records of the atmospheric greenhouse gases (GHGs) CO_2, CH_4, and N_2O are necessary input data for transient climate simulations, and their associated radiative forcing represents important components in analyses of climate sensitivity and feedbacks. Since the available data from ice cores are discontinuous and partly ambiguous, a well-documented decision process during data compilation followed by some interpolating post-processing is necessary to obtain those desired time series. Here, we document our best possible data compilation of published ice core records and recent measurements on firn air and atmospheric samples spanning the interval from the penultimate glacial maximum ($\sim$ 156 kyr BP) to the beginning of the year 2016 CE. We use the most recent age scales for the ice core data and apply a smoothing spline method to translate the discrete and irregularly spaced data points into continuous time series. These splines are then used to compute the radiative forcing for each GHG using well-established, simple formulations. We compile only a Southern Hemisphere record of CH_4 and discuss how much larger a Northern Hemisphere or global CH_4 record might have been due to its interpolar difference. The uncertainties of the individual data points are considered in the spline procedure. Based on the given data resolution, time-dependent cutoff periods of the spline, defining the degree of smoothing, are prescribed, ranging from 5000 years for the less resolved older parts of the records to 4 years for the densely sampled recent years. The computed splines seamlessly describe the GHG evolution on orbital and millennial timescales for glacial and glacial–interglacial variations and on centennial and decadal timescales for anthropogenic times. Data connected with this paper, including raw data and final splines, are available at https://doi.org/10.1594/PANGAEA.871273.

1 Introduction

Our knowledge of changes in the atmospheric mixing ratios of the important greenhouse gases (GHGs) CO_2, CH_4, and N_2O beyond the instrumental record is mainly based on discrete data points derived from gas extractions in polar ice cores. While there are recent developments towards continuous CH_4 records using gas extraction and measurement systems coupled to continuous-flow analysis systems (Schüpbach et al., 2009; Chappellaz et al., 2013; Rhodes et al., 2013, 2015), this approach has not yet been developed for the other two important GHGs, CO_2 and N_2O. To obtain the continuous GHG records, necessary for transient climate simulations, these discrete data have to be processed in order to extract those variabilities that have climatological significance and to account for measurement uncertainties.

All three GHG records have special features which need some attention during data compilation:

- For some of the CO_2 records obtained from different ice cores, there exist significant and as yet unexplained offsets (Ahn et al., 2012; Bereiter et al., 2012; Marcott et al., 2014; Bauska et al., 2015). These offsets need to be addressed in our data compilation.
- Due to the dominance of CH_4 sources in the Northern Hemisphere, the CH_4 concentrations are higher in records from Greenland than from Antarctica (referred to as interpolar difference; e.g. Baumgartner et al., 2012).
- In situ production of N_2O connected to high mineral dust values leads to unreliable N_2O concentrations (e.g. Schilt et al., 2010a), particularly during glacial peak times and in records from Greenland, for which special care has to be taken during data selection.

Rapid changes are most pronounced in CH_4 and N_2O (and to some extent also in CO_2) during millennial-scale climate variability, or the so-called Dansgaard–Oeschger (D/O) events. Therefore, only well synchronised ice cores from Greenland and Antarctica can be used if records from the Northern and the Southern Hemisphere are to be merged into one global record. However, even with the recent efforts on ice core age scale development, there remain issues with this north–south synchronisation. For example, inconsistencies in the timing of abrupt changes in CH_4 concentration in the North Greenland Ice Core Project (NGRIP), EPICA Dronning Maud Land (EDML), and Talos Dome (TALDICE) ice cores have been identified for several D/O event transitions (Baumgartner et al., 2014) if based on AICC2012, the Antarctic Ice Core Chronology of four major Antarctic ice cores (Veres et al., 2013). Furthermore, when comparing data from the West Antarctic Ice Sheet Divide ice core (WDC) on its most recent age scale, WD2014, with data from Greenlandic ice cores, the chronology of the latter (GICC05) has been stretched by 0.63 % in order to find the best match to the absolute U/Th-dated paleo record of Hulu Cave (WAIS Divide Project Members, 2015).

In order for these issues to be overcome, careful data selection and processing are required. Here, we document our assumptions during data compilation and calculate continuous time series of CO_2, CH_4, and N_2O via spline-smoothing (Enting, 1987; Bruno and Joos, 1997) with a nominal temporal resolution Δt of 1 year from the penultimate glacial maximum until present, the time window of interest for PALMOD, the German Paleo Modelling Project (www.palmod.de). Note, however, that this Δt represents not the true resolution but only the typical spline average for each year and that the ice core information represents a low-pass filtered signal of atmospheric variability concentrations by the slow bubble enclosure process. Furthermore, the resulting spline is of restricted use for in-depth analysis with a focus on the rates of changes in the three GHGs, since the spline smoothing suppresses the most abrupt changes in the GHGs. Here, we extend the ice-core-based paleo records using instrumental data up until the beginning of the year 2016 CE, including several decades of overlap between the ice core and instrumental data. The resulting continuous GHG records might also be of interest and may be used in the Last Deglaciation experiment within PMIP4 (Paleoclimate Modelling Intercomparison Project phase 4) (Ivanovic et al., 2016). Note that different GHG data sets have so far been chosen to force transient simulations for the last 21 kyr in Ivanovic et al. (2016), but well-motivated different set-ups (e.g. using the GHG splines compiled here) are possible within PMIP4.

Previous splines (similar to our approach here but not identical in detail) have also been proposed to be used in interglacial experiments of the Holocene within PMIP4 (Otto-Bliesner et al., 2016). Within the most recent model intercomparison project, the Coupled Model Intercomparison Project Phase 6 (CMIP6), a slightly different compilation of GHGs for historical times, or the Common Era, has been presented (Meinshausen et al., 2017). While this alternative approach has its focus on the time since 1850 CE, its data compilation nevertheless extends back until the year 0 CE, based solely on the Law Dome ice core in non-instrumental times (MacFarling-Meure et al., 2006; Rubino et al., 2013). We will finally compare our splines with these forcing data sets proposed by Meinshausen et al. (2017) to be used within CMIP6.

As will be seen in detail in the next section, the mathematical formulation of the spline smoothing method needs information on the uncertainties or errors in the data points supporting the spline. These data uncertainties represent the precisions of individual measurements (1σ errors) and are of the order of a few parts per million for CO_2 or a few parts per billion for CH_4 and N_2O. The uncertainty in the final spline, however, is larger, since the applied smoothing, which depends on the chosen cutoff periods, adds some additional uncertainty. Furthermore, the estimates of the radiative forcing based on these three GHGs given here are even more uncertain, since the calculations of the radiative forcing themselves are based on models (Myhre et al., 1998) with an embedded intrinsic uncertainty of about $\sim 10\,\%$ (Forster et al., 2007). Note that the calculations of the GHG radiative forcing provided here are just a first-order approximation, since we use the simplified expressions of Myhre et al. (1998), while full climate models calculate radiative forcing internally, when forced with variable GHG concentrations.

In the following, ages are either given in years CE (Common Era) or in years BP (before present), where present is defined as 1950 CE. We define the onset of anthropogenic activities at 1750 CE (or 200 BP), based on the timing of the increase in CO_2 and CH_4 in our final splines, although we acknowledge that the onset of the Anthropocene is still debated (e.g. Lewis and Maslin, 2015; Steffen et al., 2016; Williams et al., 2016).

2 Details on the spline smoothing method

The numerical code for spline smoothing is based on Enting (1987), but see also Bruno and Joos (1997) and Enting et al. (2006) for further details, discussions, and applications. It offers the possibility to select different cutoff periods for different time intervals or parts of the input data set, which is needed when data spacing is variable throughout the data set.

In a smoothing spline a cost function is minimised. This cost function includes two terms: (i) the error-weighted deviation between the spline value and the actual data value and (ii) the curvature of the spline, represented by its second derivative. A parameter λ defines how much weight is given to the curvature. For a large λ, the optimisation results in low curvature, i.e. a very smooth spline and relatively large deviations from the original data. Similarly, increasing errors in the data results in a smoother spline for a given λ. In other words, the smoothing of the spline depends on both the assumed errors in the data and the parameter λ.

According to Fourier, each time series can be represented by a sum of sine functions. Since a smoothing spline acts as a low-pass filter, high frequencies are dampened in the spline. The period at which the amplitude is attenuated to 50 % is defined as the cutoff period P_c (e.g. Enting et al., 2006). The parameter λ is linked to P_c as described in detail in Eq. (1) below.

Let us assume input data are t_j, y_j, and v_j corresponding to time, value, and error (1σ). For a given interval of the input data, an average error, v, and an average data spacing, Δt, can be computed. The link between the cutoff period (P_c), the data spacing (Δt), and the 1σ error in the input data (v) is

$$P_c = 2\pi \cdot \left(\lambda \cdot \Delta t \cdot v^2\right)^{0.25}. \quad (1)$$

In the following, we prescribe P_c and can calculate λ following the given relationship in Eq. (1). We choose a P_c value such that it is much larger than the temporal resolution of the data, Δt, to avoid overfitting. However, since the choice of P_c is also partially subjective, we investigate its influence on the final spline by sensitivity studies, in which P_c is varied by ±50 %. One aspect of Eq. (1) is that P_c depends only weakly on Δt.

Let us now assume we have a data set with variable data spacing, for which we would like to apply different smoothing depending on Δt. We proceed by modifying λ to follow the predefined individual P_c for each interval of the input data set as follows.

- **Reference interval:** We take the most recent time window, consisting of instrumental measurements, as reference interval. λ is computed using Eq. (1) for the given cutoff period, average data spacing, and average error for this first interval.

- **Other intervals:** A modified $\lambda' = \lambda \cdot s^2$, with λ taken from the reference interval, is used for other intervals, implying that for the reference interval $s = 1$ and $\lambda' = \lambda$. The scaling factor s is chosen to gain the desired P_c' after

$$s = \frac{\left(\frac{P_c'}{2\pi}\right)^2}{\sqrt{(\lambda \cdot \Delta t')} \cdot v'}, \quad (2)$$

where P_c', $\Delta t'$, and v' are the cutoff period, the mean data spacing, and the mean error for the interval under consideration.

An intermediate product with t_j, y_j, and v_j' is calculated, in which the revised uncertainty v_j' is defined by Eq. (2) using the cutoff-related scaling factor s. From this intermediate product, the final spline with time-dependent P_c is calculated. In doing so, the approach abstains from any further merging of partial time series to a final spline. The resulting spline follows the prescribed cutoff periods throughout the whole time series. However, for every change in cutoff period from P_{c1} to P_{c2} a transition window around the time of change, t_{change}, exists (defined as $t_{change} \pm P_{ct}$, with P_{ct} being the smaller of P_{c1} and P_{c2}), in which the variability of the spline transits from one cutoff period to the other and does not follow the prescribed P_c exactly. The summaries of the spline calculation contained in Tables 3, 6, and 8 show the effect of this transition in a column of averaged realised cutoff periods, which are always slightly lower than the prescribed cutoff periods.

The uncertainties of the final splines are calculated from the square root of the sum of squares of three individual errors ($\sigma = \sqrt{\sigma_1^2 + \sigma_2^2 + \sigma_3^2}$).

- P_c error (σ_1): Mean difference from the standard spline by smoothing with cutoff periods P_c which are varied by ±50 %.

- Data resolution error (σ_2): The importance of uncertainty of the individual data points v_i for the spline smoothing by setting all v_i to 0.01.

- Monte Carlo error (σ_3): Repeated ($n = 500$) realisation of the data sets y_i by randomly choosing data points out of the normally distributed data using the given uncertainty ranges v_i.

3 Greenhouse gas data compilations and final splines

Our GHG data compilations are based on various data sets from 13 global distributed locations. An overview of the locations, including latitude and longitude, is provided in Table 1. Please note that CH_4 data are only included

Table 1. Locations of the different data sources, ordered north to south. Individual sites of the NOAA observational network are not explicitly mentioned here, when they only contribute to global mean calculations. SH CH_4: Southern Hemisphere CH_4.

Site	Latitude	Longitude	Data used here
NGRIP	75.10° N	42.32° W	N_2O
GRIP	72.583° N	37.633° W	comparing to SH CH_4
Barrow	71.3230° N	156.6114° W	comparing to SH CH_4
Mauna Loa	19.5362° N	155.5763° W	CO_2
Law Dome[1]	~66.73° S	~112.83° E	CO_2, SH CH_4, N_2O
Talos Dome (TALDICE)	72.817° S	159.183° E	CO_2, N_2O
EPICA Dronning Maud Land (EDML)	75.0° S	0.067° E	CO_2
EPICA Dome C (EDC)	75.1° S	123.35° E	CO_2, SH CH_4, N_2O
Taylor Glacier[2]	~77.77° S	~161.7° E	N_2O
WAIS Divide Ice Core (WDC)	79.468° S	112.086° W	CO_2, SH CH_4
Siple Dome	81.66° S	148.82° W	CO_2
South Pole[1]	90° S	59° E	CO_2, SH CH_4

Notes: [1] The data compilation of MacFarling-Meure et al. (2006) and Rubino et al. (2013) on CO_2, SH CH_4, and N_2O uses data from the Law Dome deep ice core and from various shallow ice and firn cores in its vicinity but also from atmospheric data from Cape Grim and firn core data from the South Pole. While we here state all the relevant positions, the original source of the individual data points is not marked in Tables 2, 5, and 7 or in the data files uploaded to PANGAEA, where data are only labelled with "Law Dome" as their source. Please see the original references for further details. [2] Data taken from Taylor Glacier are based on a "horizontal ice core", which does not have a point location like all other sites do.

Table 2. Data used to construct the CO_2 spline.

Time (in BP)	Time (in CE)	Source	Age scale	Citation
−66 to −8	2016 to 1958	Mauna Loa[1] (monthly)	–	Dlugokencky et al. (2016b)
−10 to 1949	1960 to 1	Law Dome[2]	as in references.	MacFarling-Meure et al. (2006); Rubino et al. (2013)
200 to 1210	1750 to 740	WDC[3]	WD2014	Ahn et al. (2012); Bauska et al. (2015)
1902 to 10 954	48 to before CE	EDC[4]	AICC2012	Monnin et al. (2001, 2004)
8807 to 22 909	–	WDC[6]	WD2014	Marcott et al. (2014); Buizert et al. (2015); Sigl et al. (2016)
21 926 to 48 720	–	Siple Dome	GICC05	Ahn and Brook (2014)
38 127 to 69 672	–	Talos Dome[5]	AICC2012	Bereiter et al. (2012)
43 205 to 113 429	–	EDML	AICC2012	Bereiter et al. (2012); Lüthi et al. (2010)
104 331 to 156 306	–	EDC	AICC2012	Schneider et al. (2013)
124 859 to 153 135	–	EDC	AICC2012	Lourantou et al. (2010a)

Notes: [1] Data taken from ftp://aftp.cmdl.noaa.gov/products/trends/co2/co2_mm_mlo.txt. [2] Law Dome data are taken from various sources; see references for details. They are available from 2001 CE to 1 CE; only data before 1960 CE are taken for the spline. [3] WDC data are available from 10 BP to 1217 BP, but not all were used here. Published WDC data have been shifted by −3.13 ppm (see Fig. A1). [4] EDC data are available from 350 BP to 22 236 BP, but not all were used here. [5] Talos Dome data exist from 34 360 BP but contain some outliers before 38 kyr BP. [6] Published WDC data have been shifted by −6.06 ppm (see Fig. A2).

from Southern Hemisphere records. These pointwise data sets are supplemented for the instrumental period by some global mean data from the National Oceanic and Atmospheric Administration (NOAA) observational network, including Radiatively Important Trace Species (RITS) nitrous oxide data from the Earth System Research Laboratory (NOAA/ESRL) halocarbons program and nitrous oxide data from the NOAA/ESRL halocarbons in situ program, which consists of globally distributed measurements. Individual data uploaded to the database PANGAEA, based on MacFarling-Meure et al. (2006) and Rubino et al. (2013) are all labelled as "Law Dome" data for simplicity, although these two studies contain data from the Law Dome deep ice core, data from various shallow cores, and atmospheric data from Cape Grim and the South Pole. Please refer to the original publications for a precise characterisation of the sample origins.

3.1 Atmospheric CO_2

There are small offsets of a few parts per million in measured CO_2 concentration between records obtained from different ice cores (e.g. Ahn et al., 2012; Bereiter et al., 2012; Marcott et al., 2014; Bauska et al., 2015). These offsets may be related to inter-laboratory differences in the calibration or po-

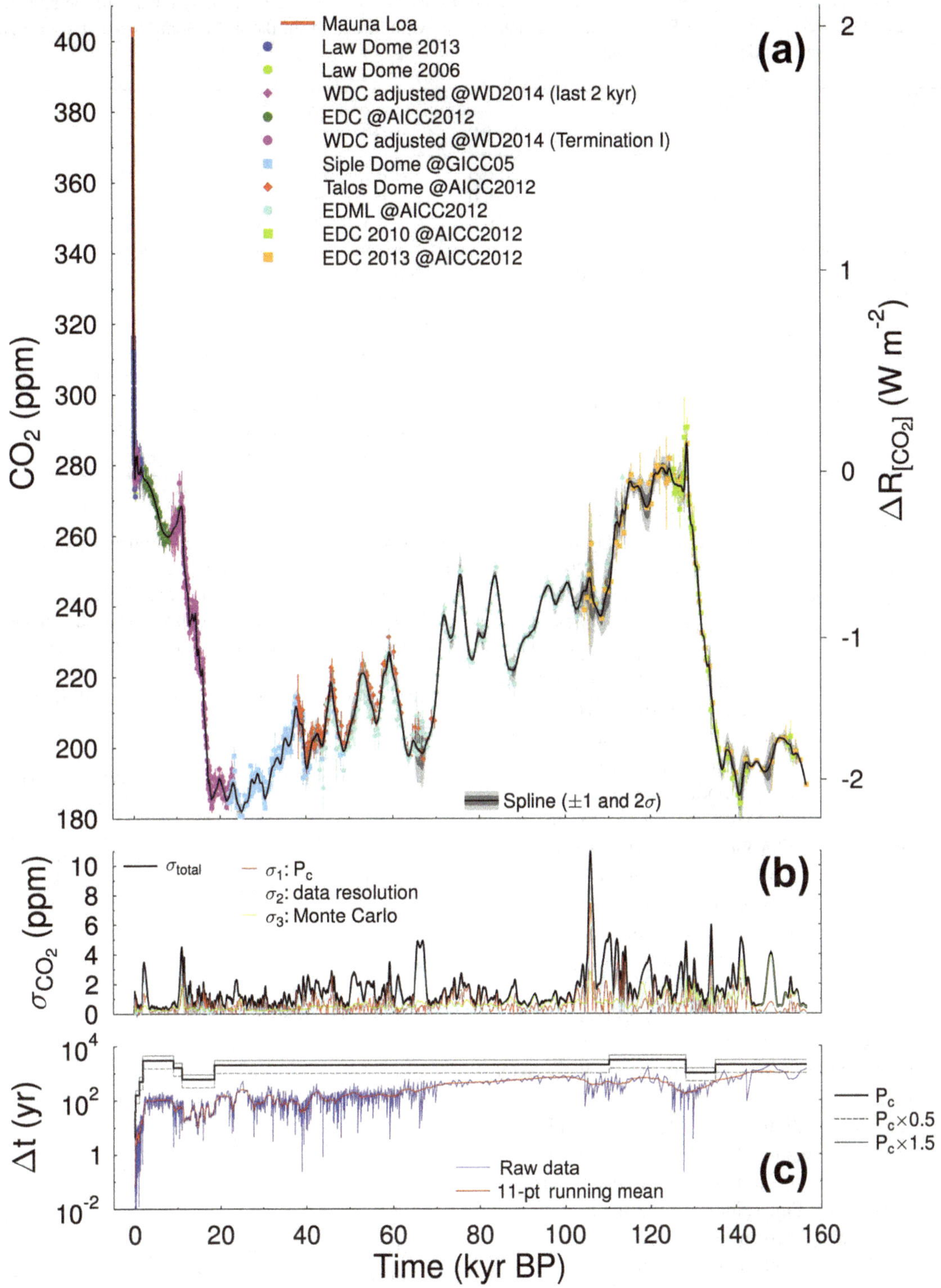

Figure 1. CO_2 spline covering all data: 2016 CE–156 307 BP. Error bars around the ice core data points are $\pm 2\sigma$. WDC data have been adjusted to reduce offsets; see text for details. In **(a)** the right axis contains the resulting radiative forcing $\Delta R_{[CO_2]} = 5.35 \cdot ln(CO_2/(278\,\text{ppm}))\,\text{W m}^{-2}$ calculated after Myhre et al. (1998). **(b)** Total uncertainty of the spline based on three individual error sources; see text for details. **(c)** Temporal resolution (Δt) of the CO_2 data points underlying the spline on a log scale. Additionally, the prescribed time-dependent cutoff period P_c is plotted, including its variation by $\pm 50\,\%$, which has been used to determine σ_1.

Table 3. Statistics of the CO_2 spline. Interval: i_{CO_2}; s: scaling factor to fulfil the constraints given by the prescribed cutoff period P_c; $\overline{P_c}$: average realised cutoff period; Δt: mean data spacing; υ: mean 1σ error – exact time framing is given by the age of the first (t_{start}) and last (t_{stop}) data point of the interval (in years BP); N: number of data points within interval. In the last column the underlying data source is briefly mentioned; see Table 2 for details and citations.

i_{CO_2}	s	P_c	$\overline{P_c}$	Δt	υ	t_{start}	t_{stop}	N	Data source
	–	yr	yr	yr	ppm	yr BP	yr BP	–	
1	1.00	4.0	4.0	0.1	0.3	−66.0	−8.1	699	Mauna Loa, Law Dome
2	2.65	20.0	18.5	0.4	1.3	−8.0	19.2	69	Law Dome
3	7.49	40.0	37.5	1.0	1.2	20.6	117.1	96	Law Dome
4	79.91	160.0	151.0	4.3	0.9	123.1	997.8	206	Law Dome, WDC
5	388.87	500.0	468.8	13.0	1.0	1006.0	1796.5	62	Law Dome, WDC
6	5377.63	3000.0	2883.1	93.3	1.0	1814.0	8994.9	78	Law Dome, WDC, EDC
7	1532.28	1600.0	1519.4	48.8	1.3	9060.2	10 962.5	40	WDC, EDC
8	357.66	600.0	567.4	28.1	1.0	11 060.3	18 463.6	264	WDC
9	1563.85	2000.0	1806.4	176.2	1.1	18 559.8	109 840.0	519	WDC, Siple D, Talos Dome, EDML, EDC
10	1690.90	3000.0	2593.3	383.9	1.5	110 555.4	127 829.0	46	EDML, EDC
11	225.60	1000.0	921.5	257.3	1.5	128 024.5	134 970.7	28	EDC
12	530.27	2000.0	1853.3	871.7	1.4	135 387.0	156 306.8	25	EDC

tentially due to in situ artefact production of CO_2 in the ice archive. For a detailed discussion, see Ahn et al. (2012) and the supplement to Bereiter et al. (2012). In addition to these offsets, the amplitudes of GHG variations can differ from one core to the next due to the site-dependent bubble enclosure characteristics, which act as low-pass filtering. Offsets require the adjustment of individual records to avoid spurious CO_2 changes when linking different records from different laboratory and ice cores. Ice core data are considered here on the best (most recent) age model available, whose details are contained in Table 2. AICC2012 refers to the most recent Antarctic Ice Cores Chronology, providing age models for EPICA Dome C (EDC), EDML, Talos Dome, Vostok, and the NGRIP record (Bazin et al., 2013; Veres et al., 2013). The CO_2 record from WDC is used here on its more recent age scale WD2014 to have the timing of CO_2 and the other two GHGs consistently on the same chronology. Using WD2014 instead of the original chronology WDC06A-7 shifts the WDC CO_2 time series towards younger ages: by about 100 years during Termination I and by about 10 years during the last 1.2 kyr.

Our CO_2 data compilation extends to ~156 kyr BP, at which point in time well-resolved CO_2 records stop. The full CO_2 spline covering the whole time window from 2016 CE to 156 kyr BP is plotted in Fig. 1, including its uncertainty estimate (b) and the temporal resolution, Δt, of the compilation of data points (c). The 11-point running mean of Δt is around 100 years in the Holocene, between 20 and 50 years during Termination I, varies between 40 and 200 years between 20 and 70 kyr BP, and rises to 1000 years prior to 70 kyr BP. Across Termination II, Δt decreases to an average of 200 years.

The CO_2 data contributing to this spline are described below (further details in Table 2):

Table 4. Comparison of our final spline data with values used for PMIP4 experiments for 21 kyr (Ivanovic et al., 2016) and 1850 CE, 6 kyr, and 127 kyr (Otto-Bliesner et al., 2016). Please note that the PMIP4 values should be millennial-scale mean numbers to serve as forcing values for time slice experiments, while the values given from our study are snapshots of the given points in time. Furthermore, we calculate SH CH_4 values, while in PMIP4 the global CH_4 is given.

GHG	Unit	1850 CE	6 kyr	21 kyr	127 kyr
Our study:					
CO_2	ppm	286.1	264.4	187	274
SH CH_4	ppb	795	553	382	660
N_2O	ppb	271	261	206	257
PMIP4:					
CO_2	ppm	284.6	264.4	190	275
global CH_4	ppb	808	597	375	685
N_2O	ppb	273	262	200	255

1. Our CO_2 data compilation uses instrumental monthly CO_2 data taken from the NOAA network up to the beginning of the year 2016 CE, or −66.0 BP (Dlugokencky et al., 2016b). We here choose to take only the data of the original so-called "Keeling curve" started by C.D. Keeling at NOAA's Mauna Loa Observatory in 1958 CE, and since 1974 CE independently measured by both the Scripps Institution of Oceanography (scrippsco2.ucsd.edu) and NOAA (www.esrl.noaa.gov/gmd/ccgg/trends/) (Fig. 2a). There is a small interpolar difference in CO_2 concentrations, with higher concentrations in the north than in the south; e.g. the 10-year averages from 2006 CE to 2015 CE were 3.5 ppm

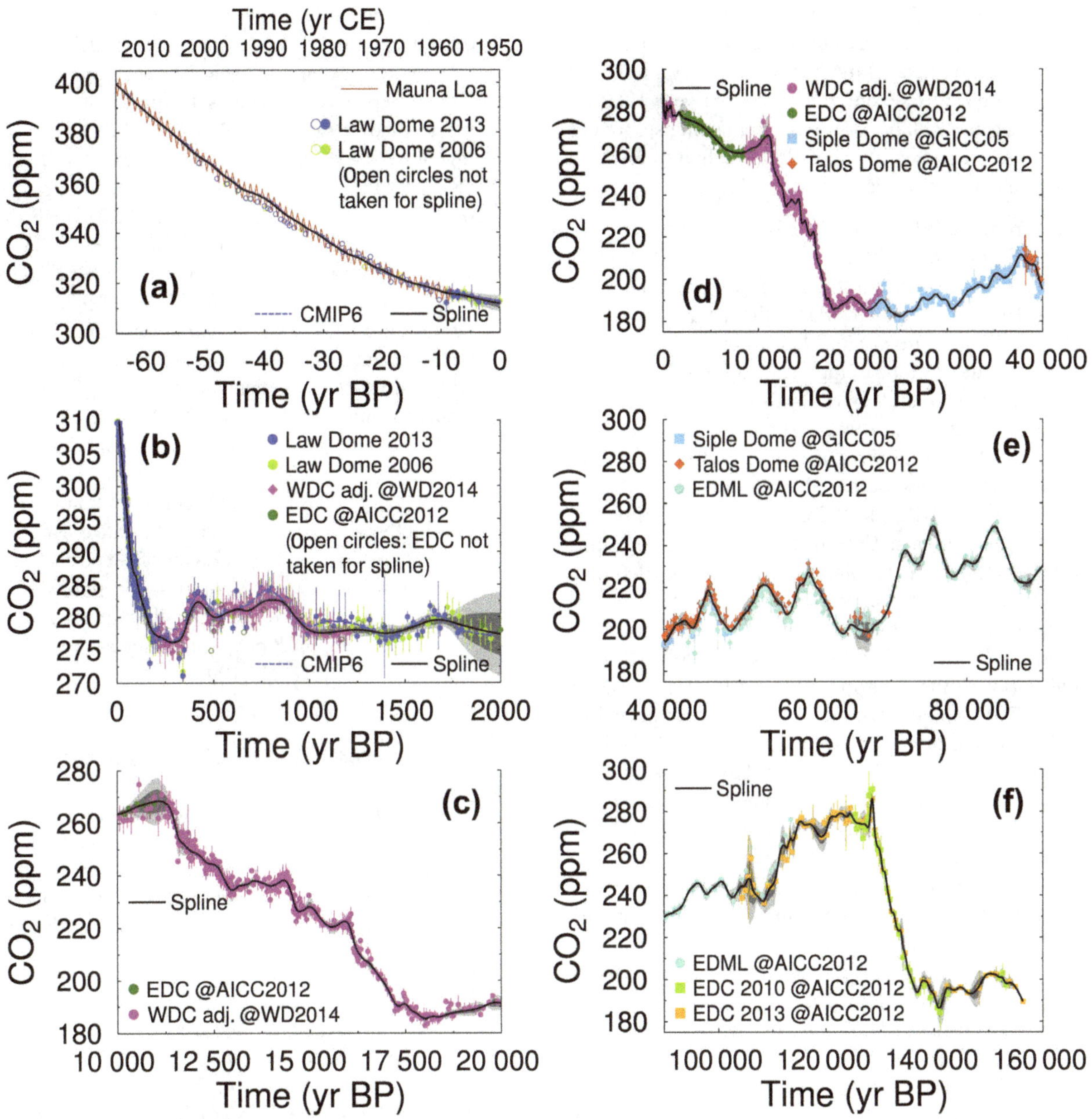

Figure 2. Details of the CO_2 spline. Light and dark grey bands around the spline represent $\pm 1\sigma$ and $\pm 2\sigma$, respectively; error bars around the ice core data points are $\pm 2\sigma$. **(a)** Instrumental times (1950–2016 CE); **(b)** 0–2000 BP; **(c)** Termination I; **(d)** 0–40 kyr BP without the Law Dome data showing the anthropogenic rise; **(e)** 40–90 kyr BP; **(f)** 90–160 kyr BP. WDC data have been adjusted to reduce offsets; see text for details. Dashed line labelled CMIP6 in panels **(a)** and **(b)** is the compiled CO_2 record to be used in CMIP6 experiments for the last 2 kyr (Meinshausen et al., 2017).

lower at the South Pole than at Mauna Loa and 1.4 ppm higher at Barrow (Alaska) than at Mauna Loa (Dlugokencky et al., 2016b). We therefore assume that CO_2 data from Mauna Loa are a good approximation of the global average concentration. However, in practice this interpolar difference cannot be determined prior to the instrumental records since CO_2 is only measured on ice cores from Antarctica, as the higher impurity content gives rise to artefacts in any CO_2 measurement based on Greenlandic ice cores (e.g. Anklin et al., 1995; Stauffer et al., 1998).

2. The firn and ice data compilation of Law Dome, which also contains some contributions from Cape Grim and the South Pole – available for the time from 1996 CE to 1 CE (−46 BP to 1949 BP) (MacFarling-Meure et al., 2006) and 2001 CE to 154 CE (−51 BP to 1796 BP)

(Rubino et al., 2013) – overlap consistently with direct atmospheric measurements. We therefore take these data as our reference time series for the Common Era (Fig. 2b) but include only the data from year 1960 CE and older in our spline compilation. In doing so, we use the more precise and temporally more highly resolved instrumental data for later times.

3. Data from the WDC ice core exist for the times of 11–1210 BP, or 1939–740 CE (Ahn et al., 2012; Bauska et al., 2015) and for Termination I (see point no. 5 below). These WDC data overlap with the Law Dome data (MacFarling-Meure et al., 2006; Rubino et al., 2013); however, the available high-resolution CO_2 records from different ice cores (Law Dome, WDC, EDML) show some apparent offsets (Ahn et al., 2012). Whilst the CO_2 data in all three ice cores converge on similar concentrations during the anthropogenic rise in CO_2 after 1750 CE, the WDC CO_2 concentrations are slightly higher than CO_2 in the other two ice cores prior to 1750 CE. In pre-anthropogenic times, EDC data not contained in the comparison of Ahn et al. (2012) also agree more with the Law Dome data than with those of WDC. We therefore choose to take WDC data only prior to the anthropogenic rise (200–1210 BP or 1750–740 CE). Furthermore, WDC data are adjusted by −3.13 ppm to bring them into agreement with the Law Dome CO_2 record. The data from Law Dome and the adjusted data from WDC contribute to our data compilation between 200–1210 BP. The mean temporal resolutions of both ice core CO_2 records within this time interval are 8 and 13 years for WDC and Law Dome, respectively. The amplitude of the CO_2 minima around 300–400 BP is controversial (Bauska et al., 2015). In our final spline, little of the large negative anomaly in CO_2 contained in the Law Dome data is preserved, since we smoothed the ice data in this time window with a cut-off period of 160 years (Fig. 2b). The time between the start of the anthropogenic rise (1750 CE) and the start of the instrumental record (1958 CE) is only supported by the Law Dome data in our compilation (Fig. 2b). Further details on this adjustment of the WDC data are covered in Fig. A1 in the Appendix.

4. EDC data exist between 350 BP and the Last Glacial Maximum (LGM) (Monnin et al., 2001, 2004) and further back in time (see point no. 7 below). They overlap with the Law Dome data between 350 and 1950 BP (Fig. 2b) without any apparent offset, and therefore no adjustment is applied to the EDC data. However, EDC data are only included in our compilation for the interval 1.9–11 kyr BP because Law Dome and WDC data provide a better resolution for the interval younger than 1.9 kyr BP, whilst the WDC data are the more highly resolved record for the interval older than 11 kyr BP (Fig. 2c).

5. Termination I is best covered by data from WDC (Marcott et al., 2014). WDC data are available for 8.8–22.9 kyr BP and are adjusted by −6.06 ppm (Fig. 2c). This difference corresponds to the duration-weighted mean offset between the WDC and EDC records during three intervals of relatively constant CO_2 (22.3–18.5 kyr BP: WDC ($n = 29$) 194.75 ± 2.44 ppm; EDC ($n = 21$) 188.22 ± 2.32 ppm; 14.5–13.0 kyr BP: WDC ($n = 45$) 243.02 ± 2.44 ppm; EDC ($n = 9$) 237.57 ± 1.42 ppm; 11.5–9.0 kyr BP: WDC ($n = 36$) 269.97 ± 3.67 ppm; EDC ($n = 27$) 264.24 ± 1.88 ppm). The intervals have been selected to minimise the influence of potential age scale differences between the two records. Only those EDC studies focusing on CO_2 measurements (Monnin et al., 2001, 2004) have been considered here, rather than those with a main focus on $\delta^{13}C$ (Lourantou et al., 2010b; Schmitt et al., 2012), which have a lower precision in CO_2 concentrations. More details on this adjustment of the WDC data during Termination I are found in Fig. A2. Our offset corrections imply an absolute CO_2 concentration uncertainty of about 5 ppm (accuracy). The corresponding uncertainty in the radiative forcing $\Delta R_{[CO_2]}$ following a simplified expression of Myhre et al. (1998),

$$\Delta R_{[CO_2]} = 5.35 \cdot \ln(CO_2/(278\,\text{ppm}))\ \text{W m}^{-2}, \quad (3)$$

is 0.15 or 0.09 W m^{-2} for a reference concentration of 180 or 280 ppm, respectively. This uncertainty is larger than the measurement uncertainty (precision) of the order of 1 ppm attached to individual data points which is used to determine the smoothing spline through the data.

6. Further back in time all ice core records used have some data overlap with their successive records. There are some small offsets between the different records (for details, see Bereiter et al., 2012). We treat them all alike, so the spline averages over all cores, and we select a large cutoff period of 2000 years for the interval 18.5–110 kyr BP to account for those uncertainties. Rapid variations in CO_2 during glacial times (Fig. 2d–f) are best recorded in the Siple Dome record between 21.9 and 48.7 kyr BP (Ahn and Brook, 2014), the Talos Dome record between 34.4 and 69.7 kyr BP (Bereiter et al., 2012), and the EDML record between 43.2 and 113.4 kyr BP (Lüthi et al., 2010; Bereiter et al., 2012). Talos Dome CO_2 data include some outliers in the interval 34–38 kyr BP that disagree with CO_2 records from other ice cores by more than 10 ppm. Therefore, Talos Dome data are only considered for the times older than 38.0 kyr BP.

7. From 104.3 to 156.3 kyr BP – the interval spanning the last glacial inception, the last interglacial, Termination

II, and the penultimate glacial maximum (Fig. 2f) – the EDC CO_2 record is used in the compilation (Schneider et al., 2013; Lourantou et al., 2010a).

For every supporting data point j a 1σ uncertainty or error v_j has to be assigned in order to be able to calculate the smoothing spline (see Sect. 2 for details). A nominal uncertainty of 0.3 ppm is assigned to the Mauna Loa data, which is for conservative reasons slightly higher than the generally stated measurement uncertainty of 0.2 ppm (https://www.esrl.noaa.gov/gmd/ccgg/about/co2_measurements.html). Uncertainties for the ice and firn data are taken either as reported or set to 0.5 ppm if the reported standard deviation is missing or less than 0.5 ppm. For Law Dome data published in MacFarling-Meure et al. (2006), we take the reported uniform uncertainty of 1.1 ppm. Note that the adjusted ice core offset between WDC and the other ice cores is not considered in our uncertainty calculation as this represents a systematic error.

The data selection as described above then leads to $n =$ 2152 data points including 20 ages with duplicate entries. These duplicates are averaged (reducing n to 2132) and the assigned uncertainties based on this averaging.

To account for the variable temporal resolution of the data points (Fig. 1c) whilst preserving as much of the abrupt changes in CO_2 during Termination I as possible (e.g. Marcott et al., 2014), the spline is divided into 12 intervals with different nominal cutoff periods that vary between 4 years (for instrumental times) and 3000 years (for the Holocene). A low P_c of 600 years was assigned to the high-resolution interval of Termination I (11–18.5 kyr BP). For the glacial interval between 18.5 and 110 kyr BP, P_c of 2000 years was chosen. For the warm interglacial between 110 and 128 kyr BP, we assign a cutoff period of 3000 years similar to the Holocene. Across Termination II (128–135 kyr BP), we use a 1000-year cutoff period, whilst for the penultimate glacial maximum a cutoff period of 2000 years was used. A summary of all details on the calculated spline is found in Table 3.

The total 1σ uncertainty of the spline is < 2 ppm on average (Fig. 1b). Across some short time windows, it rises up to 6 ppm, and around 108 kyr BP, it reaches a maximum of 11 ppm. The three different error sources contribute equally to the total uncertainty; however, time windows with large uncertainties are often dominated by one error source.

The CO_2 record of the last 2 kyr to be used within CMIP6 (Meinshausen et al., 2017) is nearly indistinguishable from our spline across the instrumental period (Fig. 2a); however, CO_2 concentrations during the pre-anthropogenic interval of the last 2 kyr are partly larger by a few parts per million than our spline (Fig. 2b). This difference is readily explained by the underlying data and the different filtering. We use a combination of Law Dome and WDC data between 200 and 1210 BP, whilst only Law Dome data are considered for CMIP6.

The CO_2 values chosen as boundary conditions for several time slice experiments within PMIP4 (Ivanovic et al., 2016; Otto-Bliesner et al., 2016) can be compared with snapshots from our splines (Table 4). However, one needs to be aware that some short-term fluctuations in our spline might offset the values from long-term averages and lead to differences between our final splines and the PMIP4 forcing data. For the mid-Holocene (6 kyr experiment), both our spline and data used in PMIP4 are based on the same EDC data and processed with the identical spline routines and cutoff frequencies, leading to identical values. Values differ by a few parts per million for the experiments 1850 CE, 21 kyr, and 127 kyr.

Since spline smoothing is a low-pass filter, abrupt changes in CO_2 are always smaller in the spline than in the original data sets. Therefore, if one wants to investigate the impact of abrupt increases in CO_2 concentration on the climate system that have been identified during three intervals (around 11.6, 14.7 or 16.2 kyr BP) across Termination I (Monnin et al., 2001; Marcott et al., 2014), an alternative continuous CO_2 record needs to be constructed. One approach might be to reduce the cutoff period so that the spline would include these pronounced jumps. For example, one might want to capture the rise in CO_2 of 12 and 13 ppm across a single century at 16.2 and 11.6 kyr BP, respectively, as identified in the WDC record (Marcott et al., 2014). For the abrupt rise in CO_2 around 14.7 kyr BP, even an increase of 15 ppm in 200 years, slightly larger than the 12 ppm of the WDC record, has been suggested to represent atmospheric changes in CO_2 potentially caused by permafrost thawing during the northern hemispheric warming into the Bølling–Allerød interstadial (Köhler et al., 2014, 2015). Transient simulations investigating these abrupt jumps in CO_2 concentration should use a CO_2 times series that contains greater details than our low-frequency spline.

3.2 Atmospheric CH_4

Our data compilation of CH_4 data and the consistently calculated CH_4 spline is restricted to the Southern Hemisphere (SH). Northern hemispheric (NH) data are shown for comparison but are not included in the spline, since for such efforts chronologies of ice cores from both hemispheres have to match perfectly during abrupt climate changes of the D/O events. However, as has been shown (Baumgartner et al., 2014), there remains some mismatch in the timing of the NH and the SH CH_4 records in the most recent chronology AICC2012. NH CH_4, and consequently global CH_4 concentrations, should, according to the estimates of the interpolar difference, be larger than our SH CH_4 values. Therefore, our SH CH_4 spline represents the lower bound of global CH_4 concentration. Baumgartner et al. (2012) found that NH CH_4 was up to +4 % (+14 ppb) and up to +10 % (+60 ppb) larger than the SH CH_4 during glacial times and the Holocene, respectively. However, new and as yet unpublished results

Table 5. Data used to construct (or compare to) the Southern Hemisphere CH_4 spline.

Time (in BP)	Time (in CE)	Source	Age scale	Spline[5]	Citation
−66 to −34	2016 to 1984	NOAA network[1] (annual)	–	no	Dlugokencky et al. (2016a)
−66 to −33	2016 to 1983	South Pole[2] (monthly)	–	yes	Dlugokencky et al. (2016a)
−66 to −33	2016 to 1983	Barrow[3] (monthly)	–	no	Dlugokencky et al. (2016a)
−32 to 168	1982 to 1782	Law Dome[4]	as in references	yes	MacFarling-Meure et al. (2006); Rubino et al. (2013)
169 to 4669	1781 to before CE	WDC discrete, OSU[6]	WD2014	yes	WAIS Divide Project Members (2015); Marcott et al. (2014); Buizert et al. (2015); Mitchell et al. (2013, 2011); Sigl et al. (2016)
4689 to 9798	–	WDC discrete, PSU[7]	WD2014	yes	WAIS Divide Project Members (2015); Marcott et al. (2014); Buizert et al. (2015); Mitchell et al. (2013, 2011); Sigl et al. (2016)
9821 to 67 233	–	WDC continuous	WD2014	yes	Rhodes et al. (2015); Buizert et al. (2015); Sigl et al. (2016)
192 to 100 469	–	GRIP	GICC05ext	no	Blunier and Brook (2001); Dällenbach et al. (2000); Flückiger et al. (2004); Landais et al. (2004)
67 401 to 15 6211	–	EDC	AICC2012	yes	Loulergue et al. (2008)

Notes: [1] Global annual mean of the NOAA network. Data taken from ftp://aftp.cmdl.noaa.gov/products/trends/ch4/ch4_annmean_gl.txt. [2] Data taken from ftp://aftp.cmdl.noaa.gov/data/trace_gases/ch4/flask/surface/ch4_spo_surface-flask_1_ccgg_month.txt. [3] Data taken from ftp://aftp.cmdl.noaa.gov/data/trace_gases/ch4/flask/surface/ch4_brw_surface-flask_1_ccgg_month.txt. [4] Law Dome data are taken from various sources; see references for details. They exist for 2005 CE to 14 CE (or −55 BP to 1936 BP), but only the part bridging the gap between instrumental data and WDC is taken for the calculation of the spline (1982 CE to 1782 CE or −32 BP to 168 BP). [5] Indicates if the data are used to construct the spline. [6] Measured in laboratory at Oregon State University (OSU). [7] Measured in laboratory at Penn State University (PSU); measured data shifted by +9.9 ppb to account for unexplained OSU and PSU laboratory offset (Rhodes et al., 2015).

point to in situ CH_4 production in Greenland ice cores during times of high dust fluxes, calling for a revision of the interpolar difference in CH_4 during glacial times. For this reason, we refrain from calculating an NH or global CH_4 spline. As CH_4 is only of secondary importance for the total greenhouse gas radiative forcing, this systematic error is of little relevance for climate simulation studies. Studies are under way to improve our knowledge of the NH CH_4 value for glacial times, too. Using an approximation of the radiative forcing

$$\Delta R_{[CH_4]} \sim 1.4 \cdot 0.036 \cdot (\sqrt{CH_4/\mathrm{ppb}} - \sqrt{742})\,\mathrm{W\,m^{-2}}, \quad (4)$$

which neglects the interacting effects of CH_4 and N_2O (Myhre et al., 1998) but which considers the approximate increase in $\Delta R_{[CH_4]}$ by 40 % through indirect effects of CH_4 on stratospheric H_2O and tropospheric O_3 (Hansen et al., 2005; Köhler et al., 2010), we estimate that our restriction of CH_4 to the SH only would underestimate the radiative forcing of CH_4 by less than 0.05 $\mathrm{W\,m^{-2}}$.

Our data compilation starts with the beginning of the year 2016 CE (−66.0 BP) and stops around 156 kyr BP to cover the same time interval as for CO_2 (Fig. 3a). The 11-point running mean temporal resolution between neighbouring data points, Δt, is less than 100 years for most of the last 67 kyr, increasing to ~ 700 years between 67 and 156 kyr BP (Fig. 3c). Our strategy here is to select one data set for each point in time and use overlapping intervals only for confirmation of data consistency. The following data sets are considered here.

1. From the NOAA network, the annual global mean concentration of CH_4 from 2016 CE to 1984 CE is available (www.esrl.noaa.gov/gmd/ccgg/trends/). These global mean concentrations lie in between the seasonally resolved CH_4 concentration measured at Barrow, Alaska (NH), and at the South Pole (SH), both reaching back in time until 1983 CE (Dlugokencky et al., 2016a) (Fig. 4a). The interpolar difference between the NH (Barrow) and the SH (South Pole) was +161 ppb or +9 % at the beginning of 2016 CE. In absolute CH_4 concentration, this most recent interpolar difference is ~ 100 ppb larger than the interpolar difference in the Holocene, while the relative interpolar difference during both time intervals is comparable (Baumgartner et al., 2012). An estimation of the radiative forcing of this interpolar difference reveals that, for the time covered by the NOAA network, the $\Delta R_{[CH_4]}$ for the NH was $< 0.1\,\mathrm{W\,m^{-2}}$ larger than for the SH. This estimate of the radiative forcing of the interpolar difference is obtained

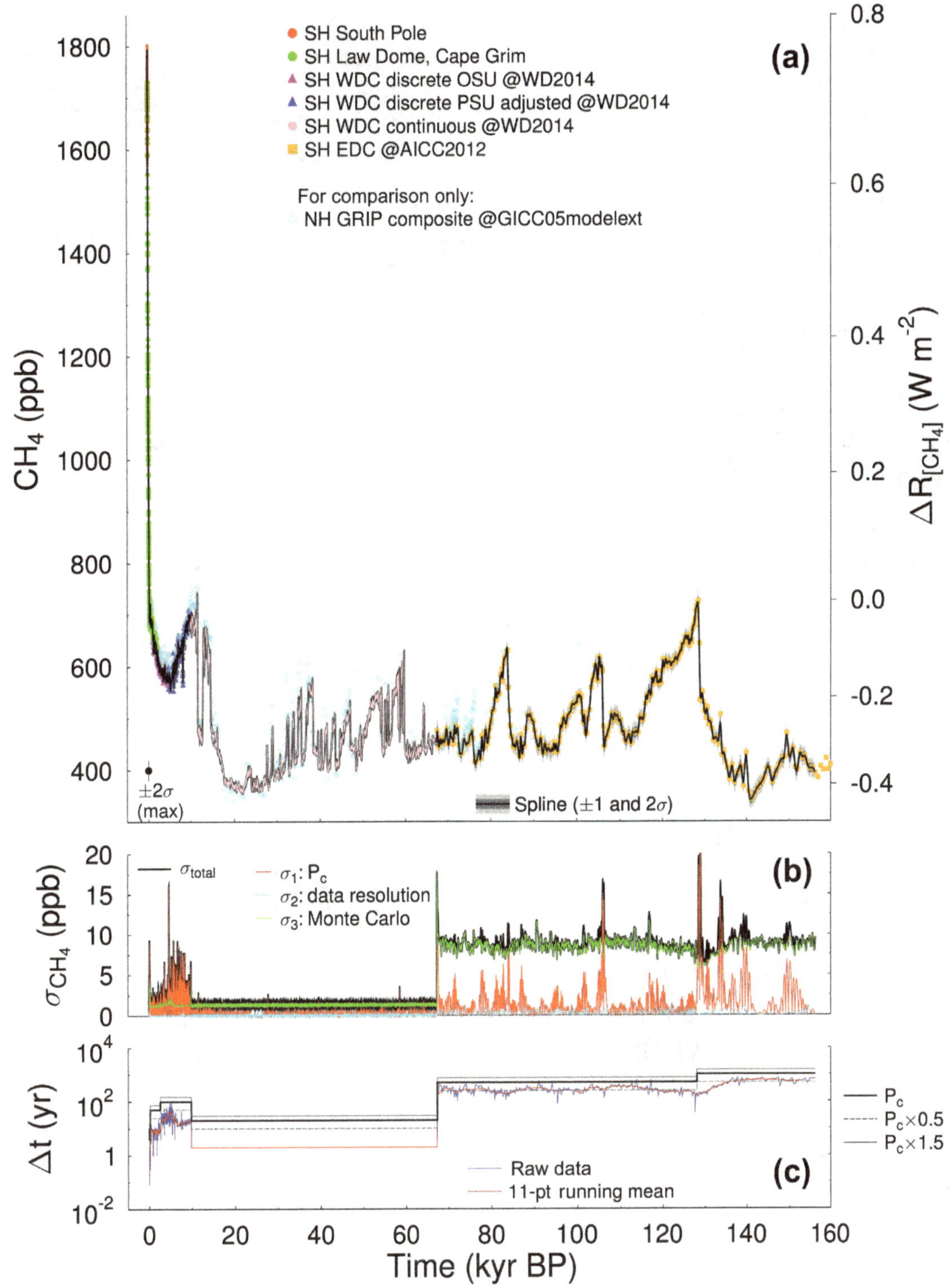

Figure 3. CH_4 spline covering all data: 2016 CE–156 211 BP. Details on plotted data are explained in the text. The maximum ice core data uncertainty ($\pm 2\sigma$) is given in the lower left corner. In **(a)** the right axis contains the resulting radiative forcing approximated with $\Delta R_{[CH_4]} \sim 1.4 \cdot 0.036 \cdot (\sqrt{CH_4/\mathrm{ppb}} - \sqrt{742})\ \mathrm{W\,m^{-2}}$ based on Myhre et al. (1998), but neglecting interacting effects of CH_4 and N_2O and considering indirect effects of CH_4 on stratospheric H_2O and tropospheric O_3 (Hansen et al., 2005; Köhler et al., 2010). The latitudinal origin of data is indicated by NH and SH, indicating Northern and Southern Hemisphere, respectively. **(b)** Total uncertainty of the spline based on three individual error sources; see text for details. **(c)** Temporal resolution (Δt) of the CH_4 data points underlying the spline on a log scale. Additionally, the prescribed time-dependent cutoff period P_c is plotted, including its variation by ±50 %, which has been used to determine σ_1.

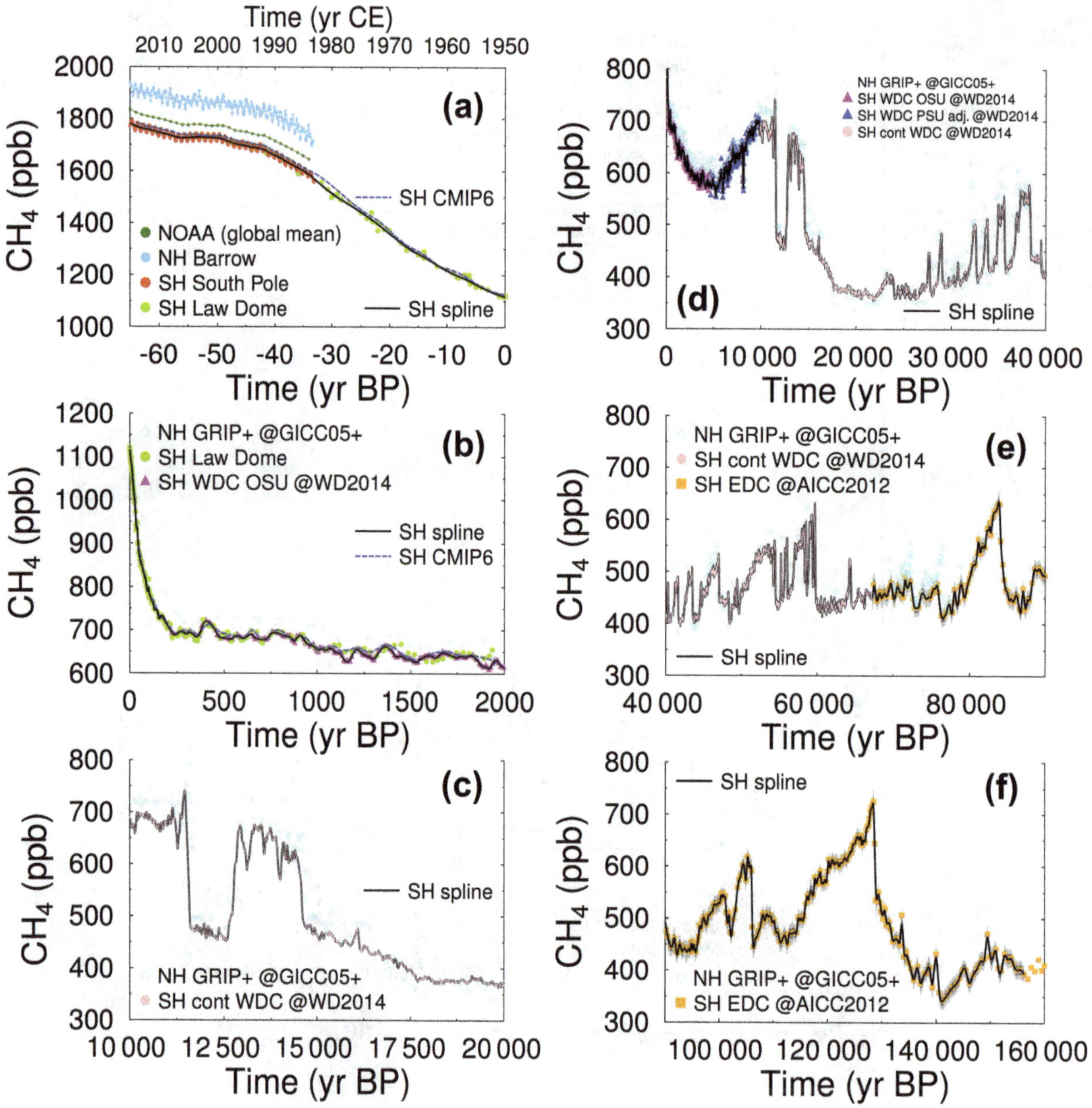

Figure 4. Details of the southern hemispheric CH_4 spline. Light and dark grey bands around the spline represents $\pm 1\sigma$ and $\pm 2\sigma$, respectively. **(a)** Instrumental times (1950–2016 CE); **(b)** 0–2000 BP; **(c)** Termination I; **(d)** 0–40 kyr BP without the Law Dome data showing the anthropogenic rise; **(e)** 40–90 kyr BP; **(f)** 90–160 kyr BP. Hemispheric origin of data is indicated by NH (north) and SH (south). WDC PSU data are adjusted by +9.9 ppb. GRIP+: Greenland composite; GICC05+: GICC05 model extended. See text for details. Dashed line labelled CMIP6 in panels **(a)** and **(b)** is the compiled CH_4 record to be used in CMIP6 experiments for the last 2 kyr (Meinshausen et al., 2017).

from Eq. (4), based on Barrow and the South Pole CH_4 data. For our SH compilation we used the South Pole data.

2. Ice core and firn air data from Law Dome and Cape Grim (SH) exist from 2005 CE back to 14 CE (= 1936 BP) (MacFarling-Meure et al., 2006; Rubino et al., 2013) with an overlap of more than 2 decades with the instrumental measurements (Fig. 4a, b). The CH_4 data from Law Dome and Cape Grim used in our compilation span the period from 1982 CE to 1782 CE (= 168 BP), bridging the gap between instrumental data and CH_4 from WDC. Where the Law Dome data overlap with the data from either the South Pole or WDC, no apparent systematic offsets between the different data sets have been identified. WDC and Law Dome data differ slightly across short intervals between 1000 and 2000 BP (Fig. 4b). However, since WDC is the more

Table 6. Statistics of the CH_4 spline. Interval: i_{CH_4}; s: scaling factor to fulfil the constraints given by the prescribed cutoff period P_c; $\overline{P_c}$: average realised cutoff period; Δt: mean data spacing; υ: mean 1σ error – exact scaling factor to fulfil the constraints given by the prescribed cutoff period P_c; $\overline{P_c}$: average realised cutoff period. Δt: mean data spacing; υ: mean 1σ error; exact time framing is given by the age of the first (t_{start}) and last (t_{stop}) data point of the interval (in years BP); N: number of data points within interval. In the last column the underlying data source is briefly mentioned; see Table 5 for details and citations.

i_{CH_4}	s	P_c	$\overline{P_c}$	Δt	υ	t_{start}	t_{stop}	N	Data source
	–	yr	yr	yr	ppb	yr BP	yr BP	–	
1	1.00	4.0	4.0	0.1	2.0	−66.0	−33.1	396	South Pole
2	0.68	10.0	9.5	1.7	4.0	−31.8	165.1	114	Law Dome
3	13.12	50.0	49.7	8.2	2.4	169.0	2591.0	296	WDC discrete
4	33.63	100.0	98.8	20.0	2.4	2602.0	9798.0	361	WDC discrete
5	3.09	20.0	20.0	2.0	3.3	9821.0	67 233.0	28 707	WDC continuous
6	56.26	500.0	497.0	257.1	10.0	67 401.0	127 831.0	236	EDC
7	171.96	1000.0	982.0	440.4	10.0	128 026.0	156 211.0	65	EDC

highly resolved record, it alone is included in the spline; no data adjustment is necessary here.

3. The discrete CH_4 data from WDC (SH) span the interval 169 BP to 67 kyr BP (WAIS Divide Project Members, 2015; Marcott et al., 2014; Buizert et al., 2015; Mitchell et al., 2013, 2011; Sigl et al., 2016). Starting with the year 9821 BP, continuous CH_4 data from WDC with higher temporal resolution are now available and are used to support our spline (Rhodes et al., 2015). These continuous CH_4 data have already been post-processed, including the support from some discrete WDC data points to improve the data set, whenever larger gaps in the continuous record appeared. The data product of the continuous CH_4 WDC data obtained at NOAA is splined to a constant temporal resolution of 2 years. For the missing part of the Holocene not contained in the continuous WDC data, discrete WDC CH_4 data are used. They have been measured in two different laboratories: at Oregon State University (OSU; 169–4669 BP) and at Pennsylvania State University (PSU; 4689–9798 BP). An unexplained inter-laboratory offset between the discrete CH_4 WDC data from OSU and PSU has been identified. To account for this offset the PSU CH_4 data have been adjusted by +9.9 ppb (Supplementary Information to Rhodes et al., 2015). To date WDC CH_4 are the temporally most highly resolved data of the last glacial, and therefore they are our reference record (Fig. 4b–e). The data not only contain the well-known abrupt CH_4 changes at the onset and end of the millennial-scale D/O events in high resolution and accuracy but also centennial-scale features that are understood to be of climatic origin (e.g. Mitchell et al., 2013; Rhodes et al., 2015).

4. We extend our SH CH_4 data compilation beyond WDC with data from EDC, spanning the period from 67 to 156 kyr BP (Loulergue et al., 2008) (Fig. 4e–f). These EDC data actually extend back to 800 kyr BP, but since our focus here is on the time since the penultimate glacial maximum (i.e. the last 156 kyr), the CH_4 record older than 156 kyr is not considered here. CH_4 data from EDML might be in part more highly resolved than in EDC because of a higher annual layer thickness between 67 kyr BP (the end of WDC) and 80 kyr BP (Ruth et al., 2007). However, a well-documented EDML CH_4 record is not available to date, and therefore none of the published EDML CH_4 data for this interval (Capron et al., 2010; Schilt et al., 2010b) are considered here.

5. The NH Greenland composite of CH_4 (Blunier and Brook, 2001; Dällenbach et al., 2000; Flückiger et al., 2004; Landais et al., 2004) is only plotted for comparison to the SH data (Fig. 4b–f).

The assigned data uncertainty (1σ error) is 2.0, 4.0, 2.4, 3.3, and 10 ppb for instrumental data, Law Dome, discrete WDC, continuous WDC, and EDC, respectively (Dlugokencky et al., 2016a; MacFarling-Meure et al., 2006; Mitchell et al., 2013; Rhodes et al., 2015; Loulergue et al., 2008). Using the approximation of $\Delta R_{[CH_4]}$, given the above 1σ errors the uncertainty in the radiative forcing is $<0.01\ W\ m^{-2}$.

Compiled data contain 30 214 data points, among which duplicate entries exist for 39 ages. These duplicates are averaged giving $n = 30175$.

The whole data set is divided into seven intervals with different assigned cutoff periods. P_c ranges from 4 years (for the interval covered by instrumental data) to 20 years (for the interval covered by the continuous WDC record). Due to lower data coverage during the Holocene and further back in time, P_c is increased to 100 years (0.2–9.8 kyr BP), 500 years (60–128 kyr BP), and 1000 years (128–156 kyr BP) (Fig. 3c). More details are shown in Table 6.

The total 1σ uncertainty of our final CH_4 spline is around 3–10 ppb in the Holocene, ~2 ppb in the time window supported by the continuous WDC CH_4 data (9.8–67 kyr BP), and around 10 ppb in earlier parts. During some short time

Table 7. Data used to construct the N_2O spline.

Time (in BP)	Time (in CE)	Source	Age scale	Citation
−66 to −49	2016 to 1999	NOAA network (monthly)	–	nitrous oxide data from the NOAA/ESRL halocarbons in situ program[1]
−50 to −38	2000 to 1988	NOAA network (monthly)	–	RITS nitrous oxide data from the NOAA/ESRL halocarbons program[2]
−33 to 1937	1983 to 13	Law Dome[3]	as in references	MacFarling-Meure et al. (2006)
1975 to 11 502	–	EDC[4]	AICC2012	Flückiger et al. (2002); Stauffer et al. (2002)
29 065 to 134 519	–	EDC	AICC2012	Spahni et al. (2005)
9858 to 15 843	–	Taylor Glacier	WD2014[5]	Schilt et al. (2014)
15 000 to 118 602	–	NGRIP[6]	AICC2012	Flückiger et al. (2004); Schilt et al. (2010b, 2013)
15 000 to 134 418	–	Talos Dome[7]	AICC2012	Schilt et al. (2010b)

Notes: [1] ftp://ftp.cmdl.noaa.gov/hats/n2o/insituGCs/CATS/global/insitu_global_N2O.txt. [2] ftp://ftp.cmdl.noaa.gov/hats/n2o/insituGCs/RITS/global/RITS_global_N2O.txt. [3] Law Dome data are taken from various sources; see references for details. They exist from 2004 CE to 13 CE (or −54 BP to 1937 BP), but only those older than the instrumental record (1986 CE and older) are taken here. [4] Data exist from 334 BP (or 1616 CE) until 11 502 BP, but only data not yet covered by the Law Dome records (13 CE or 1937 BP and older) are considered here. [5] WD2014 age model for Taylor Glacier, published in Baggenstos et al. (2017). [6] Data exist for 11 068 BP–119 555 BP, but only those older than 15 kyr BP are considered here. Five data points in the oldest part considerably disagree with SH records and are therefore rejected. [7] Data exist for 217 BP–134 418 BP, but only those older than 15 kyr BP are considered here.

windows, σ_{CH_4} reaches a maximum of 20 ppb (Fig. 3b). The uncertainty is dominated by the Monte Carlo error before 9.8 kyr BP and by the error in the cutoff period in the Holocene and during those short events in which σ_{CH_4} reached its local maxima. An abrupt jump in σ_{CH_4} appears at 67 kyr BP (transition from continuous WDC to discrete EDC data), when individual data point uncertainty rose from 3.3 to 10 ppb (changing σ_3) at the same time as Δt, and therefore P_c, increased by 2 orders of magnitude (changing σ_1).

The SH CH_4 record to be used within CMIP6 (Meinshausen et al., 2017) largely agrees with our SH CH_4 spline (Fig. 4a, b). However, during instrumental times the CMIP6 SH CH_4 record is consistently larger than our SH CH_4 spline by about 10–15 ppb, probably caused by the inclusion of different stations in the calculation of the SH CH_4 record within CMIP6, while we rely on South Pole data. Prior to the instrumental CH_4 data around 1980 CE, the maximum difference between both approaches is 30 ppb. This difference might be caused by the statistical routines within CMIP6 to account for missing stations. Further back in time (around 1150 BP, 1300 BP, and 1900 BP), higher-frequency variation contained in the WDC CH_4 record (used here but ignored within CMIP6) leads to some CH_4 variations within our SH CH_4 spline on the order 10–25 ppb that are not captured by the CMIP6 SH CH_4 record.

A comparison of our final spline with the GHG values chosen for the PMIP4 time slice experiments (Ivanovic et al., 2016; Otto-Bliesner et al., 2016) is not straightforward, since we only compile SH CH_4 data, while the PMIP4 experiments use global values. Taking the two records at face value, one finds that our SH CH_4 is 13, 44, and 25 ppb smaller than the global mean value used in PMIP4 for 1850 CE, 6 kyr, and 127 kyr, respectively. In particular, the large SH-global difference of 44 ppb around 6 kyr seems to be rather large but is readily explained by the centennial variability contained in the WDC CH_4, which leads to a local minimum in SH CH_4 around 6 ka. Similarly, our SH CH_4 spline is 7 ppb higher than the global CH_4 value chosen within PMIP4 for the 21 kyr experiment. This difference can again be explained by the centennial-scale variability contained in the WDC CH_4 record, which shows a local maximum at 21 kyr BP. A hundred years later, our SH CH_4 spline has a local minimum which is 11 ppb smaller than the global CH_4 values taken for PMIP4 (Table 4).

3.3 Atmospheric N_2O

For the data compilation of the third GHG, N_2O, one has to be aware that during times of high dust input, in situ production of N_2O occurs, leading to artefacts in the paleo record (Schilt et al., 2010a). Furthermore, the precise synchronisation of Northern and Southern Hemisphere records, as already explained for CH_4, is crucial to accurately obtain the changes in N_2O during millennial-scale D/O events.

The compiled record starts at the beginning of the year 2016 CE (−66.0 BP) but extends back in time only until ~134.5 kyr BP (Fig. 5a) because the ice cores on which the N_2O compilation is based in the older parts, Talos Dome, EDC, and NGRIP, have either no data points between 134.5 and 156 kyr BP or unreliable N_2O data containing artefacts across the penultimate glacial maximum (Schilt et al., 2010a). The latter is also the case for EDML, whose data have not been taken to support the spline because despite the agreement of the N_2O of EDML and EDC, the data from EDML have a lower temporal resolution than those of EDC (Schilt et al., 2010a).

The data sets contributing to the N_2O stack are listed below.

1. There are two contributions of N_2O data based on instrumental measurements to the NOAA network or ESRL halocarbon program: (a) in situ N_2O data are available from 2016 CE back until 1999 CE, and (b) the RITS N_2O data from 2000 CE go back until 1988 CE. Both represent global mean monthly values (Fig. 6a).

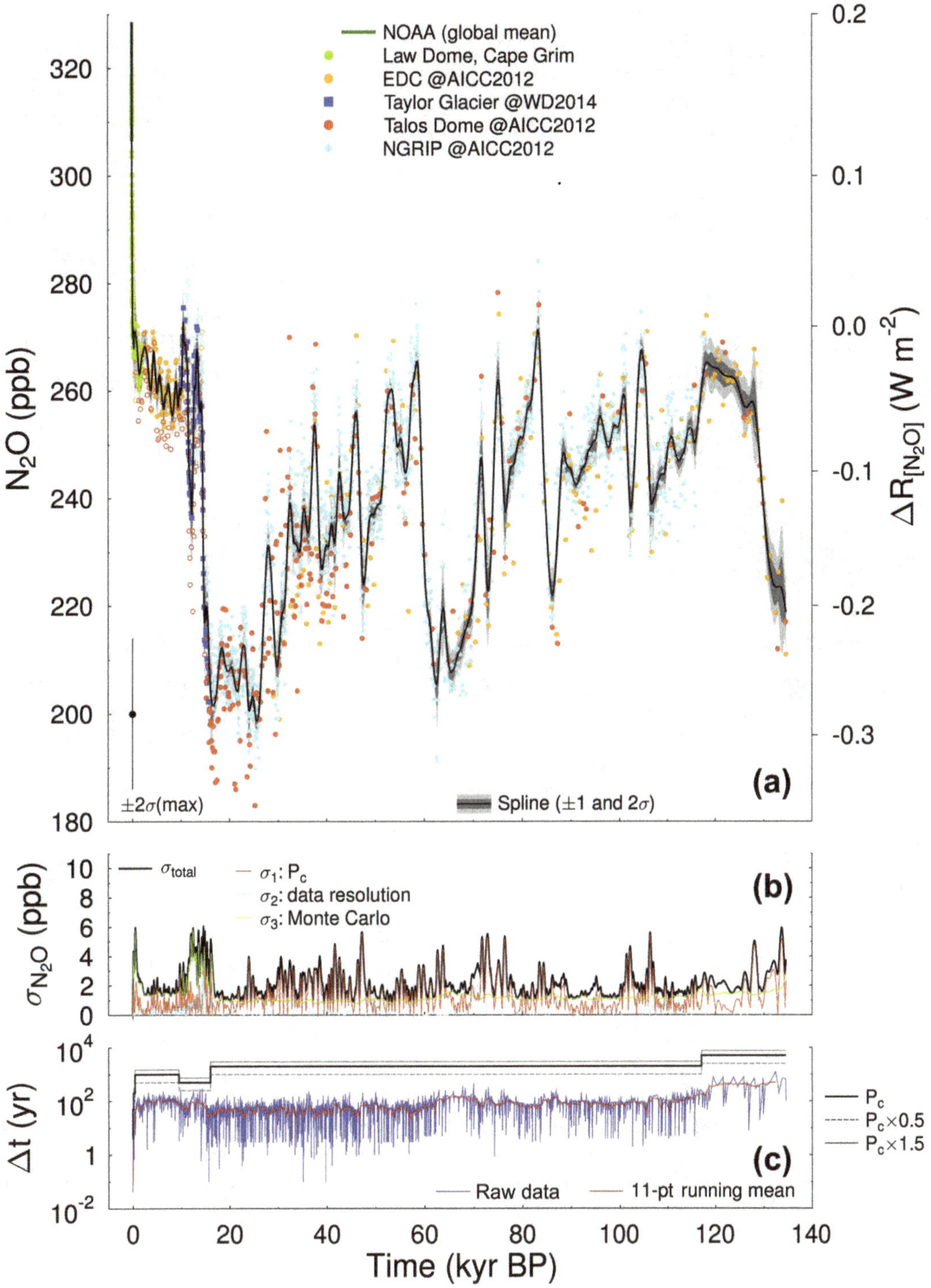

Figure 5. N_2O spline covering all data: 2016 CE–134 519 BP. Details on plotted data are explained in the text. The maximum ice core data uncertainty ($\pm 2\sigma$) is sketched in the lower left corner. In **(a)** the right axis contains the resulting radiative forcing approximated with $\Delta R_{[N_2O]} \sim 0.12 \cdot (\sqrt{N_2O/\text{ppb}} - \sqrt{272})\ \text{W m}^{-2}$ after Myhre et al. (1998), neglecting interacting effects of CH_4 and N_2O. Filled symbols: data taken for spline; open symbols: data not taken for spline. **(b)** Total uncertainty of the spline based on three individual error sources; see text for details. **(c)** Temporal resolution (Δt) of the N_2O data points underlying the spline on a log scale. Additionally, the prescribed time-dependent cutoff period P_c is plotted, including its variation by $\pm 50\,\%$, which has been used to determine σ_1.

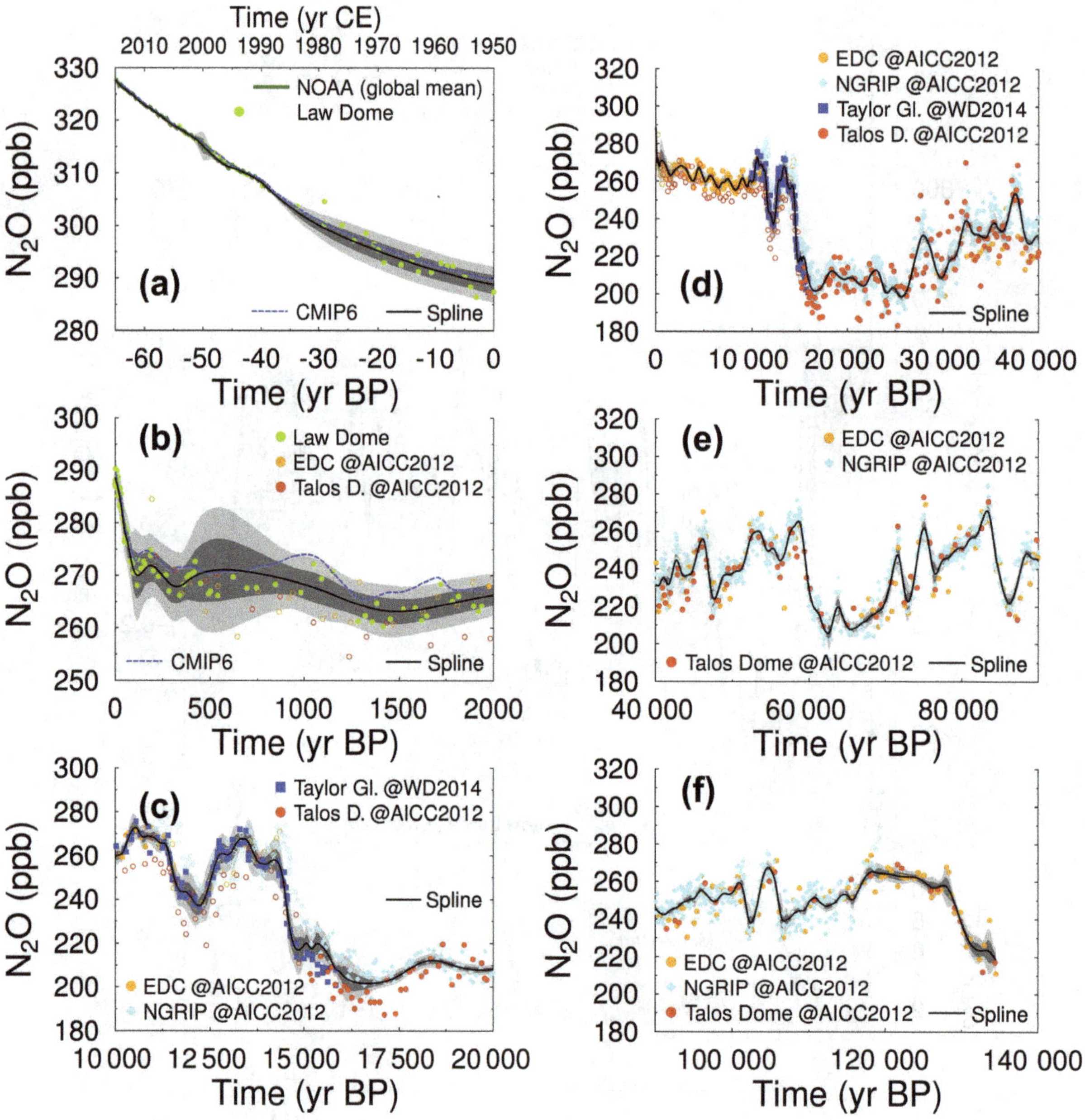

Figure 6. Details of the N_2O spline. Light and dark grey bands around the spline represent $\pm1\sigma$ and $\pm2\sigma$, respectively. **(a)** Instrumental times (1950–2016 CE); **(b)** 0–2000 BP; **(c)** Termination I; **(d)** 0–40 kyr BP without the Law Dome data showing the anthropogenic rise; **(e)** 40–90 kyr BP; **(f)** 90–140 kyr BP. Filled symbols: data taken for spline; open symbols: data not taken for spline. See text for further details. Dashed line labelled CMIP6 in panels **(a)** and **(b)** is the compiled N_2O record to be used in CMIP6 experiments for the last 2 kyr (Meinshausen et al., 2017).

Note that due to the long atmospheric lifetime of N_2O, any interpolar difference can be safely neglected.

2. Law Dome and Cape Grim N_2O data exist from 2004 CE back until 13 CE (1937 BP) (MacFarling-Meure et al., 2006) and correlate well with the instrumental data in overlapping intervals (Fig. 6a,b). Here, the Law Dome data contribute to the spline only for those years not covered by the instrumental record, i.e. 1983 CE and earlier.

3. In the Holocene, N_2O was measured at EDC (Flückiger et al., 2002) from 334 BP until 11.5 kyr BP. For the last two millennia, the EDC N_2O data points are sparser than the Law Dome data; therefore, the EDC N_2O data are only considered for times older than what is cov-

Table 8. Statistics of N_2O spline. Interval: i_{N_2O}; s: scaling factor to fulfil the constraints given by the prescribed cutoff period P_c; $\overline{P_c}$: average realised cutoff period; Δt: mean data spacing; v: mean 1σ error – exact time framing is given by the age of the first (t_{start}) and last (t_{stop}) data point of the interval (in years BP); N: number of data points within interval. In the last column the underlying data source is briefly mentioned; see Table 7 for details and citations.

i_{N_2O}	s	P_c	$\overline{P_c}$	Δt	v	t_{start}	t_{stop}	N	Data source
	–	yr	yr	yr	ppb	yr BP	yr BP	–	
1	1.00	4.0	3.4	0.1	0.8	−66.0	−38.3	334	NOAA network
2	3.01	50.0	48.7	2.5	7.2	−33.7	95.0	53	Law Dome
3	19.08	200.0	190.2	16.8	7.0	104.0	389.6	18	Law Dome
4	321.85	1000.0	952.4	85.1	4.6	400.3	9425.6	107	Law Dome, EDC
5	77.33	500.0	469.0	58.2	5.8	9517.2	15 974.7	112	EDC, Taylor Glacier, NGRIP, Talos Dome
6	1443.10	2000.0	1915.3	60.4	4.9	16 003.0	116 900.0	1672	EDC, NGRIP, Talos Dome
7	4236.11	5000.0	4792.5	370.0	4.2	117 130.0	134 519.0	48	EDC, NGRIP, Talos Dome

ered in the Law Dome N_2O record, i.e. before 1975 BP (Fig. 6b, d).

4. The most highly resolved N_2O record across large parts of Termination I is provided by the horizontal ice core from Taylor Glacier (Schilt et al., 2014) which has been linked to the chronology of the WDC ice core (WD2014) via CH_4 (Baggenstos et al., 2017). The Taylor Glacier N_2O record used in our spline covers the interval 9.6 to 15.8 kyr BP (Fig. 6c).

5. The last glacial interval is well resolved by N_2O data from the NGRIP record (Flückiger et al., 2004; Schilt et al., 2010b, 2013). While the NGRIP N_2O data cover the times between 11 kyr BP and 119.6 kyr BP, we only take those data older than 15 kyr BP due to the more highly resolved Taylor Glacier N_2O data during Termination I (Fig. 6c–f). Five data points near the bedrock in the bottom part of the NGRIP records apparently have higher N_2O values than found in ice cores from the Southern Hemisphere. These data points are rejected here, leading to the oldest NGRIP N_2O data point at 118.6 kyr BP. We are aware that due to the imperfect north–south synchronisation of gas records in AICC2012 (see Sect. 3.2 for details), the usage of N_2O data from NGRIP might introduce erroneous phasing between our global N_2O record and the purely SH-based CH_4 spline, particularly during abrupt change connected to D/O events. However, N_2O data coverage in the SH is very sparse and a spline only based on SH data would be even less reliable. This potential synchronisation problem is also addressed by large cutoff periods of the spline of 2000 to 5000 years beyond 16 kyr BP.

6. Additional N_2O data going back to 134.4 kyr BP are obtained from the Talos Dome ice core (Schilt et al., 2010b) and from further data of the EDC ice core (compilation found in Schilt et al., 2010a; data source between 29.0–134.5 kyr BP: Stauffer et al., 2002; Spahni et al., 2005). Since – besides EDML, which correlates well with EDC (Schilt et al., 2010a) – these are the only N_2O records with reliable data going back to the penultimate glacial maximum, we consider all data points from the Talos Dome and EDC ice cores here before 15 kyr BP. The data points of Talos Dome and EDC in general agree with the NGRIP data over the last glacial cycle, but NGRIP diverges from the SH records towards higher (probably biased) values in the warm previous interglacial around 115 kyr BP. As already explained above, these five NGRIP data points are rejected. However, across Termination I, Talos Dome N_2O data seems to be systematically lower than NGRIP N_2O data, with Taylor Glacier data in between both (Fig. 6c). We therefore believe that a mixture of all three records (N_2O from NGRIP, EDC, and Talos Dome) most likely represents a reasonable mean global N_2O value (Fig. 6c–f). The relatively large difference in N_2O from different ice cores during the last glacial times indicates that the uncertainty (accuracy) in N_2O is probably higher than the reported measurement errors (precision) of up to 7 ppb.

The generally assigned 1σ uncertainty of each data point is 7 ppb for the Law Dome ice core (MacFarling-Meure et al., 2006). The uncertainty of individual data points in other ice cores was in general less than 7 ppb (Flückiger et al., 2002; Schilt et al., 2014). For the instrumental measurements, we take the reported uncertainties of around 1 ppb. For 58 times, more than one data point for the same age exists. These duplicates are averaged reducing the number of N_2O data to $n = 2344$. Using an estimate of the radiative forcing of N_2O, which neglects the interacting effects of CH_4 and N_2O,

$$\Delta R_{[N_2O]} \sim 0.12 \cdot (\sqrt{N_2O/\text{ppb}} - \sqrt{272})\ \text{W m}^{-2}, \quad (5)$$

(Myhre et al., 1998), we estimate that the 1σ error in N_2O is related to an uncertainty in the radiative forcing of about $0.04\ \text{W m}^{-2}$, slightly larger than the uncertainty in ΔR related to the CH_4 data. Comparing the different values of N_2O

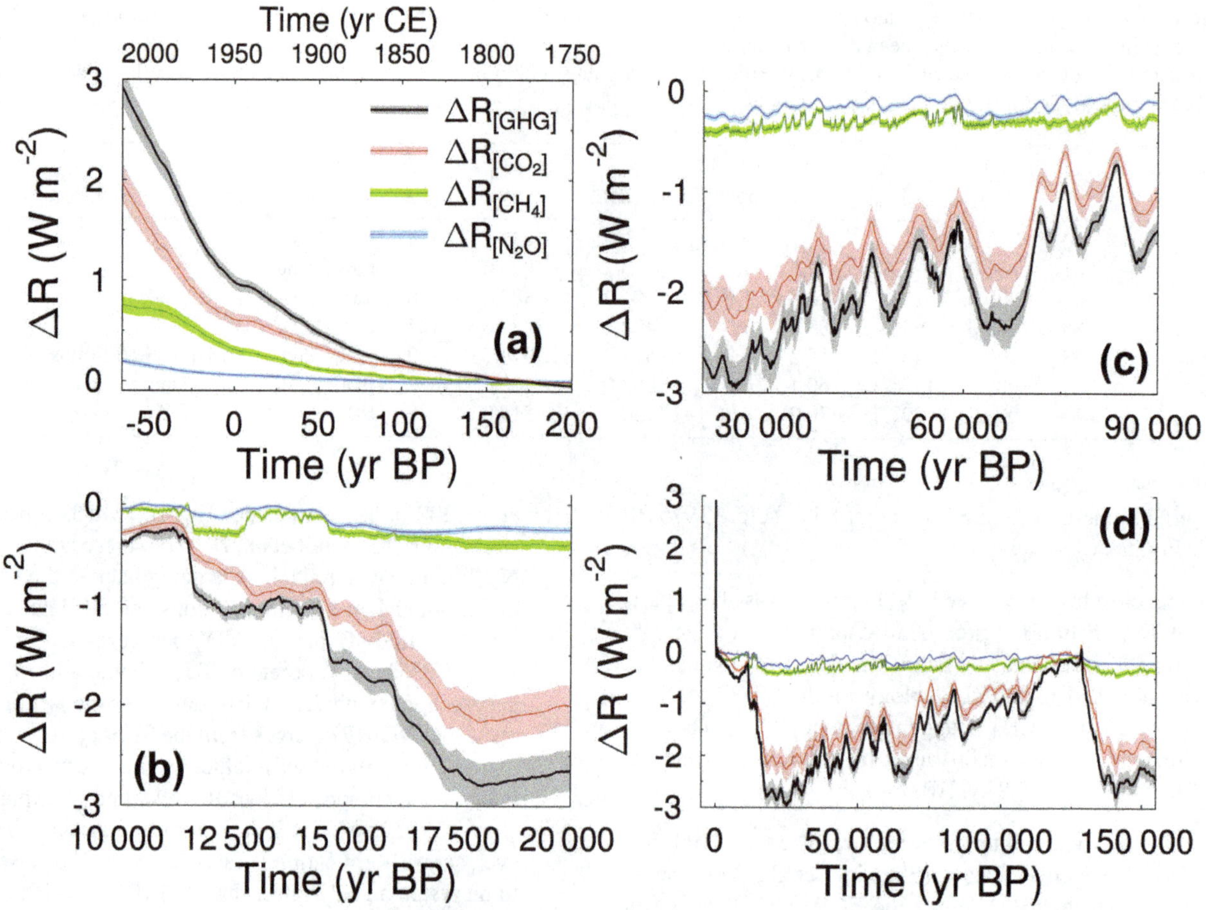

Figure 7. Calculated radiative forcing of CO_2, CH_4, N_2O, and their sum ($\Delta R_{[GHG]}$), including 1σ-uncertainty bands. In addition to the uncertainty of the spline, further uncertainties need to be considered: 10 % relative uncertainty contained in the simplified Eqs. (3)–(5); $0.2\,W\,m^{-2}$ uncertainty in $\Delta R_{[CH_4]}$ and $\Delta R_{[N_2O]}$ due to the omission of interaction effects; 5 % uncertainty in efficiency of CH_4 (for more details, see Köhler et al., 2010). The total uncertainty of the $\Delta R_{[GHG]}$ is calculated as the square root of the sum of squares of the uncertainties of the individual GHGs. The calculations are based on Eqs. (3)–(5) given in the text (following Myhre et al., 1998). Sub-figures focus on specific time windows: **(a)** anthropogenic rise since 1750 CE; **(b)** Termination I; **(c)** 20–90 kyr BP, including the abrupt changes during D/O event; **(d)** full record from 2016 CE to 156 kyr BP – here N_2O was kept constant beyond 134 kyr BP.

in Talos Dome and NGRIP for same intervals reveals differences on the order of about 10 ppb (e.g. Fig. 6c–f), suggesting that the ice-core-specific values of N_2O contain an intrinsic uncertainty which is comparable to the measurement error.

The mean temporal resolution (11-point running mean) of the underlying N_2O data is around 50 years across large parts of the last glacial cycle (15–60 kyr BP), with slightly lower resolution of 100 years in the Holocene and between 60 and 115 kyr BP. In MIS5.5, Termination II, and the penultimate glacial maximum, the mean temporal resolution rises to ~ 500 years (Fig. 5c). Based on this distribution of Δt the prescribed cutoff periods for the spline vary for seven different intervals between 4 (for the instrumental period) and 5000 years (for data older than 117 kyr BP). For the majority of the data (400 yr BP to 117 kyr BP), a P_c between 500 and 2000 years is prescribed. More details on the spline are found in Table 8.

The total 1σ uncertainty of the spline varies between 1 and 6 ppb (Fig. 5b). This uncertainty is mainly based on σ_3 (the error related to the Monte Carlo statistics) for periods younger than the LGM. For intervals older than the LGM, the main uncertainty is σ_1, the error related to the cutoff period.

If compared with the N_2O compilation used within CMIP6 (Meinshausen et al., 2017) both approaches largely agree for instrumental times (Fig. 6a). Further back in time during the last 2 kyr, both approaches rely on the same data: the published Law Dome/Cape Grim N_2O data (MacFarling-Meure et al., 2006). Interestingly, both time series differ by up to 6 ppb between 0.7 and 2.0 kyr BP (Fig. 6b). This difference is in the range of the ice core data uncertainty, and therefore still small, but we have no ready explanation. The records used in CMIP6 have a higher N_2O concentration than all data from Law Dome or other SH ice cores (Fig. 6b), for some unknown reason.

The N_2O data used as starting values in the PMIP4 experiments 1850 CE, 6 kyr, 21 kyr, and 127 kyr (Ivanovic et al., 2016; Otto-Bliesner et al., 2016) agree within 1 or 2 ppb with values based on our calculated spline; only for 21 kyr is the offset with 6 ppb greater (Table 4).

4 Data availability

Data connected with this paper are available in the scientific database PANGAEA.

In detail, for each of three GHGs the following data are available.

- Final, compiled raw data (t_j, y_j, and v_j, corresponding to time, value, and assumed 1σ error), including the data source, as described in the article.
- Preprocessed raw data (averaging of duplicate entries for similar times).
- Calculated splines with time steps of $\Delta t = 1$ year, including the 1σ total uncertainty.
- Corresponding radiative forcing based on the simplified Eqs. (3)–(5), including the 1σ total uncertainty.

When using these data, please consider citing the original publications from which the data underlying this compilation have been taken.

5 Conclusions

We have compiled available greenhouse gas records and, by calculating a smoothing spline, we were able to provide continuous records over the last glacial cycle, starting from the beginning of the year 2016 CE and going back to 134 kyr BP (for N_2O) and to 156 kyr BP (for CO_2 and CH_4). These records should serve as boundary conditions to calculate the greenhouse gas radiative forcing in transient climate simulations as planned, for example, in the German project PALMOD, or they might be used in the Last Deglaciation experiment within PMIP4 (Ivanovic et al., 2016) or in other future model intercomparison projects.

The resulting radiative forcing of the three GHGs calculated here with the simplified non-interacting Eqs. (3)–(5) has relative uncertainties of $\sim 10\,\%$ (Forster et al., 2007). These equations have also been used in Köhler et al. (2010) and Joos and Spahni (2008), while Schilt et al. (2010b) used a different equation for $\Delta R_{[CO_2]}$. The latter two studies furthermore include the interacting effects of CH_4 and N_2O (which we ignore here) but neglect the 40 % increase in CH_4 radiative forcing due to indirect effects of CH_4 (Hansen et al., 2005). Our results are very similar to recent calculations based on a complete and revised set of simplified equations, which also consider interacting effects between the three GHGs (Etminan et al., 2016), with differences between old and new expressions in $\Delta R_{[CO_2]}$, $\Delta R_{[CH_4]}$, and $\Delta R_{[N_2O]}$ of less than 0.01, 0.04, and 0.02 W m^{-2}, respectively. While the differences in $\Delta R_{[CO_2]}$ and $\Delta R_{[N_2O]}$ lie within their uncertainty bands, $\Delta R_{[CH_4]}$ is slightly higher, leading to a revised relative uncertainty of 14 % (Etminan et al., 2016). We have refrained from applying the new equations throughout the study, since the amplification in $\Delta R_{[CH_4]}$ by 40 % through indirect effects of CH_4 (Hansen et al., 2005) is not considered. We prefer to estimate the radiative forcing attributable to one of the three GHGs individually, without interacting effects. Our forcing calculations clearly illustrate the dominant contribution of CO_2, which is responsible for about two-thirds of the total radiative forcing $\Delta R_{[GHG]}$ during both the anthropogenic rise (Fig. 7a) and the reduction during the LGM (Fig. 7b). High-resolution variability in CH_4 (captured due to smaller cutoff periods during spline calculations than for CO_2) also imposes some fine-scale structure on the overall GHG radiative forcing (Fig. 7c, d); however, the dominant features are still driven by CO_2.

Appendix A

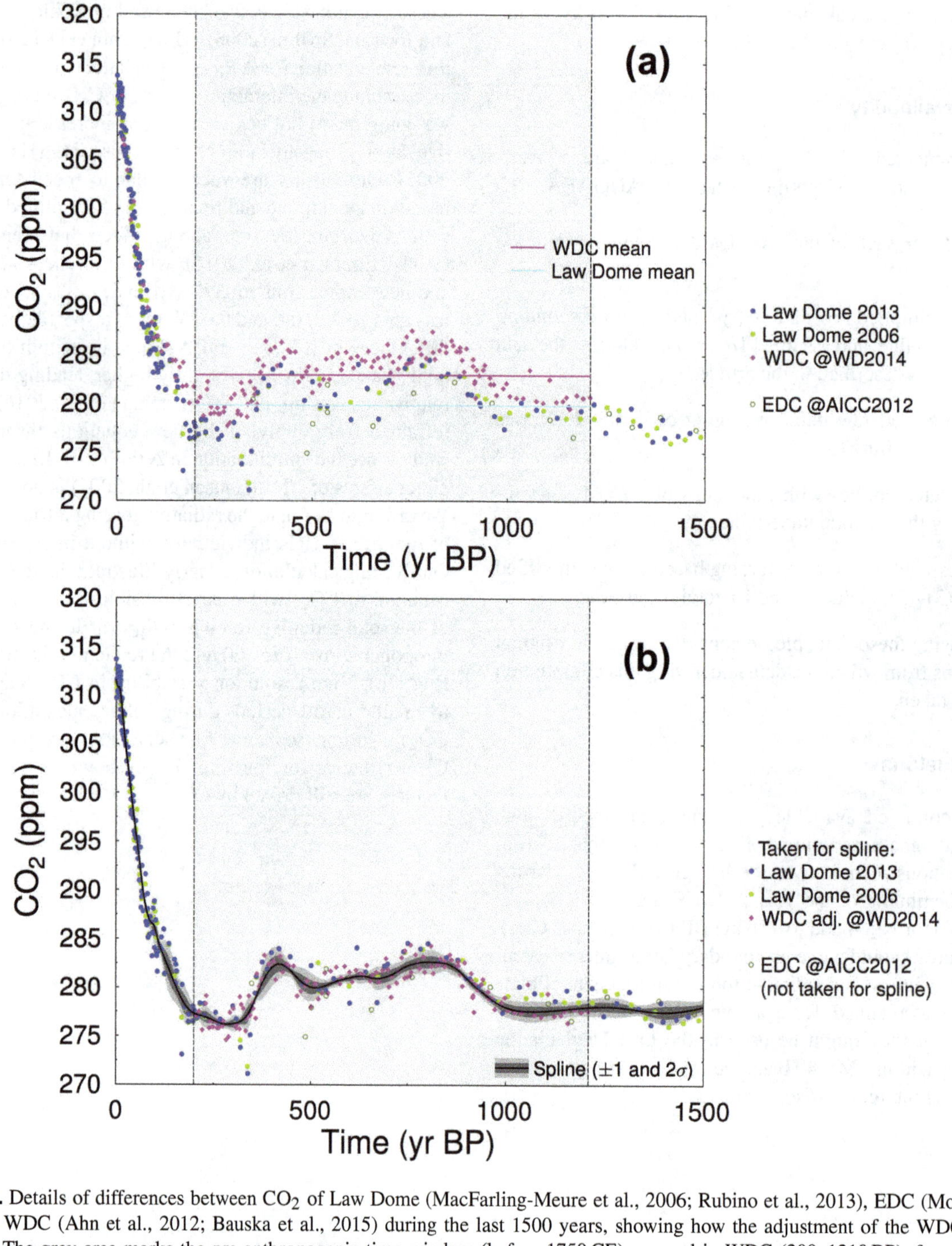

Figure A1. Details of differences between CO_2 of Law Dome (MacFarling-Meure et al., 2006; Rubino et al., 2013), EDC (Monnin et al., 2004), and WDC (Ahn et al., 2012; Bauska et al., 2015) during the last 1500 years, showing how the adjustment of the WDC has been calculated. The grey area marks the pre-anthropogenic time window (before 1750 CE) covered in WDC (200–1210 BP), from which the difference in CO_2 to WDC and Law Dome records has been determined. Horizontal lines mark the mean values for the different ice cores (cyan: Law Dome (all data); magenta: WDC). The mean offset between the WDC and Law Dome of 3.13 ppm is subtracted from the WDC data in **(b)**.

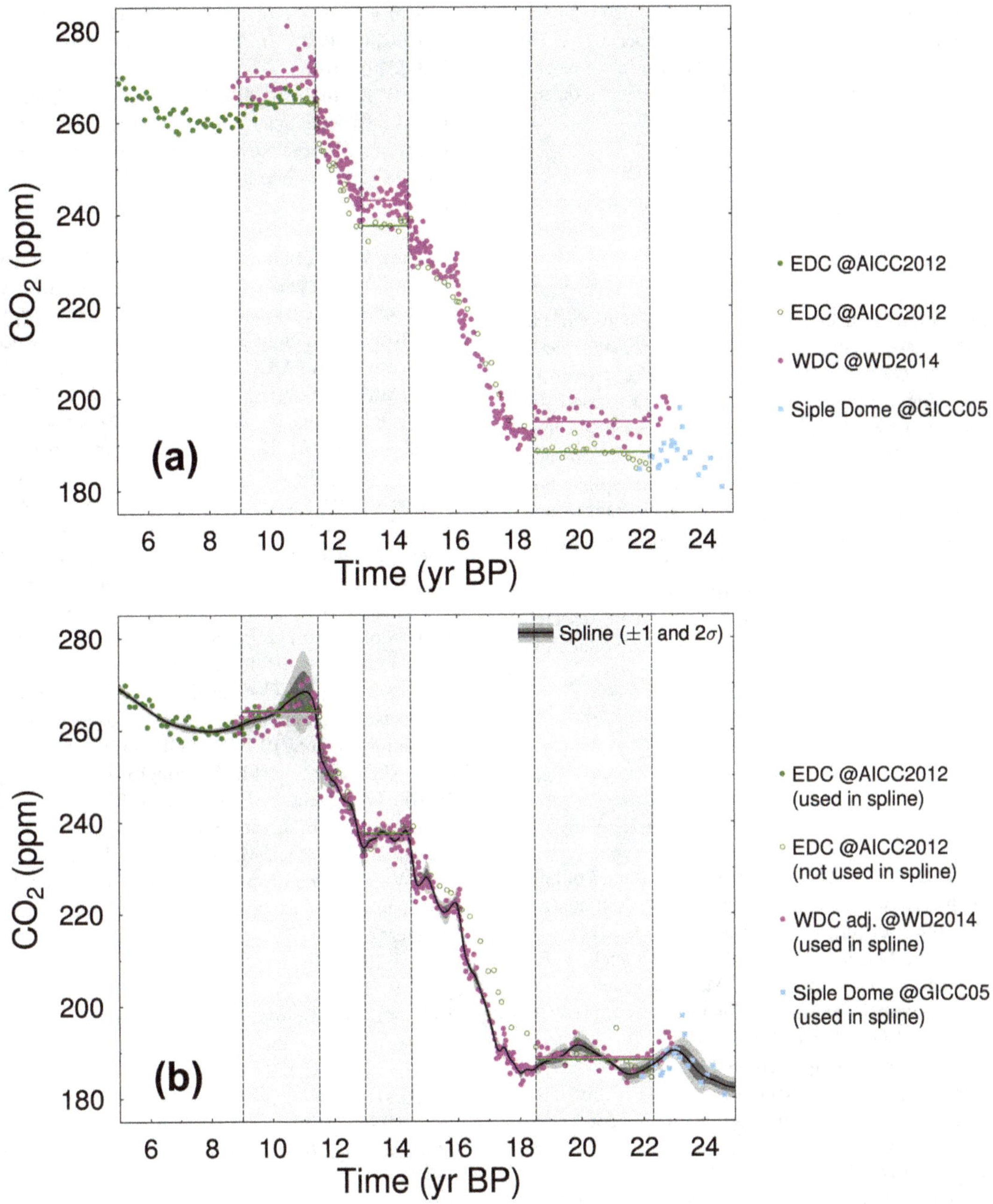

Figure A2. Details of differences between CO_2 of EDC (Monnin et al., 2001, 2004) and WDC (Marcott et al., 2014) during Termination I, showing how the adjustment of the WDC has been calculated. Grey areas mark the three time windows with relatively stable CO_2 from which the difference in CO_2 to both records has been determined. Horizontal lines mark the mean values for the different ice cores (green: EDC; magenta: WDC). The duration-weighted mean offset between the WDC and EDC of 6.06 ppm is subtracted from the WDC data in **(b)**.

Author contributions. PK initiated the work, compiled the data, calculated the spline, and led the writing of the paper. CNA, JS, TFS, and HF contributed specific insights on the data selection and advised on the spline smoothing. All co-authors commented on and improved the initial draft.

Competing interests. The authors declare that they have no conflict of interest.

Acknowledgements. We thank NOAA for the availability of the instrumental GHG data, specifically Ed Dlugokencky and Pieter Tans, NOAA/ESRL (www.esrl.noaa.gov/gmd/ccgg/trends/), and Ralph Keeling, Scripps Institution of Oceanography (scrippsco2.ucsd.edu/). Furthermore, we thank Fortunat Joos for some earlier thoughts on similar GHG data compilations and for providing the Fortran routines to calculate the spline fits; Jenny Roberts for improving the English; Thomas Bauska and two anonymous reviewers for their comments; and Rainer Sieger (PANGAEA) for comments on the uploaded data, which led to the preparation of Table 1. This is a contribution to the German BMBF project PALMOD. Long-term support of ice core research at the University of Bern by the Swiss National Science Foundation (SNF) is gratefully acknowledged.

Edited by: Alexander Kokhanovsky

References

Ahn, J. and Brook, E. J.: Siple Dome ice reveals two modes of millennial CO_2 change during the last ice age, Nature Communications, 5, 3723, https://doi.org/10.1038/ncomms4723, 2014.

Ahn, J., Brook, E. J., Mitchell, L., Rosen, J., McConnell, J. R., Taylor, K., Etheridge, D., and Rubino, M.: Atmospheric CO_2 over the last 1000 years: A high-resolution record from the West Antarctic Ice Sheet (WAIS) Divide ice core, Global Biogeochem. Cy., 26, GB2027, https://doi.org/10.1029/2011GB004247, 2012.

Anklin, M., Barnola, J. M., Schwander, J., Stauffer, B., and Raynaud, D.: Processes affecting the CO_2 concentrations measured in Greenland ice, Tellus B, 47, 461–470, https://doi.org/10.3402/tellusb.v47i4.16061, 1995.

Baggenstos, D., Bauska, T. K., Severinghaus, J. P., Lee, J. E., Schaefer, H., Buizert, C., Brook, E. J., Shackleton, S., and Petrenko, V. V.: Atmospheric gas records from Taylor Glacier, Antarctica, reveal ancient ice with ages spanning the entire last glacial cycle, Clim. Past Discuss., https://doi.org/10.5194/cp-2017-25, in review, 2017.

Baumgartner, M., Schilt, A., Eicher, O., Schmitt, J., Schwander, J., Spahni, R., Fischer, H., and Stocker, T. F.: High-resolution interpolar difference of atmospheric methane around the Last Glacial Maximum, Biogeosciences, 9, 3961–3977, https://doi.org/10.5194/bg-9-3961-2012, 2012.

Baumgartner, M., Kindler, P., Eicher, O., Floch, G., Schilt, A., Schwander, J., Spahni, R., Capron, E., Chappellaz, J., Leuenberger, M., Fischer, H., and Stocker, T. F.: NGRIP CH_4 concentration from 120 to 10 kyr before present and its relation to a $\delta^{15}N$ temperature reconstruction from the same ice core, Clim. Past, 10, 903–920, https://doi.org/10.5194/cp-10-903-2014, 2014.

Bauska, T. K., Joos, F., Mix, A. C., Roth, R., Ahn, J., and Brook, E. J.: Links between atmospheric carbon dioxide, the land carbon reservoir and climate over the past millennium, Nat. Geosci., 8, 383–387, https://doi.org/10.1038/ngeo2422, 2015.

Bazin, L., Landais, A., Lemieux-Dudon, B., Toyé Mahamadou Kele, H., Veres, D., Parrenin, F., Martinerie, P., Ritz, C., Capron, E., Lipenkov, V., Loutre, M.-F., Raynaud, D., Vinther, B., Svensson, A., Rasmussen, S. O., Severi, M., Blunier, T., Leuenberger, M., Fischer, H., Masson-Delmotte, V., Chappellaz, J., and Wolff, E.: An optimized multi-proxy, multi-site Antarctic ice and gas orbital chronology (AICC2012): 120–800 ka, Clim. Past, 9, 1715–1731, https://doi.org/10.5194/cp-9-1715-2013, 2013.

Bereiter, B., Lüthi, D., Siegrist, M., Schüpbach, S., Stocker, T. F., and Fischer, H.: Mode change of millennial CO_2 variability during the last glacial cycle associated with a bipolar marine carbon seesaw, P. Natl. Acad. Sci., 109, 9755–9760, https://doi.org/10.1073/pnas.1204069109, 2012.

Blunier, T. and Brook, E. J.: Timing of millennial-scale climate change in Antarctica and Greenland during the last glacial period, Science, 291, 109–112, https://doi.org/10.1126/science.291.5501.109, 2001.

Bruno, M. and Joos, F.: Terrestrial carbon storage during the past 200 years: A Monte Carlo analysis of CO_2 data from ice core and atmospheric measurements, Global Biogeochem. Cy., 11, 111–124, https://doi.org/10.1029/96GB03611, 1997.

Buizert, C., Cuffey, K. M., Severinghaus, J. P., Baggenstos, D., Fudge, T. J., Steig, E. J., Markle, B. R., Winstrup, M., Rhodes, R. H., Brook, E. J., Sowers, T. A., Clow, G. D., Cheng, H., Edwards, R. L., Sigl, M., McConnell, J. R., and Taylor, K. C.: The WAIS Divide deep ice core WD2014 chronology – Part 1: Methane synchronization (68–31 ka BP) and the gas age–ice age difference, Clim. Past, 11, 153–173, https://doi.org/10.5194/cp-11-153-2015, 2015.

Capron, E., Landais, A., Lemieux-Dudon, B., Schilt, A., Masson-Delmotte, V., Buiron, D., Chappellaz, J., Dahl-Jensen, D., Johnsen, S., Leuenberger, M., Loulergue, L., and Oerter, H.: Synchronising EDML and NorthGRIP ice cores using $\delta^{18}O$ of atmospheric oxygen ($\delta^{18}O_{atm}$) and CH_4 measurements over MIS5 (80-123 kyr), Quaternary Sci. Rev., 29, 222- 234, https://doi.org/10.1016/j.quascirev.2009.07.014, 2010.

Chappellaz, J., Stowasser, C., Blunier, T., Baslev-Clausen, D., Brook, E. J., Dallmayr, R., Faïn, X., Lee, J. E., Mitchell, L. E., Pascual, O., Romanini, D., Rosen, J., and Schüpbach, S.: High-resolution glacial and deglacial record of atmospheric methane by continuous-flow and laser spectrometer analysis along the NEEM ice core, Clim. Past, 9, 2579–2593, https://doi.org/10.5194/cp-9-2579-2013, 2013.

Dällenbach, A., Blunier, B., Flückiger, J., Stauffer, B., Chappellaz, J., and Raynaud, D.: Changes in the atmospheric CH_4 gradient between Greenland and Antarctica during the last glacial and the transition to the Holocene, Geophys. Res. Lett., 27, 1005–1008, https://doi.org/10.1029/1999GL010873, 2000.

Dlugokencky, E., Lang, P., Crotwell, A., Mund, J., Crotwell, M., and Thoning, K.: Atmospheric Methane Dry Air Mole Fractions from the NOAA ESRL Carbon Cycle Cooperative Global Air Sampling Network, 1983–2015, Version: 2016-07-

07, available at: ftp://aftp.cmdl.noaa.gov/data/trace_gases/ch4/flask/surface/ (last access: 17 August 2016), 2016a.

Dlugokencky, E., Lang, P., Mund, J., Crotwell, A., Crotwell, M., and Thoning, K.: Atmospheric Carbon Dioxide Dry Air Mole Fractions from the NOAA ESRL Carbon Cycle Cooperative Global Air Sampling Network, 1968–2015, Version: 2016-08-30, available at: ftp://aftp.cmdl.noaa.gov/data/trace_gases/co2/flask/surface/ (last access: 17 August 2016), 2016b.

Enting, I. G.: On the use of smoothing splines to filter CO_2 data, J. Geophys. Res.-Atmos., 92, 10977–10984, https://doi.org/10.1029/JD092iD09p10977, 1987.

Enting, I. G., Trudinger, C. M., and Etheridge, D. M.: Propoagating data uncertainty through smoothing spline fits, Tellus, 58B, 305–309, https://doi.org/10.1111/j.1600-0889.2006.00193.x, 2006.

Etminan, M., Myhre, G., Highwood, E. J., and Shine, K. P.: Radiative forcing of carbon dioxide, methane, and nitrous oxide: A significant revision of the methane radiative forcing, Geophys. Res. Lett., 43, 12614–12,23, https://doi.org/10.1002/2016GL071930, 2016.

Flückiger, J., Monnin, E., Stauffer, B., Schwander, J., and Stocker, T. F.: High-resolution Holocene N_2O ice core record and its relationship with CH_4 and CO_2, Global Biogeochem. Cy., 16, GB001417, https://doi.org/10.1029/2001GB001417, 2002.

Flückiger, J., Blunier, T., Stauffer, B., Chappellaz, J., Spahni, R., Kawamura, K., Schwander, J., Stocker, T. F., and Dahl-Jensen, D.: N_2O and CH_4 variations during the last glacial epoch: insights into global processes, Global Biogeochem. Cy., 18, GB1020, https://doi.org/10.1029/2003GB002122, 2004.

Forster, P., Ramaswamy, V., Artaxo, P., Berntsen, T., Betts, R., Fahey, D., Haywood, J., Lean, J., Lowe, D., Myhre, G., Nganga, J., Prinn, R., Raga, G., Schulz, M., and Dorland, R. V.: Changes in Atmospheric Constituents and in Radiative Forcing, in: Climate Change 2007: The Physical Science Basis. Contribution of Working Group I to the Fourth Assessment Report of the Intergovernmental Panel on Climate Change, edited by: Solomon, S., Qin, D., Manning, M., Chen, Z., Marquis, M., Averyt, K. B., Tignor, M., and Miller, H. L., Cambridge University Press, Cambridge, United Kingdom and New York, NY, USA, 129–234, 2007.

Hansen, J., Sato, M., Ruedy, R., Nazarenko, L., Lacis, A., Schmidt, G. A., Russell, G., Aleinov, I., Bauer, M., Bauer, S., Bell, N., Cairns, B., Canuto, V., Chandler, M., Cheng, Y., Genio, A. D., Faluvegi, G., Fleming, E., Friend, A., Hall, T., Jackman, C., Kelley, M., Kiang, N., Koch, D., Lean, J., Lerner, J., Lo, K., Menon, S., Miller, R., Minnis, P., Novakov, T., Oinas, V., Perlwitz, J., Perlwitz, J., Rind, D., Romanou, A., Shindell, D., Stone, P., Sun, S., Tausnev, N., Thresher, D., Wielicki, B., Wong, T., Yao, M., and Zhang, S.: Efficacy of climate forcings, J. Geophys. Res., 110, D18104, https://doi.org/10.1029/2005JD005776, 2005.

Ivanovic, R. F., Gregoire, L. J., Kageyama, M., Roche, D. M., Valdes, P. J., Burke, A., Drummond, R., Peltier, W. R., and Tarasov, L.: Transient climate simulations of the deglaciation 21–9 thousand years before present (version 1) – PMIP4 Core experiment design and boundary conditions, Geosci. Model Dev., 9, 2563–2587, https://doi.org/10.5194/gmd-9-2563-2016, 2016.

Joos, F. and Spahni, R.: Rates of change in natural and anthropogenic radiative forcing over the past 20,000 years, P. Natl. Acad. Sci. USA, 105, 1425–1430, https://doi.org/10.1073/pnas.0707386105, 2008.

Köhler, P., Bintanja, R., Fischer, H., Joos, F., Knutti, R., Lohmann, G., and Masson-Delmotte, V.: What caused Earth's temperature variations during the last 800,000 years? Data-based evidences on radiative forcing and constraints on climate sensitivity, Quaternary Sci. Rev., 29, 129–145, https://doi.org/10.1016/j.quascirev.2009.09.026, 2010.

Köhler, P., Knorr, G., and Bard, E.: Permafrost thawing as a possible source of abrupt carbon release at the onset of the Bølling/Allerød, Nature Communications, 5, 5520, https://doi.org/10.1038/ncomms6520, 2014.

Köhler, P., Völker, C., Knorr, G., and Bard, E.: High Latitude Impacts on Deglacial CO_2: Southern Ocean Westerly Winds and Northern Hemisphere Permafrost Thawing, Nova Acta Leopoldina, 121, 135–140, 2015.

Landais, A., Caillon, N., Goujon, C., Grachev, A. M., Barnola, J. M., Chappellaz, J., Jouzel, J., Masson-Delmotte, V., and Leuenberger, M.: Quantification of rapid temperature change during DO event 12 and phasing with methane inferred from air isotopic measurements, Earth Planet. Sc. Lett., 225, 221–232, https://doi.org/10.1016/j.epsl.2004.06.009, 2004.

Lewis, S. L. and Maslin, M. A.: Defining the Anthropocene, Nature, 519, 171–180, https://doi.org/10.1038/nature14258, 2015.

Loulergue, L., Schilt, A., Spahni, R., Masson-Delmotte, V., Blunier, T., Lemieux, B., Barnola, J.-M., Raynaud, D., Stocker, T. F., and Chappellaz, J.: Orbital and millennial-scale features of atmospheric CH_4 over the past 800,000 years, Nature, 453, 383–386, https://doi.org/10.1038/nature06950, 2008.

Lourantou, A., Chappellaz, J., Barnola, J.-M., Masson-Delmotte, V., and Raynaud, D.: Changes in atmospheric CO_2 and its carbon isotopic ratio during the penultimate deglaciation, Quaternary Sci. Rev., 29, 1983–1992, https://doi.org/10.1016/j.quascirev.2010.05.002, 2010a.

Lourantou, A., Lavrič, J. V., Köhler, P., Barnola, J.-M., Michel, E., Paillard, D., Raynaud, D., and Chappellaz, J.: Constraint of the CO_2 rise by new atmospheric carbon isotopic measurements during the last deglaciation, Global Biogeochem. Cy., 24, GB2015, https://doi.org/10.1029/2009GB003545, 2010b.

Lüthi, D., Bereiter, B., Stauffer, B., Winkler, R., Schwander, J., Kindler, P., Leuenberger, M., Kipfstuhl, S., Capron, E., Landais, A., Fischer, H., and Stocker, T. F.: CO_2 and O_2/N_2 variations in and just below the bubble-clathrate transformation zone of Antarctic ice cores, Earth Planet. Sc. Lett., 297, 226–233, https://doi.org/10.1016/j.epsl.2010.06.023, 2010.

MacFarling-Meure, C., Etheridge, D., Trudinger, C., Langenfelds, R., van Ommen, T., Smith, A., and Elkins, J.: Law Dome CO_2, CH_4 and N_2O ice core records extended to 2000 years BP, Geophys. Res. Lett., 33, L14810, https://doi.org/10.1029/2006GL026152, 2006.

Marcott, S. A., Bauska, T. K., Buizert, C., Steig, E. J., Rosen, J. L., Cuffey, K. M., Fudge, T. J., Severinghaus, J. P., Ahn, J., Kalk, M. L., McConnell, J. R., Sowers, T., Taylor, K. C., White, J. W., and Brook, E. J.: Centennial Scale Changes in the Global Carbon Cycle During the Last Deglaciation, Nature, 514, 616–619, https://doi.org/10.1038/nature13799, 2014.

Meinshausen, M., Vogel, E., Nauels, A., Lorbacher, K., Meinshausen, N., Etheridge, D. M., Fraser, P. J., Montzka, S. A., Rayner, P. J., Trudinger, C. M., Krummel, P. B., Beyerle, U., Canadell, J. G., Daniel, J. S., Enting, I. G., Law, R. M., Lunder, C. R., O'Doherty, S., Prinn, R. G., Reimann, S., Rubino,

M., Velders, G. J. M., Vollmer, M. K., Wang, R. H. J., and Weiss, R.: Historical greenhouse gas concentrations for climate modelling (CMIP6), Geosci. Model Dev., 10, 2057–2116, https://doi.org/10.5194/gmd-10-2057-2017, 2017.

Mitchell, L., Brook, E., Lee, J. E., Buizert, C., and Sowers, T.: Constraints on the Late Holocene Anthropogenic Contribution to the Atmospheric Methane Budget, Science, 342, 964–966, https://doi.org/10.1126/science.1238920, 2013.

Mitchell, L. E., Brook, E. J., Sowers, T., McConnell, J. R., and Taylor, K.: Multidecadal variability of atmospheric methane, 1000–1800 C.E., J. Geophys. Res., 116, G02007, https://doi.org/10.1029/2010JG001441, 2011.

Monnin, E., Indermühle, A., Dällenbach, A., Flückiger, J., Stauffer, B., Stocker, T. F., Raynaud, D., and Barnola, J.-M.: Atmospheric CO_2 concentrations over the last glacial termination, Science, 291, 112–114, https://doi.org/10.1126/science.291.5501.112, 2001.

Monnin, E., Steig, E. J., Siegenthaler, U., Kawamura, K., Schwander, J., Stauffer, B., Stocker, T. F., Morse, D. L., Barnola, J.-M., Bellier, B., Raynaud, D., and Fischer, H.: Evidence for sustantial accumulation rate variability in Antarctica during the Holocene, through synchronization of CO_2 in the Taylor Dome, Dome C and DML ice cores, Earth Planet. Sc. Lett., 224, 45–54, https://doi.org/10.1016/j.epsl.2004.05.007, 2004.

Myhre, G., Highwood, E. J., Shine, K. P., and Stordal, F.: New estimates of radiative forcing due to well mixed greenhouse gases, Geophys. Res. Lett., 25, 2715–2718, https://doi.org/10.1029/98GL01908, 1998.

Otto-Bliesner, B. L., Braconnot, P., Harrison, S. P., Lunt, D. J., Abe-Ouchi, A., Albani, S., Bartlein, P. J., Capron, E., Carlson, A. E., Dutton, A., Fischer, H., Goelzer, H., Govin, A., Haywood, A., Joos, F., Legrande, A. N., Lipscomb, W. H., Lohmann, G., Mahowald, N., Nehrbass-Ahles, C., Pausata, F. S.-R., Peterschmitt, J.-Y., Phipps, S., and Renssen, H.: The PMIP4 contribution to CMIP6 – Part 2: Two Interglacials, Scientific Objective and Experimental Design for Holocene and Last Interglacial Simulations, Geosci. Model Dev. Discuss., https://doi.org/10.5194/gmd-2016-279, in review, 2016.

Rhodes, R. H., Faïn, X., Stowasser, C., Blunier, T., Chappellaz, J., McConnell, J. R., Romanini, D., Mitchell, L. E., and Brook, E. J.: Continuous methane measurements from a late Holocene Greenland ice core: Atmospheric and in-situ signals, Earth Planet. Sc. Lett., 368, 9–19, https://doi.org/10.1016/j.epsl.2013.02.034, 2013.

Rhodes, R. H., Brook, E. J., Chiang, J. C. H., Blunier, T., Maselli, O. J., McConnell, J. R., Romanini, D., and Severinghaus, J. P.: Enhanced tropical methane production in response to iceberg discharge in the North Atlantic, Science, 348, 1016–1019, https://doi.org/10.1126/science.1262005, 2015.

Rubino, M., Etheridge, D. M., Trudinger, C. M., Allison, C. E., Battle, M. O., Langenfelds, R. L., Steele, L. P., Curran, M., Bender, M., White, J. W. C., Jenk, T. M., Blunier, T., and Francey, R. J.: A revised 1000-year atmospheric $\delta^{13}C$-CO_2 record from Law Dome and South Pole, Antarctica, J. Geophys. Res.-Atmos., 118, 8482–8499, https://doi.org/10.1002/jgrd.50668, 2013.

Ruth, U., Barnola, J.-M., Beer, J., Bigler, M., Blunier, T., Castellano, E., Fischer, H., Fundel, F., Huybrechts, P., Kaufmann, P., Kipfstuhl, S., Lambrecht, A., Morganti, A., Oerter, H., Parrenin, F., Rybak, O., Severi, M., Udisti, R., Wilhelms, F., and Wolff, E.: "EDML1": a chronology for the EPICA deep ice core from Dronning Maud Land, Antarctica, over the last 150 000 years, Clim. Past, 3, 475–484, https://doi.org/10.5194/cp-3-475-2007, 2007.

Schilt, A., Baumgartner, M., Blunier, T., Schwander, J., Spahni, R., Fischer, H., and Stocker, T. F.: Glacial-interglacial and millennial-scale variations in the atmospheric nitrous oxide concentration during the last 800 000 years, Quaternary Sci. Rev., 29, 182–192, https://doi.org/10.1016/j.quascirev.2009.03.011, 2010a.

Schilt, A., Baumgartner, M., Schwander, J., Buiron, D., Capron, E., Chappellaz, J., Loulergue, L., Schüpbach, S., Spahni, R., Fischer, H., and Stocker, T. F.: Atmospheric nitrous oxide during the last 140,000 years, Earth Planet. Sci.e Lett., 300, 33–43, https://doi.org/10.1016/j.epsl.2010.09.027, 2010b.

Schilt, A., Baumgartner, M., Eicher, O., Chappellaz, J., Schwander, J., Fischer, H., and Stocker, T. F.: The response of atmospheric nitrous oxide to climate variations during the last glacial period, Geophys. Res. Lett., 40, 1888–1893, https://doi.org/10.1002/grl.50380, 2013.

Schilt, A., Brook, E. J., Bauska, T. K., Baggenstos, D., Fischer, H., Joos, F., Petrenko, V. V., Schaefer, H., Schmitt, J., Severinghaus, J. P., Spahni, R., and Stocker, T. F.: Isotopic constraints on marine and terrestrial N_2O emissions during the last deglaciation, Nature, 516, 234–237, https://doi.org/10.1038/nature13971, 2014.

Schmitt, J., Schneider, R., Elsig, J., Leuenberger, D., Lourantou, A., Chappellaz, J., Köhler, P., Joos, F., Stocker, T. F., Leuenberger, M., and Fischer, H.: Carbon isotope constraints on the deglacial CO_2 rise from ice cores, Science, 336, 711–714, https://doi.org/10.1126/science.1217161, 2012.

Schneider, R., Schmitt, J., Köhler, P., Joos, F., and Fischer, H.: A reconstruction of atmospheric carbon dioxide and its stable carbon isotopic composition from the penultimate glacial maximum to the last glacial inception, Clim. Past, 9, 2507–2523, https://doi.org/10.5194/cp-9-2507-2013, 2013.

Schüpbach, S., Federer, U., Kaufmann, P. R., Hutterli, M. A., Buiron, D., Blunier, T., Fischer, H., and Stocker, T. F.: A New Method for High-Resolution Methane Measurements on Polar Ice Cores Using Continuous Flow Analysis, Environ. Sci. Technol., 43, 5371–5376, https://doi.org/10.1021/es9003137, 2009.

Sigl, M., Fudge, T. J., Winstrup, M., Cole-Dai, J., Ferris, D., McConnell, J. R., Taylor, K. C., Welten, K. C., Woodruff, T. E., Adolphi, F., Bisiaux, M., Brook, E. J., Buizert, C., Caffee, M. W., Dunbar, N. W., Edwards, R., Geng, L., Iverson, N., Koffman, B., Layman, L., Maselli, O. J., McGwire, K., Muscheler, R., Nishiizumi, K., Pasteris, D. R., Rhodes, R. H., and Sowers, T. A.: The WAIS Divide deep ice core WD2014 chronology – Part 2: Annual-layer counting (0–31 ka BP), Clim. Past, 12, 769-786, https://doi.org/10.5194/cp-12-769-2016, 2016.

Spahni, R., Chappellaz, J., Stocker, T. F., Loulergue, L., Hausammann, G., Kawamura, K., Flückiger, J., Schwander, J., Raynaud, D., Masson-Delmotte, V., and Jouzel, J.: Atmospheric methane and nitrous oxide of the late Pleistocene from Antarctic ice cores, Science, 310, 1317–1321, https://doi.org/10.1126/science.1120132, 2005.

Stauffer, B., Blunier, T., Dällenbach, A., Indermühle, A., Schwander, J., Stocker, T. F., Tschumi, J., Chappellaz, J., Raynaud, D., Hammer, C. U., and Claussen, H. B.: Atmospheric CO_2 concentration and millennial-scale climate change during the last glacial period, Nature, 392, 59–62, https://doi.org/10.1038/32133, 1998.

Stauffer, B., Flückiger, J., Monnin, E., Schwander, J., Barnola, J.-M., and Chappellaz, J.: Atmospheric CO_2, CH_4 and N_2O records over the past 60000 years based on the comparison of different polar ice cores, Ann. Glaciol., 35, 202–208, 2002.

Steffen, W., Leinfelder, R., Zalasiewicz, J., Waters, C. N., Williams, M., Summerhayes, C., Barnosky, A. D., Cearreta, A., Crutzen, P., Edgeworth, M., Ellis, E. C., Fairchild, I. J., Galuszka, A., Grinevald, J., Haywood, A., Ivar do Sul, J., Jeandel, C., McNeill, J., Odada, E., Oreskes, N., Revkin, A., Richter, D. d., Syvitski, J., Vidas, D., Wagreich, M., Wing, S. L., Wolfe, A. P., and Schellnhuber, H.: Stratigraphic and Earth System approaches to defining the Anthropocene, Earth's Future, 4, 324–345 https://doi.org/10.1002/2016EF000379, 2016.

Veres, D., Bazin, L., Landais, A., Toyé Mahamadou Kele, H., Lemieux-Dudon, B., Parrenin, F., Martinerie, P., Blayo, E., Blunier, T., Capron, E., Chappellaz, J., Rasmussen, S. O., Severi, M., Svensson, A., Vinther, B., and Wolff, E. W.: The Antarctic ice core chronology (AICC2012): an optimized multi-parameter and multi-site dating approach for the last 120 thousand years, Clim. Past, 9, 1733–1748, https://doi.org/10.5194/cp-9-1733-2013, 2013.

WAIS Divide Project Members: Precise interpolar phasing of abrupt climate change during the last ice age, Nature, 520, 661–665, https://doi.org/10.1038/nature14401, 2015.

Williams, M., Zalasiewicz, J., Waters, C. N., Edgeworth, M., Bennett, C., Barnosky, A. D., Ellis, E. C., Ellis, M. A., Cearreta, A., Haff, P. K., Ivar do Sul, J. A., Leinfelder, R., McNeill, J. R., Odada, E., Oreskes, N., Revkin, A., Richter, D. d., Steffen, W., Summerhayes, C., Syvitski, J. P., Vidas, D., Wagreich, M., Wing, S. L., Wolfe, A. P., and Zhisheng, A.: The Anthropocene: a conspicuous stratigraphical signal of anthropogenic changes in production and consumption across the biosphere, Earth's Future, 4, 34–53, https://doi.org/10.1002/2015EF000339, 2016.

7

The SISAL database: a global resource to document oxygen and carbon isotope records from speleothems

Kamolphat Atsawawaranunt[1], Laia Comas-Bru[2], Sahar Amirnezhad Mozhdehi[2], Michael Deininger[2,3], Sandy P. Harrison[1], Andy Baker[4], Meighan Boyd[5], Nikita Kaushal[6], Syed Masood Ahmad[7,24], Yassine Ait Brahim[8], Monica Arienzo[9], Petra Bajo[10], Kerstin Braun[11], Yuval Burstyn[12,13], Sakonvan Chawchai[14], Wuhui Duan[15], István Gábor Hatvani[16], Jun Hu[17], Zoltán Kern[16], Inga Labuhn[18], Matthew Lachniet[19], Franziska A. Lechleitner[6], Andrew Lorrey[20], Carlos Pérez-Mejías[21], Robyn Pickering[22], Nick Scroxton[23], and SISAL Working Group Members*

[1]Centre for Past Climate Change and School of Archaeology, Geography & Environmental Sciences, Reading University, Whiteknights, Reading, RG6 6AH, UK
[2]School of Earth Sciences, University College Dublin, Belfield, Dublin 4, Ireland
[3]Institute of Geosciences, Johannes-Gutenberg-University Mainz, Johann-Joachim-Becher-Weg 21, 55128 Mainz, Germany
[4]School of Biological, Earth and Environmental Sciences, University of New South Wales, Kensington 2052, Australia
[5]Department of Earth Sciences, Royal Holloway University of London, Egham, Surrey, TW20 0EX, UK
[6]Department of Earth Sciences, University of Oxford, South Parks Road, Oxford, OX1 3AN, UK
[7]CSIR – National Geophysical Research Institute, Uppal Road, 500 007 Hyderabad, India
[8]Institute of Global Environmental Change, Xi'an Jiaotong University, Xi'an, Shaanxi, China
[9]Division of Hydrologic Sciences, Desert Research Institute, 2215 Raggio Parkway, 89512 Reno, NV, USA
[10]School of Geography, The University of Melbourne, 3010 Victoria, Australia
[11]Institute of Human Origins, Arizona State University, P.O. Box 874101, 85287 Tempe, AZ, USA
[12]Institute of Sciences, Hebrew University of Jerusalem, Edmond J. Safra Campus, Givat Ram, 91904 Jerusalem, Israel
[13]Geological Survey of Israel, 30 Malkhe Israel, 95501 Jerusalem, Israel
[14]MESA Research Unit, Department of Geology, the Faculty of Science, Chulalongkorn University, 10330 Bangkok, Thailand
[15]Institute of Geology and Geophysics, Chinese Academy of Sciences, University of Chinese Academy of Sciences, No. 19 Beitucheng West Road, Chaoyang District, Beijing, China
[16]Institute for Geological and Geochemical Research, Research Centre for Astronomy and Earth Sciences, Hungarian Academy of Sciences, Budaörsi út 45, 1112 Budapest, Hungary
[17]Department of Earth Sciences, University of Southern California, 3651 Trousdale Parkway, 90089 Los Angeles, CA, USA
[18]Institute of Geography, University of Bremen, Bremen, Germany
[19]Dept. of Geoscience, University of Nevada, Box 4010, 89154 Las Vegas, NV, USA
[20]Climate, Atmosphere and Hazards Centre, National Institute of Water & Atmospheric Research, 41 Market Place, Central Business District, Auckland, New Zealand
[21]Department of Geoenvironmental Processes and Global Change, Pyrenean Institute of Ecology (IPE-CSIC), Avda. Montañana 1005, 50059 Zaragoza, Spain
[22]Department of Geological Sciences, Human Evolutionary Research Institute, University of Cape Town, 7701 Rondebosch, Cape Town, South Africa
[23]Department of Geosciences, University of Massachusetts Amherst, 611 North Pleasant Street, 01003-9297 Amherst, MA, USA
[24]Department of Geography, Faculty of Natural Sciences, Jamia Millia Islamia, New Delhi 110025, India

*A full list of authors and their affiliations appears at the end of the paper.

Correspondence: Laia Comas-Bru (laia.comas-bru@ucdconnect.ie)

Abstract. Stable isotope records from speleothems provide information on past climate changes, most particularly information that can be used to reconstruct past changes in precipitation and atmospheric circulation. These records are increasingly being used to provide "out-of-sample" evaluations of isotope-enabled climate models. SISAL (Speleothem Isotope Synthesis and Analysis) is an international working group of the Past Global Changes (PAGES) project. The working group aims to provide a comprehensive compilation of speleothem isotope records for climate reconstruction and model evaluation. The SISAL database contains data for individual speleothems, grouped by cave system. Stable isotopes of oxygen and carbon ($\delta^{18}O$, $\delta^{13}C$) measurements are referenced by distance from the top or bottom of the speleothem. Additional tables provide information on dating, including information on the dates used to construct the original age model and sufficient information to assess the quality of each data set and to erect a standardized chronology across different speleothems. The metadata table provides location information, information on the full range of measurements carried out on each speleothem and information on the cave system that is relevant to the interpretation of the records, as well as citations for both publications and archived data. The compiled data are available at https://doi.org/10.17864/1947.147.

1 Introduction

Speleothems are inorganic carbonate deposits (mostly calcite and aragonite) that grow in caves and form from drip water supersaturated with respect to $CaCO_3$. Speleothems are highly suitable for radiometric dating using uranium-series disequilibrium techniques. Since they form through continuous accretion, speleothems can provide a highly resolved record of environmental conditions, generally with a temporal resolution ranging from seasonal scale to 100 years, depending on sampling resolution.

Speleothem records are one of the types of record widely used to reconstruct past climate changes (see Bradley, 2015 for an overview of other methods). Speleothem growth is, in itself, an indicator of precipitation availability (Ayliffe et al., 1998; Wang et al., 2004), and variations in annual growth increments have been interpreted as an index of precipitation amount (Fleitmann et al., 2004; Polyak and Asmerom, 2001; Trouet et al., 2009). Many different types of measurements have been made on speleothems, but the most common are the stable isotopes of oxygen and carbon ($\delta^{18}O$, $\delta^{13}C$). Although the interpretation of such records can be complicated, for samples that are deposited close to equilibrium, changes in $\delta^{18}O$ are primarily a signal of changes in precipitation amount and source, precipitation temperature, and cave temperature (Affek et al., 2014; Hu et al., 2008; McDermott, 2004; Wang et al., 2008) and have been widely used to reconstruct changing atmospheric circulation patterns (e.g. Bar-Matthews et al., 1999; Cai et al., 2012, 2015; Luetscher et al., 2015; Spötl and Mangini, 2002; Trouet et al., 2009). Changes in $\delta^{13}C$ are a more indirect signal of precipitation changes. If not affected by non-equilibrium deposition (Baker et al., 1997), $\delta^{13}C$ can reflect the changing abundance of C_3 and C_4 plants above the cave (Baldini et al., 2008; Dorale et al., 1998) or the availability of soil CO_2 during the dissolution of limestone (Genty et al., 2003; Hendy, 1971; Salomons and Mook, 1986). Speleothem records are widely distributed geographically, and this makes them an ideal type of archive for regional climate reconstructions.

An increasing number of climate models explicitly simulate water isotopes as a tool for characterizing and diagnosing the atmospheric hydrological cycle (Schmidt et al., 2007; Steen-Larsen et al., 2017; Sturm et al., 2010; Werner et al., 2011; Haese et al., 2013). Such models are evaluated against modern observations of the isotopic composition of rainwater (see for example Yoshimura et al., 2008; Steen-Larsen et al., 2017). However, evaluations against palaeo-records such as the $\delta^{18}O$ records from speleothems can be used to provide an "out-of-sample" test (Schmidt et al., 2014) of these models. Thus, in addition to their use for climate reconstruction, speleothem records are a useful addition to the tools that are used for climate-model evaluation.

More than 500 speleothem data sets have been published to date, 70 % of which have been published in the decade since 2007. There have been some attempts to provide syntheses of speleothem data, particularly in the context of providing climate reconstructions or data sets for model evaluation (e.g. Bolliet et al., 2016; Caley et al., 2014; Harrison et al., 2014; Shah et al., 2013). However, these compilations generally lack sufficient information to allow careful screening of the records to ensure the reliability of the climate interpretation or the quality of the dating of the record. Furthermore, none of them provide comprehensive coverage of the globe.

Table 1. Information on speleothem records (entities) in the SISAL_v1 database. Elevation (Elv) is given in metres above sea level and latitude (Lat) and longitude (Long) in decimal degrees. For convenience, we have given the location of each record, although this information is not available in the database itself. Note that the latitude and longitude of Abaliget, Brown's cave and Uamh an Tartair are given correctly here but are incorrect in the SISAL_v1 database.

Entity name	Site name	Elv (m)	Lat (°)	Long (°)	Region	Citations
AB-DC-01, AB-DC-03, AB-DC-12	Abaco Island cave	−45	26.2300	−77.1600	Bahamas	Arienzo et al. (2017)
AB-DC-09	Abaco Island cave	−45	26.2300	−77.1600	Bahamas	Arienzo et al. (2015, 2017)
ABA_1, ABA_2	Abaliget cave	209	46.1333	18.1666	Hungary	Koltai et al. (2017)
Abissal, Ale-1	Abissal cave	100	−5.6000	−37.7330	Brazil	Cruz et al. (2009)
Ach-1	Achere cave	1534	8.6036	40.3729	Ethiopia	Asrat et al. (2006, 2008)
AB2	Anjohibe cave	131	−15.5300	46.8800	Madagascar	Scroxton et al. (2017)
AB3	Anjohibe cave	131	−15.5300	46.8800	Madagascar	Burns et al. (2016)
ANJB-2	Anjohibe cave	131	−15.5300	46.8800	Madagascar	Voarintsoa et al. (2017c)
MA3	Anjohibe cave	131	−15.5300	46.8800	Madagascar	Voarintsoa et al. (2017a)
MAJ-5	Anjokipoty cave		−15.5784	46.7344	Madagascar	Voarintsoa et al. (2017c)
CC-1_2004	Antro del Corchia	840	43.9800	10.2200	Italy	Drysdale et al. (2004)
CC-1_2009, CC-5_2009, CC-7	Antro del Corchia	840	43.9800	10.2200	Italy	Drysdale et al. (2009)
CC-28	Antro del Corchia	840	43.9800	10.2200	Italy	Drysdale et al. (2007)
CC-5_2005	Antro del Corchia	840	43.9800	10.2200	Italy	Drysdale et al. (2005)
POM2	Ascunsa cave	1050	45.0000	22.6000	Romania	Drăguşin et al. (2014)
BGC6, BGC11, BGC14	Ball Gown cave	100	−17.3000	124.1000	Australia	Denniston et al. (2013b)
BAR-II#L, BAR-II#B	Baradla cave	375	48.4667	20.5000	Hungary	Demény et al. (2017b)
BA-1b, BA-1, BA-2	Baschg cave	780	47.2501	9.6667	Austria	Boch et al. (2011)
Bero-1	Bero cave	1363	8.4241	40.3073	Ethiopia	Asrat et al. (2008), Baker et al. (2010)
Keklik1	Bir-Uja cave	1435	40.4833	72.5833	Kyrgyzstan	Fohlmeister et al. (2017)
BT-1, BT-2.1, BT-2.2, BT-2.3, BT-2.4, BT-2.5, BT-4, BT-6, BT-8, BT-9	Bittoo cave	3000	30.7903	77.7764	India	Kathayat et al. (2016)
BT-2	Botuverá	180	−27.2247	−49.1569	Brazil	Cruz et al. (2005)
BTV21a	Botuverá	180	−27.2247	−49.1569	Brazil	Bernal et al. (2016)
BDinf	Bourgeois–Delaunay cave	100	45.6678	0.5133	France	Couchoud et al. (2009)
BC01-07	Brown's cave	25	27.8894	−82.5186	USA	Pollock et al. (2016)
Boss, BFM-9, F2	Brown's Folly mine	150	51.3800	−2.3700	England (UK)	Baldini (2001), Baldini et al. (2005)
RL4_2006	Buca della Renella	300	44.0800	10.2100	Italy	Drysdale et al. (2006)
RL4_2016	Buca della Renella	300	44.0800	10.2100	Italy	Zanchetta et al. (2016)
RL4_2018	Buca della Renella	300	44.0800	10.2100	Italy	Drysdale et al. (unpublished)
BCC-2, BCC-4, BCC-6, BCC_composite	Buckeye creek	600	37.9825	−79.5894	USA	Hardt et al. (2010)
BCC-8, BCC-10	Buckeye creek	600	37.9825	−79.5894	USA	Springer et al. (2014)
BMS1	Bue Marino cave	0	40.2467	9.6228	Italy	Columbu et al. (2017)
Buffalo Cave Flowstone	Buffalo cave	1140	−24.1428	29.1770	South Africa	Hopley et al. (2007a, b)
BA02	Bukit Assam cave	150	4.0300	114.8000	Malaysia	Carolin et al. (2013)
BA03	Bukit Assam cave	150	4.0300	114.8000	Malaysia	Chen et al. (2016)
BA04	Bukit Assam cave	150	4.0300	114.8000	Malaysia	Partin et al. (2007)
Bu1, Bu2, Bu4, Bu6, BuStack	Bunker cave	184	51.3675	7.6647	Germany	Fohlmeister et al. (2012)
Calcite	Calcite cave		−46.0172	167.7431	New Zealand	Lorrey et al. (2008)
V3	Cango cave		−33.3925	22.2147	South Africa	Vogel (1983), Talma and Vogel (1992), Vogel and Kronfeld (1997)
COB-01-02	Cave of the Bells		31.7500	−110.7500	USA	Wagner et al. (2010)
CWN4	Cave Without a Name	377	29.8852	−98.6208	USA	Feng et al. (2014)

Table 1. Continued.

Entity name	Site name	Elv (m)	Lat (°)	Long (°)	Region	Citations
CC-1	Ceremosnja cave	530	44.4000	21.6500	Republic of Serbia	Kacanski et al. (2001)
Chau-stm6	Chauvet cave	240	44.2300	4.2600	France	Genty et al. (2006)
CHIL-1	Chilibrillo cave	60	9.1741	−79.6164	Panama	Lachniet (2004)
CL26	Clamouse cave	110	43.7100	3.5500	France	McDermott et al. (1999)
Cla4	Clamouse cave	110	43.7100	3.5500	France	Plagnes et al. (2002)
FC12-12, FC12-14, FC12-15	Clearwater cave	120	4.1000	114.8333	Malaysia	Carolin et al. (2016)
Squeeze1	Clearwater/Wind caves connection		4.1000	114.8300	Malaysia	Meckler et al. (2012)
T5	Cold Air cave	1420	−24.0000	29.1833	South Africa	Repinski et al. (1999)
T7_1999	Cold Air cave	1420	−24.0000	29.1833	South Africa	Holmgren et al. (1999), Stevenson et al. (1999)
T7_2001	Cold Air cave	1420	−24.0000	29.1833	South Africa	Lee-Thorp et al. (2001)
T7_2013	Cold Air cave	1420	−24.0000	29.1833	South Africa	Sundqvist et al. (2013)
T8	Cold Air cave	1420	−24.0000	29.1833	South Africa	Holmgren et al. (2003)
ESP03	Cova da Arcoia	1240	42.6100	−7.0900	Spain	Railsback et al. (2011)
CC3	Crag cave	60	52.2500	−9.4300	Ireland	McDermott et al. (1999), McDermott (2001)
ASR, ASM	Cueva de Asiul	285	43.3200	−3.5900	Spain	Smith et al. (2016)
CBD-2	Cueva del Diablo	1030	18.1920	−99.9210	Mexico	Bernal et al. (2011)
CUR4	Curupira cave	420	−15.2002	−56.7839	Brazil	Novello et al. (2016)
DAN-D	Dandak cave	400	19.0000	82.0000	India	Berkelhammer et al. (2010), Sinha et al. (2007)
DP1_2013	Dante cave		−19.4000	17.8833	Namibia	Sletten et al. (2013)
DP1_2016	Dante cave		−19.4000	17.8833	Namibia	Voarintsoa et al. (2017b)
DY-1	Dayu cave	870	33.1330	106.3000	China	Tan et al. (2009)
S3	Defore cave	150	17.1667	54.0833	Oman	Burns (2002)
DSSG-4	DeSoto caverns	150	33.3722	−86.3667	USA	Aharon et al. (2013)
DH2, DH2-D, DH2-E Terminal1, DH2-E Terminal2	Devils Hole	719	36.4254	−116.2920	USA	Moseley et al. (2016)
Dim-E2, Dim-E3, Dim-E4	Dim cave	232	36.5340	32.1056	Turkey	Ünal-İmer et al. (2015)
DV2	Diva cave	680	−12.3667	−41.5667	Brazil	Novello et al. (2012)
D3, D4	Dongge cave	680	25.2800	108.0800	China	Yuan (2004)
D8	Dongge cave	680	25.2800	108.0800	China	Cheng et al. (2016b)
Doubtful	Doubtful Xanadu	960	−45.3735	167.0476	New Zealand	Lorrey et al. (2008)
ARTEMISA	Ejulve cave	1240	40.4500	−0.3500	Spain	Pérez-Mejías et al. (2017)
HOR	Ejulve cave	1240	40.4500	−0.3500	Spain	Moreno et al. (2017)
TKS	Entrische Kirche cave	2119	47.1600	13.1500	Austria	Meyer et al. (2008)
GEX-SPA	Excentrica cave	100	37.1000	−7.7700	Portugal	Ponte et al. (2017)
ED1	Exhaleair cave	685	−41.2833	172.6330	New Zealand	Hellstrom et al. (1998)
FS2_2010	Fort Stanton cave	1864	33.5067	−105.4430	USA	Asmerom et al. (2010)
FS2_2012	Fort Stanton cave	1864	33.5067	−105.4430	USA	Polyak et al. (2012)
FG01	Fukugaguchi cave	170	36.9917	137.8000	Japan	Sone et al. (2013)
FR-0510	Furong cave	480	29.2289	107.9036	China	H.-C. Li et al. (2011)
FR-5	Furong cave	480	29.2289	107.9036	China	T.-Y. Li et al. (2011)
GG1, GG2	Gardener's Gut	120	−37.7394	175.1033	New Zealand	Williams et al. (2004)
GC08	Green Cathedral cave		4.2333	114.9250	Malaysia	Meckler et al. (2012)
CR1	Grotta di Carburangeli	22	38.1671	13.1615	Italy	Frisia et al. (2006), Madonia et al. (2005)
ER76	Grotta di Ernesto	1167	45.9667	11.6500	Italy	Scholz et al. (2012)

Table 1. Continued.

Entity name	Site name	Elv (m)	Lat (°)	Long (°)	Region	Citations
GP2	Grotte de Piste	1260	33.8400	−4.0900	Morocco	Wassenburg et al. (2016)
stm2, stm4	Gueldaman cave	507	36.4333	4.5667	Algeria	Ruan et al. (2016)
GT05-5	Guillotine cave	740	−42.3108	172.2178	New Zealand	Whittaker (2008)
SSC01, SCH02	Gunung-buda cave (snail shell cave)	150	4.0330	114.8000	Malaysia	Cobb et al. (2007), Moerman et al. (2013, 2014), Partin et al. (2007, 2013b)
Han-9	Han-sur-Lesse cave	180	50.1164	5.1884	Belgium	Vansteenberge et al. (2016)
Han-stm1	Han-sur-Lesse cave	180	50.1164	5.1884	Belgium	Genty et al. (1999)
Han-stm5b	Han-sur-Lesse cave	180	50.1164	5.1884	Belgium	Genty et al. (1998)
HS4_2008	Heshang cave	294	30.4500	110.4167	China	Hu et al. (2008)
HS4_2013	Heshang cave	294	30.4500	110.4167	China	Liu et al. (2013)
HOL-10	Hölloch im Mahdtal	1240	47.3781	10.1506	Germany	Moseley et al. (2015)
HOL-7, HOL-16, HOL-17, HOL-18, HOL-16-17, HOL-comp	Hölloch im Mahdtal	1240	47.3781	10.1506	Germany	Moseley et al. (2014)
HW3	Hollywood cave	130	−41.9500	171.4667	New Zealand	Whittaker et al. (2011)
H5	Hoti cave	800	23.0833	57.3500	Oman	Neff et al. (2001)
HY1, HY2, HY3	Huangye cave	1650	33.5833	105.1167	China	Tan et al. (2010)
MSD, MSL, PD, YT, H82	Hulu cave	90	32.5000	119.1700	China	Wang (2001)
IFK1	Ifoulki cave	1250	30.7080	−9.3275	Morocco	Ait Brahim et al. (2017)
JAR7, JAR14, JAR13	Jaraguá cave	570	−21.0830	−56.5830	Brazil	Novello et al. (2017)
Jeita-1, Jeita-2, Jeita-3	Jeita cave	100	33.9500	35.6500	Lebanon	Cheng et al. (2015)
AF12	Jerusalem west cave	700	31.7833	35.1500	Israel	Frumkin et al. (1999, 2000)
JHU-1	Jhumar cave	600	18.8667	81.8667	India	Sinha et al. (2011)
C996-1, C996-2	Jiuxian cave	1495	33.5667	109.1000	China	Cai et al. (2010b)
JX-2, JX-10	Juxtlahuaca cave	934	17.4000	−99.2000	Mexico	Lachniet et al. (2013)
JX-6	Juxtlahuaca cave	934	17.4000	−99.2000	Mexico	Lachniet et al. (2012)
JX-7	Juxtlahuaca cave	934	17.4000	−99.2000	Mexico	Lachniet et al. (2017)
KL 3	Kalakot cave	826	33.2219	74.4258	India	Kotlia et al. (2016)
Kanaan_MIS5	Kanaan cave	98	33.9069	35.6069	Lebanon	Nehme et al. (2015)
Kanaan_MIS6	Kanaan cave	98	33.9069	35.6069	Lebanon	Nehme et al. (2018)
GK-09-02	Kapsia cave	700	37.6233	22.3539	Greece	Finné et al. (2014)
K1, K3	Katerloch cave	900	47.0833	15.5500	Austria	Boch et al. (2009)
KS06-A-H, KS06-A, KS06-B, KS08-1-H, KS08-1, KS08-2-H, KS08-2	Kesang cave	2000	42.8700	81.7500	China	Cheng et al. (2012, 2016a)
KS08-2-MIS3, KS08-6	Kesang cave	2000	42.8700	81.7500	China	Cheng et al.(2016a)
KC-1, KC-3, KC-Composite	Kinderlinskaya cave	240	54.1500	56.8500	Russia	Baker et al. (2017)
PFU6	Klapferloch cave	1140	46.9500	10.5500	Austria	Boch et al. (2011)
SPA_126, SPA_49	Kleegruben cave	2165	47.0800	11.6700	Austria	Spötl et al. (2006)
KNI-51-0, KNI-51-3, KNI-51-4, KNI-51-7, KNI-51-10, KNI-51-A2-side1, KNI-51-A2-side2, KNI-51-C, KNI-51-F, KNI-51-G, KNI-51-H, KNI-51-I, KNI-51-J, KNI-51-N, KNI-51-O	KNI-51 cave	100	−15.1800	128.3700	Australia	Denniston et al. (2013a)

Table 1. Continued.

Entity name	Site name	Elv (m)	Lat (°)	Long (°)	Region	Citations
KNI-51-11	KNI-51 cave	100	−15.1800	128.3700	Australia	Denniston et al. (2015, 2016)
K11	Korallgrottan cave	540	64.8800	14.0000	Sweden	Sundqvist et al. (2010)
BW-1	Kulishu cave	610	39.6800	115.6500	China	Ma et al. (2012)
Min-stm1	La Mine cave	975	36.0300	9.6800	Tunisia	Genty et al. (2006)
L4	Labyrintgrottan cave	730	66.0600	14.6800	Sweden	Sundqvist et al. (2007)
LH-70s-1	Lancaster Hole	294	54.2209	−2.5168	England (UK)	Atkinson and Hopley (2013)
LH-70s-2, LH-70s-3	Lancaster Hole	294	54.2209	−2.5168	England (UK)	Atkinson and Hoffman (unpublished)
LD12	Lapa Doce cave	680	−12.3667	−41.5667	Brazil	Novello et al. (2012)
LG11, LG3	Lapa grande cave	590	−14.4200	−44.3660	Brazil	Strikis et al. (2011)
LSF16, LSF3	Lapa sem fim cave	341	−16.1503	−44.6281	Brazil	Stríkis et al. (2015)
L03	Larshullet cave	400	66.0000	14.0000	Norway	Linge et al. (2009b)
Leany	Leány cave	420	47.7000	18.8400	Hungary	Demény et al. (2013)
LC-2	Lehman caves	2080	39.0100	−114.2200	USA	Shakun et al. (2011)
LMC-14, LMC-21	Lehman caves	2080	39.0100	−114.2200	USA	Lachniet et al. (2014)
LC-1	Leviathan cave	2400	37.8900	−115.5800	USA	Lachniet et al. (2014)
LR06-B1_2009, LR06-B3_2009	Liang Luar cave	550	−8.5300	120.4300	Indonesia	Griffiths et al. (2009)
LR06-B1_2016, LR06-B3_2016	Liang Luar cave	550	−8.5300	120.4300	Indonesia	Griffiths et al. (2016)
LR06-B3_2013, LR06-C2, LR06-C3_2013, LR06-C5, LR06-C6, LL_Comp_2013	Liang Luar cave	550	−8.5300	120.4300	Indonesia	Ayliffe et al. (2013)
LR07-A8, LR07-A9, LR07-E11	Liang Luar cave	550	−8.5300	120.4300	Indonesia	Griffiths et al. (2013)
LR07-E1, LR06-C3_2011, LR07-E1-D	Liang Luar cave	550	−8.5300	120.4300	Indonesia	Lewis et al. (2011)
LII4, LII4-KH	Lobatse cave	1200	−25.2100	25.6800	Botswana	Holmgren et al. (1994, 1995)
ME-12	Ma'ale Efrayim cave	250	32.0660	35.3952	West Bank	Vaks et al. (2003)
MC01	Macal Chasm	530	16.8830	−89.1080	Belize	Akers et al. (2016), Webster et al. (2007)
MC-S1, MC-S2	Mairs cave	475	−32.1600	138.8300	Australia	Treble et al. (2017)
S1	Mavri Trypa cave	70	36.7360	21.7596	Greece	Finné et al. (2017)
KM-A	Mawmluh cave	1160	25.2622	91.8817	India	Berkelhammer et al. (2013), Breitenbach et al. (2015)
MAW-6	Mawmluh cave	1160	25.2622	91.8817	India	Lechleitner et al. (2017)
MWS-1	Mawmluh cave	1160	25.2622	91.8817	India	Breitenbach et al. (2015), Dutt et al. (2015)
MAXS	Max's cave	325	−37.7394	175.1033	New Zealand	Williams et al. (2004)
ML1	McLean's cave	300	38.0700	−120.4200	USA	Oster et al. (2015)
MB-2, MB-3, MB-5, MB-6	Milchbach cave	1840	46.6167	8.0830	Switzerland	Luetscher et al. (2011)
MC3	Moaning cave	520	38.0717	−120.4660	USA	Oster et al. (2009, 2015)
MOD-22	Modric cave	32	44.2568	15.5372	Croatia	Rudzka et al. (2012)
MO-1	Molinos cave	1050	40.7925	−0.4492	Spain	Moreno et al. (2017)
MO-7	Molinos cave	1050	40.7925	−0.4492	Spain	Moreno et al. (2017), Muñoz et al. (2015)
M1-5	Moomi cave	400	12.5000	54.0000	Yemen	Shakun et al. (2007)

Table 1. Continued.

Entity name	Site name	Elv (m)	Lat (°)	Long (°)	Region	Citations
Mun-stm2, Mun-stm1	Munagamanu cave	475	15.1500	77.9200	India	Genty et al. (unpublished)
NBJ	Natural Bridge caverns	306	29.6900	−98.3400	USA	Wong et al. (2015)
MD3	Nettlebed cave	390	−41.2500	172.6330	New Zealand	Hellstrom et al. (1998)
Gib04a	New St Michael's cave	325	36.1261	−5.3455	Gibraltar (UK)	Mattey et al. (2008, 2010)
FM3, Oks82	Okshola cave	165	67.0000	15.0000	Norway	Linge et al. (2009a)
OCNM02-1	Oregon caves national monument	1300	42.0981	−123.4070	USA	Ersek et al. (2012)
PX7	Paixão cave		−12.6182	−41.0184	Brazil	Strîkis et al. (2015)
PAL3, PAL4	Palestina cave	870	−5.9200	−77.3500	Peru	Apaéstegui et al. (2014)
PAR01, PAR03, PAR06, PAR07, PAR08, PAR16, PAR24	Paraiso cave	60	−4.0667	−55.4500	Brazil	Wang et al. (2017)
ALHO6	Pau d'Alho cave	340	−15.2055	−56.8000	Brazil	Jaqueto et al. (2016), Novello et al. (2016)
PDR-1	Perdida cave	350	18.0000	−67.0000	Puerto Rico	Winter et al. (2011)
Candela	Pindal cave	24	43.4000	−4.5300	Spain	Moreno et al. (2010), Rudzka et al. (2011)
PC-1	Pinnacle cave	1792	35.9700	−115.5000	USA	Lachniet et al. (2011)
YD01	Pippikin Pot cave	320	54.2143	−2.5123	England (UK)	Atkinson and Hopley (2013), Daley et al. (2011)
POS-STM-4	Postojna cave	529	45.7700	14.2000	Slovenia	Genty et al. (1998)
Q5	Qunf cave	650	17.1667	54.3000	Oman	Fleitmann et al. (2007)
RN1, RN4	Rainha cave	100	−5.6000	−37.7330	Brazil	Cruz et al. (2009)
Ruakuri C	Ruakuri cave	80	−38.2667	175.0667	New Zealand	Williams et al. (2004)
Merc-1, Asfa-3	RukieSSa cave	1618	8.6036	40.3772	Ethiopia	Asrat et al. (2008), Baker et al. (2007)
SAH-A, SAH-B, SAH-AB	Sahiya cave	1190	30.6000	77.8667	India	Sinha et al. (2015)
SB-10, SB-26, SB-27, SB-43, SB-44, SB-49	Sanbao cave	1900	31.6670	110.4330	China	Dong et al. (2010)
SB-12, SB-14, SB-32, SB-58	Sanbao cave	1900	31.6670	110.4330	China	Cheng et al. (2016b)
MF-3	Schafsloch cave	1890	47.2333	9.3833	Switzerland	Häuselmann et al. (2015)
SCH-5	Schneckenloch cave	1285	47.4333	9.8667	Austria	Moseley et al. (2015)
SCH-7	Schneckenloch cave	1285	47.4333	9.8667	Austria	Boch et al. (2011)
SC02, SC03	Secret cave	250	4.0848	114.8503	Malaysia	Carolin et al. (2013)
SE09-6	Seso cave	794	42.4600	0.0400	Spain	Bartolomé et al. (2015)
7H, 7H-2, 7H-3	Sieben Hengste cave	1955	46.7500	7.8100	Switzerland	Luetscher et al. (2015)
MAR_L	Skala Marion cave	41	40.6387	24.5144	Greece	Psomiadis et al. (2018)
So-1	Sofular cave	700	41.4200	31.9300	Turkey	Fleitmann et al. (2009)
2-6	Soreq cave	400	31.7558	35.0224	Israel	Orland et al. (2009)
2N	Soreq cave	400	31.7558	35.0224	Israel	Orland et al. (2012)
Soreq-composite	Soreq cave	400	31.7558	35.0224	Israel	Grant et al. (2012)
SG95	Soylegrotta cave	280	66.0000	14.0000	Norway	Linge et al. (2001)
SPA12, SPA70, SPA128, SPA127, COMNISPA II, SPA133	Spannagel cave	2310	47.0800	11.6700	Austria	Fohlmeister et al. (2013)
SPA121	Spannagel cave	2310	47.0800	11.6700	Austria	Spötl et al. (2008)
SZ2	Suozi cave	700	32.4300	107.1700	China	Zhou et al. (2008)
TM0, TM2	Tamboril cave	200	−16.0000	−47.0000	Brazil	Wortham et al. (2017)
Taurius	Taurius cave	230	−15.5333	167.0167	Vanuatu	Partin et al. (2013a)

Table 1. Continued.

Entity name	Site name	Elv (m)	Lat (°)	Long (°)	Region	Citations
Aurora	Te Anau Fiordland	320	−45.2800	167.7000	New Zealand	Lorrey et al. (2008)
Te Reinga A, Te Reinga B	Te Reinga cave		−38.8200	177.5200	New Zealand	Lorrey et al. (2008)
TM-18a, TM-18b	Tianmen	4800	30.9167	90.0667	China	Cai et al. (2012)
TM-2, TM-5	Tianmen	4800	30.9167	90.0667	China	Cai et al. (2010a)
T1	Timta cave	1900	29.8381	82.0336	India	Sinha et al. (2005)
TC1	Tityana cave	1470	30.6419	77.6521	India	Joshi et al. (2017)
TON-1, TON-2	Tonnel'naya cave	3226	38.4000	67.2300	Uzbekistan	Cheng et al. (2016a)
TR5	Torrinha cave	680	−12.3667	−41.5667	Brazil	Novello et al. (2012)
Trio	Trió cave	275	46.1100	18.1500	Hungary	Demény et al. (2017a), Siklósy et al. (2009)
Chaac	Tzabnah cave	20	20.7300	−89.7160	Mexico	Medina-Elizalde et al. (2010)
SU032	Uamh an Tartair	220	58.1400	−4.9300	Scotland (UK)	Baker et al. (2012)
SU967	Uamh an Tartair	220	58.1400	−4.9300	Scotland (UK)	Baker et al. (2011)
PU-2	Ursilor cave	482	46.5537	22.5695	Romania	Onac et al. (2002)
VSPM1	Valmiki cave	420	15.1500	77.8167	India	Raza et al. (2017)
VSPM4	Valmiki cave	420	15.1500	77.8167	India	Lone et al. (2014)
Vil-car1	Villars cave	175	45.4300	0.7800	France	Wainer et al. (2011)
Vil-stm1	Villars cave	175	45.4300	0.7800	France	Labuhn et al. (2015)
Vil-stm11	Villars cave	175	45.4300	0.7800	France	Genty et al. (2006)
Vil-stm14	Villars cave	175	45.4300	0.7800	France	Genty et al. (2010), Wainer et al. (2009)
Vil-stm27	Villars cave	175	45.4300	0.7800	France	Genty et al. (2003)
Vil-stm6	Villars cave	175	45.4300	0.7800	France	Genty (unpublished)
Vil-stm9	Villars cave	175	45.4300	0.7800	France	Genty et al. (2003, 2010)
WS-B	Wah Shikhar cave	1290	25.2500	91.8667	India	Sinha et al. (2011)
Waiau	Waiau cave	100	−46.0000	167.7300	New Zealand	Lorrey et al. (2008)
WP-1	Wazpretti cave	100	−42.3108	171.4000	New Zealand	Williams et al. (2005)
WSC-97-10-5	White Scar cave	255	54.1656	−2.4419	England (UK)	Atkinson and Hopley (2013), Daley et al. (2011)
WR5	Whiterock cave		4.1500	114.8600	Malaysia	Meckler et al. (2012)
W5	Wolkberg cave	1450	−24.1000	29.8800	South Africa	Holzkämper et al. (2009)
XBL-3, XBL-4, XBL-7, XBL-26, XBL-27, XBL-29, XBL-48, XBL-65	Xiaobailong cave	1500	24.2000	103.3500	China	Cai et al. (2015)
XL-1	Xinglong cave	710	40.5000	117.5000	China	Duan et al. (2016)
XY07-8	Xinya cave	1250	30.7500	109.4700	China	J. Y. Li et al. (2017)
XY-2	Xinya cave	1250	30.7500	109.4700	China	Li et al. (2007)
JFYK7	Yangkou cave	2140	29.0333	107.1833	China	Han et al. (2016), T.-Y. Li et al. (2017), Zhang et al. (2017)
YK5, YK12, YK23, YK47, YK61	Yangkou cave	2140	29.0333	107.1833	China	Li et al. (2014)
YOKG	Yok Balum cave	336	16.2086	−89.0735	Belize	Ridley et al. (2015)
YOKI	Yok Balum cave	336	16.2086	−89.0735	Belize	Kennett et al. (2012)
ZLP1, ZLP2	Zhuliuping cave	1217	26.0167	104.0950	China	Huang et al. (2016)

SISAL (Speleothem Isotope Synthesis and Analysis) is an international working group set up in 2017 under the auspices of the Past Global Changes (PAGES) programme (http://pastglobalchanges.org/ini/wg/sisal, last access: 4 September 2018). The aim of the working group is to compile the many hundreds of speleothem isotopic records worldwide, paying due attention to careful screening and metadata documentation, the construction of standardized age models, and age-model uncertainties, in order to produce a public-access database that can be used for palaeoclimate reconstruction and for climate-model evaluation. In this paper, we document the first publicly available version of the SISAL database, focusing on describing its structure and contents including the information that has been included to facilitate quality control.

2 Data and methods

2.1 Compilation of data

The database contains stable carbon and oxygen isotope measurements made on speleothems, as well as supporting metadata to facilitate the interpretation of these records. All available speleothem data are included, and no attempt was made to screen records on the basis of the time period covered, the resolution of the records, or the quality of the data or age models. Adequate metadata are provided to allow database users to select the records that are suitable for a particular type of analysis. The raw data were either provided by members of the SISAL working group or extracted from data lodged in PANGAEA or from the National Centres for Environmental Information. Additional information on the records was compiled from publications. All the records in the current version of the database (SISAL_v1) are listed and described in Table 1.

2.2 Structure of the database

The data are stored in a relational database (MySQL), which consists of 14 linked tables, specifically site, entity, sample, dating, dating lamina, gap, hiatus, original chronology, $\delta^{13}C$, $\delta^{18}O$, entity link reference, reference, composite link entity and notes. Figure S1 shows the relationships between these tables. A detailed description of the structure and content of each of the tables is given below. The details of the predefined lists for all fields can be found in Table S1.

2.2.1 Site metadata (table name: site)

A site is defined as the cave or cave system from which speleothem records have been obtained. A site may therefore be linked to several speleothem records, where each record is treated as a separate entity. The site table contains basic metadata about the cave or cave system, including site ID, site name, latitude, longitude, elevation, geology, rock age and monitoring (see Table S2). The elevation is that of the cave itself, not the elevation of the land surface above the cave. Since the elevation of the land surface can be obtained from other sources, we include the cave elevation to facilitate making additional lapse rate corrections for oxygen isotopes for high elevation sites (Bowen and Wilkinson, 2002). This also allows an estimation of the depth of the overburden above the speleothem site, and hence an estimate of the time taken for water to reach the cave. The description of the geology and the age of the rock formation (rock_age) is given because this is important for understanding the degree of permeability of the material above the cave. Primary porosity decreases and fracture flow increases as rocks age, which in turn affects the likely speed at which water flows through the host rock and reaches the cave system. The geology field is also useful because it gives an indication of whether the cave is formed in Mg-rich rocks (e.g. dolomite) and thus whether the speleothems are likely to be formed of aragonite, which would require special consideration in terms of oxygen and carbon isotope comparisons with that of calcite (see also Table S3). Only a limited number of descriptive terms are allowed for each field. The age of rock formation follows the standard era, period, epoch terminology as defined by the International Commission on Stratigraphy in 2015 (Cohen et al., 2015). The database includes information on whether the cave site has been monitored: positive returns in this field mean that monitoring of in-cave environmental parameters (e.g. cave air temperature) and/or cave drip chemistry has been carried out periodically for at least one entire season (as opposed to one-off measurements of in-cave conditions when the speleothem was collected). The database does not contain monitoring data, but inclusion of this field facilitates researchers being able to contact the original data providers about monitoring information, which can be useful in understanding if a cave is likely to contain speleothems that have been deposited close to isotopic equilibrium.

2.2.2 Entity metadata (table name: entity)

Each speleothem (or composite speleothem record) has a unique identifier and a unique name. The entity metadata table (Table S3) provides information on the cave environment that can affect speleothem formation. This includes the thickness of the cover above the speleothem, which might affect the time taken for water to reach the drip site feeding the speleothem and hence the responsiveness of the record to individual rain events or seasonal patterns of precipitation (Fairchild and Baker, 2012). The distance of the speleothem from the cave entrance is provided, which, depending on the morphology of a cave, can be a useful indicator of cave ventilation (direct air advection). Ventilation is important as it can control cave air temperature, humidity, evaporation and pCO_2 levels (Fairchild and Baker, 2012; Frisia et al., 2011; Spötl et al., 2005; Tremaine et al., 2011). The entity table also contains a field to document whether any tests have been carried out to establish whether there is oxygen and carbon isotope quasi-equilibrium between the drip water (CO_2–H_2O system) and the speleothem ($CaCO_3$). There are several such tests (see for example Hendy, 1971; Johnston et al., 2013; Mickler et al., 2006; Tremaine et al., 2011), but no attempt is made to identify which test has been applied in the database. The drip type (e.g. seepage flow, seasonal drip, vadose flow; Smart and Friederich, 1987) also provides useful hydrological information: seepage flow shows a small inter-annual variability of discharge and the speleothem record will therefore more likely reflect a long-term average state over several years; other drip types, such as seasonal drip, will indicate the potential to record seasonal or individual rainfall events.

The main focus of the SISAL database is stable isotope measurements, but the entity metadata table also contains information on the kinds of measurements that have been made for a specific speleothem. Only the stable isotope measurements are currently archived in the database. However, listing the range of data available from any speleothem will facilitate future updates of the database to include other types of measurements apart from stable isotopes (i.e. trace elements) and will help researchers in locating speleothems where multiple types of measurements have been done. The entity metadata table contains four fields to facilitate data traceability. The first two fields give the name of the person who was responsible for collating the data, and a DOI or URL for the original data. Some records have been partially or entirely updated since first being published. Although these records are included for data traceability, the entity_status field indicates whether they have been partially updated (e.g. with additional samples and/or improved age models) or completely superseded by a new record. The field corresponding_current indicates which entity (or entities) provides updated information. Information on original publications on specific speleothems is given in the reference table (see Sect. 2.2.11).

2.2.3 Sample metadata (table name: sample)

The sample metadata table (Table S4) contains information on the location of the sample with respect to a reference point, where the reference point can be either the top or the base of the speleothem. It also provides information on the thickness and mineralogy (calcite, secondary calcite, aragonite, vaterite, mixed, not known) of each sample. Since some samples may have mixed mineralogy, it also provides information on whether a correction for aragonite has been applied to $\delta^{18}O$ or $\delta^{13}C$, due to different phosphoric acid fractionation factors.

2.2.4 Dating information (table name: dating)

The dating information table (Table S5) provides information on the radiometric dates used to construct the original age model for each of the speleothem entities, including type of radiometric date (e.g. U series), depth of dated sample, thickness of dated sample and sample weight. Dates that are used to anchor sequences that are dated by lamina counting (see Sect. 2.2.5) are included in the dating information table and identified in date type as an event (i.e. the start or end of a laminated sequence). The degree of precision varies between different dating methods and techniques, for example mass spectrometric U/Th dating generally produces a more precise age than the alpha spectrometry U/Th data method. So the inclusion of the dating method provides a basic measure of the reliability, in terms of analytical precision, of any given date. Sample thickness also influences the dating uncertainty, because thicker samples will integrate more material of different age. Similarly, sample weight can influence precision: samples younger than a few thousand years may contain very low levels of the daughter isotope ^{230}Th (whose accumulation by radioactive decay provides the measure of the sample's age), and so require more material to provide an accurate and precise result. The content of ^{232}Th is included in the dating information table because this value is used for the detrital correction of initial ^{230}Th. Sample mineralogy is also included because this affects the reliability of individual dates (e.g. samples from re-crystallized secondary calcite are not reliable because of the loss of uranium; Bajo et al., 2016).

We provide both the original uncorrected age and the corrected age for each date. The corrected U/Th age is adjusted for detrital contamination; the corrected calibrated ^{14}C age is adjusted for dead carbon. The correction factors used to derive the corrected U/Th or ^{14}C age are included in the dating information table. The decay constant used to calculate the U/Th ages is given because the values used have changed through time (Cheng et al., 2000, 2013; Edwards et al., 1987; Ivanovich and Harmon, 1992). The calibration curve used to convert radiocarbon ages to calendar years in the original publication is also given. Several different standards have been used in the original publications for the modern reference state (e.g. BP(1950), b2k, CE / BCE or the year when U/Th chemical separation was performed) but all of these have been converted to BP(1950) in the database.

Some of the dates listed for a given entity were not used in the original age model, for example because the dating sample was contaminated with organic material or because of age inversions. The dates excluded from the original age model are flagged in the database (date_used) but the other information on these dates is nevertheless included in the dating information table to ensure transparency.

The geochemical characteristics of the sample provide information that is required to assess the quality or reliability of these dates. The ratio of $^{230}Th / ^{232}Th$, for example, is a measure of detrital thorium concentration in the sample and thus provides an initial quality control on each date. A $^{230}Th / ^{232}Th$ activity ratio > 300 is considered a good indicator of a reliable date (Hellstrom, 2006); a higher ratio indicates a cleaner sample with higher accuracy. The thorium corrected errors are also included to provide an indication of the magnitude of the correction related to detrital thorium contamination.

2.2.5 Lamina dating information (table name: dating_lamina)

Variations in the drip-water geochemistry and/or quantity or cave conditions may occur at regular intervals, forming laminae of a range of thicknesses usually linked to surface seasonal climate variations (Fairchild and Baker, 2012). A high-resolution chronology can be established for such records by lamina counting, provided an absolute date is available for either the start or the end of the laminated sequence (e.g. because U/Th dates have been obtained or because the stalagmite was actively growing when collected). The identification of individual lamina can be difficult if they are very thin or of varying width, so best practice is to provide an estimate of the counting uncertainty that propagates from the absolute anchor dates. The lamina dating information table (Table S6) provides the age of each lamina in the sequence and the uncertainty on this dating; the absolute dates used as anchor points are given in the dating information table and identified in the date type field there as an event (see Sect. 2.2.4).

It should be noted that laminae can be formed on a variety of timescales, depending on the frequency that the thresholds for the formation of specific fabrics and mineralogies are crossed. Annual laminations are more likely in regions where there is a clear seasonality in climate or cave environment. In other regions, the lamination may be a result of lower- or higher-frequency variations in, for example, hydrologically effective precipitation (e.g. infiltrated waters) or soil CO_2 production. It is imperative to demonstrate that the laminations are annual (see Table S9) before using lamina counting for dating.

2.2.6 Hiatus place mark information (table name: hiatus)

A prolonged cessation of speleothem growth can occur under unfavourable environmental conditions leading to, for example, undersaturation of drip water or cessation of dripping. Growth hiatuses can often be recognized from structural or mineralogical features, or inferred based on absolute dating. Growth hiatuses have to be taken into account in the construction of age models and thus the hiatus place mark table (Table S7) provides information on the location of such features. The hiatus is referenced to the specific depth at which it occurs, and this depth is considered as an imaginary sample that then appears with a specific sample_id in the sample table. There are some cases in which the hiatus depth was not recorded; in these cases the depth was specified as the imaginary mid-point depth between bracketing samples.

2.2.7 Gap place mark information (table name: gap)

When a composite record is created based on more than one individual speleothem from the same cave system, there may be discontinuities in the overlapping time of the individual records. These gaps are not growth hiatuses, but must be identified to facilitate plotting of the records. The gap place mark information table (Table S8) provides information on the location of sample gaps. The gap is referenced to the specific depth at which it occurs, and this depth is considered as an imaginary sample which then appears with a specific sample_id in the sample table. In composite records where sample depths are not given, the location of a gap can be derived from the sample ordering and the absence of isotopic information for a given sample. In point of fact, this table is empty in version 1 of the database.

2.2.8 Original chronology (table name: original_chronology)

The original chronology table (Table S9) provides an estimate of the age and age uncertainty, according to the original published age model for each sample on which stable isotope measurements have been made. The table also provides information on the type of age model (e.g. linear interpolation between dates, polynomial fit, Bayesian, StalAge; Scholz and Hoffmann, 2011, COPRA; Breitenbach et al., 2012, OxCal; Bronk Ramsey, 2001, 2008) used in the original publication. The fields ann_lam_check and dep_rate_check are included for quality assurance purposes, since they indicate that the assumption that laminae are truly annual has been explicitly tested.

2.2.9 Carbon isotope data (table name: d13C)

The carbon isotope data table (Table S10) contains the carbon isotope measurements. It also provides information on the laboratory precision of each measurement and the standard (PDB or Vienna-PDB) used as a reference.

2.2.10 Oxygen isotope data (table name: d18O)

The oxygen isotope data table (Table S11) contains the oxygen isotope measurements. It also provides information on the laboratory precision of each measurement and the standard (PDB or Vienna-PDB) used as a reference.

2.2.11 Publication information (table name: reference)

This table (Table S12) provides full bibliographic citations for the original references documenting the speleothems, their isotopic records and/or their age models. References on monitoring of the cave may also be provided. There may be multiple publications for a single speleothem record, and all of these references are listed. For convenience, there is also a table (Table S13) that links the publications to the specific entity.

2.2.12 Link composite and entity information (table name: composite_link_entity)

Multiple speleothem records showing a temporal overlap (and a similar signal) can be combined to create a composite record of changes through time. The composite record is treated as a distinct entity in the database. The link composite and entity information table (Table S14) is provided in order to be able to link this composite record to the individual speleothem records from which it was derived. Thus any single composite entity (composite_entity_id) is linked to multiple single entities (single_entity_id)

2.2.13 Notes (table name: notes)

The notes table (Table S15) is provided in order to record additional information regarding the site which cannot be recorded in the fields of the table; this may also include entity specific information.

2.3 Quality control

Individual records in the SISAL database were compiled either by the original authors or from published and open-access material by specialists in the collection and interpretation of speleothem records. In this latter case, the data compilers made every attempt to contact original authors to check that the compiled data were correct. The name of the person who compiled the data is included in the database (entity table, contact) so that they can be consulted in the future about queries or corrections. Individual records for the database were subsequently checked by a small number of regional coordinators, to ensure that records were being entered in a consistent way. Prior to entry in the database, the records were automatically checked using specially designed database scripts (in Python) to ensure that the entries to individual fields were in the format expected (e.g. text, decimal numeric, positive integers) or were selected from the pre-defined lists provided for specific fields. In defining both the formats and the pre-defined lists, the SISAL working group has taken special care to ensure that the entries are unambiguous. Null values for metadata fields were identified during the checking procedure, and checks were made with the data contributors whenever possible to ensure that null fields genuinely corresponded to missing information.

The database contains information designed to allow an assessment of the quality of an individual record. Thus, the entity metadata table contains information on, for example, the distance of the speleothem from the cave entrance in order to allow the user to assess whether cave temperatures are driven by advection of air or conduction through the bedrock. There are several other factors that can affect ventilation, for example the contrast between the cave and external climates, and cave morphology such as the size of the entrance or the number of entrances. Information on these factors is only rarely given in publications; we assume that this information would be more likely to be available if the original authors thought that ventilation was a significant influence on the speleothem record. Including information on distance from the cave entrance is therefore being regarded as a minimal indicator for record quality. Other fields that are included to allow the user to select appropriate records include geology, rock age, speleothem type and drip type.

The database also contains information to allow an assessment of the reliability of the dates used in constructing the original age model. The most important of these fields are those with information on the sample geochemistry (see Sect. 2.2.4), which allows the user to determine whether the samples were sufficiently large and sufficiently pure to yield good U/Th dates. The database also gives information on sample weight, which also addresses this issue. The information on the corrections employed, dating uncertainties and whether the original authors considered the date reliable (and therefore used it in constructing an age model) also provide insights into the reliability of individual chronologies.

The SISAL database is an ongoing effort and continuing efforts to update the records will include updating missing data fields for individual records. Analysis of the data is also useful for verification purposes and may result in corrections of some data. Any such changes to sites and entities included in version 1 of the database will be documented in subsequent updates. The SISAL working group also aims to provide new chronologies in future versions of the database based on Bayesian approaches, namely OxCal (Bronk Ramsey, 1995, 2008), COPRA (Breitenbach et al., 2012) and StalAge (Scholz and Hoffmann, 2011).

2.4 Overview of contents

The first version of the SISAL database contains 211 022 $\delta^{18}O$ measurements, 127 115 $\delta^{13}C$ measurements from 371 speleothem records and 10 composites from 174 cave systems. This represents approximately 58 % of published speleothem records we have identified. Of the 371 speleothem records, 8 have been superseded, 7 provide information for which there are also updates or additional information recorded as separate entities, 6 have dating information but no isotopic records because the individual entities were only used to construct composite records, and 15 do not have age models. The database also contains 6 records that have not been published.

The distribution of sites is global in extent (Fig. 1). The majority (30 %) of the sites are from Europe (53 sites) and there is currently less good representation of sites from other regions. The temporal distribution of records is excellent for the past 2000 years (Fig. 2a) and good for the past 22 000 years (Fig. 2b). Altogether, 142 entities record some

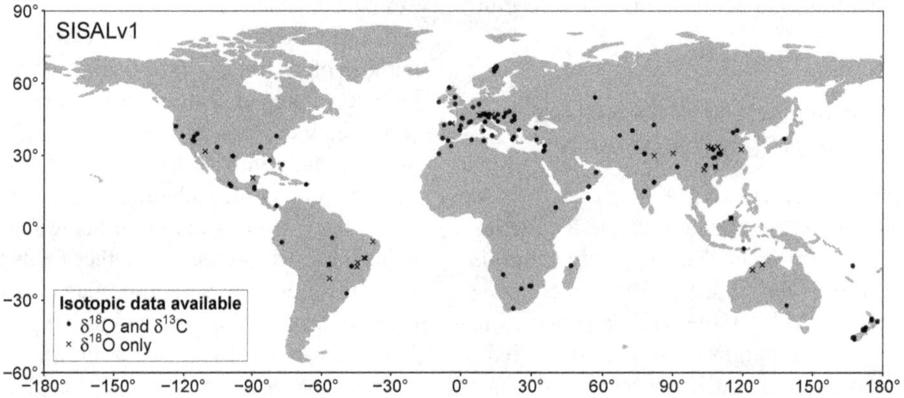

Figure 1. Map of the location of sites in the database. Note that some sites include records for multiple individual speleothems, which are treated as separate entities in the database itself. The sites are coded with different shapes to indicate whether they provide records only for oxygen isotopes, or for both oxygen and carbon isotopes.

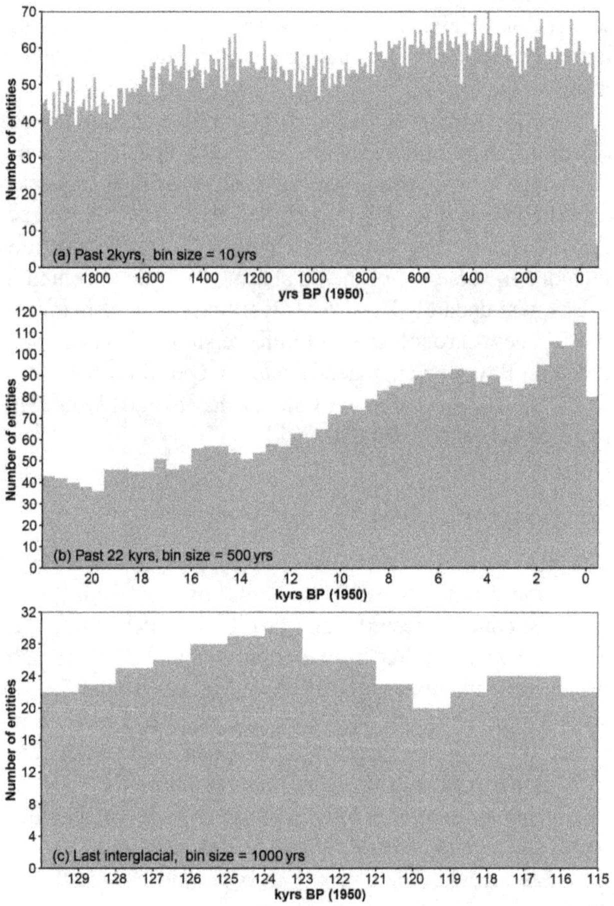

Figure 2. Plot showing the temporal coverage of individual entities in the database. Panel **(a)** shows records covering the past 2000 years (2 kyr BP), **(b)** shows records covering the past 22 000 years (22 kyr BP), and **(c)** shows records that cover the LIG (130–115 kyr BP).

part of the past 2000 years, 87 of which have a resolution $\leq$ 10 years between samples on average. There are 253 entities recording some part of the past 22 000 years, including 153 with a resolution of $\leq$ 100 years between samples on average. The database contains 42 entities from the last interglacial period (115 000 to 130 000 years before present), 41 of which have a resolution of $\leq$ 1000 years between samples on average (Fig. 2c).

3 Conclusions

The SISAL database is based on a community effort to compile isotopic measurements from speleothems to facilitate palaeoclimate analysis. Considerable effort has been made to ensure that there is adequate metadata and quality control information to allow the selection of records appropriate to answer specific questions and to document the uncertainties in the interpretation of these records. The database is publicly available.

The first version of the SISAL database contains 211 022 $\delta^{18}O$ measurements and 127 115 $\delta^{13}C$ measurements from 371 individual speleothem records, and 10 composites from 174 cave systems. The distribution of sites is global in extent. The temporal distribution is excellent for the past 2000 years and good for the past 22 000 years. There are also records that span the last interglacial period.

The format of the database is designed to facilitate the use of the data for regional- to continental-scale analyses, and in particular to facilitate comparisons with and evaluation of isotope-enabled climate model simulations. The SISAL working group will continue to expand the coverage of the SISAL database and will provide new chronologies based on standardized age models; subsequent versions of the database will be made freely available to the community.

Team list. The following authors are members of the SISAL Working Group and contributed to the database: Tim Atkinson (Departments of Earth Sciences & Geography, University College London, WC1E 6BT, UK), Avner Ayalon (Geological Survey of Israel, 30 Malkei Israel St., Jerusalem, Israel), James Baldini (Department of Earth Sciences, Durham University, DH1 3LE, UK), Miriam Bar-Matthews (Geological Survey of Israel, 30 Malkei Israel St., Jerusalem, Israel), Juan Pablo Bernal (Centro de Geociencias, Universidad Nacional Autónoma de México, Campus UNAM Juriquilla, Querétaro 76230, Querétaro, Mexico), Sebastian Breitenbach (Institute of Geology, Mineralogy & Geophysics, Ruhr-Universität Bochum, Universitätsstr. 150, NABF04/751, 44801 Bochum, Germany), Ronny Boch (Graz University of Technology, Institute of Applied Geosciences, Rechbauerstrasse 12, 8010 Graz, Austria), Andrea Borsato (School of Environmental and Life Science, University of Newcastle, 2308 NSW, Australia), Yanjun Cai (State Key Lab of Loess and Quaternary Geology, Institute of Earth Environment, Chinese Academy of Sciences, Xi'an 710061, China), Stacy Carolin (Institute of Geology, University of Innsbruck, Innrain 52, 6020 Innsbruck, Austria), Hai Cheng (Institute of Global Environmental Change, Xi'an Jiaotong University, China), Andrea Columbu (Department of Biological, Geological and Environmental Sciences, Via Zamboni 67, 40126, Bologna, Italy), Isabelle Couchoud (EDYTEM, UMR 5204 CNRS, Université Savoie Mont Blanc, Université Grenoble Alpes, 73370 Le Bourget du Lac, France), Francisco Cruz (Instituto de Geociências, Universidade de São Paulo, São Paulo, Brazil), Attila Demény (Institute for Geological and Geochemical Research, Research Centre for Astronomy and Earth Sciences, Hungarian Academy of Sciences, Budaörsi út 45, 1112 Budapest, Hungary), David Dominguez-Villar (School of Geography, Earth and Environmental Sciences, University of Birmingham, Edgbaston B15 2TT Birmingham, UK), Virgil Drăgusin (Emil Racoviță Institute of Speleology, Romanian Academy, Frumoasă St. 31, Bucharest, Romania), Russell Drysdale (School of Geography, University of Melbourne, 3010 Victoria, Australia), Vasile Ersek (Department of Geography and Environmental Sciences, Northumbria University, Newcastle upon Tyne, UK), Martin Finné (Department of Archaeology and Ancient History, Uppsala University, Sweden), Dominik Fleitmann (Department of Archaeology, School of Archaeology, Geography and Environmental Science, Whiteknights, University of Reading, RG6 6AB, UK), Jens Fohlmeister (Institute for Earth and Environmental Sciences, University of Potsdam, Karl-Liebknecht-Str. 24–25, 14476 Potsdam, Germany), Amy Frappier (Department of Geosciences, Skidmore College, Saratoga Springs, NY 12866, USA), Dominique Genty (Laboratoire des Science du Climat et de l'Environment, CNRS, L'Orme des Merisiers, 91191 Gif-sure-Yvette CEDEX, France), Steffen Holzkämper (Department of Physical Geography, Stockholm University, 106 91 Stockholm, Sweden), Philip Hopley (Department of Earth and Planetary Sciences, Birkbeck, University of London, Malet St, London, WC1E 7HX, UK), Gayatri Kathayat (Institute of Global Environmental Change, Xi'an Jiaotong University, China), Duncan Keenan-Jones (School of Historical and Philosophical Inquiry, The University of Queensland, St Lucia QLD 4072, Australia), Gabriella Koltai (Institute of Geology, University of Innsbruck, Innrain 52, 6020 Innsbruck, Austria), Marc Luetscher (Swiss Institute for Speleology and Karst Studies (SISKA), Serre 68, 2300 La Chaux-de-Fonds, Switzerland), Ting-Yong Li (1: Swiss Institute for Speleology and Karst Studies (SISKA), Serre 68, 2300 La Chaux-de-Fonds, Switzerland; 2: Field Scientific Observation & Research Base of Karst Eco-environments at Nanchuan in Chongqing, Ministry of Land and Resources of China, Chongqing, 408435, China), Mahjoor Ahmad Lone (1: High-Precision Mass Spectrometry and Environment Change Laboratory (HISPEC), Department of Geosciences, National Taiwan University, Taipei 10617, Taiwan, China; 2: Research Center for Future Earth, National Taiwan University, Taipei 10617, Taiwan, China), Monika Markowska (University of Tübingen, Hölderlinstr. 12, 72074 Tübingen, Germany), Dave Mattey (Department of Earth Sciences, Royal Holloway University of London, Egham, Surrey, TW20 0EX, UK), Frank McDermott (UCD School of Earth Sciences, University College Dublin, Belfield, Dublin 4, Ireland), Ana Moreno (Dpto. de Procesos Geoambientales y Cambio Global, Instituto Pirenaico de Ecología-CSIC, Zaragoza, Spain), Gina Moseley (Institute of Geology, University of Innsbruck, Innrain 52, 6020 Innsbruck, Austria), Carole Nehme (IDEES UMR 6266 CNRS, Geography department, University of Rouen Normandie, Mont Saint Aignan, France), Valdir F. Novello (Instituto de Geociências, Universidade de São Paulo, São Paulo, Brazil), David Psomiadis (Imprint Analytics GmbH, Werner von Siemens Str. 1, 7343 Neutal, Austria), Kira Rehfeld (1: British Antarctic Survey, High Cross, Madingley Road, CB3 0ET, Cambridge, UK; 2: Institute of Environmental Physics, Ruprecht-Karls-Universität Heidelberg, Im Neuenheimer Feld 229, 69120 Heidelberg, Germany), Jiaoyang Ruan (Guangdong Provincial Key Lab of Geodynamics and Geohazards, School of Earth Sciences and Engineering, Sun Yat-sen University, Guangzhou 510275, China), Natasha Sekhon (Department of Geological Sciences, Jackson School of Geosciences, University of Texas at Austin, Austin, TX, USA), Lijuan Sha (Institute of Global Environmental Change, Xi'an Jiaotong University, Xi'an, Shaanxi, China), Denis Sholz (Institute for Geosciences, University of Mainz, Johann-Joachim-Becher-Weg 21, 55128 Mainz, Germany), Yavor Shopov (University Center for Space Research and Technologies, Faculty of Physics, University of Sofia, James Baucher 5, Sofia 1164, Bulgaria), Andrew Smith (NERC Isotope Geoscience Facility, British Geological Survey, Nottingham, UK), Nicolás Strikis (Departamento de Geoquímica, Universidade Federal Fluminense, Niterói, Brazil), Pauline Treble (ANSTO, Lucas Heights NSW, Australia), Ezgi Ünal-İmer (Department of Geological Engineering, Hacettepe University, Turkey), Anton Vaks (Geological Survey of Israel, 30 Malkhei Israel St., Jerusalem, Israel), Stef Vansteenberge (Analytical, Environmental & Geo-Chemistry, Department of Chemistry, Vrije Universiteit Brussel, Brussels, Belgium), Cristina Veiga-Pires (University of Algarve – FCT, CIMA Research Center, Campus de Gambelas, 8005139 Faro, Portugal), Ny Riavo Voarintsoa (Institute of Earth Sciences, The Hebrew University, Jerusalem, Israel), Xianfeng Wang (Earth Observatory of Singapore, Nanyang Technological University, Singapore 636798), Corinne Wong (Environmental Science Institute, The University of Texas at Austin, 2275 Speedway, Austin, TX 78712, USA), Barbara Wortham (Department of Earth and Planetary Science, University of California, Davis, California, USA), Jennifer Wurtzel (1: Research School of Earth Sciences, Australian National University, Canberra, ACT, Australia; 2: ARC Centre of Excellence for Climate System Science, Australian National University, Canberra, ACT, Australia; 3: now at: NSW Department of Primary Industries, Orange Agricultural Institute, Orange, NSW, Australia) and Baoyun Zong (Institute of Global Environmental Change, Xi'an Jiaotong University, Xi'an, Shaanxi, China).

Author contributions. LCB is the coordinator of the SISAL working group. KA, SPH, LCB, MD and AB designed the database, drawing on discussions with participants at the first SISAL workshop. KA, SPH, LCB and SAM were responsible for the database construction. NK, SMA, MA, YAB, PB, KB, YB, SC, WD, IGH, JH, ZK, IL, ML, FL, AL, CP-M, RP and NS coordinated the regional data collection. All authors listed as "SISAL Working Group members" provided data or helped to complete data entry. TA and DG contributed original unpublished data to the database. The first draft of the paper was written by KA, SPH, LCB, MB, AB and NK and all authors contributed to the final version.

Competing interests. The authors declare that they have no conflict of interest.

Acknowledgements. SISAL (Speleothem Isotopes Synthesis and Analysis) is a working group of the Past Global Changes (PAGES) programme. We thank PAGES for their support for this activity. Additional financial support for SISAL activities has been provided by the European Geosciences Union (EGU TE Winter call, grant number W2017/413), Irish Centre for Research in Applied Geosciences (iCRAG), European Association of Geochemistry (Early Career Ambassadors program 2017), Geological Survey Ireland, Quaternary Research Association UK, Navarino Environmental Observatory, Stockholm University, Savillex, John Cantle, University of Reading and University College Dublin (Seed Funding award, grant number SF1428). The design and creation of the database has been supported by funding to Sandy P. Harrison from the ERC-funded project GC2.0 (Global Change 2.0: Unlocking the past for a clearer future, grant number 694481) and by funding to Laia Comas-Bru from the Geological Survey Ireland Short Call 2017 (Developing a toolkit for model evaluation using speleothem isotope data, grant number 2017-SC-056). Sandy P. Harrison also acknowledges funding from the JPI-Belmont project "PAlaeo-Constraints on Monsoon Evolution and Dynamics (PACMEDY)" through the UK Natural Environmental Research Council (NERC). We thank SISAL members who contributed their published data to the database and provided additional information when necessary.

Edited by: David Carlson

References

Affek, H. P., Matthews, A., Ayalon, A., Bar-Matthews, M., Burstyn, Y., Zaarur, S., and Zilberman, T.: Accounting for kinetic isotope effects in Soreq Cave (Israel) speleothems, Geochim. Cosmochim. Acta, 143, 303–318, https://doi.org/10.1016/j.gca.2014.08.008, 2014.

Aharon, P., Aldridge, D., and Hellstrom, J.: Rainfall Variability and the Rise and Collapse of the Mississippian Chiefdoms: Evidence From a Desoto Caverns Stalagmite, in: Climates, Landscapes, and Civilizations, American Geophysical Union, 35–42, 2013.

Ait Brahim, Y., Cheng, H., Sifeddine, A., Wassenburg, J. A., Cruz, F. W., Khodri, M., Sha, L., Pérez-Zanón, N., Beraaouz, E. H., Apaéstegui, J., Guyot, J.-L., Jochum, K. P., and Bouchaou, L.: Speleothem records decadal to multidecadal hydroclimate variations in southwestern Morocco during the last millennium, Earth Planet. Sci. Lett., 476, 1–10, https://doi.org/10.1016/j.epsl.2017.07.045, 2017.

Akers, P. D., Brook, G. A., Railsback, L. B., Liang, F., Iannone, G., Webster, J. W., Reeder, P. P., Cheng, H., and Edwards, R. L.: An extended and higher-resolution record of climate and land use from stalagmite MC01 from Macal Chasm, Belize, revealing connections between major dry events, overall climate variability, and Maya sociopolitical changes, Palaeogeogr. Palaeocl., 459, 268–288, https://doi.org/10.1016/j.palaeo.2016.07.007, 2016.

Apaéstegui, J., Cruz, F. W., Sifeddine, A., Vuille, M., Espinoza, J. C., Guyot, J. L., Khodri, M., Strikis, N., Santos, R. V., Cheng, H., Edwards, L., Carvalho, E., and Santini, W.: Hydroclimate variability of the northwestern Amazon Basin near the Andean foothills of Peru related to the South American Monsoon System during the last 1600 years, Clim. Past, 10, 1967–1981, https://doi.org/10.5194/cp-10-1967-2014, 2014.

Arienzo, M. M., Swart, P. K., Pourmand, A., Broad, K., Clement, A. C., Murphy, L. N., Vonhof, H. B., and Kakuk, B.: Bahamian speleothem reveals temperature decrease associated with Heinrich stadials, Earth Planet. Sci. Lett., 430, 377–386, https://doi.org/10.1016/j.epsl.2015.08.035, 2015.

Arienzo, M. M., Swart, P. K., Broad, K., Clement, A. C., Pourmand, A. and Kakuk, B.: Multi-proxy evidence of millennial climate variability from multiple Bahamian speleothems, Quaternary Sci. Rev., 161, 18–29, https://doi.org/10.1016/j.quascirev.2017.02.004, 2017.

Asmerom, Y., Polyak, V. J., and Burns, S. J.: Variable winter moisture in the southwestern United States linked to rapid glacial climate shifts, Nat. Geosci., 3, 114–117, https://doi.org/10.1038/ngeo754, 2010.

Asrat, A., Baker, A., Mohammed, M. U., Leng, M. J., Van Calsteren, P., and Smith, C.: A high-resolution multi-proxy stalagmite record from Mechara, Southeastern Ethiopia: palaeohydrological implications for speleothem palaeoclimate reconstruction, J. Quaternary Sci., 22, 53–63, https://doi.org/10.1002/jqs.1013, 2006.

Asrat, A., Baker, A., Leng, M., Gunn, J., and Umer, M.: Environmental monitoring in the Mechara caves, Southeastern Ethiopia: implications for speleothem palaeoclimate studies, Int. J. Speleol., 37, 207–220, https://doi.org/10.5038/1827-806x.37.3.5, 2008.

Atkinson, T. C. and Hopley, P. J.: Speleothems and Palaeoclimates, chap. 5, in: Caves and Karst of the Yorkshire Dales, esite by: Waltham, T. and Lowe, D., British Cave Research Association, Buxton, 181–186, 2013.

Atsawawaranunt, K., Harrison, S. P., and Comas-Bru, L.: SISAL (Speleothem Isotopes Synthesis and AnaLysis Working Group) database Version 1.0, University of Reading, Dataset, https://doi.org/10.17864/1947.147, last access: 4 September 2018.

Ayliffe, L. K., Marianelli, P. C., Moriarty, K. C., Wells, R. T., McCulloch, M. T., Mortimer, G. E., and Hellstrom, J. C.: 500 ka precipitation record from southeastern Australia: Evidence for interglacial relative aridity, Geology, 26, 147–150, https://doi.org/10.1130/0091-7613(1998)026<0147:KPRFSA>2.3.CO;2, 1998.

Ayliffe, L. K., Gagan, M. K., Zhao, J., Drysdale, R. N., Hellstrom, J. C., Hantoro, W. S., Griffiths, M. L., Scott-Gagan, H., Pierre, E. S., Cowley, J. A., and Suwargadi, B. W.: Rapid interhemispheric climate links via the Australasian monsoon during the last deglaciation, Nat. Commun., 4, 2908, https://doi.org/10.1038/ncomms3908, 2013.

Bajo, P., Hellstrom, J., Frisia, S., Drysdale, R., Black, J., Woodhead, J., Borsato, A., Zanchetta, G., Wallace, M. W., Regattieri, E., and Haese, R.: "Cryptic" diagenesis and its implications for speleothem geochronologies, Quaternary Sci. Rev., 148, 17–28, https://doi.org/10.1016/j.quascirev.2016.06.020, 2016.

Baker, A., Ito, E., Smart, P. L., and McEwan, R. F.: Elevated and variable values of ^{13}C in speleothems in a British cave system, Chem. Geol., 136, 263–270, https://doi.org/10.1016/S0009-2541(96)00129-5, 1997.

Baker, A., Asrat, A., Fairchild, I. J., Leng, M. J., Wynn, P. M., Bryant, C., Genty, D., and Umer, M.: Analysis of the climate signal contained within δ^{18}O and growth rate parameters in two Ethiopian stalagmites, Geochim. Cosmochim. Acta, 71, 2975–2988, https://doi.org/10.1016/j.gca.2007.03.029, 2007.

Baker, A., Asrat, A., Fairchild, I. J., Leng, M. J., Thomas, L., Widmann, M., Jex, C. N., Dong, B., van Calsteren, P., and Bryant, C.: Decadal-scale rainfall variability in Ethiopia recorded in an annually laminated, Holocene-age, stalagmite, The Holocene, 20, 827–836, https://doi.org/10.1177/0959683610365934, 2010.

Baker, A., Wilson, R., Fairchild, I. J., Franke, J., Spötl, C., Mattey, D., Trouet, V., and Fuller, L.: High resolution δ^{18}O and δ^{13}C records from an annually laminated Scottish stalagmite and relationship with last millennium climate, Global Planet. Change, 79, 303–311, https://doi.org/10.1016/j.gloplacha.2010.12.007, 2011.

Baker, A., Bradley, C., Phipps, S. J., Fischer, M., Fairchild, I. J., Fuller, L., Spötl, C., and Azcurra, C.: Millennial-length forward models and pseudoproxies of stalagmite δ^{18}O: an example from NW Scotland, Clim. Past, 8, 1153–1167, https://doi.org/10.5194/cp-8-1153-2012, 2012.

Baker, J. L., Lachniet, M. S., Chervyatsova, O., Asmerom, Y., and Polyak, V. J.: Holocene warming in western continental Eurasia driven by glacial retreat and greenhouse forcing, Nat. Geosci., 10, 430–435, https://doi.org/10.1038/ngeo2953, 2017.

Baldini, J., McDermott, F., Baker, A., Baldini, L., Mattey, D., and Railsback, L.: Biomass effects on stalagmite growth and isotope ratios: A 20th century analogue from Wiltshire, England, Earth Planet. Sci. Lett., 240, 486–494, https://doi.org/10.1016/j.epsl.2005.09.022, 2005.

Baldini, J. U. L.: Morphologic and dimensional linkage between recently deposited speleothems and drip water from Browns Folly Mine, Wiltshire, England, J. Cave Karst Stud., 63, 83–90, available at: https://caves.org/pub/journal/PDF/V63/cave_63-03-fullr.pdf (last access: 30 May 2018), 2001.

Baldini, J. U. L., McDermott, F., Hoffmann, D. L., Richards, D. A., and Clipson, N.: Very high-frequency and seasonal cave atmosphere P_{CO_2} variability: Implications for stalagmite growth and oxygen isotope-based paleoclimate records, Earth Planet. Sci. Lett., 272, 118–129, https://doi.org/10.1016/j.epsl.2008.04.031, 2008.

Bar-Matthews, M., Ayalon, A., Kaufman, A. and Wasserburg, G. J.: The Eastern Mediterranean paleoclimate as a reflection of regional events: Soreq cave, Israel, Earth Planet. Sci. Lett., 166, 85–95, https://doi.org/10.1016/S0012-821X(98)00275-1, 1999.

Bartolomé, M., Moreno, A., Sancho, C., Stoll, H. M., Cacho, I., Spötl, C., Belmonte, Á., Edwards, R. L., Cheng, H., and Hellstrom, J. C.: Hydrological change in Southern Europe responding to increasing North Atlantic overturning during Greenland Stadial 1, P. Natl. Acad. Sci. USA, 112, 6568–6572, https://doi.org/10.1073/pnas.1503990112, 2015.

Berkelhammer, M., Sinha, A., Mudelsee, M., Cheng, H., Edwards, R. L., and Cannariato, K.: Persistent multidecadal power of the Indian Summer Monsoon, Earth Planet. Sci. Lett., 290, 166–172, https://doi.org/10.1016/j.epsl.2009.12.017, 2010.

Berkelhammer, M., Sinha, A., Stott, L., Cheng, H., Pausata, F. S. R., and Yoshimura, K.: An Abrupt Shift in the Indian Monsoon 4000 Years Ago, in: Climates, Landscapes, and Civilizations, American Geophysical Union, 75–88, 2013.

Bernal, J. P., Lachniet, M., McCulloch, M., Mortimer, G., Morales, P., and Cienfuegos, E.: A speleothem record of Holocene climate variability from southwestern Mexico, Quaternary Res., 75, 104–113, https://doi.org/10.1016/j.yqres.2010.09.002, 2011.

Bernal, J. P., Cruz, F. W., Strı́kis, N. M., Wang, X., Deininger, M., Catunda, M. C. A., Ortega-Obregón, C., Cheng, H., Edwards, R. L., and Auler, A. S.: High-resolution Holocene South American monsoon history recorded by a speleothem from Botuverá Cave, Brazil, Earth Planet. Sci. Lett., 450, 186–196, https://doi.org/10.1016/j.epsl.2016.06.008, 2016.

Boch, R. and Spötl, C.: Reconstructing palaeoprecipitation from an active cave flowstone, J. Quaternary Sci., 26, 675–687, https://doi.org/10.1002/jqs.1490, 2011.

Boch, R., Spötl, C., and Kramers, J.: High-resolution isotope records of early Holocene rapid climate change from two coeval stalagmites of Katerloch Cave, Austria, Quaternary Sci. Rev., 28, 2527–2538, https://doi.org/10.1016/j.quascirev.2009.05.015, 2009.

Boch, R., Cheng, H., Spötl, C., Edwards, R. L., Wang, X., and Häuselmann, Ph.: NALPS: a precisely dated European climate record 120–60 ka, Clim. Past, 7, 1247–1259, https://doi.org/10.5194/cp-7-1247-2011, 2011.

Bolliet, T., Brockmann, P., Masson-Delmotte, V., Bassinot, F., Daux, V., Genty, D., Landais, A., Lavrieux, M., Michel, E., Ortega, P., Risi, C., Roche, D. M., Vimeux, F., and Waelbroeck, C.: Water and carbon stable isotope records from natural archives: a new database and interactive online platform for data browsing, visualizing and downloading, Clim. Past, 12, 1693–1719, https://doi.org/10.5194/cp-12-1693-2016, 2016.

Bowen, G. J. and Wilkinson, B.: Spatial distribution of $\delta^{18}O$ in meteoric precipitation, Geology, 30, 315–318, https://doi.org/10.1130/0091-7613(2002)030<0315:SDOOIM>2.0.CO;2, 2002.

Bradley, R. S.: Paleoclimatology: Reconstructing Climates of the Quaternary, 3rd edn., Academic Press, https://doi.org/10.1016/C2009-0-18310-1, 2015.

Breitenbach, S. F. M., Rehfeld, K., Goswami, B., Baldini, J. U. L., Ridley, H. E., Kennett, D. J., Prufer, K. M., Aquino, V. V., Asmerom, Y., Polyak, V. J., Cheng, H., Kurths, J., and Marwan, N.: COnstructing Proxy Records from Age models (COPRA), Clim. Past, 8, 1765–1779, https://doi.org/10.5194/cp-8-1765-2012, 2012.

Breitenbach, S. F. M., Lechleitner, F. A., Meyer, H., Diengdoh, G., Mattey, D., and Marwan, N.: Cave ventilation and rainfall signals in dripwater in a monsoonal setting – a monitoring study from NE India, Chem. Geol., 402, 111–124, https://doi.org/10.1016/j.chemgeo.2015.03.011, 2015.

Bronk Ramsey, C.: Radiocarbon Calibration and Analysis of Stratigraphy: The OxCal Program, Radiocarbon, 37, 425–430, https://doi.org/10.1017/S0033822200030903, 1995.

Bronk Ramsey, C.: Development of the Radiocarbon Calibration Program, Radiocarbon, 43, 355–363, https://doi.org/10.1017/S0033822200038212, 2001.

Bronk Ramsey, C.: Deposition models for chronological records, Quaternary Sci. Rev., 27, 42–60, https://doi.org/10.1016/j.quascirev.2007.01.019, 2008.

Burns, S. J.: A 780-year annually resolved record of Indian Ocean monsoon precipitation from a speleothem from south Oman, J. Geophys. Res., 107, 4434, https://doi.org/10.1029/2001jd001281, 2002.

Burns, S. J., Godfrey, L. R., Faina, P., McGee, D., Hardt, B., Ranivoharimanana, L., and Randrianasy, J.: Rapid human-induced landscape transformation in Madagascar at the end of the first millennium of the Common Era, Quaternary Sci. Rev., 134, 92–99, https://doi.org/10.1016/j.quascirev.2016.01.007, 2016.

Cai, Y., Cheng, H., An, Z., Edwards, R. L., Wang, X., Tan, L., and Wang, J.: Large variations of oxygen isotopes in precipitation over south-central Tibet during Marine Isotope Stage 5, Geology, 38, 243–246, https://doi.org/10.1130/g30306.1, 2010a.

Cai, Y., Tan, L., Cheng, H., An, Z., Edwards, R. L., Kelly, M. J., Kong, X., and Wang, X.: The variation of summer monsoon precipitation in central China since the last deglaciation, Earth Planet. Sci. Lett., 291, 21–31, https://doi.org/10.1016/j.epsl.2009.12.039, 2010b.

Cai, Y., Zhang, H., Cheng, H., An, Z., Lawrence Edwards, R., Wang, X., Tan, L., Liang, F., Wang, J., Kelly, M., Edwards, R. L., Wang, X., Tan, L., Liang, F., Wang, J., and Kelly, M.: The Holocene Indian monsoon variability over the southern Tibetan Plateau and its teleconnections, Earth Planet. Sci. Lett., 335–336, 135–144, https://doi.org/10.1016/j.epsl.2012.04.035, 2012.

Cai, Y., Fung, I. Y., Edwards, R. L., An, Z., Cheng, H., Lee, J.-E., Tan, L., Shen, C.-C., Wang, X., Day, J. A., Zhou, W., Kelly, M. J., and Chiang, J. C. H.: Variability of stalagmite-inferred Indian monsoon precipitation over the past 252,000 y, P. Natl. Acad. Sci. USA, 112, 2954–2959, https://doi.org/10.1073/pnas.1424035112, 2015.

Caley, T., Roche, D. M., Waelbroeck, C., and Michel, E.: Oxygen stable isotopes during the Last Glacial Maximum climate: perspectives from data–model (iLOVECLIM) comparison, Clim. Past, 10, 1939–1955, https://doi.org/10.5194/cp-10-1939-2014, 2014.

Carolin, S. A., Cobb, K. M., Adkins, J. F., Clark, B., Conroy, J. L., Lejau, S., Malang, J., and Tuen, A. A.: Varied Response of Western Pacific Hydrology to Climate Forcings over the Last Glacial Period, Science, 340, 1564–1566, https://doi.org/10.1126/science.1233797, 2013.

Carolin, S. A., Cobb, K. M., Lynch-Stieglitz, J., Moerman, J. W., Partin, J. W., Lejau, S., Malang, J., Clark, B., Tuen, A. A., and Adkins, J. F.: Northern Borneo stalagmite records reveal West Pacific hydroclimate across MIS 5 and 6, Earth Planet. Sci. Lett., 439, 182–193, https://doi.org/10.1016/j.epsl.2016.01.028, 2016.

Chen, S., Hoffmann, S. S., Lund, D. C., Cobb, K. M., Emile-Geay, J., and Adkins, J. F.: A high-resolution speleothem record of western equatorial Pacific rainfall: Implications for Holocene ENSO evolution, Earth Planet. Sci. Lett., 442, 61–71, https://doi.org/10.1016/j.epsl.2016.02.050, 2016.

Cheng, H., Edwards, R., Hoff, J., Gallup, C., Richards, D., and Asmerom, Y.: The half-lives of uranium-234 and thorium-230, Chem. Geol., 169, 17–33, https://doi.org/10.1016/S0009-2541(99)00157-6, 2000.

Cheng, H., Zhang, P. Z., Spötl, C., Edwards, R. L., Cai, Y. J., Zhang, D. Z., Sang, W. C., Tan, M., and An, Z. S.: The climatic cyclicity in semiarid-arid central Asia over the past 500,000 years, Geophys. Res. Lett., 39, L01705, https://doi.org/10.1029/2011GL050202, 2012.

Cheng, H., Lawrence Edwards, R., Shen, C.-C., Polyak, V. J., Asmerom, Y., Woodhead, J., Hellstrom, J., Wang, Y., Kong, X., Spötl, C., Wang, X., and Calvin Alexander, E.: Improvements in ^{230}Th dating, ^{230}Th and ^{234}U half-life values, and U–Th isotopic measurements by multi-collector inductively coupled plasma mass spectrometry, Earth Planet. Sci. Lett., 371–372, 82–91, https://doi.org/10.1016/j.epsl.2013.04.006, 2013.

Cheng, H., Sinha, A., Verheyden, S., Nader, F. H., Li, X. L., Zhang, P. Z., Yin, J. J., Yi, L., Peng, Y. B., Rao, Z. G., Ning, Y. F., and Edwards, R. L.: The climate variability in northern Levant over the past 20,000 years, Geophys. Res. Lett., 42, 8641–8650, https://doi.org/10.1002/2015GL065397, 2015.

Cheng, H., Spötl, C., Breitenbach, S. F. M., Sinha, A., Wassenburg, J. A., Jochum, K. P., Scholz, D., Li, X., Yi, L., Peng, Y., Lv, Y., Zhang, P., Votintseva, A., Loginov, V., Ning, Y., Kathayat, G., and Edwards, R. L.: Climate variations of Central Asia on orbital to millennial timescales, Sci. Rep., 6, 36975, https://doi.org/10.1038/srep36975, 2016a.

Cheng, H., Edwards, R. L., Sinha, A., Spötl, C., Yi, L., Chen, S., Kelly, M., Kathayat, G., Wang, X., Li, X., Kong, X., Wang, Y., Ning, Y., and Zhang, H.: The Asian monsoon over the past 640,000 years and ice age terminations, Nature, 534, 640–646, https://doi.org/10.1038/nature18591, 2016b.

Cobb, K. M., Adkins, J. F., Partin, J. W., and Clark, B.: Regional-scale climate influences on temporal variations of rainwater and cave dripwater oxygen isotopes in northern Borneo, Earth Planet. Sci. Lett., 263, 207–220, https://doi.org/10.1016/j.epsl.2007.08.024, 2007.

Cohen, K. M., Finney, S., and Gibbard, P. L.: International Chronostratigraphic Chart, Int. Comm. Stratigr., available at: http://www.stratigraphy.org (last access: 30 May 2018), 2015.

Columbu, A., Drysdale, R., Capron, E., Woodhead, J., Waele, J. De, Sanna, L., Hellstrom, J., and Bajo, P.: Early last glacial intra-interstadial climate variability recorded in a Sardinian speleothem, Quaternary Sci. Rev., 169, 391–397, https://doi.org/10.1016/j.quascirev.2017.05.007, 2017.

Couchoud, I., Genty, D., Hoffmann, D., Drysdale, R., and Blamart, D.: Millennial-scale climate variability during the Last Interglacial recorded in a speleothem from southwestern France, Quaternary Sci. Rev., 28, 3263–3274, https://doi.org/10.1016/j.quascirev.2009.08.014, 2009.

Cruz, F. W., Burns, S. J., Karmann, I., Sharp, W. D., Vuille, M., Cardoso, A. O., Ferrari, J. A., Dias, P. L. S., and Viana, O.: Insolation-driven changes in atmospheric circulation over the past 116,000 years in subtropical Brazil, Nature, 434, 63–66, https://doi.org/10.1038/nature03365, 2005.

Cruz, F. W., Vuille, M., Burns, S. J., Wang, X., Cheng, H., Werner, M., Lawrence Edwards, R., Karmann, I., Auler, A. S., and Nguyen, H.: Orbitally driven east–west antiphasing of South American precipitation, Nat. Geosci., 2, 210–214, https://doi.org/10.1038/ngeo444, 2009.

Daley, T. J., Thomas, E. R., Holmes, J. A., Street-Perrott, F. A., Chapman, M. R., Tindall, J. C., Valdes, P. J., Loader, N. J., Marshall, J. D., Wolff, E. W., Hopley, P. J., Atkinson, T., Barber, K. E., Fisher, E. H., Robertson, I., Hughes, P. D. M., and Roberts, C. N.: The 8200 yr BP cold event in stable isotope records from the North Atlantic region, Global Planet. Change, 79, 288–302, https://doi.org/10.1016/j.gloplacha.2011.03.006, 2011.

Demény, A., Czuppon, G., Siklósy, Z., Leél-Őssy, S., Lin, K., Shen, C.-C., and Gulyás, K.: Mid-Holocene climate conditions and moisture source variations based on stable H, C and O isotope compositions of speleothems in Hungary, Quatern. Int., 293, 150–156, https://doi.org/10.1016/j.quaint.2012.05.035, 2013.

Demény, A., Kern, Z., Czuppon, G., Németh, A., Schöll-Barna, G., Siklósy, Z., Leél-Őssy, S., Cook, G., Serlegi, G., Bajnóczi, B., Sümegi, P., Király, Á., Kiss, V., Kulcsár, G., and Bondár, M.: Middle Bronze Age humidity and temperature variations, and societal changes in East-Central Europe, Quatern. Int., https://doi.org/10.1016/j.quaint.2017.11.023, online first, 2017a.

Demény, A., Kern, Z., Czuppon, G., Németh, A., Leél-Őssy, S., Siklósy, Z., Lin, K., Hu, H.-M., Shen, C.-C., Vennemann, T. W., and Haszpra, L.: Stable isotope compositions of speleothems from the last interglacial – Spatial patterns of climate fluctuations in Europe, Quaternary Sci. Rev., 161, 68–80, https://doi.org/10.1016/j.quascirev.2017.02.012, 2017b.

Denniston, R. F., Wyrwoll, K.-H., Polyak, V. J., Brown, J. R., Asmerom, Y., Wanamaker, A. D., LaPointe, Z., Ellerbroek, R., Barthelmes, M., Cleary, D., Cugley, J., Woods, D., and Humphreys, W. F.: A Stalagmite record of Holocene Indonesian–Australian summer monsoon variability from the Australian tropics, Quaternary Sci. Rev., 78, 155–168, https://doi.org/10.1016/j.quascirev.2013.08.004, 2013a.

Denniston, R. F., Wyrwoll, K.-H., Asmerom, Y., Polyak, V. J., Humphreys, W. F., Cugley, J., Woods, D., LaPointe, Z., Peota, J., and Greaves, E.: North Atlantic forcing of millennial-scale Indo-Australian monsoon dynamics during the Last Glacial period, Quaternary Sci. Rev., 72, 159–168, https://doi.org/10.1016/j.quascirev.2013.04.012, 2013b.

Denniston, R. F., Villarini, G., Gonzales, A. N., Wyrwoll, K.-H., Polyak, V. J., Ummenhofer, C. C., Lachniet, M. S., Wanamaker, A. D., Humphreys, W. F., Woods, D., and Cugley, J.: Extreme rainfall activity in the Australian tropics reflects changes in the El Niño/Southern Oscillation over the last two millennia, P. Natl. Acad. Sci. USA, 112, 4576–4581, https://doi.org/10.1073/pnas.1422270112, 2015.

Denniston, R. F., Ummenhofer, C. C., Wanamaker, A. D., Lachniet, M. S., Villarini, G., Asmerom, Y., Polyak, V. J., Passaro, K. J., Cugley, J., Woods, D., and Humphreys, W. F.: Expansion and Contraction of the Indo-Pacific Tropical Rain Belt over the Last Three Millennia, Sci. Rep., 6, 34485, https://doi.org/10.1038/srep34485, 2016.

Dong, J., Wang, Y., Cheng, H., Hardt, B., Edwards, R. L., Kong, X., Wu, J., Chen, S., Liu, D., Jiang, X., and Zhao, K.: A high-resolution stalagmite record of the Holocene East Asian monsoon from Mt Shennongjia, central China, The Holocene, 20, 257–264, https://doi.org/10.1177/0959683609350393, 2010.

Dorale, J. A., Edwards, R. L., Ito, E., and González, L. A.: Climate and vegetation history of the midcontinent from 75 to 25 ka: a speleothem record from Crevice Cave, Missouri, USA, Science, 282, 1871–1874, https://doi.org/10.1126/science.282.5395.1871, 1998.

Drăguşin, V., Staubwasser, M., Hoffmann, D. L., Ersek, V., Onac, B. P., and Veres, D.: Constraining Holocene hydrological changes in the Carpathian–Balkan region using speleothem $\delta^{18}O$ and pollen-based temperature reconstructions, Clim. Past, 10, 1363–1380, https://doi.org/10.5194/cp-10-1363-2014, 2014.

Drysdale, R., Zanchetta, G., Hellstrom, J., Maas, R., Fallick, A., Pickett, M., Cartwright, I., and Piccini, L.: Late Holocene drought responsible for the collapse of Old World civilizations is recorded in an Italian cave flowstone, Geology, 34, 101–104, https://doi.org/10.1130/g22103.1, 2006.

Drysdale, R. N., Zanchetta, G., Hellstrom, J. C., Fallick, A. E., Zhao, J., Isola, I., and Bruschi, G.: Palaeoclimatic implications of the growth history and stable isotope ($\delta^{18}O$ and $\delta^{13}C$) geochemistry of a Middle to Late Pleistocene stalagmite from central-western Italy, Earth Planet. Sci. Lett., 227, 215–229, https://doi.org/10.1016/j.epsl.2004.09.010, 2004.

Drysdale, R. N., Zanchetta, G., Hellstrom, J. C., Fallick, A. E., and Zhao, J.: Stalagmite evidence for the onset of the Last Interglacial in southern Europe at 129 ± 1 ka, Geophys. Res. Lett., 32, L24708, https://doi.org/10.1029/2005GL024658, 2005.

Drysdale, R. N., Zanchetta, G., Hellstrom, J. C., Fallick, A. E., McDonald, J., and Cartwright, I.: Stalagmite evidence for the precise timing of North Atlantic cold events during the early last glacial, Geology, 35, 77–80, https://doi.org/10.1130/G23161A.1, 2007.

Drysdale, R. N., Hellstrom, J. C., Zanchetta, G., Fallick, A. E., Sanchez Goni, M. F., Couchoud, I., McDonald, J., Maas, R., Lohmann, G., and Isola, I.: Evidence for Obliquity Forcing of Glacial Termination II, Science, 325, 1527–1531, https://doi.org/10.1126/science.1170371, 2009.

Duan, W., Cheng, H., Tan, M., and Edwards, R. L.: Onset and duration of transitions into Greenland Interstadials 15.2 and 14 in northern China constrained by an annually laminated stalagmite, Sci. Rep., 6, 20844, https://doi.org/10.1038/srep20844, 2016.

Dutt, S., Gupta, A. K., Clemens, S. C., Cheng, H., Singh, R. K., Kathayat, G., and Edwards, R. L.: Abrupt changes in Indian summer monsoon strength during 33,800 to 5500 years B.P., Geophys. Res. Lett., 42, 5526–5532, https://doi.org/10.1002/2015GL064015, 2015.

Edwards, R. L., Chen, J. H., and Wasserburg, G. J.: ^{238}U–^{234}U–^{230}Th–^{232}Th systematics and the precise measurement of time over the past 500,000 years, Earth Planet. Sci. Lett., 81, 175–192, https://doi.org/10.1016/0012-821X(87)90154-3, 1987.

Ersek, V., Clark, P. U., Mix, A. C., Cheng, H., and Edwards, R. L.: Holocene winter climate variability in mid-latitude western North America, Nat. Commun., 3, 1219, https://doi.org/10.1038/ncomms2222, 2012.

Fairchild, I. J. and Baker, A.: Speleothem Science, John Wiley & Sons, Ltd, Chichester, UK, 2012.

Feng, W., Hardt, B. F., Banner, J. L., Meyer, K. J., James, E. W., Musgrove, M., Edwards, R. L., Cheng, H., and Min, A.: Changing amounts and sources of moisture in the U.S. southwest since the Last Glacial Maximum in response to global climate change, Earth Planet. Sci. Lett., 401, 47–56, https://doi.org/10.1016/j.epsl.2014.05.046, 2014.

Finné, M., Bar-Matthews, M., Holmgren, K., Sundqvist, H. S., Liakopoulos, I., and Zhang, Q.: Speleothem evidence for late Holocene climate variability and floods in Southern Greece, Quaternary Res., 81, 213–227, https://doi.org/10.1016/j.yqres.2013.12.009, 2014.

Finné, M., Holmgren, K., Shen, C.-C., Hu, H.-M., Boyd, M., and Stocker, S.: Late Bronze Age climate change and the destruction of the Mycenaean Palace of Nestor at Pylos, PLoS One, 12, e0189447, https://doi.org/10.1371/journal.pone.0189447, 2017.

Fleitmann, D., Burns, S. J., Neff, U., Mudelsee, M., Mangini, A., and Matter, A.: Palaeoclimatic interpretation of high-resolution oxygen isotope profiles derived from annually laminated speleothems from Southern Oman, Quaternary Sci. Rev., 23, 935–945, https://doi.org/10.1016/j.quascirev.2003.06.019, 2004.

Fleitmann, D., Burns, S. J., Mangini, A., Mudelsee, M., Kramers, J., Villa, I., Neff, U., Al-Subbary, A. A., Buettner, A., Hippler, D., and Matter, A.: Holocene ITCZ and Indian monsoon dynamics recorded in stalagmites from Oman and Yemen (Socotra), Quaternary Sci. Rev., 26, 170–188, https://doi.org/10.1016/j.quascirev.2006.04.012, 2007.

Fleitmann, D., Cheng, H., Badertscher, S., Edwards, R. L., Mudelsee, M., Göktürk, O. M., Fankhauser, A., Pickering, R., Raible, C. C., Matter, A., Kramers, J., and Tüysüz, O.: Timing and climatic impact of Greenland interstadials recorded in stalagmites from northern Turkey, Geophys. Res. Lett., 36, L19707, https://doi.org/10.1029/2009gl040050, 2009.

Fohlmeister, J., Schröder-Ritzrau, A., Scholz, D., Spötl, C., Riechelmann, D. F. C., Mudelsee, M., Wackerbarth, A., Gerdes, A., Riechelmann, S., Immenhauser, A., Richter, D. K., and Mangini, A.: Bunker Cave stalagmites: an archive for central European Holocene climate variability, Clim. Past, 8, 1751–1764, https://doi.org/10.5194/cp-8-1751-2012, 2012.

Fohlmeister, J., Vollweiler, N., Spötl, C., and Mangini, A.: COMNISPA II: Update of a mid-European isotope climate record, 11 ka to present, The Holocene, 23, 749–754, https://doi.org/10.1177/0959683612465446, 2013.

Fohlmeister, J., Plessen, B., Dudashvili, A. S., Tjallingii, R., Wolff, C., Gafurov, A., and Cheng, H.: Winter precipitation changes during the Medieval Climate Anomaly and the Little Ice Age in arid Central Asia, Quaternary Sci. Rev., 178, 24–36, https://doi.org/10.1016/j.quascirev.2017.10.026, 2017.

Frisia, S., Borsato, A., Mangini, A., Spötl, C., Madonia, G., and Sauro, U.: Holocene Climate Variability in Sicily from a Discontinuous Stalagmite Record and the Mesolithic to Neolithic Transition, Quaternary Res., 66, 388–400, https://doi.org/10.1016/j.yqres.2006.05.003, 2006.

Frisia, S., Fairchild, I. J., Fohlmeister, J., Miorandi, R., Spotl, C., and Borsato, A.: Carbon mass-balance modelling and carbon isotope exchange processes in dynamic caves, Geochim. Cosmochim. Acta, 75, 380–400, https://doi.org/10.1016/j.gca.2010.10.021, 2011.

Frumkin, A., Ford, D. C., and Schwarcz, H. P.: Continental Oxygen Isotopic Record of the Last 170,000 Years in Jerusalem, Quaternary Res., 51, 317–327, https://doi.org/10.1006/qres.1998.2031, 1999.

Frumkin, A., Ford, D. C., and Schwarcz, H. P.: Paleoclimate and vegetation of the Last Glacial Cycles in Jerusalem from a Speleothem Record, Global Biogeochem. Cy., 14, 863–870, https://doi.org/10.1029/1999gb001245, 2000.

Genty, D., Vokal, B., Obelic, B., and Massault, M.: Bomb ^{14}C time history recorded in two modern stalagmites – importance for soil organic matter dynamics and bomb ^{14}C distribution over continents, Earth Planet. Sci. Lett., 160, 795–809, https://doi.org/10.1016/S0012-821X(98)00128-9, 1998.

Genty, D., Massault, M., Gilmour, M., Baker, A., Verheyden, S., and Kepens, E.: Calculation of Past Dead Carbon Proportion and Variability by the Comparison of AMS ^{14}C and Tims U/TH Ages on Two Holocene Stalagmites, Radiocarbon, 41, 251–270, https://doi.org/10.1017/S003382220005712X, 1999.

Genty, D., Blamart, D., Ouahdi, R., Gilmour, M., Baker, A., Jouzel, J., and Van-Exter, S.: Precise dating of Dansgaard–Oeschger climate oscillations in western Europe from stalagmite data, Nature, 421, 833–837, https://doi.org/10.1038/nature01391, 2003.

Genty, D., Blamart, D., Ghaleb, B., Plagnes, V., Causse, C., Bakalowicz, M., Zouari, K., Chkir, N., Hellstrom, J., and Wainer, K.: Timing and dynamics of the last deglaciation from European and North African δ^{13}C stalagmite profiles—comparison with Chinese and South Hemisphere stalagmites, Quaternary Sci. Rev., 25, 2118–2142, https://doi.org/10.1016/j.quascirev.2006.01.030, 2006.

Genty, D., Combourieu-Nebout, N., Peyron, O., Blamart, D., Wainer, K., Mansuri, F., Ghaleb, B., Isabello, L., Dormoy, I., and von Grafenstein, U.: Isotopic characterization of rapid climatic events during OIS3 and OIS4 in Villars Cave stalagmites (SW-France) and correlation with Atlantic and Mediterranean pollen records, Quaternary Sci. Rev., 29, 2799–2820, https://doi.org/10.1016/j.quascirev.2010.06.035, 2010.

Grant, K. M., Rohling, E. J., Bar-Matthews, M., Ayalon, A., Medina-Elizalde, M., Ramsey, C. B., Satow, C., and Roberts, A. P.: Rapid coupling between ice volume and polar temperature over the past 150,000 years, Nature, 491, 744–747, https://doi.org/10.1038/nature11593, 2012.

Griffiths, M. L., Drysdale, R. N., Gagan, M. K., Zhao, J.-X., Ayliffe, L. K., Hellstrom, J. C., Hantoro, W. S., Frisia, S., Feng, Y.-X., Cartwright, I., Pierre, E. St., Fischer, M. J., and Suwargadi, B. W.: Increasing Australian–Indonesian monsoon rainfall linked to early Holocene sea-level rise, Nat. Geosci., 2, 636–639, https://doi.org/10.1038/ngeo605, 2009.

Griffiths, M. L., Drysdale, R. N., Gagan, M. K., Hellstrom, J. C., Couchoud, I., Ayliffe, L. K., Vonhof, H. B., and Hantoro, W. S.: Australasian monsoon response to Dansgaard–Oeschger event 21 and teleconnections to higher latitudes, Earth Planet. Sci. Lett., 369–370, 294–304, https://doi.org/10.1016/j.epsl.2013.03.030, 2013.

Griffiths, M. L., Kimbrough, A. K., Gagan, M. K., Drysdale, R. N., Cole, J. E., Johnson, K. R., Zhao, J.-X., Cook, B. I., Hellstrom, J. C., and Hantoro, W. S.: Western Pacific hydroclimate linked to global climate variability over the past two millennia, Nat. Commun., 7, 11719, https://doi.org/10.1038/ncomms11719, 2016.

Haese, B., Werner, M., and Lohmann, G.: Stable water isotopes in the coupled atmosphere–land surface model ECHAM5-JSBACH, Geosci. Model Dev., 6, 1463–1480, https://doi.org/10.5194/gmd-6-1463-2013, 2013.

Han, L.-Y., Li, T.-Y., Cheng, H., Edwards, R. L., Shen, C.-C., Li, H.-C., Huang, C.-X., Li, J.-Y., Yuan, N., Wang, H.-B., Zhang, T.-T., and Zhao, X.: Potential influence of temperature changes in the Southern Hemisphere on the evolution of the Asian summer monsoon during the last glacial period, Quatern. Int., 392, 239–250, https://doi.org/10.1016/j.quaint.2015.05.068, 2016.

Hardt, B., Rowe, H. D., Springer, G. S., Cheng, H., and Edwards, R. L.: The seasonality of east central North American precipitation based on three coeval Holocene speleothems from southern West Virginia, Earth Planet. Sci. Lett., 295, 342–348, https://doi.org/10.1016/j.epsl.2010.04.002, 2010.

Harrison, S. P., Bartlein, P. J., Brewer, S., Prentice, I. C., Boyd, M., Hessler, I., Holmgren, K., Izumi, K., and Willis, K.: Climate model benchmarking with glacial and mid-Holocene climates, Clim. Dynam., 43, 671–688, https://doi.org/10.1007/s00382-013-1922-6, 2014.

Häuselmann, A. D., Fleitmann, D., Cheng, H., Tabersky, D., Günther, D., and Edwards, R. L.: Timing and nature of the penultimate deglaciation in a high alpine stalagmite from Switzerland, Quaternary Sci. Rev., 126, 264–275, https://doi.org/10.1016/j.quascirev.2015.08.026, 2015.

Hellstrom, J., McCulloch, M., and Stone, J.: A Detailed 31,000-Year Record of Climate and Vegetation Change, from the Isotope Geochemistry of Two New Zealand Speleothems, Quaternary Res., 50, 167–178, https://doi.org/10.1006/qres.1998.1991, 1998.

Hellstrom, J. C.: U–Th dating of speleothems with high initial ^{230}Th using stratigraphical constraint, Quat. Geochronol., 1, 289–295, https://doi.org/10.1016/j.quageo.2007.01.004, 2006.

Hendy, C.: The isotopic geochemistry of speleothems – I. The calculation of the effects of different modes of formaion on the isotopic composition of speleothems and their applicability as paleoclimatic indicators, Geochim. Cosmochim. Acta, 35, 801–824, https://doi.org/10.1016/0016-7037(71)90127-X, 1971.

Holmgren, K., Lauritzen, S.-E., and Possnert, G.: ^{230}Th^{234}U and ^{14}C dating of a late Pleistocene stalagmite in Lobatse II Cave, Botswana, Quaternary Sci. Rev., 13, 111–119, https://doi.org/10.1016/0277-3791(94)90036-1, 1994.

Holmgren, K., Karlén, W., and Shaw, P. A.: Paleoclimatic Significance of the Stable Isotopic Composition and Petrology of a Late Pleistocene Stalagmite from Botswana, Quaternary Res., 43, 320–328, https://doi.org/10.1006/qres.1995.1038, 1995.

Holmgren, K., Karlén, W., Lauritzen, S. E., Lee-Thorp, J. A., Partridge, T. C., Piketh, S., Repinski, P., Stevenson, C., Svanered, O., and Tyson, P. D.: A 3000-year high-resolution stalagmitebased record of palaeoclimate for northeastern South Africa, The Holocene, 9, 295–309, https://doi.org/10.1191/095968399672625464, 1999.

Holmgren, K., Lee-Thorp, J. A., Cooper, G. R. J., Lundblad, K., Partridge, T. C., Scott, L., Sithaldeen, R., Talma, A. S., and Tyson, P. D.: Persistent millennial-scale climatic variability over the past 25,000 years in Southern Africa, Quaternary Sci. Rev., 22, 2311–2326, https://doi.org/10.1016/s0277-3791(03)00204-x, 2003.

Holzkämper, S., Holmgren, K., Lee-Thorp, J., Talma, S., Mangini, A., and Partridge, T.: Late Pleistocene stalagmite growth in Wolkberg Cave, South Africa, Earth Planet. Sci. Lett., 282, 212–221, https://doi.org/10.1016/j.epsl.2009.03.016, 2009.

Hopley, P. J., Weedon, G. P., Marshall, J. D., Herries, A. I. R., Latham, A. G., and Kuykendall, K. L.: High- and low-latitude orbital forcing of early hominin habitats in South Africa, Earth Planet. Sci. Lett., 256, 419–432, https://doi.org/10.1016/j.epsl.2007.01.031, 2007a.

Hopley, P. J., Marshall, J. D., Weedon, G. P., Latham, A. G., Herries, A. I. R., and Kuykendall, K. L.: Orbital forcing and the spread of C4 grasses in the late Neogene: stable isotope evidence from South African speleothems, J. Hum. Evol., 53, 620–634, https://doi.org/10.1016/j.jhevol.2007.03.007, 2007b.

Hu, C., Henderson, G. M., Huang, J., Xie, S., Sun, Y., and Johnson, K. R.: Quantification of Holocene Asian monsoon rainfall from spatially separated cave records, Earth Planet. Sci. Lett., 266, 221–232, https://doi.org/10.1016/j.epsl.2007.10.015, 2008.

Huang, W., Wang, Y., Cheng, H., Edwards, R. L., Shen, C.-C., Liu, D., Shao, Q., Deng, C., Zhang, Z., and Wang, Q.: Multi-scale Holocene Asian monsoon variability deduced from a twin-stalagmite record in southwestern China, Quaternary Res., 86, 34–44, https://doi.org/10.1016/j.yqres.2016.05.001, 2016.

Ivanovich, M. and Harmon, R.: Uranium-series disequilibrium: applications to earth, marine, and environmental sciences. 2, available at: https://inis.iaea.org/search/search.aspx?orig_q=RN:25065862 (last access: 30 January 2018), 1992.

Jaqueto, P., Trindade, R. I. F., Hartmann, G. A., Novello, V. F., Cruz, F. W., Karmann, I., Strauss, B. E., and Feinberg, J. M.: Linking speleothem and soil magnetism in the Pau d'Alho cave (central South America), J. Geophys. Res.-Sol. Ea., 121, 7024–7039, https://doi.org/10.1002/2016JB013541, 2016.

Johnston, V. E., Borsato, A., Spötl, C., Frisia, S., and Miorandi, R.: Stable isotopes in caves over altitudinal gradients: fractionation behaviour and inferences for speleothem sensitivity to climate change, Clim. Past, 9, 99–118, https://doi.org/10.5194/cp-9-99-2013, 2013.

Joshi, L. M., Kotlia, B. S., Ahmad, S. M., Wu, C.-C., Sanwal, J., Raza, W., Singh, A. K., Shen, C.-C., Long, T., and Sharma, A. K.: Reconstruction of Indian monsoon precipitation variability between 4.0 and 1.6 ka BP using speleothem δ^{18}O records from the Central Lesser Himalaya, India, Arab. J. Geosci., 10, 356, https://doi.org/10.1007/s12517-017-3141-7, 2017.

Kacanski, A., Carmi, I., Shemesh, A., Kronfeld, J., Yam, R., and Flexer, A.: Late Holocene Climatic Change in the Balkans: Speleothem Isotopic Data from Serbia, Radiocarbon, 43, 647–658, https://doi.org/10.1017/s0033822200041308, 2001.

Kathayat, G., Cheng, H., Sinha, A., Spötl, C., Edwards, R. L., Zhang, H., Li, X., Yi, L., Ning, Y., Cai, Y., Lui, W. L., and Breitenbach, S. F. M.: Indian monsoon variability on millennial-orbital timescales, Sci. Rep., 6, 24374, https://doi.org/10.1038/srep24374, 2016.

Kennett, D. J., Breitenbach, S. F. M., Aquino, V. V, Asmerom, Y., Awe, J., Baldini, J. U. L., Bartlein, P., Culleton, B. J., Ebert, C., Jazwa, C., Macri, M. J., Marwan, N., Polyak, V., Prufer, K. M., Ridley, H. E., Sodemann, H., Winterhalder, B., and Haug, G. H.: Development and Disintegration of Maya Political Systems in Response to Climate Change, Science, 338, 788–791, https://doi.org/10.1126/science.1226299, 2012.

Koltai, G., Spötl, C., Shen, C.-C., Wu, C.-C., Rao, Z., Palcsu, L., Kele, S., Surányi, G., and Bárány-Kevei, I.: A penultimate glacial climate record from southern Hungary, J. Quaternary Sci., 32, 946–956, https://doi.org/10.1002/jqs.2968, 2017.

Kotlia, B. S., Singh, A. K., Sanwal, J., Raza, W., Ahmad, S. M., Joshi, L. M., Sirohi, M., Sharma, A. K., and Sagar, N.: Stalagmite Inferred High Resolution Climatic Changes through Pleistocene-Holocene Transition in Northwest Indian Himalaya, J. Earth Sci. Clim. Change, 7, https://doi.org/10.4172/2157-7617.1000338, 2016.

Labuhn, I., Genty, D., Vonhof, H., Bourdin, C., Blamart, D., Douville, E., Ruan, J., Cheng, H., Edwards, R. L., Pons-Branchu, E., and Pierre, M.: A high-resolution fluid inclusion δ^{18}O record from a stalagmite in SW France: modern calibration and comparison with multiple proxies, Quaternary Sci. Rev., 110, 152–165, https://doi.org/10.1016/j.quascirev.2014.12.021, 2015.

Lachniet, M. S.: A 1500-year El Niño/Southern Oscillation and rainfall history for the Isthmus of Panama from speleothem calcite, J. Geophys. Res., 109, D20117, https://doi.org/10.1029/2004jd004694, 2004.

Lachniet, M. S., Asmerom, Y., and Polyak, V.: Deglacial paleoclimate in the southwestern United States: an abrupt 18.6 ka cold event and evidence for a North Atlantic forcing of Termination I, Quaternary Sci. Rev., 30, 3803–3811, https://doi.org/10.1016/j.quascirev.2011.09.022, 2011.

Lachniet, M. S., Bernal, J. P., Asmerom, Y., Polyak, V., and Piperno, D.: A 2400 yr Mesoamerican rainfall reconstruction links climate and cultural change, Geology, 40, 259–262, https://doi.org/10.1130/g32471.1, 2012.

Lachniet, M. S., Asmerom, Y., Bernal, J. P., Polyak, V. J., and Vazquez-Selem, L.: Orbital pacing and ocean circulation-induced collapses of the Mesoamerican monsoon over the past 22,000 y, P. Natl. Acad. Sci. USA, 110, 9255–9260, https://doi.org/10.1073/pnas.1222804110, 2013.

Lachniet, M. S., Denniston, R. F., Asmerom, Y., and Polyak, V. J.: Orbital control of western North America atmospheric circulation and climate over two glacial cycles, Nat. Commun., 5, 3805, https://doi.org/10.1038/ncomms4805, 2014.

Lachniet, M. S., Asmerom, Y., Polyak, V., and Bernal, J. P.: Two millennia of Mesoamerican monsoon variability driven by Pacific and Atlantic synergistic forcing, Quaternary Sci. Rev., 155, 100–113, https://doi.org/10.1016/j.quascirev.2016.11.012, 2017.

Lechleitner, F. A., Breitenbach, S. F. M., Cheng, H., Plessen, B., Rehfeld, K., Goswami, B., Marwan, N., Eroglu, D., Adkins, J., and Haug, G.: Climatic and in-cave influences on δ^{18}O and δ^{13}C in a stalagmite from northeastern India through the last deglaciation, Quaternary Res., 88, 458–471, https://doi.org/10.1017/qua.2017.72, 2017.

Lee-Thorp, J. A., Holmgren, K., Lauritzen, S.-E., Linge, H., Moberg, A., Partridge, T. C., Stevenson, C., and Tyson, P. D.: Rapid climate shifts in the southern African interior throughout the Mid to Late Holocene, Geophys. Res. Lett., 28, 4507–4510, https://doi.org/10.1029/2000gl012728, 2001.

Lewis, S. C., Gagan, M. K., Ayliffe, L. K., Zhao, J., Hantoro, W. S., Treble, P. C., Hellstrom, J. C., LeGrande, A. N., Kelley, M., Schmidt, G. A., and Suwargadi, B. W.: High-resolution stalagmite reconstructions of Australian–Indonesian monsoon rainfall variability during Heinrich stadial 3 and Greenland interstadial 4, Earth Planet. Sci. Lett., 303, 133–142, https://doi.org/10.1016/j.epsl.2010.12.048, 2011.

Li, H.-C., Lee, Z.-H., Wan, N.-J., Shen, C.-C., Li, T.-Y., Yuan, D.-X., and Chen, Y.-H.: The δ^{18}O and δ^{13}C records in an aragonite stalagmite from Furong Cave, Chongqing, China: A-2000-year record of monsoonal climate, J. Asian Earth Sci., 40, 1121–1130, https://doi.org/10.1016/j.jseaes.2010.06.011, 2011.

Li, J.-Y., Li, H.-C., Li, T.-Y., Mii, H.-S., Yu, T.-L., Shen, C.-C., and Xu, X.: High-resolution δ^{18}O and δ^{13}C records of an AMS ^{14}C and ^{230}Th/U dated stalagmite from Xinya Cave in Chongqing: Climate and vegetation change during the late Holocene, Quatern. Int., 447, 75–88, https://doi.org/10.1016/j.quaint.2017.06.075, 2017.

Li, T., Yuan, D., Li, H., Yang, Y., Wang, J., Wang, X., Li, J., Qin, J., Zhang, M., and Lin, Y.: High-resolution climate variability of southwest China during 57–70 ka reflected in a stalagmite δ^{18}O record from Xinya Cave, Sci. China Earth Sci., 50, 1202–1208, https://doi.org/10.1007/s11430-007-0059-z, 2007.

Li, T.-Y., Shen, C.-C., Li, H.-C., Li, J.-Y., Chiang, H.-W., Song, S.-R., Yuan, D.-X., Lin, C. D.-J., Gao, P., Zhou, L., Wang, J.-L., Ye, M.-Y., Tang, L.-L., and Xie, S.-Y.: Oxygen and carbon isotopic systematics of aragonite speleothems and water in Furong Cave, Chongqing, China, Geochim. Cosmochim. Acta, 75, 4140–4156, https://doi.org/10.1016/j.gca.2011.04.003, 2011.

Li, T.-Y., Shen, C.-C., Huang, L.-J., Jiang, X.-Y., Yang, X.-L., Mii, H.-S., Lee, S.-Y., and Lo, L.: Stalagmite-inferred variability of the Asian summer monsoon during the penultimate glacial–interglacial period, Clim. Past, 10, 1211–1219, https://doi.org/10.5194/cp-10-1211-2014, 2014.

Li, T.-Y., Han, L.-Y., Cheng, H., Edwards, R. L., Shen, C.-C., Li, H.-C., Li, J.-Y., Huang, C.-X., Zhang, T.-T., and Zhao, X.: Evolution of the Asian summer monsoon during Dansgaard/Oeschger events 13–17 recorded in a stalagmite constrained by high-precision chronology from southwest China, Quaternary Res., 88, 121–128, https://doi.org/10.1017/qua.2017.22, 2017.

Linge, H., Lauritzen, S.-E., Lundberg, J., and Berstad, I. M.: Stable isotope stratigraphy of Holocene speleothems: examples from a cave system in Rana, northern Norway, Palaeogeogr. Palaeocl., 167, 209–224, https://doi.org/10.1016/s0031-0182(00)00225-x, 2001.

Linge, H., Lauritzen, S.-E., Andersson, C., Hansen, J. K., Skoglund, R. Ø., and Sundqvist, H. S.: Stable isotope records for the last 10 000 years from Okshola cave (Fauske, northern Norway) and regional comparisons, Clim. Past, 5, 667–682, https://doi.org/10.5194/cp-5-667-2009, 2009a.

Linge, H., Baker, A., Andersson, C., and Lauritzen, S.-E.: Variability in luminescent lamination and initial 230 Th/232 Th activity ratios in a late Holocene stalagmite from northern Norway, Quat. Geochronol., 4, 181–192, 2009b.

Liu, Y.-H., Henderson, G. M., Hu, C.-Y., Mason, A. J., Charnley, N., Johnson, K. R., and Xie, S.-C.: Links between the East Asian monsoon and North Atlantic climate during the 8,200 year event, Nat. Geosci., 6, 117–120, https://doi.org/10.1038/ngeo1708, 2013.

Lone, M. A., Ahmad, S. M., Dung, N. C., Shen, C.-C., Raza, W., and Kumar, A.: Speleothem based 1000-year high resolution record of Indian monsoon variability during the last deglaciation, Palaeogeogr. Palaeocl., 395, 1–8, https://doi.org/10.1016/j.palaeo.2013.12.010, 2014.

Lorrey, A., Williams, P., Salinger, J., Martin, T., Palmer, J., Fowler, A., Zhao, J., and Neil, H.: Speleothem stable isotope records interpreted within a multi-proxy framework and implications for New Zealand palaeoclimate reconstruction, Quatern. Int., 187, 52–75, https://doi.org/10.1016/j.quaint.2007.09.039, 2008.

Luetscher, M., Hoffmann, D. L., Frisia, S., and Spötl, C.: Holocene glacier history from alpine speleothems, Milchbach cave, Switzerland, Earth Planet. Sci. Lett., 302, 95–106, https://doi.org/10.1016/j.epsl.2010.11.042, 2011.

Luetscher, M., Boch, R., Sodemann, H., Spötl, C., Cheng, H., Edwards, R. L., Frisia, S., Hof, F., and Müller, W.: North Atlantic storm track changes during the Last Glacial Maximum recorded by Alpine speleothems, Nat. Commun., 6, 6344, https://doi.org/10.1038/ncomms7344, 2015.

Ma, Z.-B., Cheng, H., Tan, M., Edwards, R. L., Li, H.-C., You, C.-F., Duan, W.-H., Wang, X., and Kelly, M. J.: Timing and structure of the Younger Dryas event in northern China, Quaternary Sci. Rev., 41, 83–93, https://doi.org/10.1016/j.quascirev.2012.03.006, 2012.

Madonia, G., Frisia, S., Borsato, A., Macaluso, T., Mangini, A., Paladini, M., Piccini, L., Miorandi, R., Spötl, C., Sauro, U., Agnesi, V., Di Pietro, R., Palmeri, A., and Vattano, M.: La Grotta di Carburangeli–ricostruzione climatica dell'Olocene per la piana costiera della Sicilia nord-occidentale, Stud. Trent Sci Nat Acta Geol, 80, 153–167 available at: http://www2.muse.it/pubblicazioni/6/actaG80/Vol_ACTA_80_2003_153-167.pdf (last access: 30 May 2018), 2005.

Mattey, D., Lowry, D., Duffet, J., Fisher, R., Hodge, E., and Frisia, S.: A 53 year seasonally resolved oxygen and carbon isotope record from a modern Gibraltar speleothem: Reconstructed drip water and relationship to local precipitation, Earth Planet. Sci. Lett., 269, 80–95, https://doi.org/10.1016/j.epsl.2008.01.051, 2008.

Mattey, D. P., Fairchild, I. J., Atkinson, T. C., Latin, J.-P., Ainsworth, M., and Durell, R.: Seasonal microclimate control of calcite fabrics, stable isotopes and trace elements in modern speleothem from St Michaels Cave, Gibraltar, Geol. Soc. London, Spec. Publ., 336, 323–344, https://doi.org/10.1144/sp336.17, 2010.

McDermott, F.: Centennial-Scale Holocene Climate Variability Revealed by a High-Resolution Speleothem $\delta^{18}O$ Record from SW Ireland, Science, 294, 1328–1331, https://doi.org/10.1126/science.1063678, 2001.

McDermott, F.: Palaeo-climate reconstruction from stable isotope variations in speleothems: a review, Quaternary Sci. Rev., 23, 901–918, https://doi.org/10.1016/j.quascirev.2003.06.021, 2004.

McDermott, F., Frisia, S., Huang, Y., Longinelli, A., Spiro, B., Heaton, T. H. E., Hawkesworth, C. J., Borsato, A., Keppens, E., Fairchild, I. J., van der Borg, K., Verheyden, S., and Selmo, E.: Holocene climate variability in Europe: Evidence from $\delta^{18}O$, textural and extension-rate variations in three speleothems, Quaternary Sci. Rev., 18, 1021–1038, https://doi.org/10.1016/S0277-3791(98)00107-3, 1999.

Meckler, A. N., Clarkson, M. O., Cobb, K. M., Sodemann, H., and Adkins, J. F.: Interglacial Hydroclimate in the Tropical West Pacific Through the Late Pleistocene, Science, 336, 1301–1304, https://doi.org/10.1126/science.1218340, 2012.

Medina-Elizalde, M., Burns, S. J., Lea, D. W., Asmerom, Y., von Gunten, L., Polyak, V., Vuille, M., and Karmalkar, A.: High resolution stalagmite climate record from the Yucatán Peninsula spanning the Maya terminal classic period, Earth Planet. Sci. Lett., 298, 255–262, https://doi.org/10.1016/j.epsl.2010.08.016, 2010.

Meyer, M. C., Spötl, C., and Mangini, A.: The demise of the Last Interglacial recorded in isotopically dated speleothems from the Alps, Quaternary Sci. Rev., 27, 476–496, https://doi.org/10.1016/j.quascirev.2007.11.005, 2008.

Mickler, P. J., Stern, L. A., and Banner, J. L.: Large kinetic isotope effects in modern speleothems Large kinetic isotope effects in modern speleothems, Geol. Soc. Am. Bull., 118, 65–81, https://doi.org/10.1130/B25698.1, 2006.

Moerman, J. W., Cobb, K. M., Adkins, J. F., Sodemann, H., Clark, B., and Tuen, A. A.: Diurnal to interannual rainfall $\delta^{18}O$ variations in northern Borneo driven by regional hydrology, Earth Planet. Sci. Lett., 369–370, 108–119, https://doi.org/10.1016/j.epsl.2013.03.014, 2013.

Moerman, J. W., Cobb, K. M., Partin, J. W., Meckler, A. N., Carolin, S. A., Adkins, J. F., Lejau, S., Malang, J., Clark, B., and Tuen, A. A.: Transformation of ENSO-related rainwater to dripwater δ 18 O variability by vadose water mixing, Geophys. Res. Lett., 41, 7907–7915, https://doi.org/10.1002/2014GL061696, 2014.

Moreno, A., Stoll, H., Jiménez-Sánchez, M., Cacho, I., Valero-Garcés, B., Ito, E., and Edwards, R. L.: A speleothem record of glacial (25–11.6kyr BP) rapid climatic changes from northern Iberian Peninsula, Global Planet. Change, 71, 218–231, https://doi.org/10.1016/j.gloplacha.2009.10.002, 2010.

Moreno, A., Pérez-Mejías, C., Bartolomé, M., Sancho, C., Cacho, I., Stoll, H., Delgado-Huertas, A., Hellstrom, J., Edwards, R. L., and Cheng, H.: New speleothem data from Molinos and Ejulve caves reveal Holocene hydrological variability in northeast Iberia, Quaternary Res., 88, 223–233, https://doi.org/10.1017/qua.2017.39, 2017.

Moseley, G. E., Spötl, C., Svensson, A., Cheng, H., Brandstätter, S., and Edwards, R. L.: Multi-speleothem record reveals tightly coupled climate between central Europe and Greenland during Marine Isotope Stage 3, Geology, 42, 1043–1046, https://doi.org/10.1130/g36063.1, 2014.

Moseley, G. E., Spötl, C., Cheng, H., Boch, R., Min, A., and Edwards, R. L.: Termination-II interstadial/stadial climate change recorded in two stalagmites from the north European Alps, Quaternary Sci. Rev., 127, 229–239, https://doi.org/10.1016/j.quascirev.2015.07.012, 2015.

Moseley, G. E., Edwards, R. L., Wendt, K. A., Cheng, H., Dublyansky, Y., Lu, Y., Boch, R., and Spotl, C.: Reconciliation of the

Devils Hole climate record with orbital forcing, Science, 351, 165–168, https://doi.org/10.1126/science.aad4132, 2016.

Muñoz, A., Bartolomé, M., Muñoz, A., Sancho, C., Moreno, A., Hellstrom, J. C., Osácar, M. C., and Cacho, I.: Solar influence and hydrological variability during the Holocene from a speleothem annual record (Molinos Cave, NE Spain), Terra Nov., 27, 300–311, https://doi.org/10.1111/ter.12160, 2015.

Neff, U., Burns, S. J., Mangini, A., Mudelsee, M., Fleitmann, D., and Matter, A.: Strong coherence between solar variability and the monsoon in Oman between 9 and 6 kyr ago, Nature, 411, 290–293, https://doi.org/10.1038/35077048, 2001.

Nehme, C., Verheyden, S., Noble, S. R., Farrant, A. R., Sahy, D., Hellstrom, J., Delannoy, J. J., and Claeys, P.: Reconstruction of MIS 5 climate in the central Levant using a stalagmite from Kanaan Cave, Lebanon, Clim. Past, 11, 1785–1799, https://doi.org/10.5194/cp-11-1785-2015, 2015.

Nehme, C., Verheyden, S., Breitenbach, S. F. M., Gillikin, D. P., Verheyden, A., Cheng, H., Edwards, R. L., Hellstrom, J., Noble, S. R., Farrant, A. R., Sahy, D., Goovaerts, T., Salem, G., and Claeys, P.: Climate dynamics during the penultimate glacial period recorded in a speleothem from Kanaan Cave, Lebanon (central Levant), Quaternary Res., 90, 10–25, https://doi.org/10.1017/qua.2018.18, 2018.

Novello, V. F., Cruz, F. W., Karmann, I., Burns, S. J., Stríkis, N. M., Vuille, M., Cheng, H., Lawrence Edwards, R., Santos, R. V, Frigo, E., and Barreto, E. A. S.: Multidecadal climate variability in Brazil's Nordeste during the last 3000 years based on speleothem isotope records, Geophys. Res. Lett., 39, L23706, https://doi.org/10.1029/2012GL053936, 2012.

Novello, V. F., Vuille, M., Cruz, F. W., Stríkis, N. M., de Paula, M. S., Edwards, R. L., Cheng, H., Karmann, I., Jaqueto, P. F., Trindade, R. I. F., Hartmann, G. A., and Moquet, J. S.: Centennial-scale solar forcing of the South American Monsoon System recorded in stalagmites, Sci. Rep., 6, 24762, https://doi.org/10.1038/srep24762, 2016.

Novello, V. F., Cruz, F. W., Vuille, M., Stríkis, N. M., Edwards, R. L., Cheng, H., Emerick, S., de Paula, M. S., Li, X., Barreto, E. de S., Karmann, I., and Santos, R. V: A high-resolution history of the South American Monsoon from Last Glacial Maximum to the Holocene, Sci. Rep., 7, 44267, https://doi.org/10.1038/srep44267, 2017.

Onac, B. P., Constantin, S., Lundberg, J., and Lauritzen, S.-E.: Isotopic climate record in a Holocene stalagmite from Ursilor Cave (Romania), J. Quaternary Sci., 17, 319–327, https://doi.org/10.1002/jqs.685, 2002.

Orland, I. J., Bar-Matthews, M., Kita, N. T., Ayalon, A., Matthews, A., and Valley, J. W.: Climate deterioration in the Eastern Mediterranean as revealed by ion microprobe analysis of a speleothem that grew from 2.2 to 0.9 ka in Soreq Cave, Israel, Quaternary Res., 71, 27–35, https://doi.org/10.1016/j.yqres.2008.08.005, 2009.

Orland, I. J., Bar-Matthews, M., Ayalon, A., Matthews, A., Kozdon, R., Ushikubo, T., and Valley, J. W.: Seasonal resolution of Eastern Mediterranean climate change since 34 ka from a Soreq Cave speleothem, Geochim. Cosmochim. Acta, 89, 240–255, https://doi.org/10.1016/j.gca.2012.04.035, 2012.

Oster, J. L., Montañez, I. P., Sharp, W. D., and Cooper, K. M.: Late Pleistocene California droughts during deglaciation and Arctic warming, Earth Planet. Sci. Lett., 288, 434–443, https://doi.org/10.1016/j.epsl.2009.10.003, 2009.

Oster, J. L., Montañez, I. P., Santare, L. R., Sharp, W. D., Wong, C., and Cooper, K. M.: Stalagmite records of hydroclimate in central California during termination 1, Quaternary Sci. Rev., 127, 199–214, https://doi.org/10.1016/j.quascirev.2015.07.027, 2015.

Partin, J. W., Cobb, K. M., Adkins, J. F., Clark, B., and Fernandez, D. P.: Millennial-scale trends in west Pacific warm pool hydrology since the Last Glacial Maximum, Nature, 449, 452–455, https://doi.org/10.1038/nature06164, 2007.

Partin, J. W., Quinn, T. M., Shen, C.-C., Emile-Geay, J., Taylor, F. W., Maupin, C. R., Lin, K., Jackson, C. S., Banner, J. L., Sinclair, D. J., and Huh, C.-A.: Multidecadal rainfall variability in South Pacific Convergence Zone as revealed by stalagmite geochemistry, Geology, 41, 1143–1146, https://doi.org/10.1130/g34718.1, 2013a.

Partin, J. W., Cobb, K. M., Adkins, J. F., Tuen, A. A., and Clark, B.: Trace metal and carbon isotopic variations in cave dripwater and stalagmite geochemistry from northern Borneo, Geochemistry, Geophys. Geosystems, 14, 3567–3585, https://doi.org/10.1002/ggge.20215, 2013b.

Pérez-Mejías, C., Moreno, A., Sancho, C., Bartolomé, M., Stoll, H., Cacho, I., Cheng, H., and Edwards, R. L.: Abrupt climate changes during Termination III in Southern Europe, P. Natl. Acad. Sci. USA, 114, 10047–10052, https://doi.org/10.1073/pnas.1619615114, 2017.

Plagnes, V., Causse, C., Genty, D., Paterne, M., and Blamart, D.: A discontinuous climatic record from 187 to 74 ka from a speleothem of the Clamouse Cave (south of France), Earth Planet. Sci. Lett., 201, 87–103, https://doi.org/10.1016/s0012-821x(02)00674-x, 2002.

Pollock, A. L., van Beynen, P. E., DeLong, K. L., Polyak, V., and Asmerom, Y.: A speleothem-based mid-Holocene precipitation reconstruction for West-Central Florida, The Holocene, 27, 987–996, https://doi.org/10.1177/0959683616678463, 2016.

Polyak, V. J. and Asmerom, Y.: Late Holocene Climate and Cultural Changes in the Southwestern United States, Science, 294, 148–151, https://doi.org/10.1126/science.1062771, 2001.

Polyak, V. J., Asmerom, Y., Burns, S. J., and Lachniet, M. S.: Climatic backdrop to the terminal Pleistocene extinction of North American mammals, Geology, 40, 1023–1026, https://doi.org/10.1130/g33226.1, 2012.

Ponte, J. M., Font, E., Veiga-Pires, C., Hillaire-Marcel, C., and Ghaleb, B.: The effect of speleothem surface slope on the remanent magnetic inclination, J. Geophys. Res.-Sol. Ea., 122, 4143–4156, https://doi.org/10.1002/2016jb013789, 2017.

Psomiadis, D., Dotsika, E., Albanakis, K., Ghaleb, B., and Hillaire-Marcel, C.: Speleothem record of climatic changes in the northern Aegean region (Greece) from the Bronze Age to the collapse of the Roman Empire, Palaeogeogr. Palaeocl., 489, 272–283, https://doi.org/10.1016/j.palaeo.2017.10.021, 2018.

Railsback, L. B., Liang, F., Vidal Romaní, J. R., Grandal-d'Anglade, A., Vaqueiro Rodríguez, M., Santos Fidalgo, L., Fernández Mosquera, D., Cheng, H., and Edwards, R. L.: Petrographic and isotopic evidence for Holocene long-term climate change and shorter-term environmental shifts from a stalagmite from the Serra do Courel of northwestern Spain, and implications for climatic history across Europe and the Mediterranean, Palaeogeogr. Palaeocl., 305, 172–184, https://doi.org/10.1016/j.palaeo.2011.02.030, 2011.

Raza, W., Ahmad, S. M., Lone, M. A., Shen, C.-C., Sarma, D. S., and Kumar, A.: Indian summer monsoon variability in southern India during the last deglaciation: Evidence from a high resolution stalagmite $\delta^{18}O$ record, Palaeogeogr. Palaeocl., 485, 476–485, https://doi.org/10.1016/j.palaeo.2017.07.003, 2017.

Repinski, P., Holmgren, K., Lauritzen, S. E., and Lee-Thorp, J. A.: A late Holocene climate record from a stalagmite, Cold Air Cave, Northern Province, South Africa, Palaeogeogr. Palaeocl., 150, 269–277, https://doi.org/10.1016/s0031-0182(98)00223-5, 1999.

Ridley, H. E., Asmerom, Y., Baldini, J. U. L., Breitenbach, S. F. M., Aquino, V. V, Prufer, K. M., Culleton, B. J., Polyak, V., Lechleitner, F. A., Kennett, D. J., Zhang, M., Marwan, N., Macpherson, C. G., Baldini, L. M., Xiao, T., Peterkin, J. L., Awe, J., and Haug, G. H.: Aerosol forcing of the position of the intertropical convergence zone since ad 1550, Nat. Geosci., 8, 195–200, https://doi.org/10.1038/ngeo2353, 2015.

Ruan, J., Kherbouche, F., Genty, D., Blamart, D., Cheng, H., Dewilde, F., Hachi, S., Edwards, R. L., Régnier, E., and Michelot, J.-L.: Evidence of a prolonged drought ca. 4200 yr BP correlated with prehistoric settlement abandonment from the Gueldaman GLD1 Cave, Northern Algeria, Clim. Past, 12, 1–14, https://doi.org/10.5194/cp-12-1-2016, 2016.

Rudzka, D., McDermott, F., Baldini, L. M., Fleitmann, D., Moreno, A., and Stoll, H.: The coupled $\delta^{13}C$-radiocarbon systematics of three Late Glacial/early Holocene speleothems; insights into soil and cave processes at climatic transitions, Geochim. Cosmochim. Acta, 75, 4321–4339, https://doi.org/10.1016/j.gca.2011.05.022, 2011.

Rudzka, D., Mcdermott, F., and Surić, M.: A late Holocene climate record in stalagmites from Modrič Cave (Croatia), J. Quaternary Sci., 27, 585–596, https://doi.org/10.1002/jqs.2550, 2012.

Salomons, W. and Mook, W. G.: Isotope geochemistry of carbonates in the weathering zone, in Handbook of environmental isotope geochemistry, 239–269, 1986.

Schmidt, G. A., LeGrande, A. N., and Hoffmann, G.: Water isotope expressions of intrinsic and forced variability in a coupled ocean-atmosphere model, J. Geophys. Res., 112, D10103, https://doi.org/10.1029/2006JD007781, 2007.

Schmidt, G. A., Annan, J. D., Bartlein, P. J., Cook, B. I., Guilyardi, E., Hargreaves, J. C., Harrison, S. P., Kageyama, M., LeGrande, A. N., Konecky, B., Lovejoy, S., Mann, M. E., Masson-Delmotte, V., Risi, C., Thompson, D., Timmermann, A., Tremblay, L.-B., and Yiou, P.: Using palaeo-climate comparisons to constrain future projections in CMIP5, Clim. Past, 10, 221–250, https://doi.org/10.5194/cp-10-221-2014, 2014.

Scholz, D. and Hoffmann, D. L.: StalAge – An algorithm designed for construction of speleothem age models, Quat. Geochronol., 6, 369–382, https://doi.org/10.1016/J.QUAGEO.2011.02.002, 2011.

Scholz, D., Frisia, S., Borsato, A., Spötl, C., Fohlmeister, J., Mudelsee, M., Miorandi, R., and Mangini, A.: Holocene climate variability in north-eastern Italy: potential influence of the NAO and solar activity recorded by speleothem data, Clim. Past, 8, 1367–1383, https://doi.org/10.5194/cp-8-1367-2012, 2012.

Scroxton, N., Burns, S. J., McGee, D., Hardt, B., Godfrey, L. R., Ranivoharimanana, L., and Faina, P.: Hemispherically in-phase precipitation variability over the last 1700 years in a Madagascar speleothem record, Quaternary Sci. Rev., 164, 25–36, https://doi.org/10.1016/j.quascirev.2017.03.017, 2017.

Shah, A. M., Morrill, C., Gille, E. P., Gross, W. S., Anderson, D. M., Bauer, B. A., Buckner, R., and Hartman, M.: Global Speleothem Oxygen Isotope Measurements Since the Last Glacial Maximum, Dataset Pap. Geosci., 2013, 1–9, https://doi.org/10.7167/2013/548048, 2013.

Shakun, J. D., Burns, S. J., Fleitmann, D., Kramers, J., Matter, A., and Al-Subary, A.: A high-resolution, absolute-dated deglacial speleothem record of Indian Ocean climate from Socotra Island, Yemen, Earth Planet. Sci. Lett., 259, 442–456, https://doi.org/10.1016/j.epsl.2007.05.004, 2007.

Shakun, J. D., Burns, S. J., Clark, P. U., Cheng, H., and Edwards, R. L.: Milankovitch-paced Termination II in a Nevada speleothem?, Geophys. Res. Lett., 38, L18701, https://doi.org/10.1029/2011GL048560, 2011.

Siklósy, Z., Demény, A., Vennemann, T. W., Pilet, S., Kramers, J., Leél-Össy, S., Bondár, M., Shen, C.-C., and Hegner, E.: Bronze Age volcanic event recorded in stalagmites by combined isotope and trace element studies, Rapid Commun. Mass Spectrom., 23, 801–808, https://doi.org/10.1002/rcm.3943, 2009.

Sinha, A., Cannariato, K. G., Stott, L. D., Li, H.-C., You, C.-F., Cheng, H., Edwards, R. L., and Singh, I. B.: Variability of Southwest Indian summer monsoon precipitation during the Bølling-Ållerød, Geology, 33, 813–816, https://doi.org/10.1130/G21498.1, 2005.

Sinha, A., Cannariato, K. G., Stott, L. D., Cheng, H., Edwards, R. L., Yadava, M. G., Ramesh, R., and Singh, I. B.: A 900-year (600 to 1500 A.D.) record of the Indian summer monsoon precipitation from the core monsoon zone of India, Geophys. Res. Lett., 34, L16707, https://doi.org/10.1029/2007gl030431, 2007.

Sinha, A., Berkelhammer, M., Stott, L., Mudelsee, M., Cheng, H., and Biswas, J.: The leading mode of Indian Summer Monsoon precipitation variability during the last millennium, Geophys. Res. Lett., 38, L15703, https://doi.org/10.1029/2011gl047713, 2011.

Sinha, A., Kathayat, G., Cheng, H., Breitenbach, S. F. M., Berkelhammer, M., Mudelsee, M., Biswas, J., and Edwards, R. L.: Trends and oscillations in the Indian summer monsoon rainfall over the last two millennia, Nat. Commun., 6, 6309, https://doi.org/10.1038/ncomms7309, 2015.

Sletten, H. R., Railsback, L. B., Liang, F., Brook, G. A., Marais, E., Hardt, B. F., Cheng, H., and Edwards, R. L.: A petrographic and geochemical record of climate change over the last 4600years from a northern Namibia stalagmite, with evidence of abruptly wetter climate at the beginning of southern Africa's Iron Age, Palaeogeogr. Palaeocl., 376, 149–162, https://doi.org/10.1016/j.palaeo.2013.02.030, 2013.

Smart, P. L. and Friederich, H.: Water movement and storage in the unsaturated zone of a maturely karstified carbonate aquifer, Mendip Hills, England, in Proceedings of the Environmental Problems in Karst Terranes and their Solutions Conference, 57–87, available at: https://ci.nii.ac.jp/naid/10012582404/ (last access: 29 January 2018), 1987.

Smith, A. C., Wynn, P. M., Barker, P. A., Leng, M. J., Noble, S. R., and Tych, W.: North Atlantic forcing of moisture delivery to Europe throughout the Holocene, Sci. Rep., 6, 24745, https://doi.org/10.1038/srep24745, 2016.

Sone, T., Kano, A., Okumura, T., Kashiwagi, K., Hori, M., Jiang, X., and Shen, C.-C.: Holocene stalagmite oxygen isotopic record from the Japan Sea side of the Japanese Islands, as a new proxy of the East Asian winter monsoon, Quaternary Sci. Rev., 75, 150–160, https://doi.org/10.1016/j.quascirev.2013.06.019, 2013.

Spötl, C. and Mangini, A.: Stalagmite from the Austrian Alps reveals Dansgaard-Oeschger events during isotope stage 3: Implications for the absolute chronology of Greenland ice cores, Earth Planet. Sci. Lett., 203, 507–518, https://doi.org/10.1016/S0012-821X(02)00837-3, 2002.

Spötl, C., Fairchild, I. J., and Tooth, A. F.: Cave air control on dripwater geochemistry, Obir Caves (Austria): Implications for speleothem deposition in dynamically ventilated caves, Geochim. Cosmochim. Acta, 69, 2451–2468, https://doi.org/10.1016/j.gca.2004.12.009, 2005.

Spötl, C., Mangini, A., and Richards, D. A.: Chronology and paleoenvironment of Marine Isotope Stage 3 from two high-elevation speleothems, Austrian Alps, Quaternary Sci. Rev., 25, 1127–1136, https://doi.org/10.1016/j.quascirev.2005.10.006, 2006.

Spötl, C., Scholz, D., and Mangini, A.: A terrestrial U/Th-dated stable isotope record of the Penultimate Interglacial, Earth Planet. Sci. Lett., 276, 283–292, https://doi.org/10.1016/j.epsl.2008.09.029, 2008.

Springer, G. S., Rowe, H. D., Hardt, B., Cheng, H., and Edwards, R. L.: East central North America climates during marine isotope stages 3-5, Geophys. Res. Lett., 41, 3233–3237, https://doi.org/10.1002/2014gl059884, 2014.

Steen-Larsen, H. C., Risi, C., Werner, M., Yoshimura, K., and Masson-Delmotte, V.: Evaluating the skills of isotope-enabled general circulation models against in situ atmospheric water vapor isotope observations, J. Geophys. Res.-Atmos., 122, 246–263, https://doi.org/10.1002/2016JD025443, 2017.

Stevenson, C., Lee-Thorp, J. A., and Holmgren, K.: A 3000-year isotopic record from a stalagmite in Cold Air Cave, Makapansgat Valley, Northern Province, S. Afr. J. Sci., 95, 46–48, available at: https://journals.co.za/content/sajsci/95/1/AJA00382353_7827 (last access: 30 May 2018), 1999.

Strikis, N. M., Cruz, F. W., Cheng, H., Karmann, I., Edwards, R. L., Vuille, M., Wang, X., de Paula, M. S., Novello, V. F., and Auler, A. S.: Abrupt variations in South American monsoon rainfall during the Holocene based on a speleothem record from central-eastern Brazil, Geology, 39, 1075–1078, https://doi.org/10.1130/g32098.1, 2011.

Strĺkis, N. M., Chiessi, C. M., Cruz, F. W., Vuille, M., Cheng, H., de Souza Barreto, E. A., Mollenhauer, G., Kasten, S., Karmann, I., Edwards, R. L., Bernal, J. P., and Sales, H. dos R.: Timing and structure of Mega-SACZ events during Heinrich Stadial 1, Geophys. Res. Lett., 42, 5477–5484A, https://doi.org/10.1002/2015GL064048, 2015.

Sturm, C., Zhang, Q., and Noone, D.: An introduction to stable water isotopes in climate models: benefits of forward proxy modelling for paleoclimatology, Clim. Past, 6, 115–129, https://doi.org/10.5194/cp-6-115-2010, 2010.

Sundqvist, H. S., Holmgren, K., and Lauritzen, S.-E.: Stable isotope variations in stalagmites from northwestern Sweden document climate and environmental changes during the early Holocene, The Holocene, 17, 259–267, https://doi.org/10.1177/0959683607073292, 2007.

Sundqvist, H. S., Holmgren, K., Moberg, A., Spötl, C., and Mangini, A.: Stable isotopes in a stalagmite from NW Sweden document environmental changes over the past 4000 years, Boreas, 39, 77–86, https://doi.org/10.1111/j.1502-3885.2009.00099.x, 2010.

Sundqvist, H. S., Holmgren, K., Fohlmeister, J., Zhang, Q., Matthews, M. B., Spötl, C., and Körnich, H.: Evidence of a large cooling between 1690 and 1740 AD in southern Africa, Sci. Rep., 3, 1767, https://doi.org/10.1038/srep01767, 2013.

Talma, A. S. and Vogel, J. C.: Late Quaternary Paleotemperatures Derived from a Speleothem from Cango Caves, Cape Province, South Africa, Quaternary Res., 37, 203–213, https://doi.org/10.1016/0033-5894(92)90082-t, 1992.

Tan, L., Cai, Y., Cheng, H., An, Z., and Edwards, R. L.: Summer monsoon precipitation variations in central China over the past 750years derived from a high-resolution absolute-dated stalagmite, Palaeogeogr. Palaeocl., 280, 432–439, https://doi.org/10.1016/j.palaeo.2009.06.030, 2009.

Tan, L., Cai, Y., An, Z., Edwards, R. L., Cheng, H., Shen, C.-C., and Zhang, H.: Centennial- to decadal-scale monsoon precipitation variability in the semi-humid region, northern China during the last 1860 years: Records from stalagmites in Huangye Cave, The Holocene, 21, 287–296, https://doi.org/10.1177/0959683610378880, 2010.

Treble, P. C., Baker, A., Ayliffe, L. K., Cohen, T. J., Hellstrom, J. C., Gagan, M. K., Frisia, S., Drysdale, R. N., Griffiths, A. D., and Borsato, A.: Hydroclimate of the Last Glacial Maximum and deglaciation in southern Australia's arid margin interpreted from speleothem records (23–15 ka), Clim. Past, 13, 667–687, https://doi.org/10.5194/cp-13-667-2017, 2017.

Tremaine, D. M., Froelich, P. N., and Wang, Y.: Speleothem calcite farmed in situ: Modern calibration of $\delta^{18}O$ and $\delta^{13}C$ paleoclimate proxies in a continuously-monitored natural cave system, Geochim. Cosmochim. Acta, 75, 4929–4950, https://doi.org/10.1016/J.GCA.2011.06.005, 2011.

Trouet, V., Esper, J., Graham, N. E., Baker, A., Scourse, J. D., and Frank, D. C.: Persistent Positive North Atlantic Oscillation Mode Dominated the Medieval Climate Anomaly, Science, 324, 78–80, https://doi.org/10.1126/science.1166349, 2009.

Ünal-İmer, E., Shulmeister, J., Zhao, J.-X., Tonguç Uysal, I., Feng, Y.-X., Duc Nguyen, A., and Yüce, G.: An 80 kyr-long continuous speleothem record from Dim Cave, SW Turkey with paleoclimatic implications for the Eastern Mediterranean, Sci. Rep., 5, 13560, https://doi.org/10.1038/srep13560, 2015.

Vaks, A., Bar-Matthews, M., Ayalon, A., Schilman, B., Gilmour, M., Hawkesworth, C. J., Frumkin, A., Kaufman, A., and Matthews, A.: Paleoclimate reconstruction based on the timing of speleothem growth and oxygen and carbon isotope composition in a cave located in the rain shadow in Israel, Quaternary Res., 59, 182–193, https://doi.org/10.1016/s0033-5894(03)00013-9, 2003.

Vansteenberge, S., Verheyden, S., Cheng, H., Edwards, R. L., Keppens, E., and Claeys, P.: Paleoclimate in continental northwestern Europe during the Eemian and early Weichselian (125–97 ka): insights from a Belgian speleothem, Clim. Past, 12, 1445–1458, https://doi.org/10.5194/cp-12-1445-2016, 2016.

Voarintsoa, N. R. G., Wang, L., Railsback, L. B., Brook, G. A., Liang, F., Cheng, H., and Edwards, R. L.: Multiple proxy analyses of a U/Th-dated stalagmite to reconstruct paleoenvironmental changes in northwestern Madagascar between 370CE and 1300CE, Palaeogeogr. Palaeocl., 469, 138–155, https://doi.org/10.1016/j.palaeo.2017.01.003, 2017a.

Voarintsoa, N. R. G., Brook, G. A., Liang, F., Marais, E., Hardt, B., Cheng, H., Edwards, R. L., and Railsback, L. B.: Stalagmite multi-proxy evidence of wet and dry intervals in northeastern Namibia: Linkage to latitudinal shifts of the Inter-Tropical Convergence Zone and changing solar activity from AD 1400 to 1950, The Holocene, 27, 384–396, https://doi.org/10.1177/0959683616660170, 2017b.

Voarintsoa, N. R. G., Railsback, L. B., Brook, G. A., Wang, L., Kathayat, G., Cheng, H., Li, X., Edwards, R. L., Rakotondrazafy, A. F. M., and Madison Razanatseheno, M. O.: Three distinct Holocene intervals of stalagmite deposition and nondeposition revealed in NW Madagascar, and their paleoclimate implications, Clim. Past, 13, 1771–1790, https://doi.org/10.5194/cp-13-1771-2017, 2017c.

Vogel, J. C.: ^{14}C Variations During the Upper Pleistocene, Radiocarbon, 25, 213–218, https://doi.org/10.1017/s0033822200005506, 1983.

Vogel, J. C. and Kronfeld, J.: Calibration of Radiocarbon Dates for the Late Pleistocene Using U/Th Dates on Stalagmites, Radiocarbon, 39, 27–32, https://doi.org/10.1017/s003382220004087x, 1997.

Wagner, J. D. M., Cole, J. E., Beck, J. W., Patchett, P. J., Henderson, G. M., and Barnett, H. R.: Moisture variability in the southwestern United States linked to abrupt glacial climate change, Nat. Geosci., 3, 110–113, https://doi.org/10.1038/ngeo707, 2010.

Wainer, K., Genty, D., Blamart, D., Hoffmann, D., and Couchoud, I.: A new stage 3 millennial climatic variability record from a SW France speleothem, Palaeogeogr. Palaeocl., 271, 130–139, https://doi.org/10.1016/j.palaeo.2008.10.009, 2009.

Wainer, K., Genty, D., Blamart, D., Daëron, M., Bar-Matthews, M., Vonhof, H., Dublyansky, Y., Pons-Branchu, E., Thomas, L., van Calsteren, P., Quinif, Y., and Caillon, N.: Speleothem record of the last 180 ka in Villars cave (SW France): Investigation of a large $\delta^{18}O$ shift between MIS6 and MIS5, Quaternary Sci. Rev., 30, 130–146, https://doi.org/10.1016/j.quascirev.2010.07.004, 2011.

Wang, X., Auler, A. S., Edwards, R. L., Cheng, H., Cristalli, P. S., Smart, P. L., Richards, D. A., and Shen, C.-C.: Wet periods in northeastern Brazil over the past 210 kyr linked to distant climate anomalies, Nature, 432, 740–743, https://doi.org/10.1038/nature03067, 2004.

Wang, X., Edwards, R. L., Auler, A. S., Cheng, H., Kong, X., Wang, Y., Cruz, F. W., Dorale, J. A., and Chiang, H.-W.: Hydroclimate changes across the Amazon lowlands over the past 45,000 years, Nature, 541, 204–207, https://doi.org/10.1038/nature20787, 2017.

Wang, Y., Cheng, H., Edwards, R. L., Kong, X., Shao, X., Chen, S., Wu, J., Jiang, X., Wang, X., and An, Z.: Millennial- and orbital-scale changes in the East Asian monsoon over the past 224,000 years, Nature, 451, 1090–1093, https://doi.org/10.1038/nature06692, 2008.

Wang, Y. J.: A High-Resolution Absolute-Dated Late Pleistocene Monsoon Record from Hulu Cave, China, Science, 294, 2345–2348, https://doi.org/10.1126/science.1064618, 2001.

Wassenburg, J. A., Dietrich, S., Fietzke, J., Fohlmeister, J., Jochum, K. P., Scholz, D., Richter, D. K., Sabaoui, A., Spötl, C., Lohmann, G., Andreae, M., and Immenhauser, A.: Reorganization of the North Atlantic Oscillation during early Holocene deglaciation, Nat. Geosci., 9, 602–605, https://doi.org/10.1038/ngeo2767, 2016.

Webster, J. W., Brook, G. A., Railsback, L. B., Cheng, H., Edwards, R. L., Alexander, C., and Reeder, P. P.: Stalagmite evidence from Belize indicating significant droughts at the time of Preclassic Abandonment, the Maya Hiatus, and the Classic Maya collapse, Palaeogeogr. Palaeocl., 250, 1–17, https://doi.org/10.1016/j.palaeo.2007.02.022, 2007.

Werner, M., Langebroek, P. M., Carlsen, T., Herold, M., and Lohmann, G.: Stable water isotopes in the ECHAM5 general circulation model: Toward high-resolution isotope modeling on a global scale, J. Geophys. Res.-Atmos., 116, D15109, https://doi.org/10.1029/2011JD015681, 2011.

Whittaker, T. E.: High-resolution speleothem-based palaeoclimate records from New Zealand reveal robust teleconnection to North Atlantic during MIS 1-4, PhD Thesis, The University of Waikato, available at: https://hdl.handle.net/10289/2575 (last access: 2 February 2018), 2008.

Whittaker, T. E., Hendy, C. H., and Hellstrom, J. C.: Abrupt millennial-scale changes in intensity of Southern Hemisphere westerly winds during marine isotope stages 2–4, Geology, 39, 455–458, https://doi.org/10.1130/G31827.1, 2011.

Williams, P. W., King, D. N. T., Zhao, J.-X., and Collerson, K. D.: Speleothem master chronologies: combined Holocene ^{18}O and ^{13}C records from the North Island of New Zealand and their palaeoenvironmental interpretation, The Holocene, 14, 194–208, https://doi.org/10.1191/0959683604hl676rp, 2004.

Williams, P. W., King, D. N. T., Zhao, J.-X., and Collerson, K. D.: Late Pleistocene to Holocene composite speleothem ^{18}O and ^{13}C chronologies from South Island, New Zealand – did a global Younger Dryas really exist?, Earth Planet. Sci. Lett., 230, 301–317, https://doi.org/10.1016/j.epsl.2004.10.024, 2005.

Winter, A., Miller, T., Kushnir, Y., Sinha, A., Timmermann, A., Jury, M. R., Gallup, C., Cheng, H., and Edwards, R. L.: Evidence for 800 years of North Atlantic multi-decadal variability from a Puerto Rican speleothem, Earth Planet. Sci. Lett., 308, 23–28, https://doi.org/10.1016/j.epsl.2011.05.028, 2011.

Wong, C. I., Banner, J. L., and Musgrove, M.: Holocene climate variability in Texas, USA: An integration of existing paleoclimate data and modeling with a new, high-resolution speleothem record, Quaternary Sci. Rev., 127, 155–173, https://doi.org/10.1016/j.quascirev.2015.06.023, 2015.

Wortham, B. E., Wong, C. I., Silva, L. C. R., McGee, D., Montañez, I. P., Troy Rasbury, E., Cooper, K. M., Sharp, W. D., Glessner, J. J. G., and Santos, R. V: Assessing response of local moisture conditions in central Brazil to variability in regional monsoon intensity using speleothem $^{87}Sr/^{86}Sr$ values, Earth Planet. Sci. Lett., 463, 310–322, https://doi.org/10.1016/j.epsl.2017.01.034, 2017.

Yoshimura, K., Kanamitsu, M., Noone, D., and Oki, T.: Historical isotope simulation using Reanalysis atmospheric data, J. Geophys. Res.-Atmos., 113, D19108, https://doi.org/10.1029/2008JD010074, 2008.

Yuan, D.: Timing, Duration, and Transitions of the Last Interglacial Asian Monsoon, Science, 304, 575–578, https://doi.org/10.1126/science.1091220, 2004.

Zanchetta, G., Regattieri, E., Isola, I., Drysdale, R. N., Bini, M., Baneschi, I., and Hellstrom, J. C.: The so-called "4.2 event" in the central Mediterranean and its climatic teleconnections, Alp. Mediterr. Quat., 29, 5–17, available at: http://amq.aiqua.it/index.php/issues-2012-2017/alpine-and-mediterranean-quaternary-29-1-2016/12-the-so-called-4-2-event-in-the-central-mediterranean-and- its-climatic-teleconnections (last access: 15 January 2018), 2016.

Zhang, T.-T., Li, T.-Y., Cheng, H., Edwards, R. L., Shen, C.-C., Spötl, C., Li, H.-C., Han, L.-Y., Li, J.-Y., Huang, C.-X., and Zhao, X.: Stalagmite-inferred centennial variability of the Asian summer monsoon in southwest China between 58 and 79 ka BP, Quaternary Sci. Rev., 160, 1–12, https://doi.org/10.1016/j.quascirev.2017.02.003, 2017.

Zhou, H., Zhao, J., Zhang, P., Shen, C.-C., Chi, B., Feng, Y., Lin, Y., Guan, H., and You, C.-F.: Decoupling of stalagmite-derived Asian summer monsoon records from North Atlantic temperature change during marine oxygen isotope stage 5d, Quaternary Res., 70, 315–321, https://doi.org/10.1016/j.yqres.2008.04.007, 2008.

8

Speleothem stable isotope records for east-central Europe: resampling sedimentary proxy records to obtain evenly spaced time series with spectral guidance

István Gábor Hatvani[1], Zoltán Kern[1], Szabolcs Leél-Őssy[2], and Attila Demény[1]

[1]Institute for Geological and Geochemical Research, Research Centre for Astronomy and Earth Sciences, Hungarian Academy of Sciences, Budapest, Budaörsi út 45, 1112, Hungary

[2]Eötvös Loránd University, Department of Physical and Applied Geology, Budapest, Pázmány Péter stny. 1/C, 1117, Hungary

Correspondence: István Gábor Hatvani (hatvaniig@gmail.com)

Abstract. Uneven spacing is a common feature of sedimentary paleoclimate records, in many cases causing difficulties in the application of classical statistical and time series methods. Although special statistical tools do exist to assess unevenly spaced data directly, the transformation of such data into a temporally equidistant time series which may then be examined using commonly employed statistical tools remains, however, an unachieved goal. The present paper, therefore, introduces an approach to obtain evenly spaced time series (using cubic spline fitting) from unevenly spaced speleothem records with the application of a spectral guidance to avoid the spectral bias caused by interpolation and retain the original spectral characteristics of the data. The methodology was applied to stable carbon and oxygen isotope records derived from two stalagmites from the Baradla Cave (NE Hungary) dating back to the late 18th century. To show the benefit of the equally spaced records to climate studies, their coherence with climate parameters is explored using wavelet transform coherence and discussed. The obtained equally spaced time series are available at https://doi.org/10.1594/PANGAEA.875917.

1 Introduction

With more than a hundred speleothem-related studies published every year, it is trivial to state that speleothems are one of the most important objects of paleoclimate research. Compared to other continental carbonate deposits, they are especially valuable since they (i) are formed in relatively protected cave environments that render late-stage alteration less frequent; (ii) can be precisely dated by radiometric methods (e.g., U–Th dating); (iii) can be sampled at high spatial and temporal resolution, depending on the growth rate of the speleothem; (iv) are well distributed worldwide; and (v) provide multiple independent proxies (e.g., textural characteristics, trace element and stable isotope composition; for further details, see the comprehensive overview by Fairchild and Baker (2012) suitable to the reconstruction of past climate conditions.

Global, regional, and local processes collectively determine the final geochemical signature archived in a speleothem. However, some proxies are more likely to be influenced by large-scale factors (e.g., stable oxygen isotopes; Lachniet, 2009), others are clearly locally controlled (e.g., trace elements; Fairchild and Treble, 2009). Any of the previously mentioned proxies can be used in paleoclimate evaluation, assuming that a clear relationship between climate conditions and measured proxy values can be determined and described. To assist in the interpretation of the geochemical and stable isotope composition of speleothems, the most frequently used and widely accepted method is cave monitoring, in which the physical and chemical parameters of the cave environment and carbonate precipitation are determined in a multi-year study (e.g., Breitenbach et al., 2015; Mattey et al., 2008, 2016; Surić et al., 2018; Riechelmann et al., 2011). These can be used in forward modeling to quantify and understand the role of the dominant factors regulating the geochemical signal in speleothems (Baker and Bradley, 2010). The advantage of this approach is the possibility of gaining direct information on cave dynamics and speleothem formation in the course of environmental changes; the drawback is the usually short timescale, which rarely extends as far as a decade (Fairchild and Baker, 2012; Mattey et al., 2016). The appearance or absence of seasonality (i.e., the seasonality of cave air temperature, ventilation, the chemical composition of drip water) changes in the studied cave can be determined, but the effects of stronger environmental and climate changes may not be observed due to the short periods covered.

Another approach is the comparison of meteorological parameters and speleothem data that cover a sufficient calibration period (e.g., Demény et al., 2017; Jex et al., 2010). If compared with climate parameters, such data can serve as benchmark records, whose comparison may elucidate regional proxy correlations, or can be used to test complex climate models (Wackerbarth et al., 2012). The comparison can be made visually, but detailed statistical data processing (e.g., Baker et al., 2015; von Gunten et al., 2012; Wassenburg et al., 2016) can provide much more objective results.

Unfortunately, despite all efforts, uneven spacing is a common feature of sedimentary paleoclimate records. This characteristic usually biases the results of classical statistical methods and prohibits the application of time series analysis tools in many cases (Cosford et al., 2008; Mudelsee, 2010). There are excellent toolsets that can solve certain problems directly in the case of unevenly spaced data, e.g., determining whether it has a first-order autoregressive characteristic (Schulz and Mudelsee, 2002), estimating correlation with uncertainty limits between unevenly spaced autocorrelated series (Roberts et al., 2017), or conducting a variety of spectral analyses (Schulz et al., 1999; Schulz and Stattegger, 1997). However, a problem remains: the transformation of unevenly spaced data to a temporally equidistant time series is required in order for it to be suitable for the application of, for example, wavelet spectrum, multitaper analysis, and simple smoothers (Hammer, 2017). In the case of unevenly spaced sedimentary records, the required preprocessing step is most commonly performed using linear (e.g., Holmgren et al., 2003, Lachniet et al., 2004), or nonlinear interpolation techniques (e.g., Ersek et al., 2012), rescaling of the data (e.g., Deininger et al., 2017), or by insufficiently documented methods (e.g., McCabe-Glynn et al., 2013; Duan et al., 2014). A more complex approach combining spline smoothing and linear interpolation has been presented in a multi-paleoproxy study, where in addition the spectral characteristics of the data were assessed using multiple methods as well (Cosford et al., 2008).

However, any transformation which adds data or removes data from the original record unavoidably changes its spectral characteristics. This is a factor which can easily be overlooked even in advanced studies, although it deserves a greater degree of attention.

The present paper therefore aims to introduce an approach applied to selected stable carbon and oxygen isotope records from the Baradla Cave (NE Hungary for details see Sect. 2.1). The motivation was to obtain evenly spaced time series from unevenly spaced speleothem records with the application of a spectral guidance to avoid spectral bias and retain the genuine spectral characteristics, which can then be used in further proxy-climate evaluation studies. Section 2 describes the data sources and the proposed methodology for interpolation and resampling, combining the abilities of two available software packages (Björg Ólafsdóttir et al., 2016; Schulz and Mudelsee, 2002) and the development of the spectral control. Section 3 presents the evenly spaced data generated by the application of the methodology.

2 Materials and methods

2.1 Data sources

The stable carbon and oxygen isotope compositions (expressed as $\delta^{13}C$ and $\delta^{18}O$ values in ‰ relative to V-PDB) presented in this paper were derived from two stalagmites (VK1 and NU2; Fig. 1a) from the Baradla Cave (NE Hungary; 48°28′ N, 20°30′ E). Despite being 500 m apart in the cave, the two slow-growing speleothems show a clear similarity in growth rates (VK1: 0.37 and NU2: 0.32 mm yr^{-1}; Fig. 9 in Demény et al., 2017) based on lamina counting and pattern in stable isotope variability. The values, for VK1 and NU2, respectively (Demény et al., 2017), are as follows:

- $\delta^{13}C$: median, −10.47 and −9.97; range, 2.16 and 3.15; SD, 0.53 and 0.64.
- $\delta^{18}O$: median, −6.86 and −7.41; range, 1.69 and 2.39; SD, 0.4 and 0.43.

This is not surprising, since the rock cover above the two sample locations is comparable (100–150 m). Both deposits were seasonally laminated, providing a relative chronology of annual accuracy. Besides age estimation based on lamina counting, ^{14}C activity was determined in the selected stalagmites. Radiocarbon measurements clearly reflected the post-1950 elevated atmospheric signal. Thus, the best match was obtained by shifting the NU2 and VK1 records by 3 and 5 years, respectively (Demény et al., 2017), supporting the synchronization of the individual isotope records. The good match of the isotopic peaks and the similar $\delta^{13}C$ (Fig. 1a) and $\delta^{18}O$ (Fig. 1b) amplitudes both argue for comparable water storage time in the feeding karst water system. Based on lamina counting, the chronological error at the base of VK1 is estimated to be between 3 and 5 years. However,

the perfect matches found for $\delta^{13}C$ (see Fig. 1a) and $\delta^{18}O$ (see Fig. 1b) suggest that the chronological difference is much smaller. Moreover, the stable isotope records of the two stalagmites complement one another by filling the hiatuses found (Fig. 1a). To construct composite isotope records from the two stalagmites' $\delta^{13}C$ and $\delta^{18}O$ records, the original raw data were normalized using mean and standard deviation for the 1950–2000 period (an interval of simultaneous growth for the VK1 and NU2 stalagmites) and merged to a common timescale to provide a regional reference (Demény et al., 2017). Hereinafter, the composite records are expressed as $\Delta^{13}C_{\mathrm{Baradla}}$ and $\Delta^{18}O_{\mathrm{Baradla}}$.

To test the skill of the methodology, the relationship of climate, and the processed evenly spaced $\Delta^{13}C_{\mathrm{Baradla}}$ and $\Delta^{18}O_{\mathrm{Baradla}}$ products was assessed. Thus, monthly averages of temperature and monthly precipitation totals corresponding to the cave location were retrieved a global gridded climate dataset (CRU TS3.23; Harris et al., 2014) with a time span covering 1901–2014 and a grid space of 0.5°.

2.2 Methodology

2.2.1 Interpolation and resampling with spectral guidance

To achieve a regular time axis, the gaps in the $\Delta^{13}C_{\mathrm{Baradla}}$ and $\Delta^{18}O_{\mathrm{Baradla}}$ time series were filled by cubic spline fitting, and resampled to an annual resolution by averaging (Fig. 1) (De Boor, 1978) in an R statistical environment (R Core Team, 2016) using the `stringr` package (Wickham, 2015) and the spline function of the `stats` package (R Core Team, 2016).

The advantages of cubic spline fitting are that it is considered to be highly effective in the preservation of the spectral characteristics of the original record (Horowitz, 1974); it is robust and computationally efficient, thus being suitable to process large data sets (Musial et al., 2011); it outperforms linear interpolation – especially in the higher frequency domain (Schulz and Stattegger, 1997); and it has a smaller chance of overfitting as a higher-order spline.

Furthermore, spline interpolation does not introduce outliers as the Lomb–Scargle reconstruction (Hocke and Kämpfer, 2009) or the Kondrashov and Ghil technique (Kondrashov and Ghil, 2006) do when missing values are – for example – seasonally clustered, thus rendering the estimation of ranges of frequencies and the power spectrum unreliable (Musial et al., 2011). However, because of its "local" estimation characteristic (i.e., lack of predictive capacity), with the increase in a prolonged gap the accuracy of spline interpolation decreases. With the use of rigorous tests it has been documented that if the length of the gap is, for example, slightly above 10 % of the total data length, the mean absolute error of the estimation will be below 0.1. In the case of the Baradla composite stable isotope records, the maximum gap length was 4.85 years, corresponding to 2.24 %, meaning that the use of spline interpolation could be suggested in opposition to the previously mentioned methods. For further details on the benefits and drawbacks of the previously mentioned methods and the detailed documentation on the effects of gap length on mean absolute error the reader is referred to Musial et al. (2011).

It should also be noted that in the case of interpolation, regardless of the method applied, the high-frequency components in the spectrum will be underestimated; thus the interpolated time series will become "reddened" (Schulz and Stattegger, 1997).

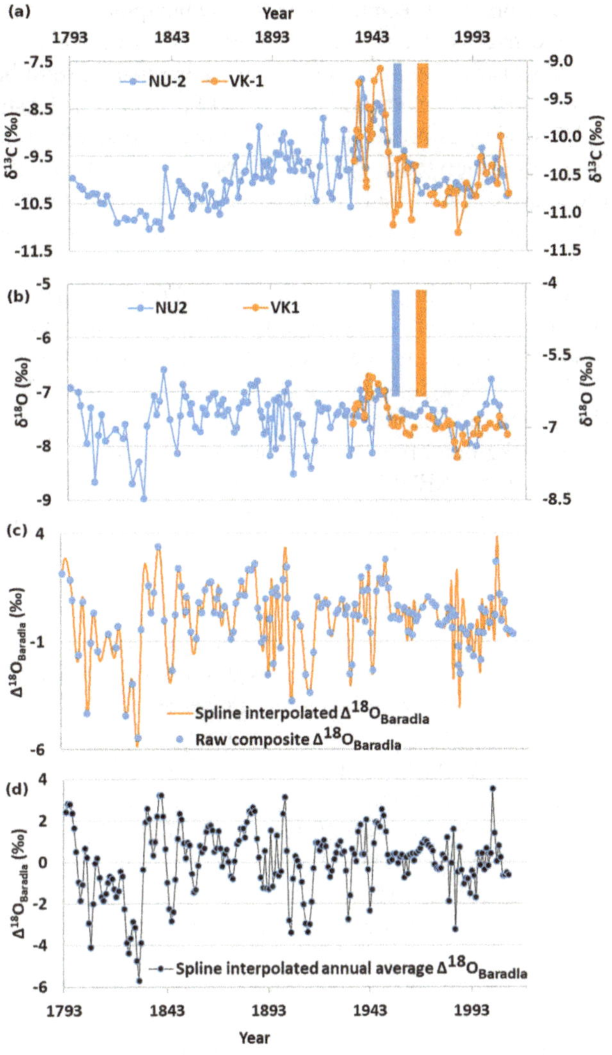

Figure 1. Raw $\delta^{13}C$ **(a)** and $\delta^{18}O$ **(b)** records of the NU2 and VK1 speleothems (hiatuses indicated by gaps in the line), cubic spline fitted to the **(c)** original unevenly spaced $\Delta^{18}O_{\mathrm{Baradla}}$ time series and the **(d)** annual averages derived from the interpolated time series. The blue and orange rectangles in panels **(a, b)** mark the hiatuses of the NU2 and VK1 speleothems, respectively, as discussed in Demény et al. (2017).

Therefore, the spectral bias caused by interpolation has to be objectively quantified and checked. The spectral characteristics of the original and the interpolated data have to be compared to detect potential spectral artifacts. Thus, a threshold frequency is needed to be quantified, beyond which (i) the interpolation (cubic spline in the present case) leaves the original spectrum mainly unchanged, and (ii) the significant powers of the original time series are in coherence with the interpolated spectrum.

To determine this threshold frequency objectively, first the spectral characteristics of the original and the interpolated time series were explored to see if the powers of the time series' Lomb–Scargle–Fourier transform periodograms (LSPs) (Lomb, 1976; Scargle, 1982) are significant against red-noise background from a first-order autoregressive (AR(1)) process using REDFIT (Schulz and Mudelsee, 2002). In the calculations, a `Welch I` window with two overlapping (50 %) segments was used, the oversampling parameter was set as `ofac`=4, as in the works of Hocke and Kämpfer (2009) and Press et al. (1996), and `hifac`=1, the number of Monte Carlo simulations to obtain the significance levels was `nsim`=1000, in line with other studies dealing with speleothem time series (Holzkämper et al., 2004; Neff et al., 2001), and the red-noise boundary was estimated as the bias-corrected 80 % chi-squared limit of a fitted AR(1) process. For the computations, the `redfit` function of the `dplR` package was used (Bunn, 2008). The obtained periodograms will be referred to hereinafter as redfit Lomb–Scargle–Fourier transform periodograms (rLSPs). The obtained rLSPs of the original and the interpolated time series were then compared visually.

A record with pronounced periodic signals was selected to test the performance and robustness of the presented approach. Seasonal averages were computed from the monthly mean total sunspot numbers (SSN; WDC SILSO, 2017) and randomly resampled. Seven data points were taken randomly from each block of 10 years to replicate the resolution of the Baradla speleothem (avg. 0.7 stable isotope data per year). This was repeated a 100 times, so an ensemble of 100 randomly resampled time series were obtained, replicating the nonuniform resolution of the Baradla $\delta^{18}O$ record. The maximum difference between the consecutive samples in the artificially resampled dataset was practically the same as in that in the Baradla speleothem record (~ 7; Fig. 2a, b). Moreover, the relatively large gaps (≥ 4 year) represented about 5–10 % of the artificial time series, which was in harmony with the original speleothem (raw and composite) record (Fig. 2b). The only slight difference between the speleothem record and the artificial time series was seen in the most frequent gaps. In the case of the VK1 record, the most frequent gap between two samples was the same as in the artificial time series – 1 year – while in one of the raw speleothems and the composite record, this was 2 years.

As a final step in line with the proposed protocol, the 100 randomly resampled time series were spline-interpolated and annual averages were calculated, and these were then processed using REDFIT to assess their spectral characteristics and compare them to the spectra of the actual annual averages of the original (sunspot) record.

All the 100 rLSPs replicated the well-known ~ 11-year sunspot cycle ($\alpha = 0.95$). Moreover, most of the rLSPs of the unequally spaced artificial time series mirrored an even smaller peak (~ 8 years). However, the noise caused by spline interpolation can be clearly observed at the higher frequencies (Fig. 2c), highlighting the necessity of controlling/checking the spectral bias caused by interpolation.

After the visual comparison of the rLSPs, the coherence of the unevenly spaced original and the interpolated time series was also computed, using REDFIT-X (Björg Ólafsdóttir et al., 2016), developed for the cross-spectral analysis of unevenly spaced paleoclimate time series. The run parameters were the same as those indicated above.

Combining the visual comparison of the original and interpolated time series' rLSPs with their quantified coherence spectrum, the smallest significant period of the original data was determined, which was also present, and in coherence with the spline-interpolated one. This could then be set as the threshold frequency above which the original spectra could be taken as unbiased. Finally, the variance below this threshold frequency was removed/filtered from the spline-interpolated time series using the bandpass function of the astrochron package (Meyers, 2014).

2.2.2 Wavelet transform coherence

Data preprocessing was necessary to ensure a time series is evenly spaced in time (Sect. 2.2.1) in order to find the areas with common powers between the speleothem stable isotope time series and the climatic data in the time–frequency space. Wavelet transform coherence (WTC) (Torrence and Webster, 1999) was used to assess and visualize the coherence and phase-angle difference of the time series on so-called power spectrum density (PSD) graphs (e.g., Fig. 4). The phase-angle differences are indicated by arrows. It should be noted that, depending on the chronological error, the phase information can be biased. For example, an error of 4 years could already reverse the phase angle of an 8-year period signal. Thus, keeping the chronological error of the input data (see Sect. 2.1) in mind, to avoid over-interpretation, only the dominantly in-phase relationships (indicated by arrows pointing to the right) and dominantly anti-phase relationships (arrows pointing to the left) are considered in the study.

WTC may be considered similar to a correlation coefficient, but with the difference that here we are dealing with a localized time–frequency space (Grinsted et al., 2004). It is based on wavelet spectrum analysis, a function localized in both frequency and time, with a mean of zero (Grinsted et al., 2004), and may be taken as the convolution of the data and the wavelet function (Kovács et al., 2010) for a time series (X_n, $n = 1, \ldots, N$) with a Δt degree of uniform resolu-

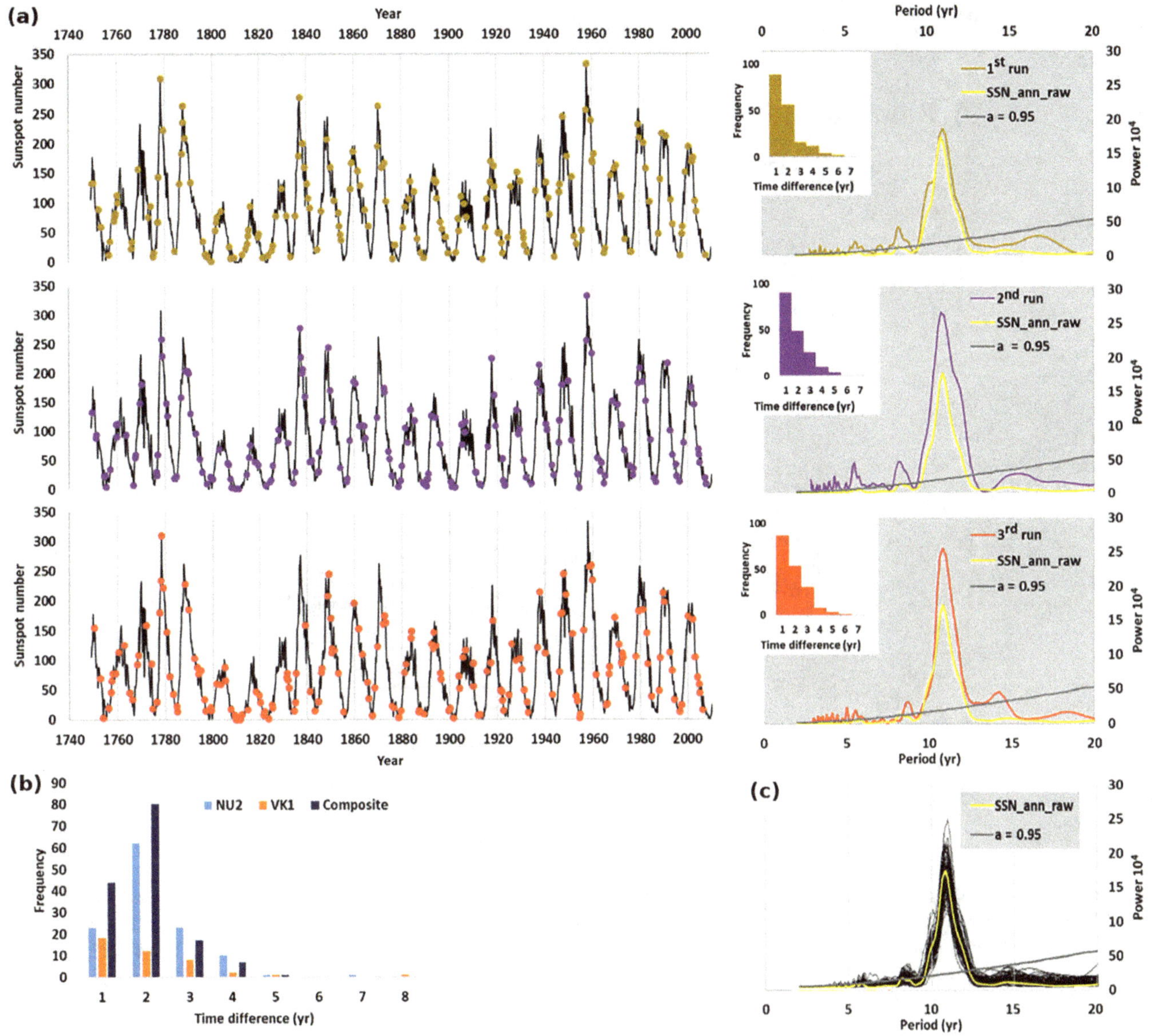

Figure 2. Illustration of the performance and robustness of the resampling protocol. **(a)** Three examples of the randomly resampled annual mean sunspot numbers (left panels), along with the histograms representing the distribution of time differences between the resampled values, and the rLSPs of the original and the corresponding resampled annual mean sunspot number data (right panels). **(b)** Histograms of the time differences between the Baradla stable isotope records' samples. **(c)** The rLSPs of the original and the 100 randomly resampled annual mean sunspot number data. The rLSP of the annual mean sunspot numbers is in yellow (right column in **a** and **c**). The rLSPs of the randomly resampled un-evenly spaced artificial (sunspot) time series, spline interpolated and annualized from the monthly mean sunspot numbers are in orange, purple and red for the three examples respectively **(a)**, and in black for all the 100 randomly resampled time series **(c)**. The bias-corrected 95 % chi-squared limit of a fitted AR(1) process for the rLSPs is shown in grey **(a, c)**.

tion Eq. (1):

$$W_n^X(s) = \sqrt{\frac{\Delta t}{S}} \sum_{n'=1}^{N} X_{n'} \Psi^* \left[(n' - n) \frac{\Delta t}{S} \right], \qquad (1)$$

where N is the length of the time series, ψ^* is the wavelet function, and s is the scale. In this study the Morlet mother wavelet (Morlet et al., 1982) was used to generate daughter wavelets, because it establishes a clear distinction between random fluctuations and periodic regions (Andreo et al., 2006).

If two time series correlate, it does not mean that their WTC will indicate a strong common periodic behavior, because the periodic component has to be present in order to find a meaningful WTC. In the course of the evaluation only those signals were considered which were significant ($\alpha = 0.1$) against an AR(1) process; for details see Roesch

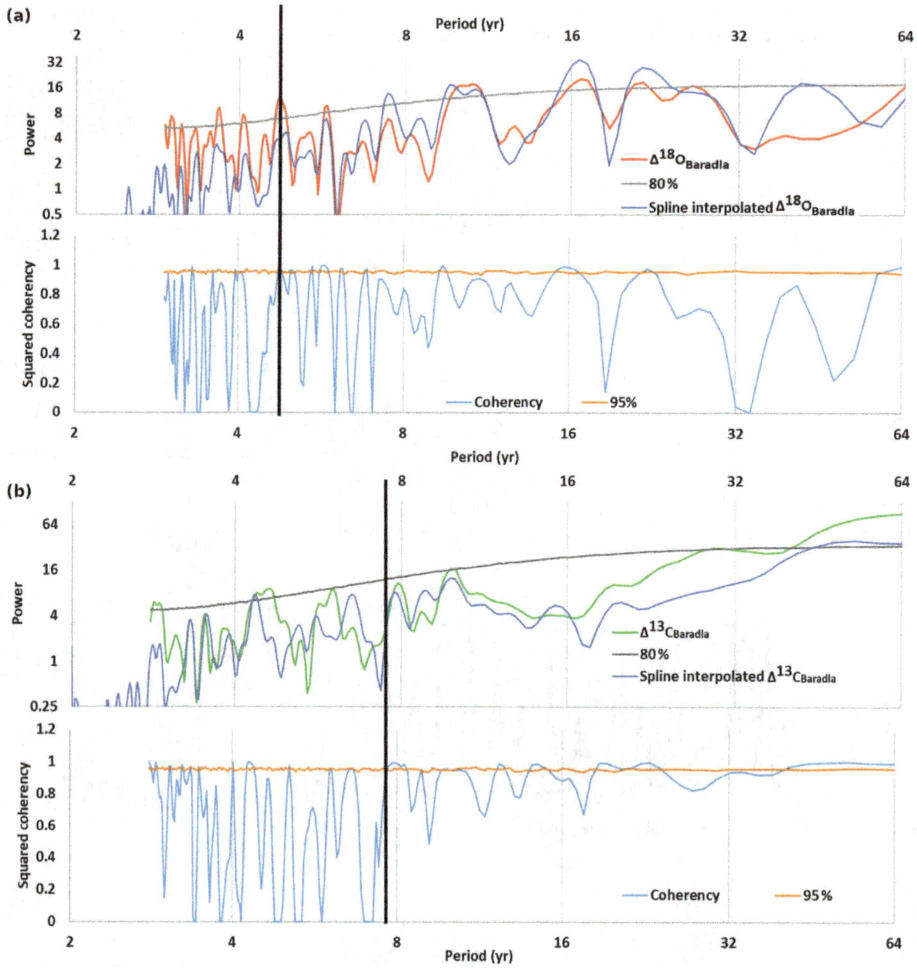

Figure 3. rLSPs (upper panels) and coherency spectra (lower panels) of the original, and the spline-interpolated **(a)** $\Delta^{18}O_{Baradla}$ and **(b)** $\Delta^{13}C_{Baradla}$ records of the Baradla speleothem. The bias-corrected 80% chi-squared limit of a fitted AR(1) process for the rLSPs is shown in grey. The coherency spectra were produced with a 95 % Monte Carlo false-alarm levels (lower panels). The vertical black line indicates the determined cut-off period.

and Schmidbauer (2014). Since the wavelet functions are normalized to have unit energy, the obtained wavelet transforms may even be compared with other time series (Torrence and Compo, 1998).

From a practical point of view, the PSD graphs of the WTC analysis between the stable isotope time series and the monthly climate data were calculated to find the months with the highest response. Then the monthly data which indicated the strongest coherence were averaged (e.g., December–January–February–March) and their WTC with the proxy time series was calculated again. Even if there was a month in between those indicating a strong relationship with the climate data (e.g., strong relationship with December and February–March; weak relationship with January), the multi-monthly averages (December–March and February–March) were calculated and assessed using WTC. Thus, consecutive multi-monthly averages (seasons) were obtained, indicating the maximum response forming the final output of the analysis. The WTC PSD graphs were generated with the `analyze.coherency` function of the `Wavelet-comp` package (Roesch and Schmidbauer, 2014) in `R`.

3 Results and discussions

3.1 Data preprocessing before WTC analysis

After obtaining the rLSPs and the coherency spectra of the original and spline-interpolated $\Delta^{13}C_{Baradla}$ and $\Delta^{18}O_{Baradla}$ time series, these were compared as discussed in Sect. 2.2.1.

As low as a period of ~ 5 years, the significant powers ($\alpha = 0.8$) of the original $\Delta^{18}O_{Baradla}$ record were all mirrored in the spline-interpolated one, and to a high degree of coherency, and the original, and the interpolated record indicated a similar pattern on their rLSPs (Fig. 3a). However, around and below the 4.5-year period, the significant powers of the original $\Delta^{18}O_{Baradla}$ record were no longer reflected

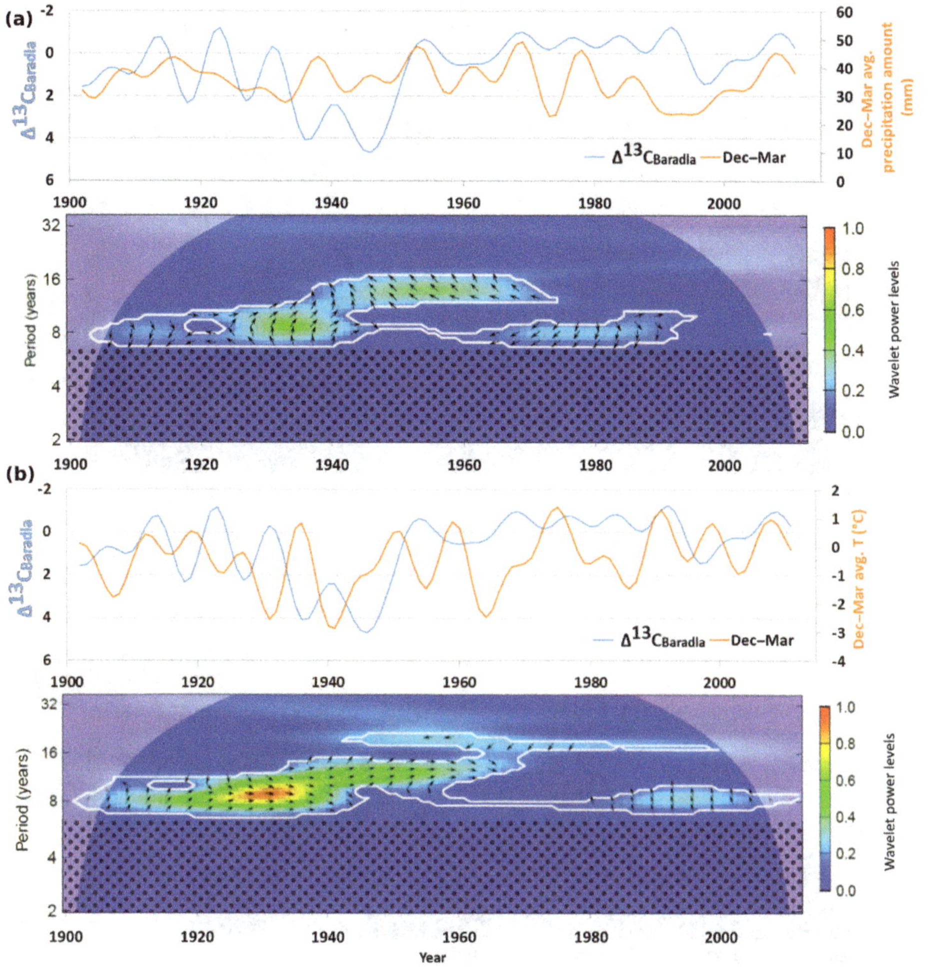

Figure 4. Time series of the 7.5-year low-passed composite $\Delta^{13}C_{Baradla}$ records (on a reversed axis) with the gridded December–March average **(a)** temperature and **(b)** precipitation amount (Harris et al., 2014) (upper panel) and their time–frequency coherency image (lower panel). The white contours in the lower panel show the 90 % confidence levels calculated based on 1000 AR(1) surrogates. The black arrows indicate the phase-angle difference, where an in-phase relationship is indicated by arrows pointing to the right, whereas anti-phase relationship is shown by arrows pointing to the left. The dotted area marks the period band below the cut-off threshold.

in the spline-interpolated one, and the two time series' level of coherency became generally low as well. Therefore, to be consistent and to take a conservative (that is, careful) approach, the period domain below 4.5 years was omitted with a low-pass filter for the $\Delta^{18}O_{Baradla}$ time series. It should be noted, however, that in cases when the storage component in the aquifer is larger than the numerically chosen cut-off period, the water storage time in the karst may amplify the auto-correlation of the isotopic time series.

The same steps were then performed on the $\Delta^{13}C_{Baradla}$ record as well. Thus, around and below the ~7.5-year period, the significant powers of the original $\Delta^{13}C_{Baradla}$ record were no longer reflected in the spline-interpolated one, and the level of the two time series' coherency became generally low as well. Above the threshold, the avg. squared coherency is 0.9 and only drops below 0.8 in exceptional cases, while below it, its average is 0.57 and frequently drops to zero (Fig. 3b). Thus, the domain below 7.5 years was cut off from the $\Delta^{13}C_{Baradla}$ spectrum to avoid the spectral bias caused by spline interpolation (Fig. 3b).

3.2 Climate signals in the stable isotope composition of carbonates of the Baradla speleothems

To present the applicability of the now equally spaced paleoproxy data in climate studies, the $\Delta^{13}C_{Baradla}$ and $\Delta^{18}O_{Baradla}$ records were compared to precipitation and temperature (primary climate variables).

The first step was to explore whether meaningful and significant coherences could be found in the time–frequency space between the equally spaced and 7.5-year low-passed speleothem $\Delta^{13}C_{Baradla}$ time series and similarly low-passed primary climate parameters (precipitation, temperature; Fig. 4). In the case of the composite $\Delta^{13}C_{Baradla}$ record,

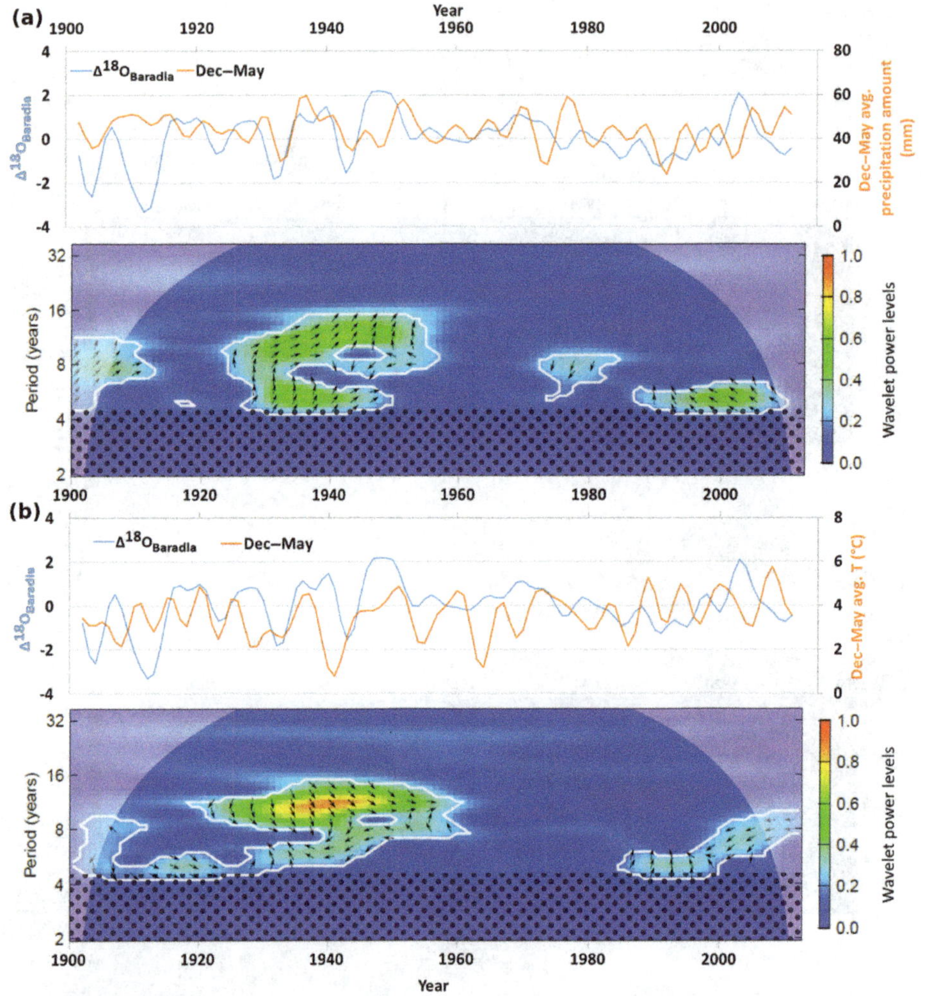

Figure 5. Time series of the 4.5-year low-passed composite $\Delta^{18}O$ records with the gridded climate data (Harris et al., 2014) December–May average **(a)** precipitation and **(b)** temperature (upper panels) and their time–frequency coherency image (lower panels). For further details see the caption of Fig. 4.

the best coherence was found with the December–March averages of temperature and precipitation amount, although a more persistent relationship was detected with precipitation than with temperature (Fig. 4).

The generally anti-phase coherence/relationship between precipitation amount and the $\Delta^{13}C_{Baradla}$ record is in accordance with the former interpretation (Demény et al., 2017), namely that a higher amount of precipitation results in enhanced biological soil activity, more biogenic CO_2 in the soil, and consequently more negative $\Delta^{13}C$ values in the speleothems of the Baradla Cave. Although the relationship between the $\Delta^{13}C$ and precipitation amount had previously been observed for individual records using visual comparisons and regression analysis (Demény et al., 2017), it was found to be weak due to dating uncertainties and the varying effect of additional factors contributing to the precipitation–sedimentary proxy relationship (e.g., kinetic fractionation, vegetation change, prior calcite precipitation; see Fairchild and Baker, 2012, for the compilation of governing factors). Nevertheless, with WTC analysis, it was not the whole spectra which was taken into account all at once (as in the case of linear regression analysis), but the relationship was mapped for each frequency band and observed over time. The phase of the coherence was observed to vary on the decadal scale. A generally anti-phase difference was revealed between $\Delta^{13}C$ and precipitation amount (Fig. 4a).

As for the relationship between $\Delta^{13}C$ and temperature, significant coherences were observed in the 8–16-year period band, but these, however, weakened between ~1960 and ~1975. Furthermore, the phase-angle differences were not persistent afterwards (Fig. 4b). The varying coherence with temperature – regarding the period and phase angle (Fig. 4b) – might be related to biological activity in the soil driven primarily by the interplay between temperature and water saturation conditions mutually affecting the decomposition of the organic matter that is the dominant source of speleothem carbon (Noronha et al., 2015).

Significant coherences were found between the Baradla speleothem's $\Delta^{18}O_{Baradla}$ records and the multi-monthly (December–May) averages of precipitation and temperature. The coherence between precipitation and $\Delta^{18}O_{Baradla}$ was patchily developed, and was significant in the intervals 1900 to ~1915 and ~1920 to ~1955 for the period range of 4–16 years, and between ~1970 and 2010 for around the ~5-year period (Fig. 5a). In the case of temperature and $\Delta^{18}O_{Baradla}$ a more developed, less patchy pattern was seen, with one longer period lacking coherence between ~1960 and the early 1980s (Fig. 5b). Coherence is primarily present in the 4–16-year period range, as in the previous case. A mutual characteristic of the relationship of both precipitation and temperature to $\Delta^{18}O_{Baradla}$ is the common discontinuity around the 1960s (Fig. 5). Such coherences were to be expected, since the oxygen isotope composition of the speleothem is driven by temperature and drip water composition (Lachniet, 2009). Drip water composition is directly related to meteoric water composition, governed by atmospheric temperature and moisture origin in the mid-latitudes (Dansgaard, 1964), and has been documented in the surroundings of the study area (Bottyán et al., 2017; Kaiser et al., 2001). In relation to the seasonal characteristic of the response, a strong contrast exists in the seasonal water balance at the Baradla Cave. Precipitation exceeds evapotranspiration from October to April (Demény et al., 2017), which is well reflected in the isotopic character of the drip water, arguing for the overwhelming dominance (~90%) of winter precipitation (Czuppon et al., 2018). Large-scale tropospheric circulation controls precipitation amount and water stable isotope composition over Europe (Comas-Bru et al., 2016), which invites future investigations of coherence with atmospheric circulation indices.

However, with regard to the phase differences between the $\Delta^{18}O_{Baradla}$ records and the climate parameters, a somewhat confusing picture can be observed in the investigated ~110 years. The phase-angle differences changed multiple times, and a dominant direction could hardly be assigned. This may be a result of the complex interplay of governing factors.

4 Conclusions

A cubic-spline-based universally applicable methodology was developed to handle the uneven spacing of sedimentary proxy records. Additionally, the bias that interpolation may cause was evaluated with spectral control. The methodology was successfully applied to the composite stable carbon and oxygen isotope records of the speleothems of the Baradla Cave, NE Hungary. The composite stable isotope records of the Baradla Cave speleothem were compared with temperature and precipitation-amount data at a monthly resolution using wavelet transform coherence analyses. A generally negative relationship between the carbon isotope record and cold season precipitation amount was revealed, in accordance with earlier assumptions. With temperature, however, a less persistent relationship was observed, and this is suspected to be related to biological activity in the soil. In the case of oxygen isotopic composition, its relationship with temperature and precipitation amount was less clear, probably due to the competing and/or superimposed factors determining the carbonate's oxygen isotopic composition. These observations provide a firm basis for the interpretation of stable isotope data obtained for the Baradla Cave system. The now evenly resampled and low-pass-filtered composite records may also serve as regional benchmarks in future proxy paleoclimate evaluations.

Author contributions. IGH developed the model code and performed the simulations. ZK preprocessed the data and provided the composite records. AD conceived the study and provided the geochemical interpretation. SL-Ő was responsible for the cave studies. All authors took part in the manuscript preparation.

Competing interests. The authors declare that they have no conflict of interest.

Acknowledgements. Financial support was received from the Hungarian Academy of Sciences (MTA "Lendület" program; LP2012-27/2012), the National Research, Development and Innovation Office (OTKA NK 101664), and the János Bolyai Research Scholarship of the Hungarian Academy of Sciences. This is contribution no. 53 of the 2ka Palæoclimatology Research Group.

Edited by: David Carlson

References

Andreo, B., Jiménez, P., Durán, J. J., Carrasco, F., Vadillo, I., and Mangin, A.: Climatic and hydrological variations during the last 117–166 years in the south of the Iberian Peninsula, from spectral and correlation analyses and continuous wavelet analyses, J. Hydrol., 324, 24–39, https://doi.org/10.1016/j.jhydrol.2005.09.010, 2006.

Baker, A. and Bradley, C.: Modern stalagmite $\delta^{18}O$: Instrumental calibration and forward modelling, Global Planet. Change, 71, 201–206, https://doi.org/10.1016/j.gloplacha.2009.05.002, 2010.

Baker, A., C. Hellstrom, J., Kelly, B. F. J., Mariethoz, G., and Trouet, V.: A composite annual-resolution stalagmite record of North Atlantic climate over the last three millennia, Sci. Rep.-UK, 5, 10307, https://doi.org/10.1038/srep10307, 2015.

Björg Ólafsdóttir, K., Schulz, M., and Mudelsee, M.: REDFIT-X: Cross-spectral analysis of unevenly spaced paleoclimate time series, Comput. Geosci., 91, 11–18, https://doi.org/10.1016/j.cageo.2016.03.001, 2016.

Bottyán, E., Czuppon, G., Weidinger, T., Haszpra, L., and Kármán, K.: Moisture source diagnostics and isotope characteristics for precipitation in east Hungary: implications for their relationship, Hydrolog. Sci. J., 62, 2049–2060, https://doi.org/10.1080/02626667.2017.1358450, 2017.

Breitenbach, S. F. M., Lechleitner, F. A., Meyer, H., Diengdoh, G., Mattey, D., and Marwan, N.: Cave ventilation and rainfall signals in dripwater in a monsoonal setting – a monitoring study from NE India, Chem. Geol., 402, 111–124, https://doi.org/10.1016/j.chemgeo.2015.03.011, 2015.

Bunn, A. G.: A dendrochronology program library in R (dplR), Dendrochronologia, 26, 115–124, https://doi.org/10.1016/j.dendro.2008.01.002, 2008.

Comas-Bru, L., McDermott, F., and Werner, M.: The effect of the East Atlantic pattern on the precipitation $\delta^{18}O$–NAO relationship in Europe, Clim. Dynam., 47, 2059–2069, https://doi.org/10.1007/s00382-015-2950-1, 2016.

Cosford, J., Qing, H., Eglington, B., Mattey, D., Yuan, D., Zhang, M., and Cheng, H.: East Asian monsoon variability since the Mid-Holocene recorded in a high-resolution, absolute-dated aragonite speleothem from eastern China, Earth Planet. Sc. Lett., 275, 296–307, https://doi.org/10.1016/j.epsl.2008.08.018, 2008.

Czuppon, G., Demény, A., Leél-Őssy, S., Óvari, M., Molnár, M., Stieber, J., Kiss, K., Kármán, K., Surányi, G., and Haszpra, L.: Cave monitoring in the Béke and Baradla caves (Northeastern Hungary): implications for the conditions for the formation cave carbonates, Int. J. Speleol., 47, 13–28, https://doi.org/10.5038/1827-806X.47.1.2110, 2018.

Dansgaard, W.: Stable isotopes in precipitation, Tellus, 16, 436–468, 1964.

De Boor, C.: A Practical Guide to Splines, Springer-Verlag, New York, 1978.

Deininger, M., McDermott, F., Mudelsee, M., Werner, M., Frank, N., and Mangini, A.: Coherency of late Holocene European speleothem $\delta^{18}O$ records linked to North Atlantic Ocean circulation, Clim. Dynam., 49, 595–618, https://doi.org/10.1007/s00382-016-3360-8, 2017.

Demény, A., Németh, A., Kern, Z., Czuppon, G., Molnár, M., Leél-Őssy, S., Óvári, M., and Stieber, J.: Recently forming stalagmites from the Baradla Cave and their suitability assessment for climate–proxy relationships, Central European Geology, 60, 1–34, https://doi.org/10.1556/24.60.2017.001, 2017.

Duan, F., Wang, Y., Shen, C.-C., Wang, Y., Cheng, H., Wu, C.-C., Hu, H.-M., Kong, X., Liu, D., and Zhao, K.: Evidence for solar cycles in a late Holocene speleothem record from Dongge Cave, China, Sci. Rep.-UK, 4, 5159, https://doi.org/10.1038/srep05159, 2014.

Ersek, V., Clark, P. U., Mix, A. C., Cheng, H., and Lawrence Edwards, R.: Holocene winter climate variability in mid-latitude western North America, Nat. Commun., 3, 1219, https://doi.org/10.1038/ncomms2222, 2012.

Fairchild, I. J. and Baker, A.: Geochemistry of Speleothems, in: Speleothem Science: From Process to Past Environments, John Wiley & Sons, Ltd, Chichester, UK, https://doi.org/10.1002/9781444361094.ch8, 2012.

Fairchild, I. J. and Treble, P. C.: Trace elements in speleothems as recorders of environmental change, Quaternary Sci. Rev., 28, 449–468, https://doi.org/10.1016/j.quascirev.2008.11.007, 2009.

Grinsted, A., Moore, J. C., and Jevrejeva, S.: Application of the cross wavelet transform and wavelet coherence to geophysical time series, Nonlin. Processes Geophys., 11, 561–566, https://doi.org/10.5194/npg-11-561-2004, 2004.

Hammer, Ø.: PAST – PAleontological STatistics, Version 3.15, Natural History Museum, University of Oslo, Oslo, Norway, 2017.

Harris, I., Jones, P. D., Osborn, T. J., and Lister, D. H.: Updated high-resolution grids of monthly climatic observations – the CRU TS3.10 Dataset, Int. J. Climatol., 34, 623–642, https://doi.org/10.1002/joc.3711, 2014.

Hocke, K. and Kämpfer, N.: Gap filling and noise reduction of unevenly sampled data by means of the Lomb-Scargle periodogram, Atmos. Chem. Phys., 9, 4197–4206, https://doi.org/10.5194/acp-9-4197-2009, 2009.

Holmgren, K., Lee-Thorp, J. A., Cooper, G. R. J., Lundblad, K., Partridge, T. C., Scott, L., Sithaldeen, R., Siep Talma, A., and Tyson, P. D.: Persistent millennial-scale climatic variability over the past 25,000 years in Southern Africa, Quaternary Sci. Rev., 22, 2311–2326, https://doi.org/10.1016/S0277-3791(03)00204-X, 2003.

Holzkämper, S., Mangini, A., Spötl, C., and Mudelsee, M.: Timing and progression of the Last Interglacial derived from a high alpine stalagmite, Geophys. Res. Lett., 31, L07201, https://doi.org/10.1029/2003GL019112, 2004.

Horowitz, L.: The effects of spline interpolation on power spectral density, IEEE T. Acoust. Speech, 22, 22–27, 1974.

Jex, C. N., Baker, A., Fairchild, I. J., Eastwood, W. J., Leng, M. J., Sloane, H. J., Thomas, L., and Bekaroğlu, E.: Calibration of speleothem $\delta^{18}O$ with instrumental climate records from Turkey, Global Planet. Change, 71, 207–217, https://doi.org/10.1016/j.gloplacha.2009.08.004, 2010.

Kaiser, A., Scheifinger, H., Kralik, M., Papesch, W., Rank, D., and Stichler, W.: Links between meteorological conditions and spatial/temporal variations in long-term isotope records from the Austrian precipitation network, Study of environmental change using isotope techniques, International Atomic Energy Agency (IAEA), IAEA-CSP–13/P, 67–77, 2001.

Kondrashov, D. and Ghil, M.: Spatio-temporal filling of missing points in geophysical data sets, Nonlin. Processes Geophys., 13, 151–159, https://doi.org/10.5194/npg-13-151-2006, 2006.

Kovács, J., Hatvani, I. G., Korponai, J., and Kovács, I. S.: Morlet wavelet and autocorrelation analysis of long-term data series of the Kis-Balaton water protection system (KBWPS), Ecol. Eng., 36, 1469–1477, https://doi.org/10.1016/j.ecoleng.2010.06.028, 2010.

Lachniet, M. S.: Climatic and environmental controls on speleothem oxygen-isotope values, Quaternary Sci. Rev., 28, 412–432, https://doi.org/10.1016/j.quascirev.2008.10.021, 2009.

Lachniet, M. S., Burns, S. J., Piperno, D. R., Asmerom, Y., Polyak, V. J., Moy, C. M., and Christenson, K.: A 1500-year El Niño/Southern Oscillation and rainfall history for the Isthmus of Panama from speleothem calcite, J. Geophys. Res., 109, D20117, https://doi.org/10.1029/2004JD004694, 2004.

Lomb, N. R.: Least-squares frequency analysis of unequally spaced data, Astrophys. Space Sci., 39, 447–462, 1976.

Mattey, D., Lowry, D., Duffet, J., Fisher, R., Hodge, E., and Frisia, S.: A 53 year seasonally resolved oxygen and carbon isotope record from a modern Gibraltar speleothem: Reconstructed drip water and relationship to local precipitation, Earth Planet. Sc. Lett., 269, 80–95, https://doi.org/10.1016/j.epsl.2008.01.051, 2008.

Mattey, D. P., Atkinson, T. C., Barker, J. A., Fisher, R., Latin, J. P., Durrell, R., and Ainsworth, M.: Carbon dioxide, ground air and carbon cycling in Gibraltar karst, Geochim. Cosmochim. Ac., 184, 88–113, https://doi.org/10.1016/j.gca.2016.01.041, 2016.

McCabe-Glynn, S., Johnson, K. R., Strong, C., Berkelhammer, M., Sinha, A., Cheng, H., and Edwards, R. L.: Variable North Pacific influence on drought in southwestern North America since AD 854, Nat. Geosci., 6, 617–621, https://doi.org/10.1038/ngeo1862, 2013.

Meyers, S. R.: astrochron: An R Package for Astrochronology, available at: https://cran.r-project.org/package=astrochron (last access: January 2018), 2017.

Morlet, J., Arens, G., Fourgeau, E., and Giard, D.: Wave propagation and sampling theory; Part I, Complex signal and scattering in multilayered media, Geophysics, 47, 203–221, https://doi.org/10.1190/1.1441328, 1982.

Mudelsee, M.: Climate Time Series Analysis: Classical Statistical and Bootstrap Methods, Springer, Dordecht, 2010.

Musial, J. P., Verstraete, M. M., and Gobron, N.: Technical Note: Comparing the effectiveness of recent algorithms to fill and smooth incomplete and noisy time series, Atmos. Chem. Phys., 11, 7905–7923, https://doi.org/10.5194/acp-11-7905-2011, 2011.

Neff, U., Burns, S. J., Mangini, A., Mudelsee, M., Fleitmann, D., and Matter, A.: Strong coherence between solar variability and the monsoon in Oman between 9 and 6 kyr ago, Nature, 411, 290–293, https://doi.org/10.1038/35077048, 2001.

Noronha, A. L., Johnson, K. R., Southon, J. R., Hu, C., Ruan, J., and McCabe-Glynn, S.: Radiocarbon evidence for decomposition of aged organic matter in the vadose zone as the main source of speleothem carbon, Quaternary Sci. Rev., 127, 37–47, https://doi.org/10.1016/j.quascirev.2015.05.021, 2015.

Press, W. H., Teukolsky, S. A., Vetterling, W. T., and Flannery, B. P.: Numerical recipes in C, Cambridge University Press, Cambridge, UK, 1996.

R Core Team: R: A Language and Environment for Statistical Computing, R Foundation for Statistical Computing, Vienna, Austria, 2016.

Riechelmann, D. F. C., Schröder-Ritzrau, A., Scholz, D., Fohlmeister, J., Spötl, C., Richter, D. K., and Mangini, A.: Monitoring Bunker Cave (NW Germany): A prerequisite to interpret geochemical proxy data of speleothems from this site, J. Hydrol., 409, 682–695, https://doi.org/10.1016/j.jhydrol.2011.08.068, 2011.

Roberts, J., Curran, M., Poynter, S., Moy, A., Ommen, T. v., Vance, T., Tozer, C., Graham, F. S., Young, D. A., Plummer, C., Pedro, J., Blankenship, D., and Siegert, M.: Correlation confidence limits for unevenly sampled data, Comput. Geosci., 104, 120–124, https://doi.org/10.1016/j.cageo.2016.09.011, 2017.

Roesch, A. and Schmidbauer, H.: WaveletComp: A guided tour through the R-package, available at: http://www.hs-stat.com/projects/WaveletComp/WaveletComp_guided_tour.pdf (last access: January 2018), 2014.

Scargle, J. D.: Studies in astronomical time series analysis. II-Statistical aspects of spectral analysis of unevenly spaced data, Astrophys. J., 263, 835–853, 1982.

Schulz, M. and Mudelsee, M.: REDFIT: estimating red-noise spectra directly from unevenly spaced paleoclimatic time series, Comput. Geosci., 28, 421–426, https://doi.org/10.1016/S0098-3004(01)00044-9, 2002.

Schulz, M. and Stattegger, K.: Spectrum: spectral analysis of unevenly spaced paleoclimatic time series, Comput. Geosci., 23, 929–945, https://doi.org/10.1016/S0098-3004(97)00087-3, 1997.

Schulz, M., Berger, W. H., Sarnthein, M., and Grootes, P. M.: Amplitude variations of 1470-year climate oscillations during the last 100,000 years linked to fluctuations of continental ice mass, Geophys. Res. Lett., 26, 3385–3388, https://doi.org/10.1029/1999GL006069, 1999.

Surić, M., Lončarić, R., Bočić, N., Lončar, N., and Buzjak, N.: Monitoring of selected caves as a prerequisite for the speleothem-based reconstruction of the Quaternary environment in Croatia, Quatern. Int., https://doi.org/10.1016/j.quaint.2017.06.042, in press, 2018.

Torrence, C. and Compo, G. P.: A Practical Guide to Wavelet Analysis, B. Am. Meteorol. Soc., 79, 61–78, https://doi.org/10.1175/1520-0477(1998)079<0061:APGTWA>2.0.CO;2, 1998.

Torrence, C. and Webster, P. J.: Interdecadal Changes in the ENSO–Monsoon System, J. Climate, 12, 2679–2690, https://doi.org/10.1175/1520-0442(1999)012<2679:ICITEM>2.0.CO;2, 1999.

von Gunten, L., D'Andrea, W. J., Bradley, R. S., and Huang, Y.: Proxy-to-proxy calibration: Increasing the temporal resolution of quantitative climate reconstructions, Sci. Rep.-UK, 2, 609, https://doi.org/10.1038/srep00609, 2012.

Wackerbarth, A., Langebroek, P. M., Werner, M., Lohmann, G., Riechelmann, S., Borsato, A., and Mangini, A.: Simulated oxygen isotopes in cave drip water and speleothem calcite in European caves, Clim. Past, 8, 1781–1799, https://doi.org/10.5194/cp-8-1781-2012, 2012.

Wassenburg, J. A., Dietrich, S., Fietzke, J., Fohlmeister, J., Jochum, K. P., Scholz, D., Richter, D. K., Sabaoui, A., Spotl, C., Lohmann, G., Andreae, M. O., and Immenhauser, A.: Reorganization of the North Atlantic Oscillation during early Holocene deglaciation, Nat. Geosci., 9, 602–605, https://doi.org/10.1038/ngeo2767, 2016.

Wickham, H.: stringr: Simple, Consistent Wrappers for Common String Operations, R package, available at: https://CRAN.R-project.org/package=stringr (last access: January 2018), 2017.

9

The ISC-GEM Earthquake Catalogue (1904–2014): status after the Extension Project

Domenico Di Giacomo[1], **E. Robert Engdahl**[2], **and Dmitry A. Storchak**[1]

[1]International Seismological Centre (ISC), Pipers Lane, Thatcham, Berkshire, RG19 4NS, UK
[2]University of Colorado, Boulder, CO, USA

Correspondence: Domenico Di Giacomo (domenico@isc.ac.uk)

Abstract. We outline the work done to extend and improve the ISC-GEM Global Instrumental Earthquake Catalogue, a dataset which was first released in 2013 (Storchak et al., 2013, 2015). In its first version (V1) the catalogue included global earthquakes selected according to time-dependent cut-off magnitudes: 7.5 and above between 1900 and 1918 (plus significant continental earthquakes 6.5 and above); 6.25 between 1918 and 1959; 5.5 between 1960 and 2009. Such selection criteria were dictated by time and resource limitations. With the Extension Project we added both pre-1960 events below the original cut-off magnitudes (if enough station data were available to perform relocation and magnitude recomputation) and added events with magnitude 5.5 and above from 2010 to 2014. The project ran over a 4-year period during which a new version of the ISC-GEM Catalogue was released each year via the ISC website (http://http://www.isc.ac.uk/iscgem/, last access: 10 October 2018). For each year, not only have we added new events to the catalogue for a given time range but also revised events already in V1 if additional data became available or location and/or magnitude reassessments were required. Here we recall the general background behind the production of the ISC-GEM Catalogue and describe the features of the different periods in which the catalogue has been extended. Compared to the 2013 release, we eliminated earthquakes during the first 4 years (1900–1903) of the catalogue (due to lack of reliable station data), added approximately 12 000 and 2500 earthquakes before 1960 and between 2010 and 2014, respectively, and improved the solution for approximately 2000 earthquakes already listed in previous versions. We expect the ISC-GEM Catalogue to continue to be one of the most useful datasets for studies of the Earth's global seismicity and an important benchmark for seismic hazard analyses, and, ultimately, an asset for the seismological community as well as other geoscience fields, education and outreach activities. The ISC-GEM Catalogue is freely available at https://doi.org/10.31905/D808B825.

1 Introduction

Earthquake catalogues are used in many activities by the seismological community. Usually these list basic focal parameters of seismic events (e.g. location, origin time, depth) along with the magnitude, and, eventually, other parameters (e.g. moment tensor or fault plane solutions). Studies concerning seismic hazard and the Earth's global seismicity often require as input an earthquake catalogue that (ideally) has been obtained using the same procedures over a long period of time. For such and other purposes, global instrumental earthquake catalogues have been produced by many authors since the beginning of the last century. Among others, catalogues from Gutenberg and Richter (1954), Båth and Duda (1979), Abe (1981, 1984), Abe and Noguchi (1983a, b) and Pacheco and Sykes (1992) have been extensively used over the past decades until Engdahl and Villaseñor (2002) and Allen et al. (2009) released the Centennial Catalogue and PAGER-CAT, respectively, both covering the period 1900–2007. Although such catalogues proved to be important resources for many years, they cover different time periods and, more importantly, are often characterised by either large heterogeneities in their parameters and/or produced with undocumented or mixed procedures and/or underlying data (e.g. Di Giacomo et al., 2015a). For example, the Centennial Catalogue lists both locations from various catalogues (including the ones mentioned above) and recomputed ones (from 1964 onwards and only for selected large earthquakes between 1918 and 1964) using the Engdahl et al. (1998) methodology (normally referred to as EHB), whereas magnitudes are not recomputed but compiled from several different sources/authors (see Di Giacomo et al., 2015a). Very similar considerations also apply to PAGER-CAT, which is based on the Centennial Catalogue up to 1973 (Allen et al., 2009). In addition, most of these catalogues terminate at different times and are no longer maintained. In this context, in 2010 the International Seismological Centre (ISC, http://www.isc.ac.uk/, last access: 10 October 2018), as requested by the GEM Foundation (https://www.globalquakemodel.org/, last access: 10 October 2018), undertook a major effort to reprocess 100+ years of instrumental seismological data to reassess both locations and magnitudes of global (i.e. having magnitude 5.5 and above in our framework) earthquakes and, consequently, to produce a new earthquake catalogue using homogeneous and documented methodologies over the longest possible period of instrumental seismology (i.e. since the early 20th century). In January 2013, after a 27-month project, the ISC and a team of international experts (http://www.isc.ac.uk/iscgem/people.php, last access: 10 October 2018) released on the ISC website (http://www.isc.ac.uk/iscgem/, last access: 10 October 2018) the first version (V1, for a general description see Storchak et al., 2013, 2015) of the ISC-GEM Global Instrumental Earthquake Catalogue (1900–2009).

Since then the ISC-GEM Catalogue has been used by many researchers investigating seismicity rates, patterns of seismicity and earthquake forecast (e.g. Cambiotti et al., 2016; Geist, 2014; Ikuta et al., 2015; Kagan, 2017; Kagan and Jackson, 2016; Katsumata, 2015; Pollitz et al., 2014; Quinteros Cartaya et al., 2016; Roth et al., 2017; Zaliapin and Kreemer, 2017; Zechar et al., 2016; Zhan and Shearer, 2015) as well as by groups working on earthquake catalogues for seismic hazard purposes (e.g. Alvarez et al., 2016; Deif et al., 2017; Ghasemi et al., 2016; Kadirioğlu et al., 2016; Markušić et al., 2015; Mikhailova et al., 2015; Poggi et al., 2017; Weatherill et al., 2016) and other seismological studies (e.g. Lange et al., 2017; Leonard, 2014; Metzger et al., 2017; Ye et al., 2016).

In recognition of the value of such a homogeneous (to the largest extent possible) instrumental catalogue, funding from public and commercial organisations (http://www.isc.ac.uk/iscgem/sponsors.php, last access: 10 October 2018) has been given to the ISC since November 2013 to work on the extension of the ISC-GEM Catalogue over a 4-year project which aimed, in a nutshell, at adding as many earthquakes as possible before 1960 and prolonging the catalogue beyond 2009. The Extension Project was also motivated by the fact that damaging pre-1960 earthquakes were below the cut-off magnitude of 6.25 (e.g. the 30 October 1930 central Italy event, which caused collapse and severe damage in various towns) and many pre-1960 events had no initial magnitude and therefore could not be selected for V1, yet they could be large enough to be part of the ISC-GEM Catalogue.

Below we detail the work done during the 4 years of the Extension Project (which ended in December 2017) and discuss features of the different time periods extended. Then we outline the overall state of the ISC-GEM Catalogue in its latest version (V5) and, finally, present the outlook for its further advancement.

2 The 4-year plan of the Extension Project

The Extension Project of the ISC-GEM Catalogue has been designed to add earthquakes smaller than magnitude 6.25 before 1960 and extend it beyond 2009 with events of magnitude 5.5 and above. In addition, many earthquakes pre-1960 with no magnitude information needed to be processed to reassess location and magnitude, if enough station data were available. Figure 1 summarises the annual number of events before 1960 included in V1 of the ISC-GEM Catalogue along with the pre-1960 events available in the International Seismological Summary (ISS, 1918–1963; see also Villaseñor and Engdahl, 2005), BAAS (1913–1917) and the Centennial Catalogue plus additional hypocentres (hereafter we refer to it as the augmented Centennial Catalogue) that were not processed for the V1 release (see also Fig. 8 in Storchak et al., 2015). Note that ISS and BAAS earthquakes are also listed in the Centennial Catalogue but throughout the paper we try to refer to the original sources as much as possible. For simplicity, in the following we refer to earthquakes in grey in Fig. 1 as extension events (i.e. not listed in V1), meaning that those are the events we digitised station data for but not necessarily all will be selected for processing and then included in the ISC-GEM Catalogue. The station data collection and the selection process will be discussed in the following sections.

The annual number of events in V1 oscillates between 4 and 12 for 1904–1917, 31 and 92 for 1918–1959 and 235 and 489 for 1960–2009. Such variations reflect the cut-off magnitudes adopted for selecting earthquakes in different time periods: 7.5 and above before 1918 (plus significant continental earthquakes 6.5 and above); 6.25 between 1918 and 1959; 5.5 from 1960 onwards (see Di Giacomo et al., 2015b, for more details on the V1 earthquake selection criteria). It is worth remembering here that the cut-off magnitudes are simply thresholds set for selection purposes (not all pre-1960 events have known or reliable magnitudes) and should not be interpreted as completeness levels (variations of the completeness over different time periods for V1 were briefly outlined by Di Giacomo et al., 2015a, and investigated in more detail by Michael, 2014).

Considering the number of pre-1960 earthquakes available (nearly 21 000, i.e. about 2000 more than the V1 release covering 1900–2009) in the ISS (1918–1959), BAAS (1913–

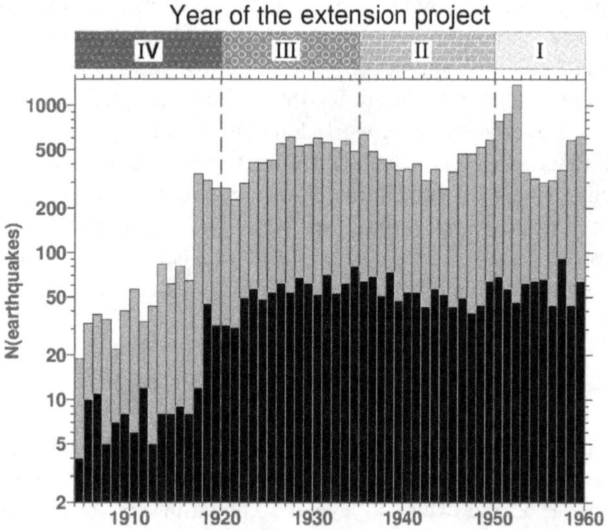

Figure 1. Annual number of pre-1960 earthquakes in V1 of the ISC-GEM Catalogue (black, total = 2439 events) and the events that are available in the ISS between 1918 and 1959 and the augmented Centennial Catalogue/BAAS between 1904 and 1917 (grey, total of 20 865 events) that were not processed for V1 (in the text referred to as extension events). The hachure patterns on top outline the period extended in each year of the Extension Project: 1950–1959, 1935–1949, 1920–1934 and 1904–1919 in Year I, II, III and IV, respectively. The period 2010–2014, not shown here, has also been progressively added during the Extension Project.

1917) and augmented Centennial Catalogue for which we had to look for station data (and, consequentially, digitise), we planned to extend the catalogue following a 4-year schedule as outlined in Fig. 1. Such a time frame was necessary to allow us to be as comprehensive as possible in the station data collection task and also to assess the ~ 60 % of extension events that had no initial magnitude information (in our database), and, therefore, could not have been selected just using any cut-off magnitude criteria (details in the next section). In addition, the extension of the catalogue beyond 2009 would benefit from the data concurrently released in the ISC Bulletin and would follow the original selection criteria (i.e. earthquakes with magnitude 5.5 and above).

At the end of each project year an upgraded version of the catalogue was made available for download at http://www.isc.ac.uk/iscgem/ (last access: 10 October 2018). The catalogue is distributed in CSV format and is composed of two parts (the Main catalogue, also available as a KMZ file for use with Google Earth, and the Supplementary catalogue, the latter including events with either poor location and/or magnitude quality; see Storchak et al., 2015). Location parameters and magnitudes (either direct or proxy moment magnitude M_w; Di Giacomo et al., 2015a) come with formal uncertainties and quality flags (from A to D, denoting well and poorly constrained parameters, respectively), followed, if available, by the solution of the Global Centroid Moment Tensor (GCMT; http://www.globalcmt.org, last access: 10 October 2018, Dziewonski et al., 1981; Ekström et al., 2012). The criteria to assign the quality flags for location, depth and magnitude are summarised in Table 1. For the location quality flag we consider the secondary azimuthal gap (largest azimuthal gap filled by a single station; Bondár et al., 2004, hereafter referred to as SGAP), the eccentricity of the error ellipses (Bondár and Storchak, 2011) and the event location accuracy if it is of high confidence to become a candidate for the IASPEI Reference Event List (GTCAND in Table 1; see Bondár and McLaughlin, 2009, and http://www.isc.ac.uk/gtevents, last access: 10 October 2018). For the depth quality flag we consider the availability of very close stations (within 10 km, NSTA10) and in the local distance range (within 150 km, NSTAlocal), the depth constrained by depth-phases (if available, depdp in Table 1) and the location accuracy (GTCAND). For the magnitude quality flag we consider the author (GCMT or literature; Lee and Engdahl, 2015) for direct M_w values, whereas for the M_w proxy based on our recomputed M_S or m_b (Di Giacomo et al., 2015a), the quality flag depends on combinations of the magnitude value, type (M_S or m_b), uncertainty, number of stations used and the uncertainty of M_w proxy.

One of the key features of the ISC-GEM Catalogue is that all events since 1904 have been reprocessed using instrumental station parametric data and the ak135 model (Kennett et al., 1995). To extend the catalogue, we followed the same steps and methodologies used to create V1, as described in the following.

- Di Giacomo et al. (2015b, and references therein) was followed for digitising from printed bulletins body-wave arrival times and amplitudes/periods (of surface waves in particular) for the pre-1960 events to allow relocations and magnitude recomputation, respectively. For the extension events, the most important source of body-wave arrival times was the ISS, whereas amplitudes and periods were retrieved from individual station or network printed bulletins.

- Bondár et al. (2015) describe the two-tier relocation approach, which benefits both from the EHB location algorithm (Engdahl et al., 1998) and the new ISC locator (Bondár and Storchak, 2011) used to constrain the depth and the epicentre, respectively. As the EHB and ISC location algorithms are also used to cross-check each other, the location consistency is checked twice.

- Di Giacomo et al. (2015a) describe the magnitude recomputation, particularly for the surface wave magnitude M_S, which, in turn, is used as the basis for M_w conversion for most of the events pre-1960.

- Lee and Engdahl (2015) are referred to for the literature search of reliable and direct computations of seismic

Table 1. Criteria to assign the location quality flags for location, depth and magnitude. SGAP is the secondary azimuthal gap, GTCAND denotes a high confidence location accuracy that makes the event a candidate for the IASPEI Reference Event List (Bondár and McLaughlin, 2009; see also http://www.isc.ac.uk/gtevents, last access: 10 October 2018), depdp is the depth constrained by depth phases (if available), NSTA10 is the number of stations within 10 km and NSTA (local) is the number of stations within 150 km (Bondár and Storchak, 2011). M_S is considered well constrained when it is obtained from more than four stations, within 5.5–7.5, and has uncertainty ≤ 0.2. See text for details.

Quality flag	Location	Depth	Magnitude (M_w, direct or proxy)
A	SGAP < 120 && eccentricity < 0.75 or GTCAND	depdp or GTCAND or NSTA10	GCMT
B	SGAP < 160	NSTAlocal > 2	Literature or Proxy based on well-constrained M_S
C	Other cases	Other cases	Literature or Proxy based on poorly constrained M_S or Proxy based on m_b
D	Manually assigned	Manually assigned	No magnitude or Proxy uncertainty > 0.7 or M_w proxy based on $M_S < 5$

moment M_0 and, therefore, of M_w, for events before the GCMT solutions start in 1976.

The data collection has been the most time-consuming task and indispensable part, not only to extend the catalogue but also to revise and better constrain solutions of events already in V1 (details in the next sections). Indeed, compared to the data collected for the V1 release, we made a significant improvement in the number of amplitude and period data digitised, particularly for M_S recomputation, thanks both to additional bulletins donated (or lent) to the ISC from various institutions and individuals (including the personal collection of Nicholas Ambraseys; more details are available at http://www.isc.ac.uk/iscgem/acknowledge.php) and station bulletins that were not processed for V1 due to time and resource limitations. Later we also show how the additional data gathered during the last 4 years helped us revise and better constrain the M_S of pre-1960 earthquakes already listed in V1. With the end of the Extension Project in December 2017, in the following we outline the improvements and features of different time periods during which the ISC-GEM Catalogue has been extended.

3 Extension for the period 1920–1959

In this section we describe the work done in the first 3 years of the Extension Project to add earthquakes in the predigital period between 1920 and 1959. Note that throughout this work we consider as predigital earthquakes those that occurred before 1964 (i.e. before the beginning of the ISC Bulletin).

3.1 Station data collection and earthquake selection

The variations in the annual number of the extension events shown in Fig. 1 are the result of various factors. For example, a significant increase in the annual number of events can be seen in 1918 coinciding with the beginning of the ISS, whereas a dip in the late 1930s to mid-1940s is associated with the disruption caused by World War II (more details later) and another dip in the mid-1950s is due to the censoring introduced by ISS procedures (more details on page 3 of the ISS, 1953) to reduce the workload. The annual variations in the number of the extension events also introduce an issue in selecting earthquakes for the ISC-GEM Catalogue. For example, between 1950 and 1952 the annual number of extension events in the ISS is between 782 and 1384, and such numbers are above the annual number of earthquakes of magnitude 5.5 and above in the ISC-GEM Catalogue in recent years (e.g. the largest annual number of earthquakes is 654 for 2011). This means that a subset of the extension events in 1950–1952 should not be part of the ISC-GEM Catalogue as it falls below the cut-off magnitude of 5.5. However, since, as mentioned earlier, about 60 % of such events

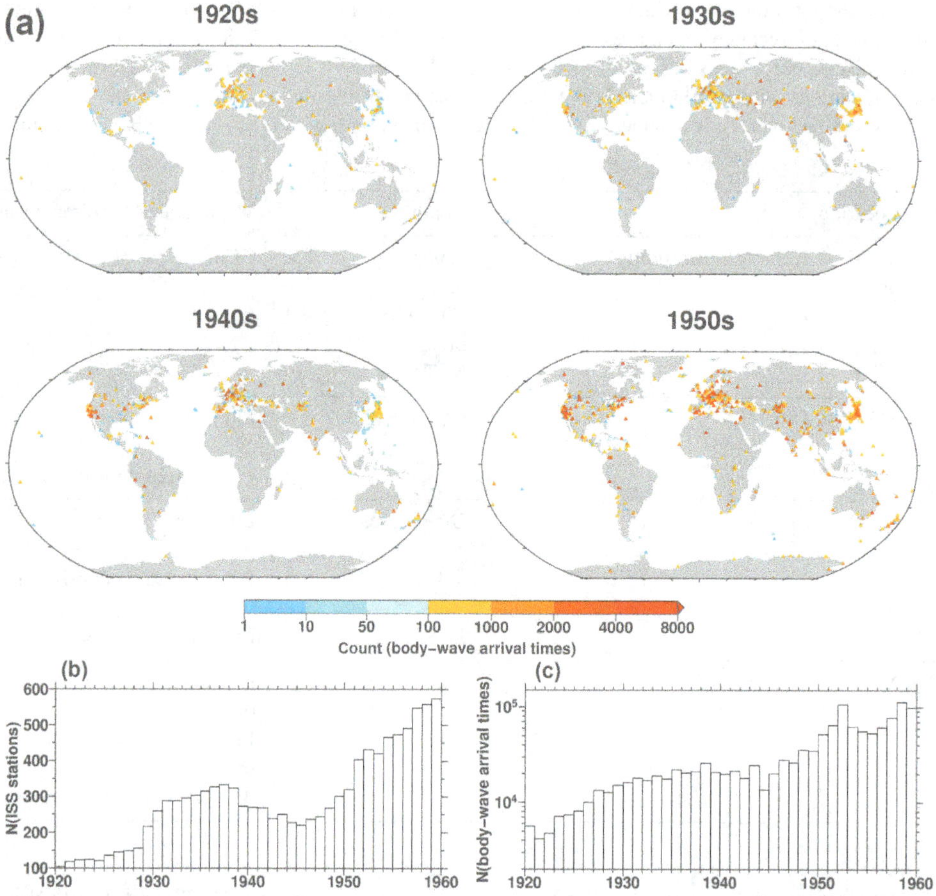

Figure 2. Distribution of stations listed in the ISS in each decade contributing body-wave arrival times to the extension events and colour-coded by number of body-wave arrival times. The annual number of stations **(b)** and body-wave arrival times **(c)** are also summarised.

have no magnitude information in our database, we could not use the original magnitude criteria of 5.5. Thus, for the extension events we decided to base our selection criteria both on the distribution of stations in the ISS and the number of stations contributing amplitudes/periods for magnitude recomputation. This required a major effort to digitise both all ISS pages (not available in any electronic format) and amplitude and period pairs (of surface waves, in particular) from the station/network printed bulletins (Di Giacomo et al., 2015b) for all extension events. Here we briefly summarise the station data collected for the extension events and highlight some features that are relevant to the ISC-GEM Catalogue.

Figure 2 shows the distribution of stations listed in the ISS for each decade (1920s, 1930s, 1940s, 1950s) colour-coded by their body-wave arrivals contribution to the extension events along with the annual number of stations and body-wave arrivals digitised from the ISS. The number of stations listed in the ISS generally increased from the 1920s to the late 1930s before World War II affected various seismic stations, and it is only around 1953 that the station contribution improved significantly. The box-and-whisker plot of Fig. 3 summarises the median number of stations per event in each year. It shows that only a limited number of stations (median number ranging from 9 to 26) are usually associated with the extension events until 1952, whereas from 1953 onwards there is a general improvement in this respect (median number of stations ranging from 66 to 99). Another relevant feature to point out is the uneven station distribution, with Europe showing the highest density particularly before the 1950s, and the lack of stations in Africa and vast parts of the Southern Hemisphere.

Figures 4 and 5, similarly to Figs. 2 and 3, show the distribution of stations contributing amplitudes for each decade and the median number of stations supplying amplitudes in each year, respectively. The number of stations reporting amplitudes increased until World War II, dropped in the 1940s and improved significantly from 1953. European and many Russian stations are the most important contributors to amplitude readings compared to stations in other continents, except for La Paz (LPZ, Observatorio San Calixto, Bolivia) and Riverview College (RIV, Sydney, Australia) from the Jesuit seismic network (Udías and Stauder, 1996). The number of

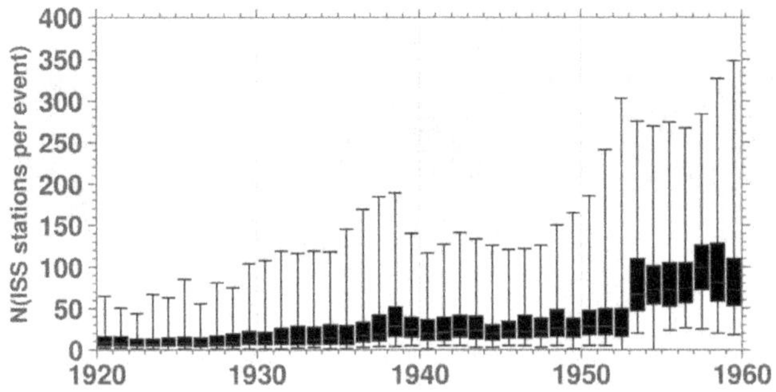

Figure 3. Box-and-whisker plot for the extension events of the median number of ISS stations per event in each year. The box represents the 25 %–75 % quantile, the band inside the box represents the median and the ends of the whiskers represent the minimum and maximum of all data.

stations per event contributing amplitudes ranges from 0 to above 40, with the median per year oscillating from 0 to 6 (Fig. 5).

We based the selection of the extension events on combinations of the number of body-wave arrival times and the number of stations supplying amplitude data. Considering that our relocation approach (Bondár et al., 2015) relies largely on teleseismic observations (i.e. above 18° distance) and the magnitude reassessment (Di Giacomo et al., 2015a) on the availability of three (or two in some case) station magnitudes, we first excluded events with no teleseismic phases and fewer than two stations contributing amplitudes. After this first cut, we further excluded earthquakes with a limited number of body-wave arrival times and fewer than two to three stations with amplitudes. These are earthquakes for which we could not obtain a reliable solution (due to poor station coverage and/or arrival times) after preliminary relocation attempts. It is worth pointing out we have tried to be as comprehensive and conservative as possible by not rejecting all poorly constrained relocations (see next section). Also, we included all extension events between 1953 and 1956 available in the ISS (due to their small number; see Fig. 1) and well-recorded earthquakes but without amplitudes. As a result, out of the 19 341 extension events between 1920 and 1959 we relocated 11 572. The annual numbers are shown in Fig. 6, where the variations are linked to the state of the global network during those years and the operational practice changes at the ISS, as mentioned earlier.

3.2 Relocations

The location reassessment of previous hypocentres (from ISS or other authors adopted by it) of the selected extension events is one of the fundamental tasks of this work. The relocations are obtained by closely following the approach described by Bondár et al. (2015). In Fig. 7 the box-and-whisker plots of the defining stations (i.e. stations with at least one arrival time that constrains the location, hereafter referred to as NDEFSTA) and SGAP for each year are shown. The NDEFSTA gradually increases from the 1920s to the 1950s (except for the slight dip in the 1940s, for reasons explained earlier), whereas the SGAP gradually improves over time. This in general leads to improved confidence in locations. Figure 8 shows the location and depth differences between the previous (ISS or authors adopted by ISS) and the ISC-GEM hypocentres. With a few large exceptions, median location differences range from about 100 km in the 1920s to about 20 km in the late 1950s. With depth differences, one must consider that for 9418 relocated extension events the original depth was unknown and nominally set to zero. Also, it is important to point out that about half of the relocated extension events have no depth phases; therefore for those the depth was assigned to a default depth resulting from the tectonic setting or nearby earthquakes. However, as already pointed out by Bondár et al. (2015), we remove the artefact of having most shallow earthquakes set at zero km depth.

Table 2. Summary of the location and depth quality flags for the extension events between 1920 and 1959.

Quality flag	Count for location	Count for depth
A	1744	1886
B	3315	479
C	6431	9208
D	83	0

We checked the reliability of the ISC-GEM relocations in terms of network coverage and deviation from the available hypocentres grouped for an event, performed a cross-check between the EHB and ISCloc algorithms and considered the nearby seismicity. At times we also used available comments in the individual station bulletins as a guide in solving uncertain cases. Obviously, relocations for events with large SGAP (> 270°) and/or small NDEFSTA are not well constrained and we decided case by case whether to manually assign location flag D (i.e. the event will be listed in the Supplementary Catalogue). A typical case in this respect (although time-dependent) is represented by earthquakes in the North Atlantic ridge where most of the phase data would come from European stations and SGAP could be even larger than 300° simply because North American stations (see Fig. 2) would not systematically report data for such earthquakes (except for large ones).

Table 2 summarises the location and depth quality flags for the relocated extension events between 1920 and 1959. The most frequent quality flag both for location and depth is C. However, despite the limitations of the global seismic network, particularly before the 1950s, it is possible to recognize the improvements of the ISC-GEM locations with respect to the original ones even on a global scale, as shown in Fig. 9. Although we do not claim that the ISC-GEM locations

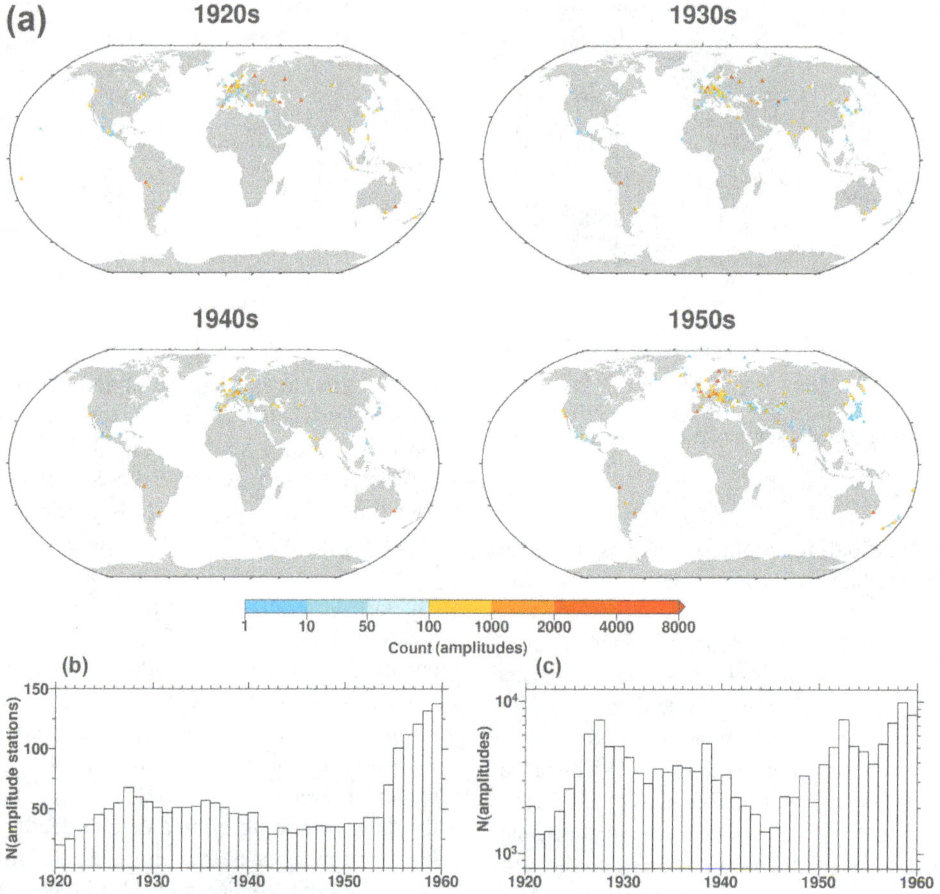

Figure 4. Distribution of the stations in each decade contributing amplitudes to the extension events and colour-coded by number of amplitudes. The annual number of stations **(b)** and amplitudes **(c)** are also summarised.

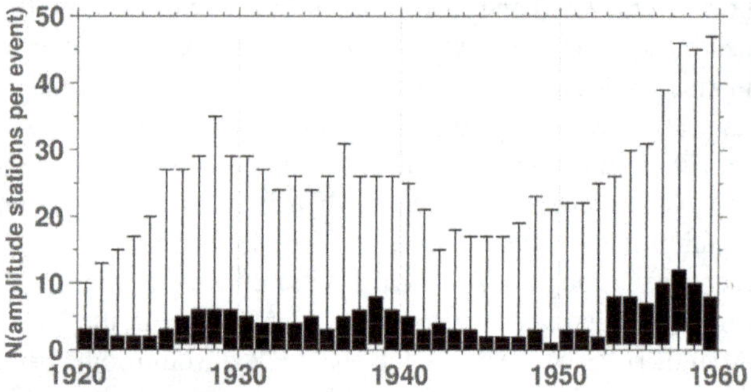

Figure 5. Box-and-whisker plot for the extension events of the median number of stations supplying amplitudes per event in each year. The box represents the 25 %–75 % quantile, the band inside the box represents the median and the ends of the whiskers represent the minimum and maximum of all data.

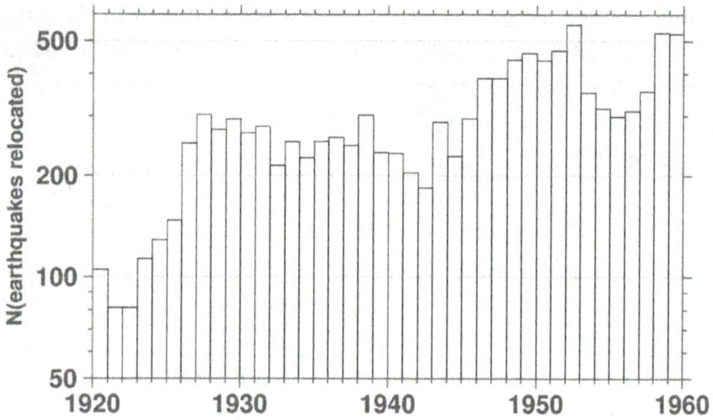

Figure 6. Annual number of the relocated extension events between 1920 and 1959. See text for details.

are the best possible solutions in this period for every single event, we recommend that any regional or focused study of predigital earthquakes instrumentally recorded should start from the ISC-GEM locations as they are obtained from (currently) the most comprehensive set of instrumental data.

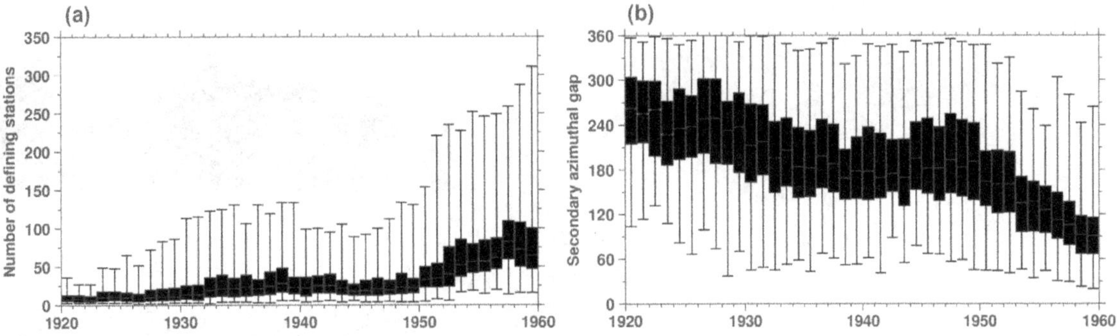

Figure 7. Box-and-whisker plots of the number of defining stations (NDEFSTA, **a**) and the secondary azimuthal gap (SGAP, **b**) in each year.

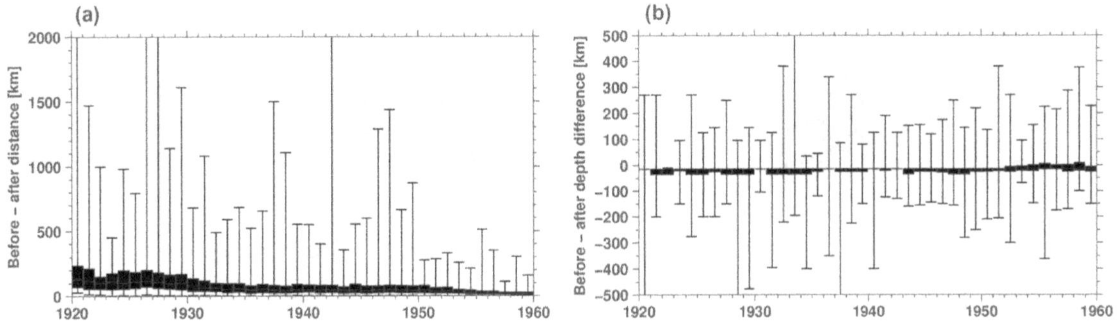

Figure 8. Box-and-whisker plots of the epicentre **(a)** and depth **(b)** differences between previous hypocentres (before) from ISS (or authors adopted by ISS) and ISC-GEM (after) locations in each year. For 9417 of the 11 572 extension events relocated between 1920 and 1959 the depth for the previous hypocentres (before) was unknown and nominally set to zero.

3.3 Magnitude reassessment

We used the approach described in Di Giacomo et al. (2015a) to reassess the magnitude of the extension events consistently with their ISC-GEM relocations. Due to the lack of short-period body-wave amplitudes before the 1960s, here we focus on recomputed M_S as the basis for the calculation of the proxy M_w. The M_S recomputation is based on the amplitudes and periods of surface waves digitised during this work (Figs. 4 and 5). Before accepting an M_S value, we checked the station distribution and, when possible, cross-checked our magnitudes with other magnitude information to investigate cases of large differences with previous results. Figure 10 shows the timeline of the recomputed M_S and their annual counts. Besides the recurrent features discussed earlier (i.e. general increase in the annual counts from the early 1920s and the dip in the 1940s), there are 2304 events with M_S below 5.5. This occurs because our selection criteria for this period, as explained earlier, had to be based on station data availability rather on magnitude. Although events with magnitude below 5.5 would not normally be part of the ISC-GEM Catalogue, we did not exclude them because of the importance of reassessing the magnitude of predigital earthquakes. Most of these events with $M_S < 5.5$ are mostly located in an area covering the mid-oceanic ridge of the North Atlantic to the European Mediterranean region. This is not surprising considering the distribution of stations contributing amplitudes (Fig. 4). Also, there are 80 earthquakes with $M_S \geq 6.5$ that should have already been in V1. These events were not originally selected because the available magnitude information was considered not reliable or it was below the cut-off value of 6.25. This further highlights the necessity of a comprehensive and systematic magnitude reassessment with homogeneous procedures.

In total, we recomputed M_S for 6575 ($\sim 57\,\%$) of the relocated extension events and obtained a magnitude (M_S or any other type) for the first time (at least to the best of our knowledge) for 3011 of them. A lack of stations reporting amplitudes is normally the cause for not having a recomputed M_S as we normally require a minimum of three stations. The only exception occurs when we have two station magnitudes from a subset of specially selected stations that do not differ more than 0.3 magnitude units (m.u.). In such circumstances we allowed M_S recomputation for 276 earthquakes and assigned M_S uncertainty of 0.5 m.u.

If no direct M_w value is available for an event, the recomputed M_S values are then used as the basis for proxy calculations of M_w and magnitude quality flags (Di Giacomo et al., 2015a). Table 3 summarises the counts for the magnitude quality flags for the relocated extension events between 1920

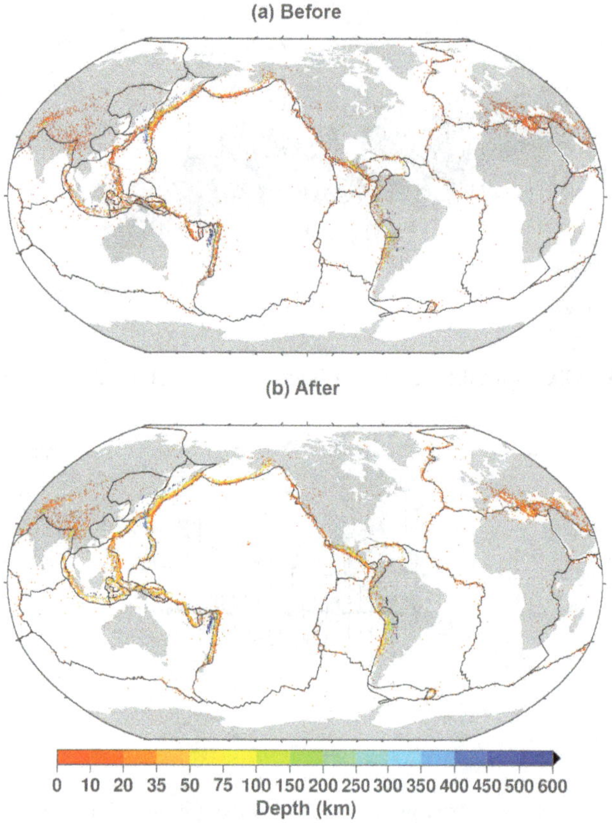

Figure 9. (a) ISS (or authors adopted by ISS) (before) and **(b)** ISC-GEM locations (after) for the extension events relocated between 1920 and 1959. Bird (2003) plate tectonic boundaries are also shown. It is possible to observe how the ISC-GEM locations better depict the seismicity of the Earth even on a global scale.

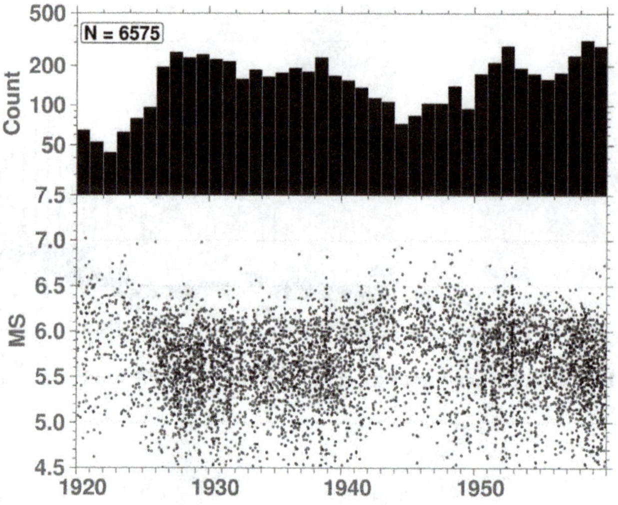

Figure 10. Timeline of the 6575 recomputed M_S (bottom) for the relocated extension events during 1920–1959 and their annual counts (top). 2304 earthquakes have $M_S < 5.5$ and 80 have $M_S \geq 6.5$.

Table 3. Summary of the magnitude quality flags for the relocated extension events between 1920 and 1959. Included in the D flag are 4984 events for which no magnitude was recomputed.

Quality flag	Count for magnitude
A	0
B	3030
C	2824
D	5719

and 1959. Only five of the extension events have direct values of M_w from the literature search (Lee and Engdahl, 2015). The high number of magnitude quality flags D is largely due to events for which no recomputed magnitude (M_S, m_b or M_w from the literature) is available and for which M_S, as the basis for M_w conversion, is below 5. Figure 11 shows the timeline of the earthquakes without recomputed magnitudes along with their annual counts and depth frequency. Although M_S is not estimated for deep earthquakes according to IASPEI (2013), the clear majority (nearly 70 %) of events without magnitude are shallow (depth $\leq$ 50 km). For such shallow earthquakes we continue to look for additional amplitudes (more details in a later section) so that we can calculate M_S and eventually move some of those events from the Supplementary to the Main catalogue.

4 Extension for the period 1904–1919

During the last year of the Extension Project we focused on the first part of 20th century and made special efforts to gather not only body-wave arrival times and amplitude of surface waves, but also known earthquakes not available in the ISC database. We did not add any station data before 1904 (basically only stations belonging to the Milne network are available; see, e.g. Adams, 1989) and, consequently, we decided to drop the 10 pre-1904 events listed before V5 from the ISC-GEM Catalogue and have the catalogue starting in 1904.

4.1 Data collection

Before the ISS was put in production starting with earthquakes that occurred in 1918, other seismic bulletins were compiled by different authors/agencies (e.g. Schweitzer and Lee, 2002; Storchak et al., 2015, and references therein). For this work we gathered station data from the following sources.

- International Seismological Associations (ISA, 1904–1908) bulletins are the most comprehensive both in terms of earthquakes and stations listed for those years. They are composed of two parts, one for the large/significant earthquakes (in German "Haupt-

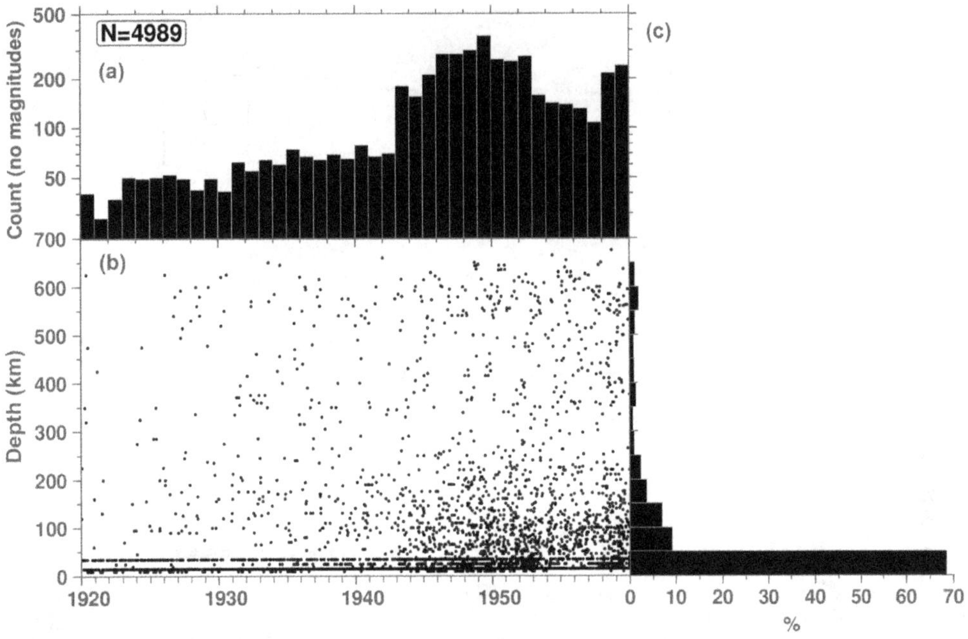

Figure 11. Timeline **(b)** of the relocated extension events without magnitude between 1920 and 1959 along with their depth frequency **(c)** and annual counts **(a)**.

beben") and one for the small ones ("Kleinere Beben"). Unfortunately, the 1908 "Hauptbeben" part was not printed (at least to the best of our knowledge). These bulletins are referred to as ISA in the following.

- Shide Circulars (BAASSC, 1900–1912) were used for 1908–1912, referred to as SHIDE.
- Gutenberg's notepads were utilized for 1908–1916 (mostly up to 1912), referred to as GUTE.
- Russian network bulletins were consulted for 1908 and 1911–1912, referred to as RUS.
- BAAS (1913–1917) bulletins (predecessor of the ISS) listed both locations and station data.
- The ISS was used for 1918–1919.
- Individual station bulletins (1904–1919, referred to as BULLETINS) were consulted, not only for what concerns the surface wave amplitudes but also for body-wave arrival times (in support of relocation rather than only magnitude reassessment) as some stations (e.g. Uppsala, Nordlingen, Munich) were partially or completely missing in the BAAS or ISS. Furthermore, the body-wave arrival times from individual station bulletins are fundamental for the newly added earthquakes that we describe later in this section.

The ISA, SHIDE and RUS bulletins are available from the supplementary material of Schweitzer and Lee (2002), whereas scanned images of GUTE notepads were kindly provided by Katsuyuki Abe. The ISA, BAAS and ISS bulletins list arrival times from most of the stations operating at that time, whereas SHIDE mostly includes data from Milne stations and the GUTE notepads only a subset of global stations. Except for ISS 1918–1919 (already electronically available), the various sources of body-wave arrival times (ISA, SHIDE, GUTE, RUS and BAAS) for the 1904–1917 extension events were all manually typed in text files and then parsed into the ISC database.

As shown in Fig. 1, the annual number of recorded earthquakes, at least up to 1917, is smaller than an approximate average rate of $\sim 100\,\mathrm{yr}^{-1}$ for events of magnitude 6 and above. Therefore, for this period we also tried to add as many known earthquakes as possible that are not listed in the augmented Centennial Catalogue, BAAS or even the ISS 1918–1919. To do that we considered the following sources:

- Catalog of Damaging Earthquakes in the World (http://iisee.kenken.go.jp/utsu/index_eng.html, last access: 10 October 2018, Utsu, 1990, 2002, 2004), referred to as UTSU in the following;
- ISA (only for 1904–1907);
- SHARE European Earthquake Catalogue (SHEEC) 1900–2006 (Grünthal et al., 2013), referred to as SHEEC in the following;
- Karnik (1971) catalogue of the European area (referred to as KAR) and Papazachos et al. (2000, 2010) cat-

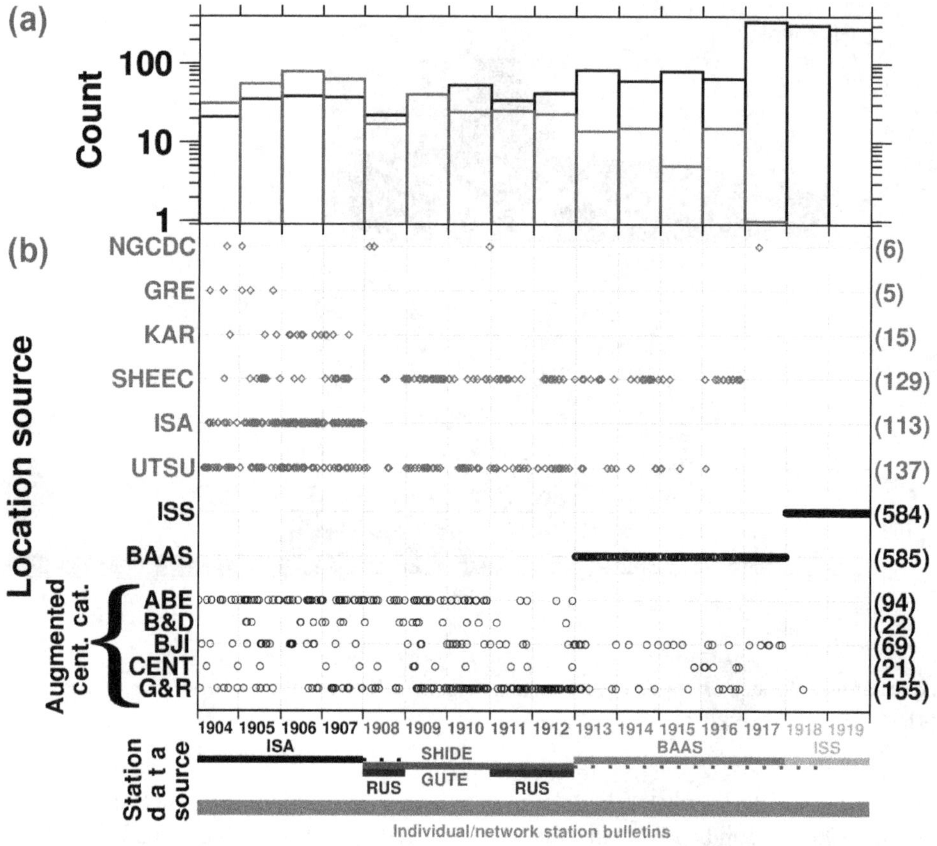

Figure 12. Timelines of the extension earthquakes already in our record (black circles) and added ones (grey diamonds) split by original location author/source. See text for the augmented Centennial Catalogue authors (black) and a brief descriptions of the additional location sources (grey). The total counts for each location source are shown on the right-hand side. The annual counts **(a)** of the extension earthquakes already known and added ones (black and grey histograms, respectively) are summarised. The station data sources **(b)** are also outlined and shown in different grey colours for the time ranges they have been used for (see text for details). For 1908 we have added station data from the "Kleinere Beben" part of ISA (black dots) and during 1913–1918 we also looked into the GUTE notepads for earthquakes not listed in the BAAS and ISS (dark grey dots). Individual/network station bulletins have been used to add both surface wave amplitudes and body-wave arrival times between 1904 and 1919.

alogue for Greece and surrounding areas (available at http://geophysics.geo.auth.gr/ss/CATALOGS/seiscat.dat (last access: 10 October 2018), referred to as GRE) for earthquakes before 1908 with station data in ISA (either not available in SHARE or for which the KAR/GRE solution would be a better starting point considering the ISA station data);

- Significant Earthquake Database of the National Geophysical Data Center/World Data Service (https://www.ngdc.noaa.gov/nndc/struts/form?t=101650&s=1&d=1, last access: 10 October 2018), referred to as NGDC.

As we have a rather mixed set of starting points for hypocentre relocations, in Fig. 12 we show the timelines of the extension earthquakes 1904–1919 split by original location author, along with their counts and the time coverage of the station data sources we digitised. The augmented Centennial Catalogue location sources G&R, B&D, ABE, CENT and BJI are from Gutenberg and Richter (1954), Båth and Duda (1979), Abe (1981, 1984) and Abe and Noguchi (1983a, b), Centennial itself and Chinese catalogue, respectively. In total we have found 405 additional earthquakes (mostly before 1917) on top of the 1530 earthquakes already listed between 1904 and 1919 in the augmented Centennial Catalogue, BAAS and ISS. Notably, between 1904 and 1907 the annual number of earthquakes we added (mostly from ISA and UTSU) is larger than the annual number of extension earthquakes previously available in our record. Between 1908 and 1912 the annual number of earthquakes added is comparable or smaller than the ones already available, whereas from the beginning of the BAAS and then ISS the annual number of newly added earthquakes drops significantly during the BAAS and then it is zero with the beginning of the ISS.

For all earthquakes outlined in Fig. 12 we tried to associate as many body-wave arrival times and surface wave am-

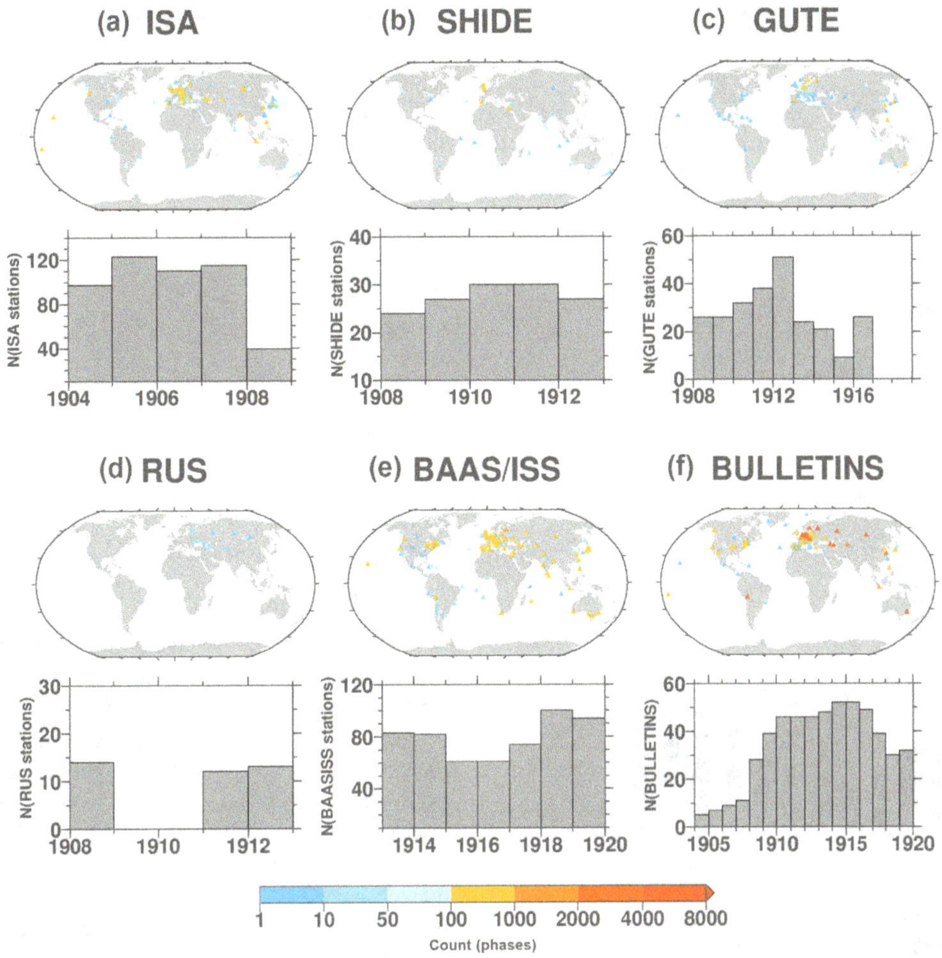

Figure 13. Station data contributions from ISA **(a)**, SHIDE **(b)**, GUTE **(c)**, RUS **(d)**, BAAS and ISS **(e)** and BULLETINS **(f)** identifying the individual/network station bulletins. The stations are colour-coded by number of phases digitised (see text for details). For each station data source, the annual station counts are shown below the corresponding map (note the different scales for each contributor).

plitudes as possible from the station data sources mentioned earlier. The contribution of each station data source is presented in Fig. 13. For the early years of the past century, ISA was comprehensive in compiling data from stations around the world, whereas the other sources only included subsets of the stations operating at that time. Unfortunately, between 1908 and 1912 (coinciding with the end of ISA, "Hauptbeben" part, in 1907 and before the beginning of BAAS in 1913) we do not have a comprehensive bulletin such as ISA in preceding years or BAAS in the following ones. Therefore, we gathered station data from SHIDE, GUTE, RUS and individual/network station bulletins. From 1913 onwards, the overall station data collection improves significantly thanks to BAAS and then ISS.

Considering all sources depicted in Fig. 13, Fig. 14 shows the overall annual counts for the number of stations, phases and, finally, the box-and-whisker plot of the annual number of stations per event. A significant dip is present in the station data between 1908 and 1912 since the station (and location) sources available to us for these years are not as comprehensive as ISA or BAAS/ISS. The box-and-whisker plot of Fig. 14 also shows that several earthquakes have none to three associated stations (59 from the augmented Centennial Catalogue, BAAS and ISS and 116 from the newly added ones). Obviously, the limitations in the collection of station data influenced the earthquakes that we finally selected for processing and the quality of the relocations/magnitude reassessment. The results are discussed in the next two subsections.

4.2 Relocations

Not all extension earthquakes have sufficient station data to perform a relocation using our approach. First, we have discarded 175 earthquakes with fewer than four stations, as pointed out earlier. We then progressively discarded another 650 as either the relocation failed or was considered unreliable. We may go back to the discarded earthquakes if ad-

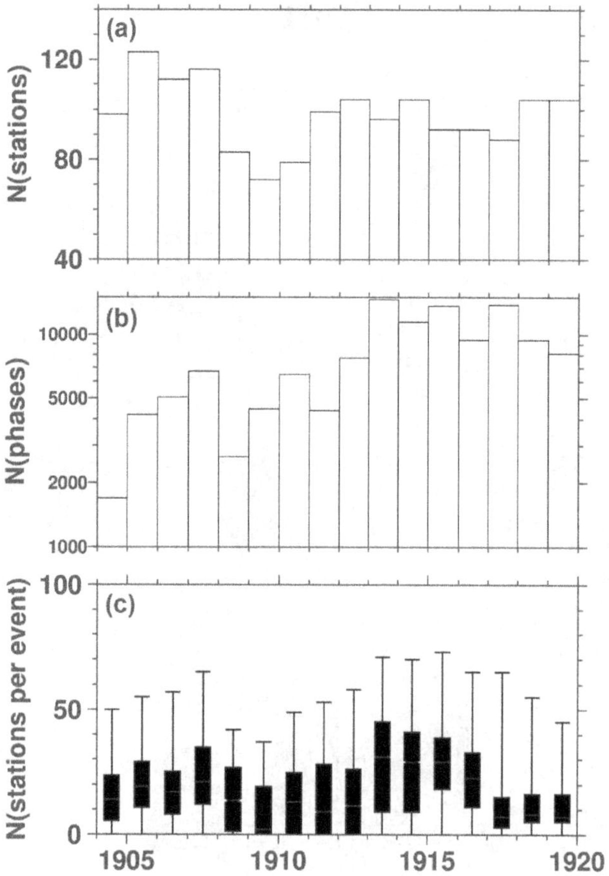

Figure 14. Annual number of station **(a)**, phases **(b)** and box-and-whisker plot **(c)** for the overall station data collected between 1904 and 1919.

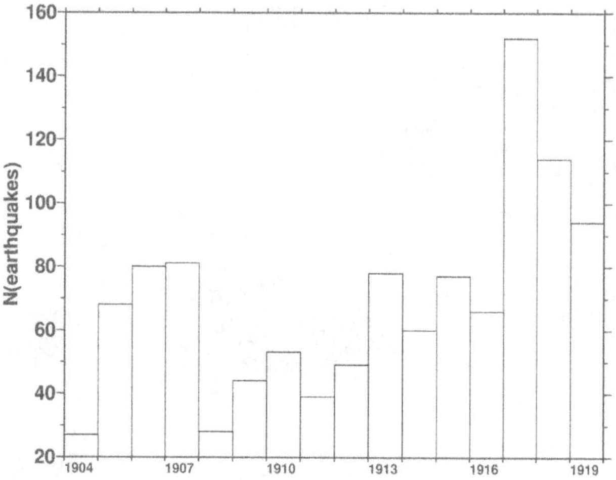

Figure 15. Annual number of relocated extension events between 1904 and 1919. See text for details.

ditional station data become available to us. In the end, we accepted the relocation for 1110 out of the 1935 extension earthquakes. Figure 15 shows the annual counts of the relocated extension earthquakes 1904–1919. Note the dip in the annual number of the relocated extension earthquakes for 1908–1912, reflecting the absence (to the best of our knowledge) of a comprehensive global bulletin between ISA and BAAS.

As in Fig. 7, Fig. 16 shows the box-and-whisker plots of NDEFSTA and SGAP. For this period the relocations are usually based on a small number of stations (median between 6 and 16) resulting in a large SGAP (median between 201 and 310°), even during the years covered by BAAS and ISS. Figure 17 shows the median location, depth and origin time differences between previous (see Fig. 12) and ISC-GEM locations. The median location differences oscillate between 70 and 205 km, with large differences above 1000 km for 46 earthquakes (16 above 2000 and 4 above 3000 km). Such large location differences can occur for various reasons (from typos in the latitude/longitude of previous locations to poorly recorded earthquakes having low confidence locations). One extreme example is the epicentre change from Bristol Bay, offshore Alaska (G&R location), to offshore Jamaica (ISC-GEM location) for an event that occurred on 22 August 1907 (~ 22 h 23 m). The reason for such a large difference originates from the fact that G&R ignored the report that the event was felt in Kingston (see, e.g. ISA, 1907, part B, p. 73) and preferred to fit the phase data to an intermediate-depth event offshore Alaska. As for 1920–1959, most of the earthquakes have no depth resolution and the previous depths were largely unknown or set to zero, and this occasionally results in large depth changes (±100 and ±300 km for 51 and 10 earthquakes, respectively). Figure 17 also shows the box-and-whisker plot of the origin time (OT) differences in each year. We show the OT differences because in this period (particularly before BAAS) the OT listed in the previous location sources was at times truncated to the minute or with some minute error that we were able to address thanks to the stations data we digitised. Although ~ 90 % of the OT differences are within 1 min, some large OT changes of ±5 min or more occur for 16 earthquakes (8 originally from ABE).

Similar to the 1920–1959 period, we assigned location quality flag D if the location was not constrained well enough. This time this task was done not only by considering the usual criteria (see Sect. 3.2) but also consulting available information on the earthquake's effects (e.g. tsunami, damage). In this respect we made systematic use of the earthquake effect information available in UTSU and NGCDC. Table 4 summarises the location and depth quality flags for the relocated extension events between 1904 and 1919. The limitations of the global network in this period are generally more prominent than for 1920–1959 and this translates in most of the earthquakes having location and depth quality C and about 246 of them have location quality D. As for the discarded earthquakes, if additional station data become available we will try to improve the location quality and eventu-

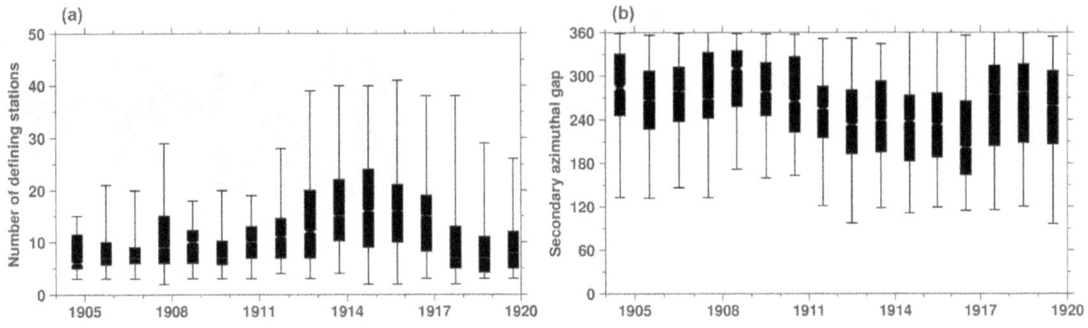

Figure 16. As for Fig. 7 but for the period 1904–1919.

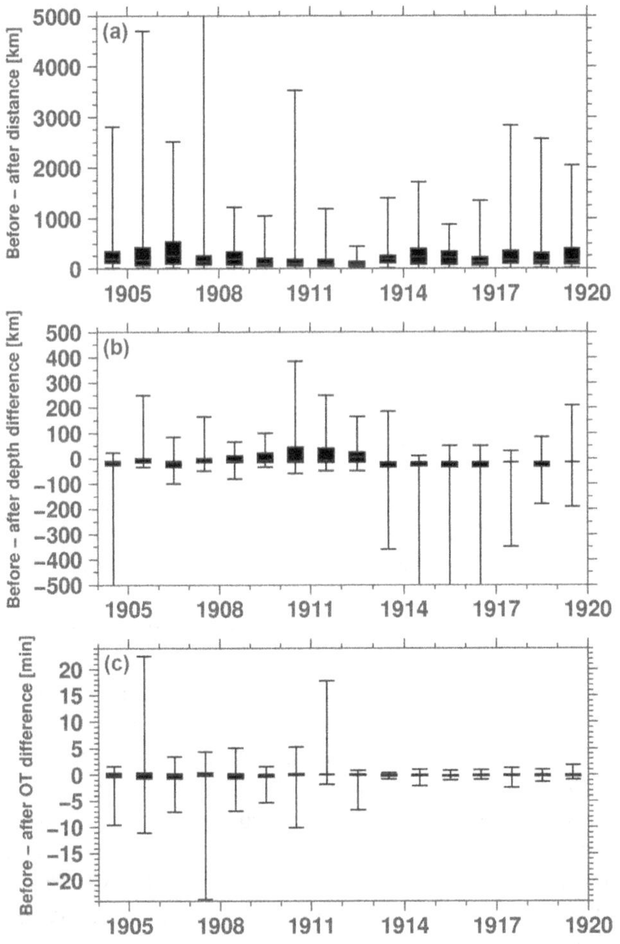

Figure 17. Box-and-whisker plots of the epicentre **(a)**, depth **(b)** and origin time (OT) differences between previous (before, see Fig. 12) and ISC-GEM hypocentres (after) in each year. For 880 of the 1110 extension events relocated between 1904 and 1919 the depth of the previous locations was unknown and nominally set to zero.

Table 4. Summary of the location and depth quality flags for the relocated extension events between 1904 and 1919.

Quality flag	Count for location	Count for depth
A	14	6
B	75	2
C	775	1102
D	246	0

ally move some of the location flag D earthquakes from the Supplementary to the Main catalogue. As for Fig. 9, Fig. 18 compares the previous (before) and ISC-GEM locations (after) on global maps for which, again, a general improvement in the earthquakes' distribution along plate boundaries is delineated. This is particularly the case for several global earthquakes along the subduction zone of the Pacific and Indian oceans whose previous locations were hundreds of kilometres away from plate boundaries.

4.3 Magnitude reassessment

Even for this period the magnitude reassessment is mostly based on our recomputed M_S. Following the same procedures described earlier, we obtained 927 M_S for the relocated extension earthquakes, as shown in Fig. 19. For 500 of them we have computed a magnitude for the first time (in our record). Notably, for 137 earthquakes $M_S < 5.5$, whereas $M_S \geq 6.5$ for 306 of them and > 7.5 for 12 of them. The latter includes six earthquakes originally from GUTE, four from ABE and two from BAAS that were not selected for V1 because the magnitudes available were not considered reliable or were below 7.5 (the original cut-off magnitude for the V1 selection before ISS started in 1918). Nearly all earthquakes with $M_S < 5.5$ occurred in the European Mediterranean area (because in this period the stations contributing surface wave amplitudes are strongly concentrated in Europe; see Fig. 13). In 1904 the collection of surface wave amplitudes is limited to two stations, GTT (Göttingen) and POT (Potsdam), until December, when we could also add

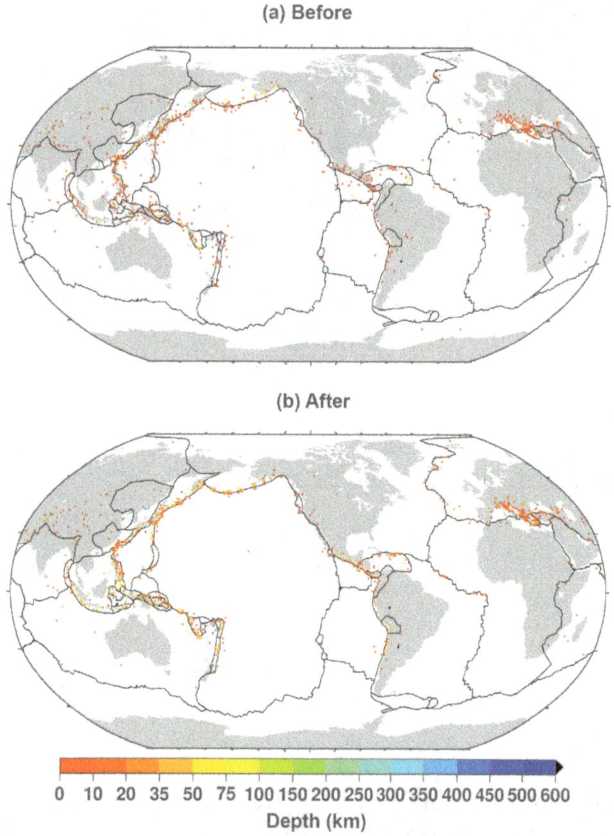

Figure 18. As for Fig. 9 but for 1904–1919. The location authors of the map in **(a)** are outlined in Fig. 12.

Table 5. As for Table 3 but for 1904–1919. Included in the D flag are 183 events for which no magnitude was recomputed.

Quality flag	Count for magnitude
A	0
B	420
C	427
D	262

data from station LEI (Leipzig). Consequently, in 1904 we were able to recompute M_S for three earthquakes only, all occurring in December. For 18 earthquakes between 1905 and 1919 we accepted M_S based on two station magnitudes. Except for four earthquakes for which we have direct M_w values from the Lee and Engdahl (2015) literature search, all recomputed M_S values are used as the basis for M_w proxy calculations (Di Giacomo et al., 2015a). Table 5 summarises the counts for the magnitude quality flags for the relocated extension events between 1904 and 1919. About 50 % of the 183 relocated extension earthquakes for which we do not have a magnitude (no direct M_w or recomputed M_S) are deep (M_S not allowed in our procedures).

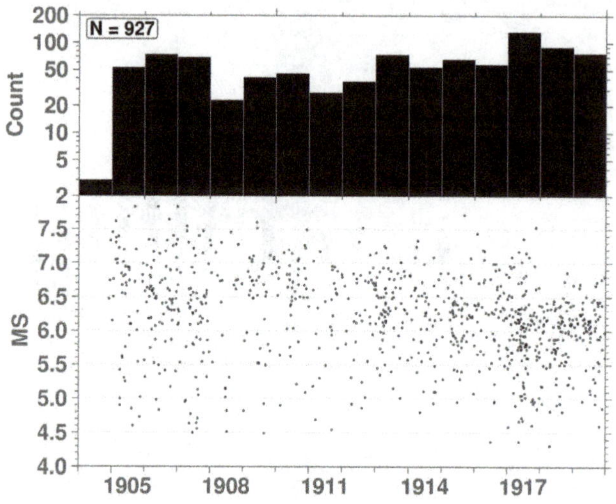

Figure 19. As for Fig. 10 but for 1904–1919.

Table 6. Number of earthquakes added between 2010 and 2014.

Year	Count
2010	504
2011	672
2012	414
2013	484
2014	521

5 Summary of the Extension for 2010–2014

The extension of the ISC-GEM Catalogue beyond 2009 (last year in V1) benefits from the data already available in the ISC Bulletin and the review of global earthquakes by ISC analysts. The earthquake selection for recent years is based on magnitude (5.5 and above). Table 6 summarises the number of earthquakes added per year during 2010–2014. The relatively high number of earthquakes in 2011 is due to the 11 March $M_w = 9.1$ Tohoku earthquake that was followed by about 120 aftershocks with magnitude 5.5 and above just in the first 24 h. In contrast to the predigital period, global earthquakes in recent years are recorded by a dense global network that usually allows us to constrain the location with hundreds of stations and a relatively small SGAP. This is shown in Fig. 20 (note the difference in scale for the plot of the number of stations compared to Figs. 7 and 16). The ISC-GEM epicentres do not move significantly from the previous ones (ISC locations), although occasional significant changes in depth occur, as shown in Fig. 21.

As to magnitude, we largely list direct M_w from GCMT (2347 earthquakes). Proxy M_w values from recomputed M_S or m_b are given for 248 earthquakes. The location and magnitudes of these earthquakes will be included in the figures of the section outlining the state of V5.

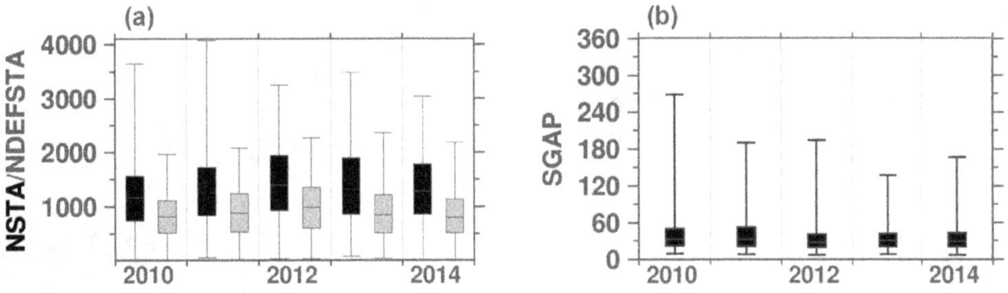

Figure 20. Box-and-whisker plots of **(a)** the number of stations (NSTA, black) and defining stations (NDEFSTA, grey) and **(b)** of the secondary azimuthal gap (SGAP) in each year for the earthquakes added for the period 2010–2014.

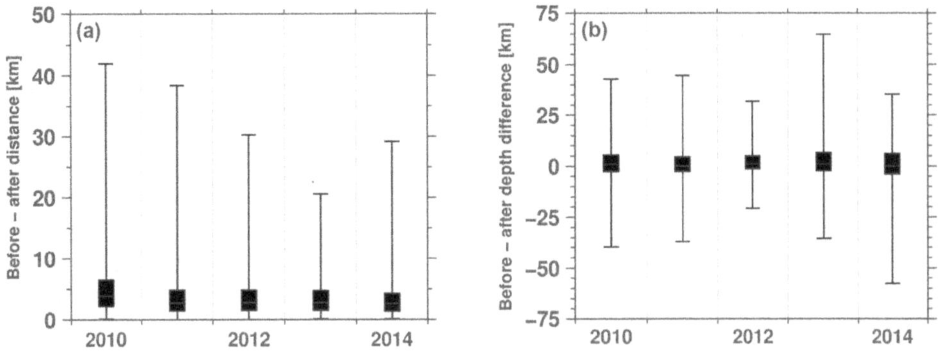

Figure 21. As for Fig. 8 but for 2010–2014. Note the different scales compared to Figs. 8 and 17.

6 Review of events that have already been part of the catalogue

The ISC-GEM Catalogue comes with a version number that keeps track of the catalogues's updates and/or additions. Even when an earthquake is listed in the catalogue, we continue to look for additional station data and information that could help us to improve, whenever necessary, the earthquakes' parameters we list in the catalogue. At the same time, we cooperate with users of the catalogue who inquire about earthquakes of their interest in different parts of the world, at times resulting in an updated location, depth and/or magnitude for one or more earthquake. Examples of updates we made thanks to users' help are available on the ISC-GEM Catalogue update log web page (http://www.isc.ac.uk/iscgem/update_log, last access: 10 October 2018). We also run internal checks as progress is made with the Rebuild of the ISC Bulletin (Storchak et al., 2017) and/or the ISC-EHB dataset (Weston et al., 2018). We try to keep the number of releases to a minimum and recommend users quote the version number when using the ISC-GEM Catalogue for their studies.

As mentioned before, during the Extension Project we gathered station data (particularly for amplitudes of surface waves) from printed station bulletins that were not available to us. Therefore, during the data collection task of the Extension Project we did not limit the search for amplitude data to extension earthquakes but also to earthquakes that were already listed in previous versions (before V5) of the catalogue. This way we revised the M_S of earthquakes already listed in the catalogue even if we added just one or two station readings. Figure 22 shows the number of stations contributing to M_S as well as the comparison between original and revised M_S for pre-1960 earthquakes already listed in previous versions of the catalogue. The increase in the number of stations contributing to the recomputation of M_S is significant: $\sim 30\,\%$ and $\sim 74\,\%$ of the original M_S were constrained using fewer than 6 and 11 stations, respectively, whereas with the revised M_S these percentages drop to $\sim 8.5\,\%$ and $31\,\%$. Also, the station data added allowed us to gain about 50 earthquakes with M_S. About 97 % of the revised M_S are within ± 0.3 m.u. of the original ones, with only five earthquakes with M_S differences above ± 0.6 m.u. (often due to originally mis-associated readings, also resulting in the loss of four original M_S values).

7 Overall status of the new (V5) ISC-GEM Catalogue

The new version (V5) of the ISC-GEM Catalogue covers the period 1904–2014 and was released on the ISC website (http://www.isc.ac.uk/iscgem/, last access: 10 October 2018) on 27 February 2018. It is composed of 35 225 earthquakes in total, with 7126 listed in the Supplementary catalogue (about 93 % of them having occurred before 1960). The an-

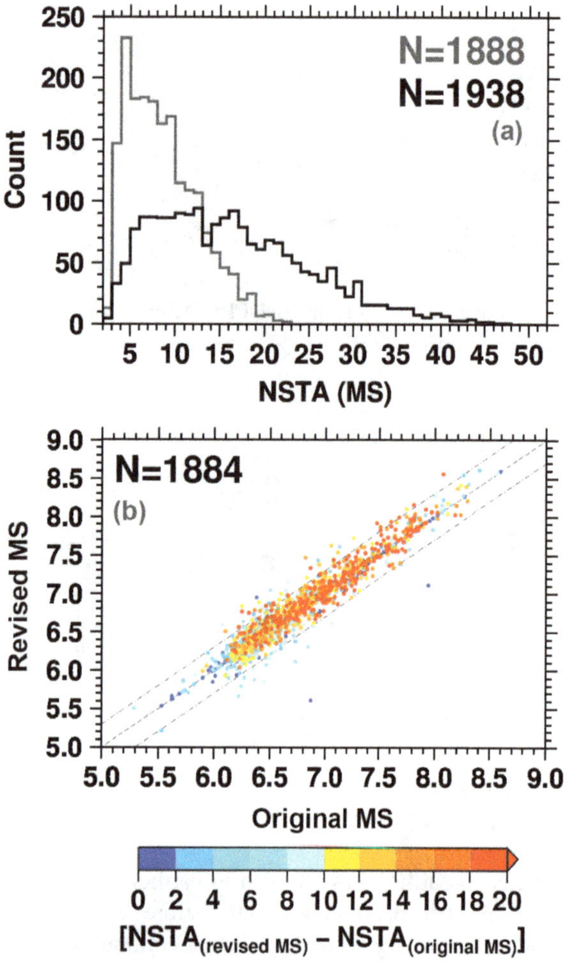

Figure 22. **(a)** Histogram distributions of the number of stations (NSTA) contributing to M_S for the pre-1960 earthquakes already in previous versions of the ISC-GEM Catalogue (grey) and after revision (black) using the newly added amplitude data. **(b)** Comparison between original and revised M_S colour-coded by the difference of NSTA used to obtain M_S. The black dashed and the dotted-dashed lines are for the 1 : 1 and the ±0.3 values, respectively. Note that during revision we dropped the M_S for four earthquakes.

nual number of events in the Main and Supplementary files is shown in Fig. 23. The magnitude sources are the same four described in Di Giacomo et al. (2015a) and the updated composition is as follows: 45.72 % for M_w from GCMT, 42.85 % for M_w proxy from recomputed M_S, 8.1 % for M_w proxy from recomputed m_b and, finally, 3.33 % for M_w from the literature search. As outlined in Di Giacomo et al. (2015a), the M_w proxy values based on M_S are mostly for pre-GCMT earthquakes (i.e. with few exceptions, before 1976).

The primary use of the ISC-GEM Catalogue is seismic hazard (including calibration of regional seismic catalogues) and Earth's seismicity pattern studies as is it the longest and most homogeneous record of natural global seismicity recorded during the instrumental period. For this reason, in

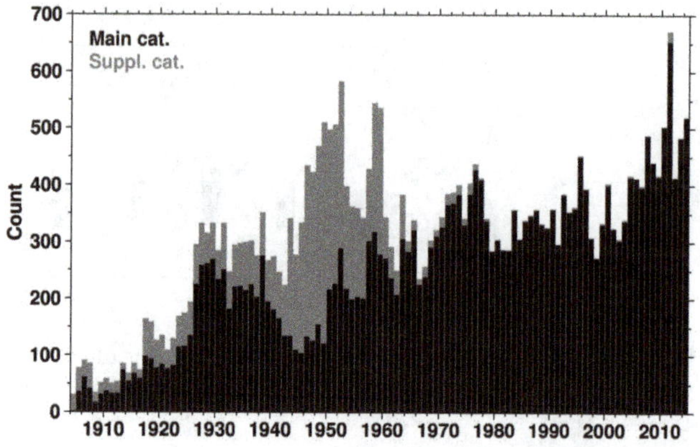

Figure 23. Annual number of earthquakes in the Main (black) and Supplementary (grey) ISC-GEM Catalogue; from the count of the Supplementary catalogue we have excluded a subset of earthquakes with M_S below 5.

Fig. 24 we plot V5 of the ISC-GEM Catalogue using Agnew (2014) symbols to emphasise the magnitude of the earthquakes in the catalogue. To produce the figure, we included earthquakes in the Main catalogue plus those earthquakes in the Supplementary catalogue that have reliable magnitude but poor location (i.e. magnitude quality flag not equal to D and location/depth quality flag equal to D). The subduction zones and the mid-oceanic ridges are well depicted as are areas where global earthquakes occur more frequently.

The current magnitude content as well as a basic magnitude completeness (M_c) assessment is shown in Fig. 25 (update on Fig. 20 of Di Giacomo et al., 2015a). It is not our aim to do a detailed completeness study as Michael (2014); here we use the magnitude content and M_c to highlight the following features of the catalogue.

- The predigital period is not as complete (average annual M_c varying between 5.7 and 6.8) as more recent decades (average annual M_c between 5.5 and 5.7 since 1964). Important fluctuations in the annual number of earthquakes/M_c are present in specific periods or years. For example, because of World War II there is a significant decrease in the number of recorded earthquakes (particularly below magnitude 6) consistent with the disruption of the global network during the 1940s; other minor fluctuations are present in almost every decade (e.g. slight rise in M_c in the early 1960s and late 1970s). The fluctuations over time of the number of earthquakes (i.e. variations of M_c) in the full catalogue (especially at the lower magnitudes, below ~ 6.5) should be checked before using it in its current status for studies concerning temporal and seismicity patterns.

- The number of intermediate-depth (between 60 and 300 km) and deep (≥ 300 km) earthquakes per year before the 1950s–1960s is significantly smaller compared to more recent decades. The reason is not fully clear and

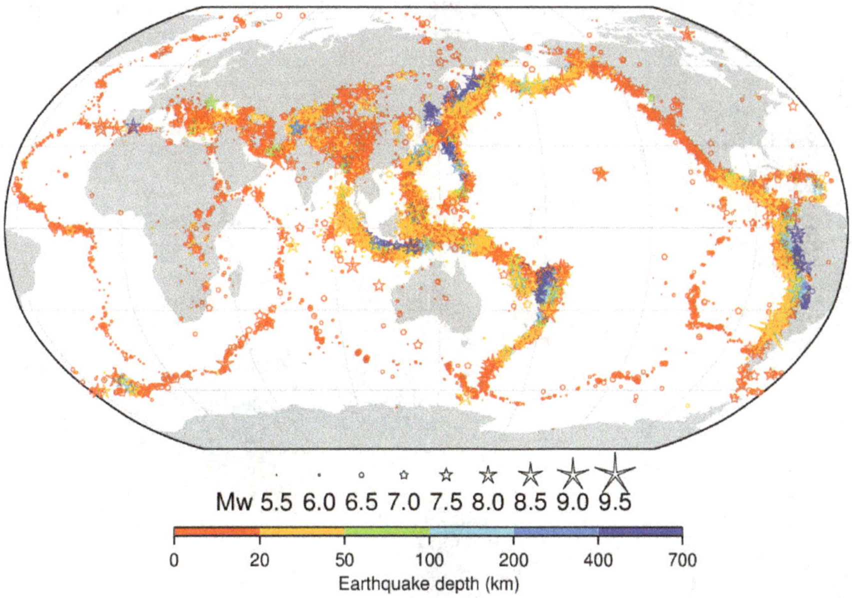

Figure 24. Map showing the earthquakes listed in V5 of the ISC-GEM Catalogue (more than 28 000 earthquakes, see Fig. 25). The symbols are plotted according to Agnew (2014) and colour-coded according to the ISC-GEM depth. The earthquakes shown are from the Main catalogue plus those earthquakes in the Supplementary catalogue that have reliable magnitude but poor location (i.e. magnitude quality flag not equal to D and location/depth quality flag equal to D).

will be a matter for further investigation (see Sect. 8). Most likely, it is the result of a combination of factors, which include the detection capability for moderate deep-focus earthquakes of analog seismographs (see, e.g. Kanamori, 1988) deployed around the world before the 1950s, the lack of stations close to subduction zones for many decades (Figs. 2 and 13) and the earthquake selection criteria. For global earthquakes, instruments such as the Wiechert, Bosh-Omori, Maika and Galitzin were able to record surface wave signals (medium period range, centred around 20 s) better than body-waves (higher frequency signals, particularly P-waves, from around 10 s and below). The effect could have been that many stations would not report station data for moderate deep-focus earthquakes and, therefore, the ISS would not compile data for such earthquakes (i.e. the earthquake would not be recorded). The selection criteria could also play a role, although the earthquakes not selected for processing either lack station data (and depth resolution) or, more importantly, are usually too small to account for the small number of deep-focus earthquakes depicted in Fig. 25.

In addition, users should be aware that the magnitude uncertainty for predigital earthquakes is inevitably larger than for earthquakes in the GCMT era (from 1976 onwards). The timeline of the M_w uncertainty in the ISC-GEM Main Catalogue is shown in Fig. 26. This is to further remind users of the full catalogue that, for patterns of seismicity studies, they should be aware of the larger magnitude uncertainty in the first part of last century.

8 Outlook

We plan to continue maintaining the ISC-GEM Catalogue for years to come and work on its advancement by

- adding recent years (2015 onwards);
- regularising the magnitude for earthquakes between 1960 and 1990 to remove as many fluctuations as possible in the M_c over those decades;
- adding earthquakes between magnitude 5 and 5.5 that have occurred in continental areas from 1960 onwards;
- improving the content for the predigital period (before 1964) by filling gaps in the station reports (particularly for what concerns surface wave amplitudes) and possibly bringing additional earthquakes and station data from the Bureau Central International de Seismologie (BCIS, 1933–1968); we will also consider any other source (if available) not considered so far that will bring useful data (station data and/or earthquake information) that will allow us to improve the catalogue; with this task we aim at moving as many earthquakes as possible from the Supplementary to the Main catalogue (see Fig. 23);

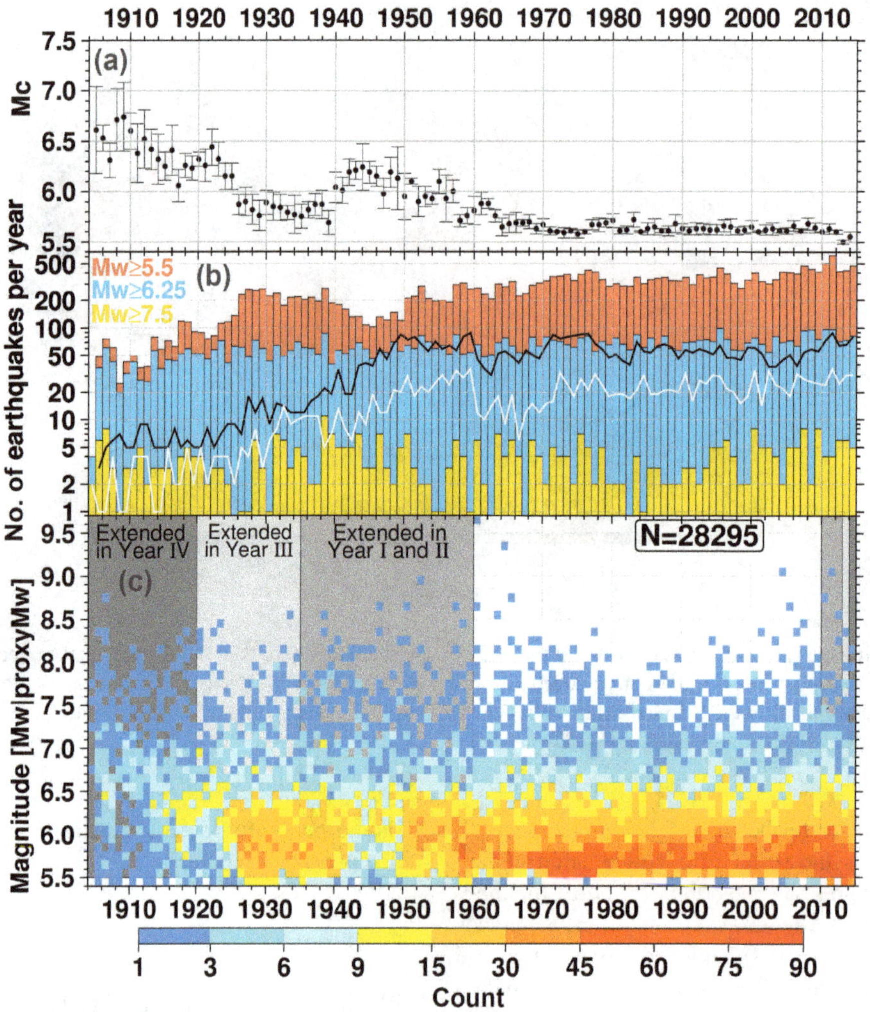

Figure 25. (c) Time–magnitude distribution colour-coded for cells of 0.1 m.u. in each year for the earthquakes used to produce Fig. 24. **(b)** Cumulative annual number of earthquakes with $M_W \geq 5.5$ (red), ≥ 6.5 (blue) and ≥ 7.5 (yellow), along with the annual counts of intermediate-depth (60 km < depth < 300 km, solid black line) and deep ($\geq$ 300 km, solid white line) earthquakes; the blue and yellow lines are obtained considering all earthquakes in V5 (i.e. both Main and Supplementary Catalogue). **(a)** Annual completeness magnitude (M_C, black circles ± 1 standard deviation) estimated with the maximum curvature method of Wiemer and Wyss (2000) implemented in the R-code of Mignan and Woessner (2012). Note that we skipped 1904 for the M_C assessment due to the small number of earthquakes.

- integrating the results from the ISC Bulletin Rebuild project (1964–2010; see Storchak et al., 2017) and the ISC-EHB reconstruction (1964 onwards, Weston et al., 2018);
- continuing and extending our literature search for new or updates of direct estimation of M_W for pre-GCMT earthquakes as well as general focal parameters; we also aim at including fault plane solutions from the literature for predigital earthquakes.

A more detailed description of the Advancement Project of the ISC-GEM Catalogue is available at http://www.isc.ac.uk/iscgem/advancement.php (last access: 10 October 2018). We will continue releasing a new version after the end of each year of the Advancement Project. In this way we will be able to provide the seismological, as well as the broader geoscience community, with the most comprehensive and homogeneous account of earthquake global seismicity recorded instrumentally at any point in time.

9 Data availability

Since 27 February 2018, V5 of the ISC-GEM Catalogue has been available for download at http://doi.org/10.31905/D808B825 (International Seismological Centre, 2018). All data used in this paper are maintained at the ISC (http://www.isc.ac.uk/, last access: 10 October 2018). The ISC-GEM Catalogue is released without the associated seismic

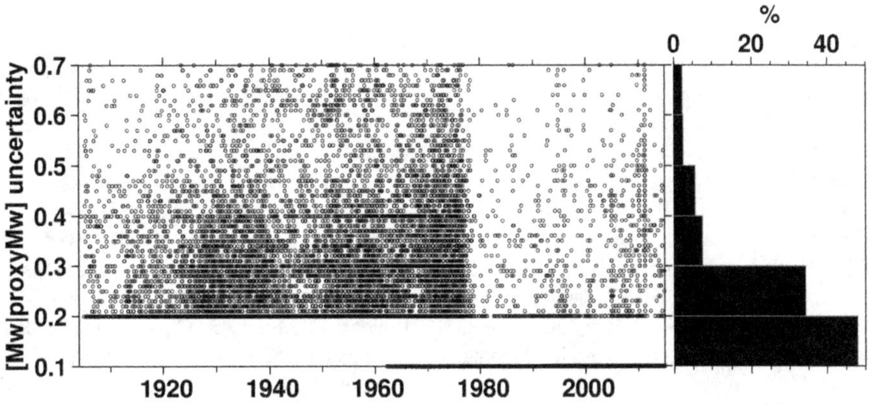

Figure 26. Timeline of the magnitude uncertainty in the Main file of the ISC-GEM Catalogue. On the right the percentage distribution is shown.

wave arrival times and amplitudes used for this work. These underlying parametric data are either already available or will be before the end of 2018 as part of corresponding events in the ISC Bulletin (http://www.isc.ac.uk/iscbulletin/, last access: 10 October 2018).

10 Conclusions

We presented the procedures and results of a 4-year project which extended and improved the ISC-GEM Catalogue first released in 2013 (Storchak et al., 2013). We have added about 12 000 more events between 1904 and 1960 and the new version (V5) ends in 2014 instead of 2009. To extend the catalogue before the 1960s we have digitised $\sim$ 650 000 phase arrival times from various sources (ISS, BAAS, ISA, Shide Circulars, Gutenberg notepads, etc.) in different periods and added $\sim$ 140 000 amplitudes from printed station bulletins. The features and limitations of the global network before 1960 have been outlined and the results show that the relocations, based on our two-tier approach (Bondár et al., 2015), provide solutions distributed along main tectonic boundaries, even though they are usually based on a small number of stations compared to relocations of earthquakes in recent years. We have recomputed over 6000 M_S values for pre-1960 earthquakes and obtained (to the best of our knowledge) a magnitude for the first time for more than 3000 of them. For the period 2010–2014 we have greatly benefited from both the station data available in the ISC Bulletin and the reviews done by ISC analysts which provide us with robust starting points for the relocations and the M_w from the GCMT.

At the same time as the digitisation from printed sources of stations supplying amplitude data (of surface waves in particular), we also looked for additional data for predigital earthquakes (pre-1960) already listed in previous versions of the ISC-GEM Catalogue. The newly added amplitude data made us revise a significant number of pre-1960 earthquakes listed in V1 and improve the magnitude solutions as the revised magnitudes are now based on a much higher number of stations.

The current state of V5 of the ISC-GEM Catalogue has been summarised and its features outlined. With the Advancement project we aim to further improve and extend the catalogue in coming years and address some of the limitations that have been pointed out here during different periods of time.

Author contributions. DDG was the leading author of the paper and responsible for the station data collection, earthquake selection, second step of the relocation task, magnitude reassessment and final checks. ERE provided scientific input and determined the depths and starting locations which were used in the second step of the relocation process. DAS obtained the funding for the project, oversaw its progress and gathered additional station bulletins. All authors contributed to the paper and approved the final version.

Competing interests. The authors declare no competing interests in the production of the ISC-GEM Catalogue.

Acknowledgements. This project was supported by the NSF (Award 1417970), USGS (Award G15AC00202), FM Global, OYO Corporation, the Lighthill Risk Network, the Aspen Re, Bundesanstalt für Geowissenschaften und Rohstoffe (BGR) and 65 members of the ISC (http://www.isc.ac.uk/members/, last access: 10 October 2018). Year I and II of the Extension Project were also supported by the GEM Foundation. We thank two anonymous reviewers for their comments that helped us to improve the manuscript. Daniela Olaru and Elizabeth Ayres were instrumental in the data collection from printed station bulletins and the ISS. We used computer codes by Antonio Villase to digitise the ISS phase data. Lynn Elms checked and streamlined the text. We are deeply indebted to various institutions and individuals that provided additional station bulletins to the ISC (more details at http://www.isc.ac.uk/iscgem/acknowledge.php, last access: 10 October 2018). We are grateful to Josep Batlló for sharing additional data on the 1919 Torremendo (Spain) series (Batlló et al., 2015) and related discussions which allowed us to correct the corresponding events originally listed in the ISS. For the M_W literature search we thank Paolo Harabaglia for pointing out the papers from Pino et al. (2000, 2008) for two significant earthquakes in Italy (12 December 1908 Messina and 23 July 1930 Irpinia) as well as the centroid solution for the 29 November 1975 Hilo (Hawaii) earthquake from Nettles and Ekström (2004). Further acknowledgements to users of the ISC-GEM Catalogue are available at http://www.isc.ac.uk/iscgem/update_log/ (last access: 10 October 2018). Figures were drawn using the Generic Mapping Tools (Wessel et al., 2013).

Edited by: David Carlson

References

Abe, K.: Magnitudes of large shallow earthquakes from 1904 to 1980, Phys. Earth Planet. In., 27, 72–92, https://doi.org/10.1016/0031-9201(81)90088-1, 1981.

Abe, K.: Complements to "Magnitudes of large shallow earthquakes from 1904 to 1980", Phys. Earth Planet. Int., 34, 17–23, https://doi.org/10.1016/0031-9201(84)90081-5, 1984.

Abe, K. and Noguchi, S.: Determination of magnitude for large shallow earthquakes 1898–1917, Phys. Earth Planet. Int., 32, 45–59, https://doi.org/10.1016/0031-9201(83)90077-8, 1983a.

Abe, K. and Noguchi, S.: Revision of magnitudes of large shallow earthquakes, 1897–1912, Phys. Earth Planet. In., 33, 1–11, https://doi.org/10.1016/0031-9201(83)90002-x, 1983b.

Adams, R. D.: The development of global earthquake seismology, in: Observatory Seismology, edited by: Liteheiser, J. J., Univeristy of California Press, Berkeley, CA, available at: http://ark.cdlib.org/ark:/13030/ft7m3nb4pj/ (last access: 10 October 2018), 1989.

Agnew, D. C.: Variable Star Symbols for Seismicity Plots, Seismol. Res. Lett., 85, 775–780, https://doi.org/10.1785/0220130214, 2014.

Allen, T. I., Marano, K. D., Earle, P. S., and Wald, D. J.: PAGER-CAT: A Composite Earthquake Catalog for Calibrating Global Fatality Models, Seismol. Res. Lett., 80, 57–62, https://doi.org/10.1785/gssrl.80.1.57, 2009.

Alvarez, L., Lindholm, C., and Villalón, M.: Seismic Hazard for Cuba: A New Approach, B. Seismol. Soc. Am., 107, 229–239, https://doi.org/10.1785/0120160074, 2016.

BAAS: British Association for the Advancement of Science, Seismological Committee, quarterly issues, 1913–1917.

BAASSC: British Association for the Advancement of Science, Seismological Committee, Circular Nos. 1–27, these circulars are generally known as the "Shide Circulars", 1900–1912.

Båth, M. and Duda, S. J.: Some aspects of global seismicity, Tectonophysics, 54, T1–T8, https://doi.org/10.1016/0040-1951(79)90105-7, 1979.

Batlló, J., Martínez-Solares, J. M., Macià, R., Stich, D., Morales, J., and Garrido, L.: The autumn 1919 Torremendo (Jacarilla) earthquake series (SE Spain), Ann. Geophys., 58, S0324, https://doi.org/10.4401/ag-6686, 2015.

BCIS: Bureau Central International de Seismologie, monthly issues, 1933–1968.

Bird, P.: An updated digital model of plate boundaries, Geochem. Geophy. Geosy., 4, 1027, https://doi.org/10.1029/2001gc000252, 2003.

Bondár, I. and McLaughlin, K. L.: A new Ground Truth data set for seismic studies, Seismol. Res. Lett., 80, 465–472, https://doi.org/10.1785/gssrl.80.3.465, 2009.

Bondár, I. and Storchak, D.: Improved location procedures at the International Seismological Centre, Geophys. J. Int., 186, 1220–1244, https://doi.org/10.1111/j.1365-246x.2011.05107.x, 2011.

Bondár, I., Myers, S. C., Engdahl, E. R., and Bergman, E. A.: Epicentre accuracy based on seismic network criteria, Geophys. J. Int., 156, 483–496, https://doi.org/10.1111/j.1365-246x.2004.02070.x, 2004.

Bondár, I., Engdahl, E. R., Villaseñor, A., Harris, J., and Storchak, D.: ISC-GEM: Global Instrumental Earthquake Catalogue (1900–2009), II. Location and seismicity patterns, Phys. Earth Planet. In., 239, 2–13, https://doi.org/10.1016/j.pepi.2014.06.002, 2015.

Cambiotti, G., Wang, X., Sabadini, R., and Yuen, D.: Residual polar motion caused by coseismic and interseismic deformations from 1900 to present, Geophys. J. Int., 205, 1165–1179, https://doi.org/10.1093/gji/ggw077, 2016.

Deif, A., Al-Shijbi, Y., El-Hussain, I., Ezzelarab, M., and Mohamed, A. M.: Compiling an earthquake catalogue for the Arabian Plate, Western Asia, J. Asian Earth Sci., 147, 345–357, https://doi.org/10.1016/j.jseaes.2017.07.033, 2017.

Di Giacomo, D., Bondár, I., Storchak, D. A., Engdahl, E. R., Bormann, P., and Harris, J.: ISC-GEM: Global Instrumental Earthquake Catalogue (1900–2009), III. Re-computed MS and mb, proxy MW, final magnitude composition and completeness assessment, Phys. Earth Planet. In., 239, 33–47, https://doi.org/10.1016/j.pepi.2014.06.005, 2015a.

Di Giacomo, D., Harris, J., Villaseñor, A., Storchak, D. A., Engdahl, E. R., and Lee, W. H.: ISC-GEM: Global Instrumental Earthquake Catalogue (1900–2009), I. Data collection from early instrumental seismological bulletins, Phys. Earth Planet. In., 239, 14–24, https://doi.org/10.1016/j.pepi.2014.06.003, 2015b.

Dziewonski, A. M., Chou, T.-A., and Woodhouse, J. H.: Determination of earthquake source parameters from waveform data for studies of global and regional seismicity, J. Geophys. Res.-Sol. Ea., 86, 2825–2852, https://doi.org/10.1029/jb086ib04p02825, 1981.

Ekström, G., Nettles, M., and Dziewoński, A.: The global CMT project 2004–2010: Centroid-moment tensors for 13,017 earthquakes, Phys. Earth Planet. In., 200-201, 1–9, https://doi.org/10.1016/j.pepi.2012.04.002, 2012.

Engdahl, E. R. and Villaseñor, A.: Global seismicity: 1900–1999, in: International Handbook of Earthquake and Engineering Seismology, edited by: Lee, W., Kanamori, H., Jennings, J., and Kisslinger, C., vol. A, chap. 41, 665–690, Academic Press, San Diego, 2002.

Engdahl, E. R., van der Hilst, R., and Buland, R.: Global teleseismic earthquake relocation with improved travel times and procedures for depth determination, B. Seismol. Soc. Am., 88, 722–743, 1998.

Geist, E. L.: Explanation of Temporal Clustering of Tsunami Sources Using the Epidemic-Type Aftershock Sequence Model, B. Seismol. Soc. Am., 104, 2091–2103, https://doi.org/10.1785/0120130275, 2014.

Ghasemi, H., McKee, C., Leonard, M., Cummins, P., Moihoi, M., Spiro, S., Taranu, F., and Buri, E.: Probabilistic seismic hazard map of Papua New Guinea, Nat. Hazards, 81, 1003–1025, https://doi.org/10.1007/s11069-015-2117-8, 2016.

Grünthal, G., Wahlström, R., and Stromeyer, D.: The SHARE European Earthquake Catalogue (SHEEC) for the time period 1900–2006 and its comparison to the European-Mediterranean Earthquake Catalogue (EMEC), J. Seismol., 17, 1339–1344, https://doi.org/10.1007/s10950-013-9379-y, 2013.

Gutenberg, B. and Richter, C.: Seismicity of the Earth and Associated Phenomena, Princeton Univ. Press, Princeton, NJ, 310 pp., 1954.

IASPEI: Summary of Magnitude Working Group recommendations on standard procedures for determining earthquake magnitudes from digital data, available at: ftp://ftp.iaspei.org/pub/commissions/CSOI/Summary_WG_recommendations_20130327.pdf (last access: 10 October 2018), 2013.

Ikuta, R., Mitsui, Y., Kurokawa, Y., and Ando, M.: Evaluation of strain accumulation in global subduction zones from seismicity data, Earth Planet. Space, 67, 192, https://doi.org/10.1186/s40623-015-0361-5, 2015.

International Seismological Centre: ISC-GEM Earthquake Catalogue, https://doi.org/10.31905/d808b825, 2018.

ISA: International Seismological Association, annual volumes, 1904–1908.

ISS: International Seismological Summary, annual volumes, 1918–1963.

Kadirioğlu, F. T., Kartal, R. F., Kılıç, T., Kalafat, D., Duman, T. Y., Eroğlu Azak, T., Özalp, S., and Emre, O.: An improved earthquake catalogue (M $\geq$ 4.0) for Turkey and its near vicinity (1900–2012), B. Earthq. Eng., 8, 3317–3338, https://doi.org/10.1007/s10518-016-0064-8, 2016.

Kagan, Y. Y.: Worldwide earthquake forecasts, Stoch. Env. Res. Risk A., 31, 1273–1290, https://doi.org/10.1007/s00477-016-1268-9, 2017.

Kagan, Y. Y. and Jackson, D. D.: Earthquake rate and magnitude distributions of great earthquakes for use in global forecasts, Geophys. J. Int., 206, 630–643, https://doi.org/10.1093/gji/ggw161, 2016.

Kanamori, H.: The importance of historical seismograms for geophysical research, in: Historical Seismograms and Earthquakes of the World, edited by: Lee, W., 16–33, Academic Press, New York, 1988.

Karnik, V.: Seismicity of the European Area, vol. Part II, Reidel Publ. Co., Dordrecht, Holland, 1971.

Katsumata, K.: A Long-Term Seismic Quiescence before the 2004 Sumatra (Mw 9.1) Earthquake, B. Seismol. Soc. Am., 105, 167–176, https://doi.org/10.1785/0120140116, 2015.

Kennett, B. L. N., Engdahl, E. R., and Buland, R.: Constraints on seismic velocities in the Earth from traveltimes, Geophys. J. Int., 122, 108–124, https://doi.org/10.1111/j.1365-246x.1995.tb03540.x, 1995.

Lange, D., Ruiz, J., Carrasco, S., and Manríquez, P.: The Chiloé Mw 7.6 earthquake of 2016 December 25 in Southern Chile and its relation to the Mw 9.5 1960 Valdivia earthquake, Geophys. J. Int., 213, 210–221, https://doi.org/10.1093/gji/ggx514, 2017.

Lee, W. H. K. and Engdahl, E. R.: Bibliographical search for reliable seismic moments of large earthquakes during 1900–1979 to compute MW in the ISC-GEM Global Instrumental Reference Earthquake Catalogue, Phys. Earth Planet. In., 239, 25–32, https://doi.org/10.1016/j.pepi.2014.06.004, 2015.

Leonard, M.: Self-Consistent Earthquake Fault-Scaling Relations: Update and Extension to Stable Continental Strike-Slip Faults, B. Seismol. Soc. Am., 104, 2953–2965, https://doi.org/10.1785/0120140087, 2014.

Markušić, S., Gülerce, Z., Kuka, N., Duni, L., Ivančić, I., Radovanović, S., Glavatović, B., Milutinović, Z., Akkar, S., Kovačević, S., Mihaljević, J., and Šalić, R.: An updated and unified earthquake catalogue for the Western Balkan Region, B. Earthq. Eng., 14, 321–343, https://doi.org/10.1007/s10518-015-9833-z, 2015.

Metzger, S., Schurr, B., Ratschbacher, L., Sudhaus, H., Kufner, S.-K., Schöne, T., Zhang, Y., Perry, M., and Bendick, R.: The 2015 Mw 7.2 Sarez Strike-Slip Earthquake in the Pamir Interior: Response to the Underthrusting of India's Western Promontory, Tectonics, 36, 2407–2421, https://doi.org/10.1002/2017tc004581, 2017.

Michael, A. J.: How Complete is the ISC-GEM Global Earthquake Catalog?, B. Seismol. Soc. Am., 104, 1829–1837, https://doi.org/10.1785/0120130227, 2014.

Mignan, A. and Woessner, J.: Estimating the magnitude of completeness for earthquake catalogs, https://doi.org/10.5078/corssa-00180805, available at: http://www.corssa.org (last access: 10 October 2018), 2012.

Mikhailova, N. N., Mukambayev, A. S., Aristova, I. L., Kulikova, G., Ullah, S., Pilz, M., and Bindi, D.: Central Asia earthquake catalogue from ancient time to 2009, Ann. Geophys., 58, S0102, https://doi.org/10.4401/ag-6681, 2015.

National Geophysical Data Center/World Data Service (NGDC/WDS): Significant Earthquake Database, National Geophysical Data Center, NOAA, https://doi.org/10.7289/V5TD9V7K, 1972.

Nettles, M. and Ekström, G.: Long-Period Source Characteristics of the 1975 Kalapana, Hawaii, Earthquake, B. Seismol. Soc. Am., 94, 422–429, https://doi.org/10.1785/0120030090, 2004.

Pacheco, J. and Sykes, L.: Seismic moment catalog of large shallow earthquakes, 1900 to 1989, B. Seismol. Soc. Am., 82, 1306–1349, 1992.

Papazachos, B., Comninakis, P., Karakaisis, G., Karakostas, B., Papaioannou, C., Papazachos, C., and Scordilis, E.: A catalogue of earthquakes in Greece and surrounding area for the period 550BC–1999, Publ. Geophys. Laboratory, University of Thessaloniki, 333 pp., 2000.

Papazachos, B., Comninakis, P., Scordilis, E., Karakaisis, G., and Papazachos, C.: A catalogue of earthquakes in the Mediterranean and surrounding area for the period 1901–2010, Publ. Geophys. Laboratory, University of Thessaloniki, 2010.

Pino, N. A., Giardini, D., and Boschi, E.: The December 28, 1908, Messina Straits, southern Italy, earthquake: Waveform modeling of regional seismograms, J. Geophys. Res.-Sol. Ea., 105, 25473–25492, https://doi.org/10.1029/2000jb900259, 2000.

Pino, N. A., Palombo, B., Ventura, G., Perniola, B., and Ferrari, G.: Waveform modeling of historical seismograms of the 1930 Irpinia earthquake provides insight on "blind" faulting in Southern Apennines (Italy), J. Geophys. Res., 113, B05303, https://doi.org/10.1029/2007jb005211, 2008.

Poggi, V., Durrheim, R., Tuluka, G. M., Weatherill, G., Gee, R., Pagani, M., Nyblade, A., and Delvaux, D.: Assessing seismic hazard of the East African Rift: a pilot study from GEM and AfricaArray, B. Earthq. Eng., 15, 4499–4529, https://doi.org/10.1007/s10518-017-0152-4, 2017.

Pollitz, F. F., Burgmann, R., Stein, R. S., and Sevilgen, V.: The Profound Reach of the 11 April 2012 M 8.6 Indian Ocean Earthquake: Short-Term Global Triggering Followed by a Longer-Term Global Shadow, B. Seismol. Soc. Am., 104, 972–984, https://doi.org/10.1785/0120130078, 2014.

Quinteros Cartaya, C. B., Nava Pichardo, F. A., Glowacka, E., Treviño, E. G., and Dmowska, R.: Forecast of Large Earthquakes Through Semi-periodicity Analysis of Labeled Point Processes, Pure Appl. Geophys., 173, 2571–2585, https://doi.org/10.1007/s00024-016-1338-4, 2016.

Roth, F., Dahm, T., and Hainzl, S.: Testing stress shadowing effects at the South American subduction zone, Geophys. J. Int., 211, 1272–1283, https://doi.org/10.1093/gji/ggx362, 2017.

Schweitzer, J. and Lee, W. H. K.: Old seismic bulletins to 1920: a collective heritage from early seismologists, in: International Handbook of Earthquake and Engineering Seismology, edited by: Lee, W., Kanamori, H., Jennings, J., and Kisslinger, C., vol. B, chap. 88, 1665–1723, Academic Press, San Diego, 2002.

Storchak, D., Giacomo, D. D., Engdahl, E., Harris, J., Bondár, I., Lee, W., Bormann, P., and Villaseñor, A.: The ISC-GEM Global Instrumental Earthquake Catalogue (1900–2009): Introduction, Phys. Earth Planet. In., 239, 48–63, https://doi.org/10.1016/j.pepi.2014.06.009, 2015.

Storchak, D. A., Di Giacomo, D., Bondár, I., Engdahl, E. R., Harris, J., Lee, W. H. K., Villaseñor, A., and Bormann, P.: Public Release of the ISC–GEM Global Instrumental Earthquake Catalogue (1900–2009), Seismol. Res. Lett., 84, 810–815, https://doi.org/10.1785/0220130034, 2013.

Storchak, D. A., Harris, J., Brown, L., Lieser, K., Shumba, B., Verney, R., Di Giacomo, D., and Korger, E. I. M.: Rebuild of the Bulletin of the International Seismological Centre (ISC), part 1: 1964–1979, Geoscience Letters, 4, 32, https://doi.org/10.1186/s40562-017-0098-z, 2017.

Udías, A. and Stauder, W.: The Jesuit contribution to seismology, Seismol. Res. Lett., 67, 10–19, 1996.

Utsu, T.: Catalog of Damaging Earthquakes in the World (Through 1989), Utsu, Tokuji, Tokyo, 243 pp., 1990 (in Japanese).

Utsu, T.: A list of deadly earthquakes in the World: 1500–2000, in: International Handbook of Earthquake and Engineering Seismology, edited by: Lee, W., Kanamori, H., Jennings, J., and Kisslinger, C., vol. A, chap. 42, 691–717, Academic Press, San Diego, 2002.

Utsu, T.: Catalog of Damaging Earthquakes in the World (Through 2002), the Dbase file distributed at the memorial party of Prof. Tokuji Utsu held in Tokyo, 2004.

Villaseñor, A. and Engdahl, E.: A digital hypocentre catalog for the International Seismological Summary, Seismol. Res. Lett., 76, 554–559, 2005.

Weatherill, G., Pagani, M., and Garcia, J.: Exploring earthquake databases for the creation of magnitude-homogeneous catalogues: tools for application on a regional and global scale, Geophys. J. Int., 206, 1652–1676, https://doi.org/10.1093/gji/ggw232, 2016.

Wessel, P., Smith, W. H. F., Scharroo, R., Luis, J., and Wobbe, F.: Generic Mapping Tools: Improved Version Released, Eos, Transactions American Geophysical Union, 94, 409–410, https://doi.org/10.1002/2013eo450001, 2013.

Weston, J., Engdahl, E. R., Harris, J., Di Giacomo, D., and Storchak, D. A.: ISC-EHB: reconstruction of a robust earthquake data set, Geophysical J. Int., 214, 474–484, https://doi.org/10.1093/gji/ggy155, 2018.

Wiemer, S. and Wyss, M.: Minimum Magnitude of Completeness in Earthquake Catalogs: Examples from Alaska, the Western United States, and Japan, B. Seismol. Soc. Am., 90, 859–869, https://doi.org/10.1785/0119990114, 2000.

Ye, L., Lay, T., Kanamori, H., and Koper, K. D.: Rapidly Estimated Seismic Source Parameters for the 16 September 2015 Illapel, Chile M w 8.3 Earthquake, Pure Appl. Geophys., 173, 321–332, https://doi.org/10.1007/s00024-015-1202-y, 2016.

Zaliapin, I. and Kreemer, C.: Systematic fluctuations in the global seismic moment release, Geophys. Res. Lett., 44, 4820–4828, https://doi.org/10.1002/2017gl073504, 2017.

Zechar, J. D., Marzocchi, W., and Wiemer, S.: Operational earthquake forecasting in Europe: progress, despite challenges, B. Earthq. Eng., 14, 2459–2469, https://doi.org/10.1007/s10518-016-9930-7, 2016.

Zhan, Z. and Shearer, P. M.: Possible seasonality in large deep-focus earthquakes, Geophys. Res. Lett., 42, 7366–7373, https://doi.org/10.1002/2015gl065088, 2015.

10

A database of paleoceanographic sediment cores from the North Pacific, 1951–2016

Marisa Borreggine[1], Sarah E. Myhre[1,2], K. Allison S. Mislan[1,3], Curtis Deutsch[1], and Catherine V. Davis[4]

[1]School of Oceanography, University of Washington, 1503 NE Boat Street, Box 357940, Seattle, WA 98195-7940, USA

[2]Future of Ice Initiative, University of Washington, Johnson Hall, Room 377A, Box 351360, Seattle, WA 98195-1360, USA

[3]eScience Institute, University of Washington, 3910 15th Ave NE, Box 351570, Seattle, WA 98195, USA

[4]School of Earth, Oceans, and the Environment, University of South Carolina, 701 Sumter Street, Earth and Water Science Building, Room 617, Columbia, SC 29208, USA

Correspondence to: Sarah E. Myhre (semyhre@uw.edu)

Abstract. We assessed sediment coring, data acquisition, and publications from the North Pacific (north of 30° N) from 1951 to 2016. There are 2134 sediment cores collected by American, French, Japanese, Russian, and international research vessels across the North Pacific (including the Pacific subarctic gyre, Alaskan gyre, Japan margin, and California margin; 1391 cores), the Sea of Okhotsk (271 cores), the Bering Sea (123 cores), and the Sea of Japan (349 cores) reported here. All existing metadata associated with these sediment cores are documented here, including coring date, location, core number, cruise number, water depth, vessel metadata, and coring technology. North Pacific sediment core age models are built with isotope stratigraphy, radiocarbon dating, magnetostratigraphy, biostratigraphy, tephrochronology, % opal, color, and lithological proxies. Here, we evaluate the iterative generation of each published age model and provide comprehensive documentation of the dating techniques used, along with sedimentation rates and age ranges. We categorized cores according to the availability of a variety of proxy evidence, including biological (e.g., benthic and planktonic foraminifera assemblages), geochemical (e.g., major trace element concentrations), isotopic (e.g., bulk sediment nitrogen, oxygen, and carbon isotopes), and stratigraphic (e.g., preserved laminations) proxies. This database is a unique resource to the paleoceanographic and paleoclimate communities and provides cohesive accessibility to sedimentary sequences, age model development, and proxies. The data set is publicly available through PANGAEA at https://doi.org/10.1594/PANGAEA.875998.

1 Introduction

Paleoceanographic sediments provide the sedimentary, geochemical, and biological evidence of past Earth system changes. Sediment cores produce robust reconstructions of large oceanographic provinces and provide insight into earth system mechanistic hypotheses. However, there is not a common repository for paleoceanographic data and publications, and this lack of centralization limits the efficacy of the earth science community in directing research efforts. For the North Pacific, such ongoing mechanistic hypotheses include deep ocean circulation (e.g., Rae et al., 2014; De Pol-Holz et al., 2006), deep water and intermediate water formation and ventilation (e.g., Knudson and Ravelo, 2015a; Zheng et al., 2000; Cook et al., 2016), and changes in the oceanic-preformed nutrient inventories (e.g., Jaccard and Galbraith, 2013; Knudson and Ravelo, 2015b), as well as more regional mechanisms such as sea ice extent (e.g., Max et al., 2012), upwelling intensity (e.g., Di Lorenzo et al., 2008; Hendy et al., 2004), local surface ocean productivity (e.g., Serno et al., 2014; Venti et al., 2017), and terrigenous and marine fluxes of iron (e.g., Davies et al., 2011; Praetorius et al., 2015).

Paleoceanographers have benefited from the use of large databases of climate data in the past, such as CLIMAP Project Members (1981) (Climate: Long-Range Investigation, Mapping, Prediction), which produced globally resolved temperature records for the Last Glacial Maximum (LGM) and aimed to determine climate system sensitivity from paleoclimate reconstructions (Hoffert and Covey, 1992). The PaleoDeepDive project employed a similar approach to the systematic extraction of archival data and constructed a new way to assess and engage with paleobiological data (Zhang, 2015). These projects are examples of international research teams approaching the same limit – extraction and organization of dark data – that arises when creating comprehensive paleoreconstructions.

1.1 Assembling a paleoceanographic database

A clear need exists for high quality paleoenvironmental reconstructions to fit the North Pacific into a climate global framework because the role this enormous ocean basin plays in earth system changes remains relatively unclear in comparison to the Southern and Atlantic oceans. To address the collective need, we present here a new database of North Pacific paleoceanographic research efforts along with the broad findings of our census of coring metadata, age model development, and proxy publications. We address the following questions in this paper to supplement and provide context for our database:

1. Where have sediment cores been extracted from the North Pacific seafloor (north of 30° N, including the Pacific subarctic gyre, Alaskan gyre, Japan margin, and California margin), the Bering Sea, the Sea of Okhotsk, and the Sea of Japan? What metadata are available in published cruise reports or peer-reviewed investigations, including core name, recovery date, recovery vessel and scientific agency, latitude and longitude, water depth, core length, and coring technology?

2. For sediment cores with published age models, what lines of evidence were used to develop the chronological age of the sediment, what is the age range from core top to core bottom, and what are the sedimentation rates?

3. What lines of sedimentary, geochemical, isotopic, and biological proxy evidence are available for each sediment core?

4. How has the state of North Pacific research efforts and reporting changed since the beginning of paleoceanographic expeditions?

1.2 Paleoceanographic age models, proxies, and nomenclature

Marine sedimentary age models tie the sedimentary depth (in meters below sediment surface) to calendar age (ka, thousands of years and/or Ma, millions of years). Not all sedimentary chronologies are of the same quality, and often age models are iteratively refined. Age models are developed with many different dating techniques, which are dependent upon the quality, preservation, and age of the sediments, as well as the investigative priorities of research teams and the proximity of other well-developed sedimentary chronologies. Paleoceanographic proxies, including biological, isotopic, geochemical, and sedimentary observations and measurements, address large thematic questions in the reconstruction of ocean environments, including ocean temperature, paleobiology, seafloor geochemistry, sea ice distribution, and additional sedimentary analyses.

Sediment cores are often represented by their cruise–core unique identifier, which has the general format "cruise name-core number". The cruise number is generally indicative of the research vessel employed and the year of the expedition. For example, L13-81 is the 13th cruise of S. P. Lee in 1981, MR06-04 is the 4th cruise of R/V *Mirai* in 2006, and YK07-12 is the 12th cruise of R/V *Yokosuka* in 2007. Often the core number will be preceded or followed by a PC (piston core), MC (multicore), TC (trigger core), or GC (gravity core) to signify the coring technology. The sediment core B37-04G is the fourth gravity core from the 37th cruise of R/V *Professor Bogorov*, and EW9504-11PC is the 11th piston core from the fourth cruise of R/V *Maurice Ewing* in 1995. However, this nomenclature is not comprehensive. For example, cores affiliated with iterations of the International Ocean Discovery Program are represented by the program abbreviation and their hole number (i.e., ODP Hole 1209A). Cruise name or number is unknown for many cores, and in these cases the core is referred to by number.

2 Methods

Here we assembled a database from peer-reviewed publications, publicly available online cruise reports, and print-only cruise reports available through the University of Washington library network. Detailed metadata were reported for cores where it is available, commonly from cruise reports, including water depth (in meters), recovery year, latitude and longitude, coring technology, and core length (in meters). Summary details regarding affiliated research vessels and institutions were gathered from publications or cruise reports. Cruise reports were commonly available for research expeditions affiliated with JAMSTEC, GEOMAR, and the Scripps Institution of Oceanography and were less commonly available for older cores. All evidence used in age model development was reported along with sedimentation rates and the sedimentary age ranges to provide investigators with the ca-

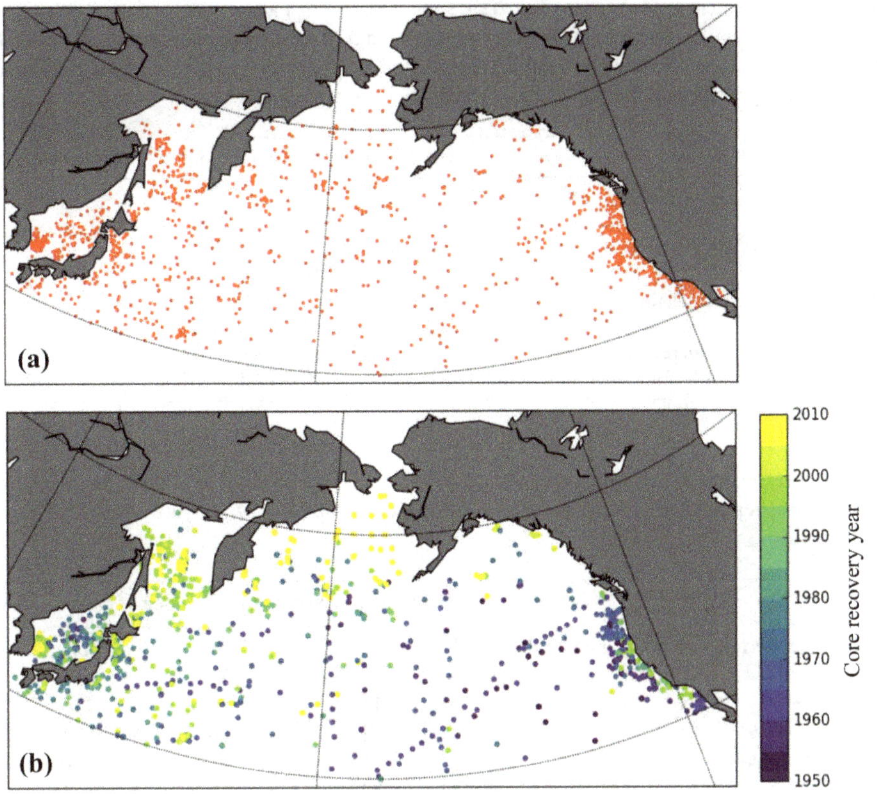

Figure 1. Location and recovery year of marine sediment cores from the North Pacific and marginal seas. **(a)** Sediment cores in the North Pacific (above 30° N) published with latitudes and longitudes (354 additional cores were documented in either cruise reports or peer-reviewed publications without latitude and longitude). **(b)** Sediment cores published with an associated core recovery year ranging from 1951 to 2010. This age range corresponds with the color bar (495 cores have been published in either cruise reports or peer-reviewed publications without a recovery year).

pacity to quickly evaluate specific cores that meet the investigative priorities. For each core, paleoceanographic proxies and associated publications are documented to provide an efficient resource for assessing the availability and quality of different lines of paleoenvironmental information. In addition, we evaluated the annual number of age models published (using any dating technique), age models published specifically with radiocarbon dating, publications generated, sediment cores collected, research cruises completed, and the mean number of proxies generated per core.

3 Results

3.1 Sediment coring and metadata

We documented 2134 sediment cores and 283 marine geology research cruises above 30° N from 1951 to 2016 in the North Pacific, the Bering Sea, the Sea of Japan, and the Sea of Okhotsk (Fig. 1, Table 1). The majority of sediment cores were extracted from the northern subarctic Pacific (1391 cores) followed by the Sea of Japan (349 cores), the Sea of Okhotsk (271 cores), and the Bering Sea (123 cores). Many of the oldest cores in this oceanographic province came from the central abyssal Pacific and were associated with the Deep Sea Drilling Project. Cores were extracted from the North Pacific from 1951 to 2010, and the oldest age models extend to 120 000 ka (Fig. 1, Table 1). Metadata associated with sediment cores or marine research cruises are frequently unavailable or omitted from publications affiliated with sediment cores. For example, 495 cores are in the literature without a recovery year, 354 sediment cores were published without latitude and longitude, and 642 cores were reported without specifying the coring technology used (Tables 1 and 2). Moreover, 1210 sediment cores reported in our database were identified in supplemental tables within publications, but no original cruise reports or peer-reviewed publications otherwise report on these cores.

3.2 Sediment chronologies

In the North Pacific, 519 marine sediment cores have published age models, and 266 of these chronologies are generated with the radiocarbon dating (^{14}C) of planktonic foraminifera, molluscs, or terrigenous material like bark or

Table 1. Regional summary of sediment core database for the Bering Sea, North Pacific, Sea of Japan, and Sea of Okhotsk, including number of cores recovered, the regional percent (%) of cores reported with latitude and longitude, number of research cruises, total regional publication count, count of cores with no peer-reviewed publications or cruise report, the range of core recovery years, the regional percent (%) of cores reported with recovery years, and the range of recovered core water depths (m b.s.l.).

Region	Bering Sea	North Pacific	Sea of Japan	Sea of Okhotsk	Total
Count of cruises	24	179	78	31	283
Count of publications	70	306	114	90	565
Count of cores with no publications* or cruise report	61	849	172	128	1210
Number of cores recovered	123	1391	349	271	2134
Percent of regional cores (%) reported with latitude and longitude	91.9	81.9	77.7	91.9	83.4
Range of core recovery years	1957–2009	1951–2009	1957–2010	1972–2009	1951–2010
Percent of regional cores (%) reported with recovery year	87	68.9	96.9	98.5	76.8
Range of recovered core water depths (m b.s.l.)	33–3930	21–9585	129–5986	105–8180	21–9585

*These cores are listed in large data tables in auxiliary publications, but the original reporting is not publicly available.

Table 2. Percent of regional cores reported with coring technology and the number of cores recovered in the North Pacific and marginal seas by coring technology. Additional reported coring technologies include the less often utilized hydrostatic cores, kasten cores, Asura cores, pressure cores, and trigger cores.

Region	Bering Sea	North Pacific	Sea of Japan	Sea of Okhotsk
% of regional cores reported with drilling technology	75.6	72.1	60.2	68.6
No. of piston cores	33	476	177	67
No. of gravity cores	11	250	15	73
No. of box cores	2	78	1	1
No. of riserless drilling cores	7	52	0	0
No. of multicores	33	39	1	39
No. of Phleger cores	0	15	0	0
No. of jumbo piston cores	4	10	0	0

wood fragments (Fig. 2). Radiocarbon dating is the most common region-wide chronological technique (51 % of age models incorporate this method). Lead dating (^{210}Pb) is used for 12 sediment chronologies. Many other lines of evidence are used in the North Pacific and marginal seas to develop paleoceanographic age models. These approaches vary regionally and include planktonic foraminifera oxygen isotope stratigraphy, diatom silica oxygen isotope stratigraphy, biostratigraphy, magnetostratigraphy, tephrochronology, chronostratigraphy, carbonate stratigraphy, opal stratigraphy, composition, lithological proxies, the presence of lamination, chlorin content, and color (a^*, b^*, and L^* values; Fig. 2, Table 3). For example, in the Sea of Japan, lithological proxies such as core laminations are often utilized as chronological proxy evidence, and 12 % of local age models incorporate this technique. Tephrochronology is also applied in 51 % of Sea of Japan age models due to regional volcanism. In the Bering Sea, peaks in silica are often used, and 13 % of the regional age models incorporated this technique. Published sedimentation rates ranged across the North Pacific (0.1–2000 cm ka^{-1}), Bering Sea (3–250 cm ka^{-1}), Sea of Okhotsk (0.7–250 cm ka^{-1}), and the Sea of Japan (0.2–74 cm ka^{-1}) with the highest rates within the Alaska Current in the North Pacific (up to 2000 cm ka^{-1}).

3.3 Paleoceanographic proxies from marine sediment cores

From all reported sediment cores in the North Pacific and marginal seas, only 40 % of cores have published proxy data (Figs. 3 and 5). Stable isotope stratigraphy was available for 293 cores, including oxygen, carbon, or nitrogen isotopes (δ^{18}O, δ^{13}C, δ^{15}N). We documented planktonic (236 cores) and benthic (178 cores) foraminiferal oxygen isotopes, planktonic (67 cores) and benthic (77) foraminifera carbon isotopes, and 34 cores with bulk sediment nitrogen isotopes. Of note, 98 cores were available with magnetostratigraphy (Table 4, Fig. 4).

We recorded paleothermometry data for 234 cores, including planktonic foraminifera oxygen isotopes, magnesium to calcium ratios from planktonic foraminifera, and alkenones TEX_{86} and U^{k}_{37} (Fig. 3). We recorded 425 cores with microfossil biostratigraphy, including foraminifera, diatom, radiolarian, ostracod, silicoflagellate, ebridian, acritarch, coccolithophore, or dinoflagellate assemblages (Fig. 3, Table 4). Biostratigraphy utilizes known microfossil assemblages and their corresponding ages to assign a geologic age range to core strata containing assemblages. Geochemical analyses are reported for 151 cores, including measurements of, for example, brassicasterol, magnesium, calcium, molybdenum,

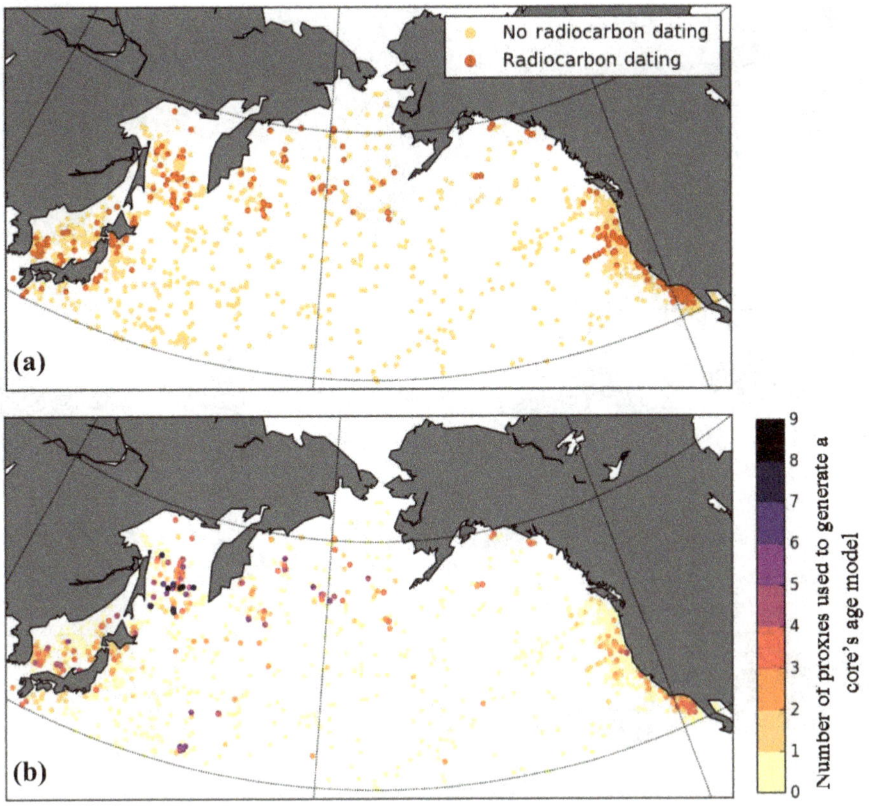

Figure 2. Age model development for sediment cores in the North Pacific and marginal seas. **(a)** Cores with age models that have been constructed with radiocarbon dating (^{14}C). **(b)** Number of paleoceanographic proxies used to generate each core's age model.

Table 3. Summary statistics on core chronology, including the number of cores with radiocarbon dating (^{14}C) and oxygen isotope stratigraphy of planktonic foraminifera ($\delta^{18}O$), as well as the regional mean core top and bottom ages, the number of cores with published sedimentation rate ranges and means, and core length minimums, maximums, and averages.

Region	Bering Sea	North Pacific	Sea of Japan	Sea of Okhotsk
^{14}C dating	28	158	38	42
$\delta^{18}O$ stratigraphy, planktonic foraminifera	20	122	29	32
Mean core top age (calendar age, ka)	6.0 ± 9.6	9.5 ± 61.6	5.9 ± 28.4	1.1 ± 2.3
Mean core bottom age (calendar age, ka)	523.7 ± 1146.2	6210.8 ± 20250.7	996.1 ± 2602.2	153.1 ± 279.2
Cores with sedimentation rate range	28	97	20	18
Cores with sedimentation rate mean	7	68	9	16
Average core length (m)	56.89	35.7	35.98	0.02
Minimum core length (m)	0.06	0.03	0.24	53.88
Maximum core length (m)	745	1180	555.3	6.50

cadmium, manganese, uranium, chromium, rhenium, chlorin, titanium, iron, zinc, and beryllium (Fig. 3, Table 3). We documented 234 cores with sea ice proxy data, including geochemical biomarker IP25, ice-rafted debris (IRD), and diatom communities (Fig. 3). The presence and concentration of these proxies are indicative of contemporary sea ice extent and volume. We recorded 521 sediment cores with lithological analyses, including the documentation of core lamina, sediment density, mass accumulation rates, biogenic opal and barium, silicon to aluminum weight ratios, carbon to nitrogen ratios, inorganic and organic calcium carbonate and carbon, inorganic nitrogen, sulfur, and clay mineral composition (Fig. 3).

3.4 Research cruise and publication rate

We cataloged 565 peer-reviewed publications and cruise reports and evaluated the progress of paleoceanographic re-

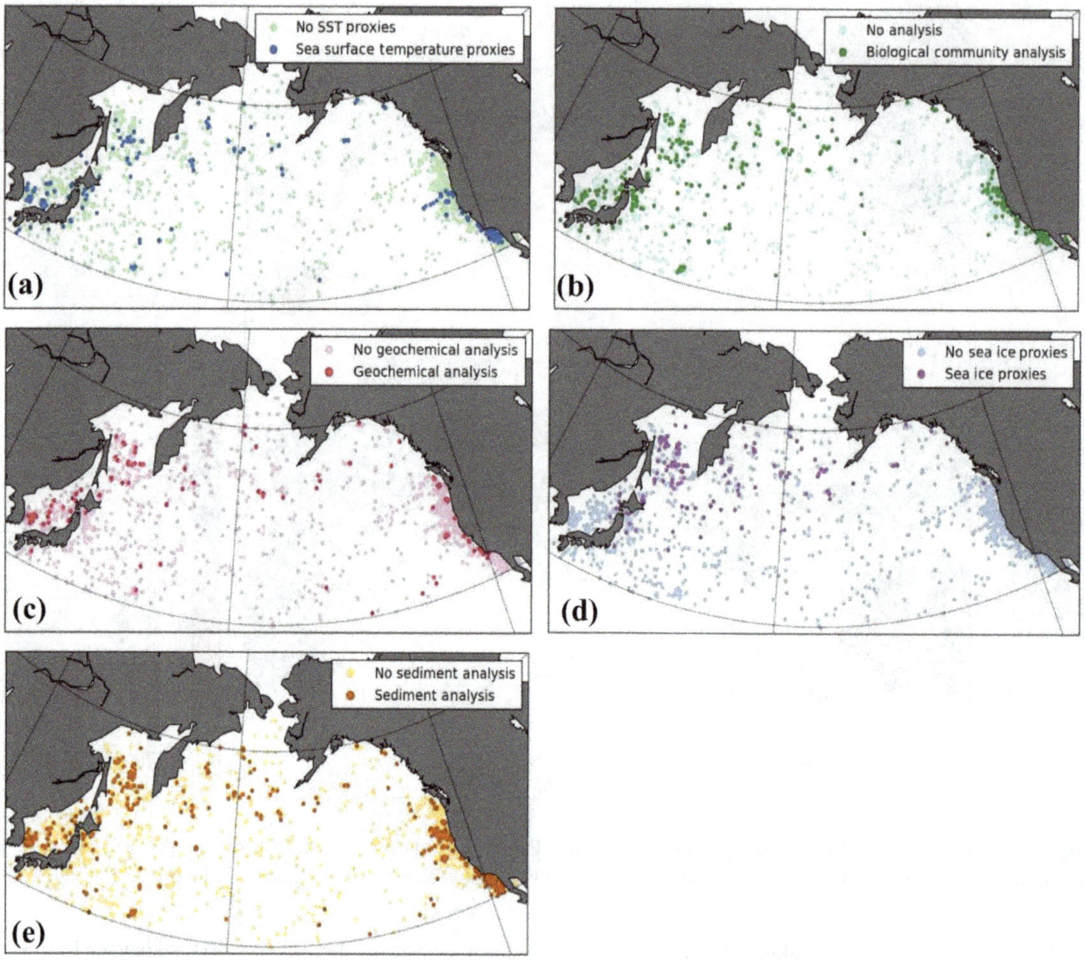

Figure 3. Published paleoceanographic proxies in the North Pacific. **(a)** Sea surface temperature reconstructions, including planktonic foraminifera oxygen isotopes ($\delta^{18}O_p$), magnesium to calcium measurements, and TEX_{86} and U^k_{37} alkenones. **(b)** Biostratigraphy of microfossils, including foraminifera, diatoms, radiolarians, ostracods, silicoflagellates, ebridians, acritarchs, coccolithophores, and dinoflagellates. **(c)** Geochemical proxy analysis, including trace metals such as brassicasterol, magnesium, calcium, molybdenum, cadmium, manganese, uranium, chromium, rhenium, chlorin, titanium, iron, zinc, and beryllium. **(d)** Presence of sea ice proxies, including geochemical biomarker IP25, ice-rafted debris (IRD), and sea-ice-related diatom communities. **(e)** Analysis of lithological core proxies, including 395 measurements of core lamina, biogenic opal and barium, silicon to aluminum weight ratio, sulfur, inorganic 20 and organic carbon weight and mass accumulation rates, mass accumulation rates of various elements, inorganic nitrogen, carbon to nitrogen ratios, sediment density, and clay mineral composition.

search using a suite of annual metrics of cruise and core metadata and publications (Fig. 5). The state of North Pacific paleoceanographic investigations has evolved incrementally in the 65-year history of research in the region, and we characterize the history into two distinct phases (before and after the early 1990s). Cruise reports were not publicly available for every cruise or core, and information on many cores cited in cruise reports was never published in peer-reviewed literature. The majority of cores (1210 cores or 57 % of all cores extracted from the North Pacific) lack any publication.

Core recovery rates were high and publication rates were low from 1951 to 1988, which is a period when expeditions were driven by individual institutions and peer-reviewed publication was not the primary research outcome (Fig. 5). Annual rates of cruise completion and sediment core extraction peaked in 1965–1968, and this is a consequence of the temporal overlap in peak research efforts by the Scripps Institute of Oceanography (1951–1988), the Lamont–Doherty Earth Observatory (1964–1975), Oregon State University (1962–1972), and the Deep Sea Drilling Project (1971–1982). Annual rates of publication (peer-reviewed and cruise reports), including those publications with age models, increased around 1995 (Fig. 5). In this later period, research

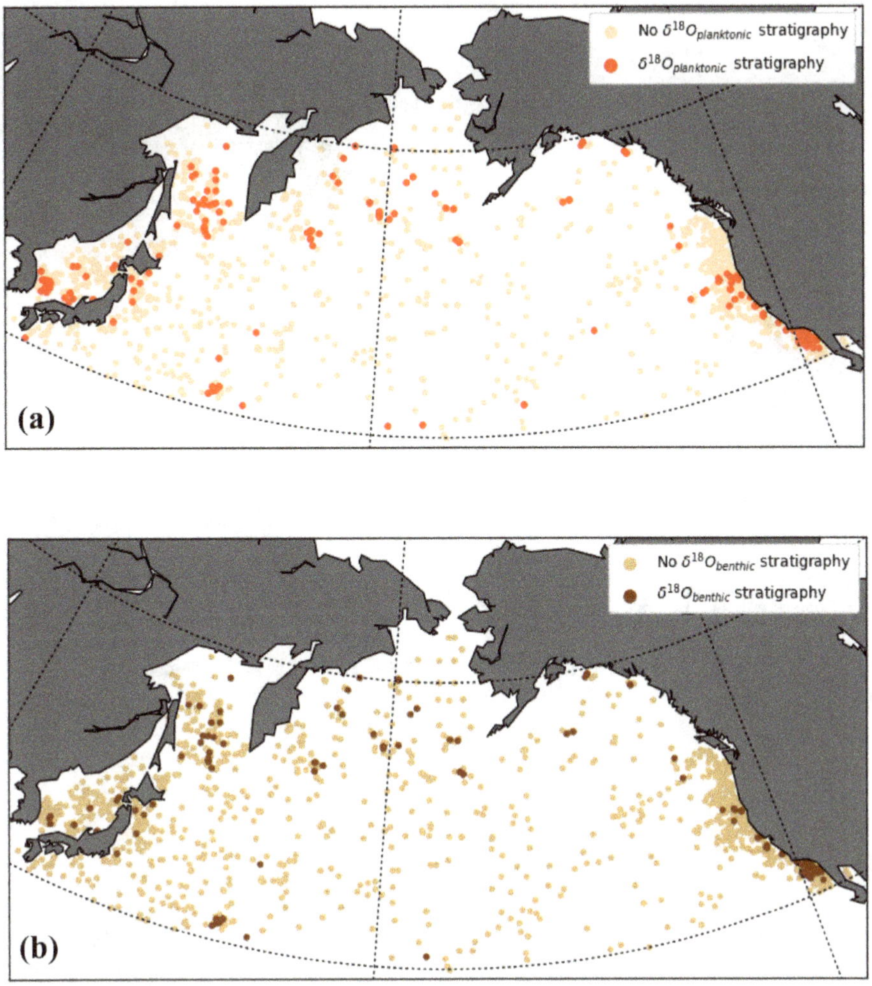

Figure 4. Published planktonic and benthic foraminiferal oxygen isotope stratigraphy for sediment core age models. **(a)** Analysis of planktonic foraminiferal oxygen isotopes and **(b)** analysis of benthic foraminiferal oxygen isotopes.

cruise efforts were dominated by international research team efforts and resulted in increased peer-reviewed publications, sediment core chronology constructions, and the proliferation of radiocarbon dating. There are 41 cores with very focused investigation (> 6 publications), and these archives are primarily located within the California Current (Fig. 6). Major peaks in cruising and coring efforts coincided with research cruises by the International Ocean Discovery Program, such as ODP Legs 145 and 146 in 1992 (North Pacific transect, Cascadia margin), ODP Leg 167 in 1996 (California margin), and IODP Expedition 323 in 2009 (Bering Sea). Despite the increase in publications around 1995, we observe no distinct temporal trend in the rate of cruise completion and coring effort (Fig. 5).

4 Discussion

4.1 Evolving state of North Pacific coring and paleoceanographic approaches

Extensive cruise and research efforts have focused on the marine geology of the North Pacific. Often, the cruise and core metadata from these efforts are unpublished, though they are integral to collaboration, continued research, and publication. Here we present a database with 2134 sediment cores, 283 research cruises, and 565 peer-reviewed publications related to paleoceanographic research (Table 1). We cataloged 519 publications with sedimentary age models, and of those age models, 266 utilized radiocarbon dating, 201 utilized planktonic foraminiferal oxygen isotope stratigraphy, and 129 utilized benthic foraminiferal oxygen isotope stratigraphy (Figs. 2 and 4). Throughout the North Pacific, Bering Sea, Sea of Okhotsk, and Sea of Japan, the techniques for reconstructing sedimentary age models var-

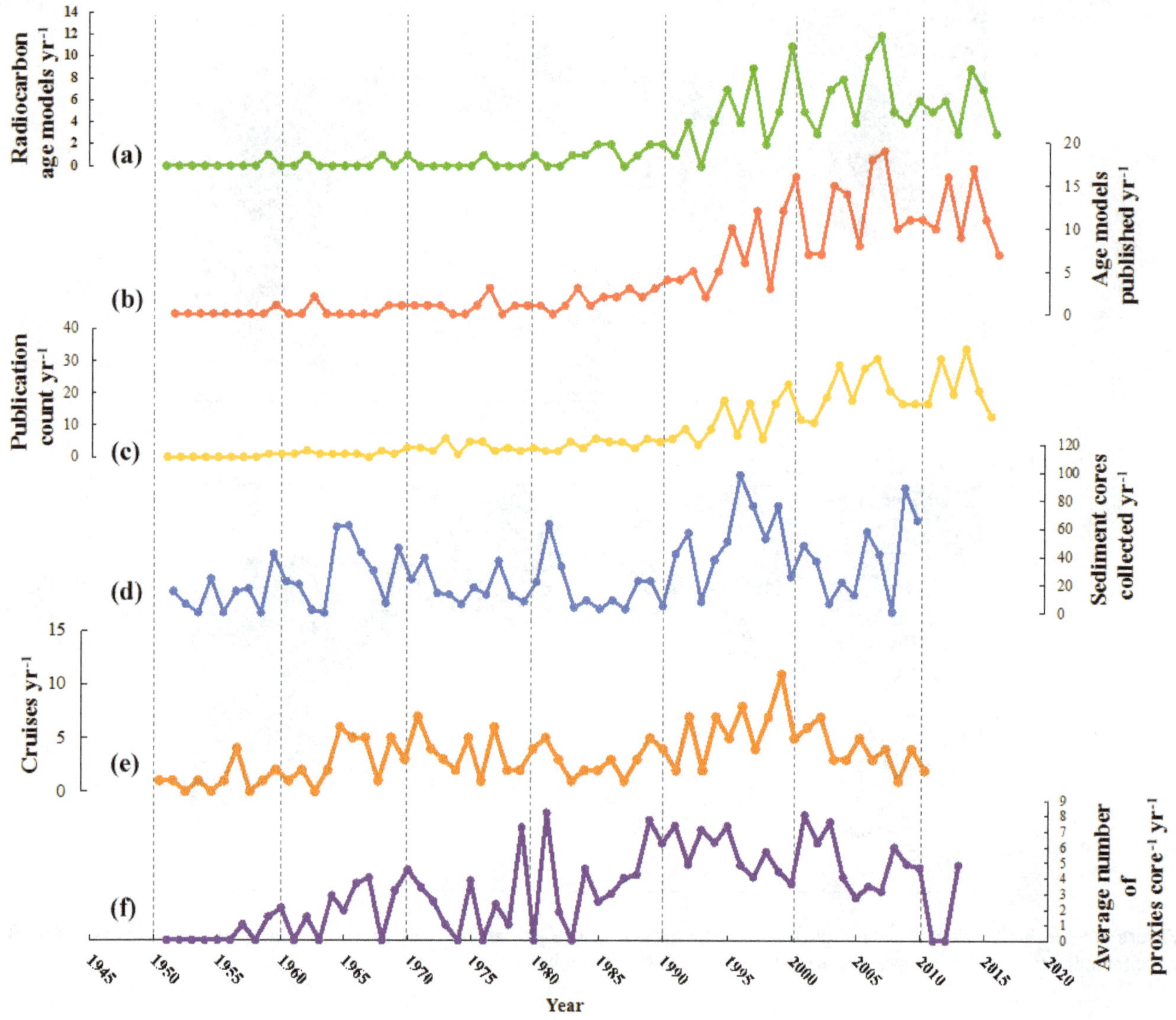

Figure 5. The marine geology cruise and paleoceanographic research progress through time for the North Pacific and marginal seas, wherein the lead–lag timing of cruise reporting and core publication is assumed for the most recent years. We utilize peer-reviewed publications to locate cores, and there is a lag between publication and core extraction. **(a)** Annual number of age models published using radiocarbon dating. **(b)** Annual number of all age models published. **(c)** Total annual number of publications and cruise reports. **(d)** Annual number of cores collected. **(e)** Marine research expedition count per year. **(f)** Annual mean number of proxy analyses published per core.

ied regionally. We documented a community-wide shift away from singular dating techniques toward age models that incorporate several techniques. Multiproxy approaches hold merit through verifying or constraining the results of a singular proxy and thereby disentangling multiple environmental drivers and providing redundancy in order to create robust records of climate and ocean conditions (Mix et al., 2000; Mann, 2002). Age model development has moved through the last 60 years to more detailed high-resolution age models constructed to investigate millennial and submillennial paleoceanographic variability (Figs. 5 and 6).

4.2 The merit of database management and open-access science

Databases are integral to facilitate efficient, hypothesis-based investigations into earth system mechanisms. Public access to databases facilitate a higher volume of research by a diverse range of scientists (Harnad and Brody, 2004). The necessity for databases to encompass a wide array of data over large oceanographic provinces is also largely recognized. Open access tools from PANGAEA support database-dependent research because hypothesis-based investigations can be more efficient through public data access. For ex-

Table 4. Regional summary of isotopic, geochemical, biological, and sedimentary proxies. Benthic and planktonic isotopic analysis is for all cores, including, but not limited to, isotope analysis used in core chronology development.

Region	Bering Sea	North Pacific	Sea of Japan	Sea of Okhotsk
$\delta^{18}O_{benthic}$	19	118	15	25
$\delta^{18}O_{planktonic}$	20	139	45	32
$\delta^{13}C_{benthic}$	7	57	7	6
$\delta^{13}C_{planktonic}$	6	50	10	1
$\delta^{15}N$	5	16	12	1
Alkenone U^{k}_{37}	7	64	15	12
TEX_{86}	0	1	0	0
Foraminiferal biostratigraphy	12	82	22	8
Foraminiferal abundance	8	53	41	18
Diatom biostratigraphy	35	87	49	25
Coccolithophore biostratigraphy	2	9	4	0
Ostracod biostratigraphy	1	1	2	0
Geochemical proxies (Mg, Mo, Cd, Mn, U, Cr, Re, Ca, T, Fe, Mg, Zn, Be, chlorin)	1	189	46	27
% opal	21	38	20	20
Total inorganic carbon (%)	16	204	49	26
Total organic carbon (%)	16	142	62	26

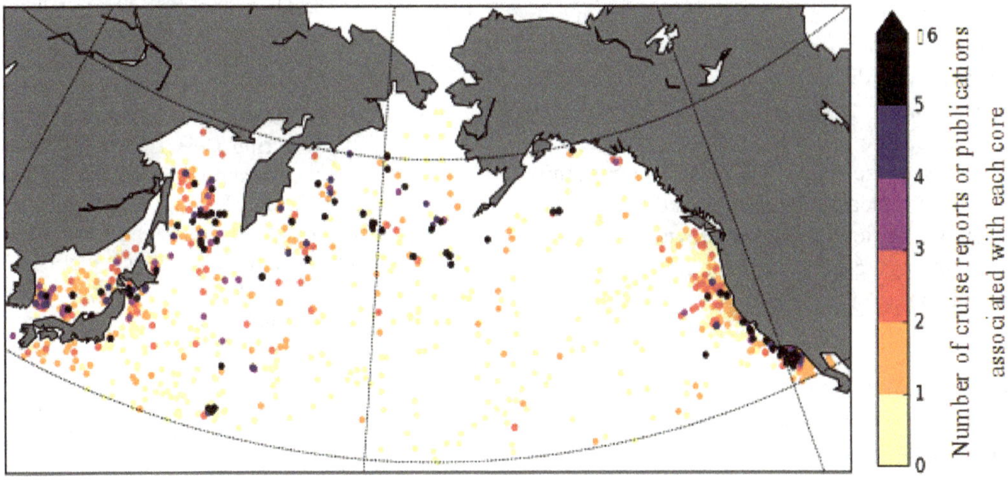

Figure 6. Number of affiliated publications and cruise reports for each core. The maximum publication count for an individual core is 23.

ample, content from online databases has contributed to research in atmospheric forcing (e.g., Shaffer et al., 2009), the Atlantic Meridional Overturning Circulation (AMOC; e.g., Schmittner and Galbraith, 2008), and Southern Ocean ventilation (e.g., Yamamoto et al., 2015). The metadata in these databases must be thorough, as data are impractical without the affiliated identifier, location, and methods (Lehnert et al., 2000). Database management should be a priority in any field that incorporates the contents of online repositories of knowledge and research. The disconnect between the research goals of the paleoceanographic community and the metadata produced here can be described as a "breakdown" that limits the progress of paleoceanographic research (Tanweer et al., 2016). These breakdowns allow us to self-assess and move forward with insight into best practices. Metadata are produced from data sets that are inherently human in design and are therefore not inerrant. Assessment of the errors in metadata reporting can directly reveal the need for community-wide improvements. As an example, all cores should be reported with latitude and longitude; the absence of these specific metadata significantly impairs further work. The database presented here and others like it consolidate the research effort of an entire community into an efficient tool for future investigative purposes.

4.3 Recommendations for the marine geology community

Ocean sediment records are one of the primary tools for understanding earth's history, and the documentation of these records benefits the entire community of earth scientists. The publications and cruise reports here represent a large body of research completed on North Pacific sediment records; however, this may not constitute the entire body of work. There was no preexisting common repository for cruise reports and coring data in the North Pacific beyond the individual institutional archival processes. This database serves as the most complete archive of publicly available cruise reports and publications in which available DOIs and URLs are reported. We demonstrate a need for more thorough, accessible documentation of marine geology and paleoceanographic research. In our examination of publications, cruise reports, and notes from research cruises, we gained insights into past inconsistencies in marine sediment record reporting. We recommend a suite of actions to ensure efficient, comprehensive sediment core collection, metadata documentation, and the publication of chronologies and proxies. We propose that each publication thoroughly reports metadata on all cores discussed and their associated cruises, including core unique identifier numbers, cruise name and number, vessel name, geographic coordinates, core recovery date, core length, core recovery water depth (m b.s.l.), sampling resolution, sampling volume, core archival repository, and the link (if existing) to public cruise reports. We also suggest summarizing each core's metadata, age model, and publication history in the methods section in order to provide a frame of reference for new findings, especially in the context of iterative age model revisions.

Author contributions. SEM and CVD initiated the building of this database. MB and SEM built the database and were joined by KASM in producing figures and analysis for this paper. All authors wrote the paper.

Competing interests. The authors declare that they have no conflict of interest.

Acknowledgements. The authors would like to thank the University of Washington Library oceanography collection, namely Louise Richards and Maureen Nolan. We also wish to acknowledge the support for this publication provided by the University of Washington Purple and Gold Scholarship, the UW School of Oceanography Lowell K. and Alice M. Barger Endowed Scholarship, the Clarence H. Campbell Endowed Lauren Donaldson Scholarship, the UW College of the Environment Student Travel Grant, and NSF grant OCE-1458967. KAS Mislan was supported by the Washington Research Foundation Fund for Innovation in Data-Intensive Discovery and the Moore/Sloan Data Science Environments Project at the University of Washington.

Edited by: Attila Demény

References

Borreggine, M., Myhre, S. E., Mislan, K. A. S., Deutsch, C., and Davis, C. V.: A database of paleoceanographic sediment cores from the North Pacific, 1951–2016, PANGAEA, https://doi.org/10.1594/PANGAEA.875998, 2017.

CLIMAP Project Members: Seasonal reconstruction of the earth's surface at the last glacial maximum, Geological Society of America, Map and Chart Series 36, 18 pp., available at: https://www.ncdc.noaa.gov/paleo/metadata/noaa-ocean-2516.html, 1981.

Cook, M. S., Ravelo, A. C., Mix, A., Nesbitt, I. M., and Miller, N. V.: Tracing subarctic Pacific water masses with benthic foraminiferal stable isotopes during the LGM and late Pleistocene, Deep-Sea Res. Pt. II, 125, 84–95, https://doi.org/10.1016/j.dsr2.2016.02.006, 2016.

Davies, M. H., Mix, A. C., Stoner, J. S., Addison, J. A., Jaeger, J., Finney, B., and Wiest, J.: The deglacial transition on the southeastern Alaska Margin: meltwater input, sea level rise, marine productivity, and sedimentary anoxia, Paleoceanography, 26, PA2223, https://doi.org/10.1029/2010pa002051, 2011.

Di Lorenzo, E., Schneider, N., Cobb, K. M., Franks, P. J. S., Chha, K., Miller, A. J., McWilliams, J. C., Bograd, S. J., Arango, H., Curchitser, E., Powell, T. M., and Rivière., P.: North Pacific Gyre Oscillation links ocean climate and ecosystem change, Geophys. Res. Lett., 35, L08607, https://doi.org/10.1029/2007GL032838, 2008.

Harnad, S. and Brody, T.: Comparing the impact of open access (OA) vs. non-OA articles in the same journals, D-lib Magazine, 10, available at: https://eprints.soton.ac.uk/260207/, 2004.

Hendy, I. L., Pedersen, T. F., Kennett, J. P., and Tada, R.: Intermittent existence of a southern Californian upwelling cell during submillennial climate change of the last 60 kyr, Paleoceanography, 19, PA3007, https://doi.org/10.1029/2003pa000965, 2004.

Hoffert, M. I. and Covey, C.: Deriving global climate sensitivity from palaeoclimate reconstructions, Nature, 360, 573–576, https://doi.org/10.1038/360573a0, 1992.

Jaccard, S. L. and Galbraith, E. D.: Direct ventilation of the North Pacific did not reach the deep ocean during the last deglaciation, Geophys. Res. Lett., 40, 199–203, https://doi.org/10.1029/2012gl054118, 2013.

Knudson, K. P. and Ravelo, A. C.: North Pacific intermediate water circulation enhanced by the closure of the Bering Strait, Paleoceanography, 30, 1287–1304, https://doi.org/10.1002/2015PA002840, 2015a.

Knudson, K. P. and Ravelo, A. C.: Enhanced subarctic Pacific stratification and nutrient utilization during glacials over the last 1.2 Myr., Geophys. Res. Lett., 42, 9870–9879, https://doi.org/10.1002/2015GL066317, 2015b.

Lehnert, K., Su, Y., Langmuir, C. H., Sarbas, B., and Nohl, U.: A global geochemical database structure for rocks, Geochem. Geophy. Geosy., 1, 1012, https://doi.org/10.1029/1999gc000026, 2000.

Mann, M. E.: Climate reconstruction: the value of multiple proxies, Science, 297, 1481–1482, https://doi.org/10.1126/science.1074318, 2002.

Max, L., Riethdorf, J.-R., Tiedemann, R., Smirnova, M., Lembke-Jene, L., Fahl, K., Nürnberg, D., Matul, A., and Mollenhauer, G.: Sea surface temperature variability and sea-ice extent in the subarctic northwest Pacific during the past 15 000 years, Paleoceanography, 27, PA3213, https://doi.org/10.1029/2012PA002292, 2012.

Mix, A. C., Bard, E., Eglinton, G., Keigwin, L. D., Ravelo, A. C., and Rosenthal, Y.: Alkenones and multiproxy strategies in paleoceanographic studies, Geochem. Geophy. Geosy., 1, 1033, https://doi.org/10.1029/2000gc000056, 2000.

Pol-Holz, R. D., Ulloa, O., Dezileau, L., Kaiser, J., Lamy, F., and Hebbeln, D.: Melting of the Patagonian Ice Sheet and deglacial perturbations of the nitrogen cycle in the eastern South Pacific, Geophys. Res. Lett., 33, L04704, https://doi.org/10.1029/2005gl024477, 2006.

Praetorius, S. K., Mix, A. C., Walczak, M. H., Wolhowe, M. D., Addison, J. A., and Prahl, F. G.: North Pacific deglacial hypoxic events linked to abrupt ocean warming, Nature, 527, 362–366, https://doi.org/10.1038/nature15753, 2015.

Rae, J. W. B., Sarnthein, M., Foster, G. L., Ridgwell, A., Grootes, P. M., and Elliott, T.: Deep water formation in the North Pacific and deglacial CO_2 rise, Paleoceanography, 29, 645–667, https://doi.org/10.1002/2013PA002570, 2014.

Schmittner, A. and Galbraith, E. D.: Glacial greenhouse-gas fluctuations controlled by ocean circulation changes, Nature, 456, 373–376, https://doi.org/10.1038/nature07531, 2008.

Serno, S., Winckler, G., Anderson, R. F., Hayes, C. T., Ren, H., Gersonde, R., and Haug, G. H.: Using the natural spatial pattern of marine productivity in the Subarctic North Pacific to evaluate paleoproductivity proxies, Paleoceanography, 29, 438–453, https://doi.org/10.1002/2013PA002594, 2014.

Shaffer, G., Olsen, S. M., and Pedersen, J. O. P.: Long-term ocean oxygen depletion in response to carbon dioxide emissions from fossil fuels, Nat. Geosci., 2, 105–109, https://doi.org/10.1038/ngeo420, 2009.

Tanweer, A., Fiore-Gartland, B., and Aragon, C.: Impediment to insight to innovation: understanding data assemblages through the breakdown–repair process, Information, Communication and Society, 19, 736–752, https://doi.org/10.1080/1369118x.2016.1153125, 2016.

Venti, N. L., Billups, K., and Herbert, T. D.: Paleoproductivity in the northwestern Pacific Ocean during the Pliocene-Pleistocene climate transition (3.0–1.8 Ma), Paleoceanography, 32, 92–103, https://doi.org/10.1002/2016PA002955, 2017.

Yamamoto, A., Abe-Ouchi, A., Shigemitsu, M., Oka, A., Takahashi, K., Ohgaito, R., and Yamanaka, Y.: Global deep ocean oxygenation by enhanced ventilation in the Southern Ocean under long-term global warming, Global Biogeochem. Cy., 29, 1801–1815, https://doi.org/10.1002/2015gb005181, 2015.

Zhang, C.: DeepDive: a data management system for automatic knowledge base construction, Thesis, available at: http://cs.stanford.edu/people/czhang/zhang.thesis.pdf, 2015.

Zheng, Y., Geen, A. V., Anderson, R. F., Gardner, J. V., and Dean, W. E.: Intensification of the Northeast Pacific oxygen minimum zone during the Bølling-Allerød Warm Period, Paleoceanography, 15, 528–536, https://doi.org/10.1029/1999pa000473, 2000.

11

Biogeochemical data from terrestrial and aquatic ecosystems in a periglacial catchment, West Greenland

Tobias Lindborg[1,2], Johan Rydberg[3], Mats Tröjbom[4], Sten Berglund[5], Emma Johansson[1,6], Anders Löfgren[7], Peter Saetre[1], Sara Nordén[1], Gustav Sohlenius[8], Eva Andersson[1], Johannes Petrone[1], Micke Borgiel[9], Ulrik Kautsky[1], and Hjalmar Laudon[2]

[1]Swedish Nuclear Fuel and Waste Management Co. (SKB), P.O. Box 250, 101 24 Stockholm, Sweden

[2]Department of Forest Ecology and Management, Swedish University of Agricultural Sciences, 901 83 Umeå, Sweden

[3]Department of Ecology and Environmental Science, Umeå University, 901 87 Umeå, Sweden

[4]Mats Tröjbom Konsult AB, Slänningevägen 28, 761 72 Norrtälje, Sweden

[5]HydroResearch AB, Stora Marknadsvägen 15, 183 34 Täby, Sweden

[6]Department of Physical Geography, Bert Bolin Centre for Climate Research, Stockholm University, 106 91 Stockholm, Sweden

[7]EcoAnalytica, Slalomvägen 28, 129 49 Hägersten, Sweden

[8]Geological Survey of Sweden (SGU), P.O. Box 670, 751 28 Uppsala, Sweden

[9]Sveriges Vattenekologer AB, Marsjö Hvilan, 643 94 Vingåker, Sweden

Correspondence to: Tobias Lindborg (tobias.lindborg@skb.se)

Abstract. Global warming is expected to be most pronounced in the Arctic where permafrost thaw and release of old carbon may provide an important feedback mechanism to the climate system. To better understand and predict climate effects and feedbacks on the cycling of elements within and between ecosystems in northern latitude landscapes, a thorough understanding of the processes related to transport and cycling of elements is required. A fundamental requirement to reach a better process understanding is to have access to high-quality empirical data on chemical concentrations and biotic properties for a wide range of ecosystem domains and functional units (abiotic and biotic pools). The aim of this study is therefore to make one of the most extensive field data sets from a periglacial catchment readily available that can be used both to describe present-day periglacial processes and to improve predictions of the future. Here we present the sampling and analytical methods, field and laboratory equipment and the resulting biogeochemical data from a state-of-the-art whole-ecosystem investigation of the terrestrial and aquatic parts of a lake catchment in the Kangerlussuaq region, West Greenland. This data set allows for the calculation of whole-ecosystem mass balance budgets for a long list of elements, including carbon, nutrients and major and trace metals. The data set is freely available and can be downloaded from PANGAEA: doi:10.1594/PANGAEA.860961.

1 Introduction

Arctic environments, including permafrost-dominated periglacial landscapes, play a fundamental role in the ongoing global climate change (Smol, 2012; Schuur et al., 2015). Currently immobile pools of organic matter in currently frozen soils are expected to be mobilised due to permafrost thaw potentially resulting in dramatic changes in the rate of release of carbon, nutrients and metals (McGuire et al., 2009; Schuur et al., 2015). Therefore, a thorough understanding of the processes related to the transport and cycling of elements within and between terrestrial and aquatic ecosystems is key to describing and quantifying the effects of climate change. This understanding is also vital for our ability to facilitate reliable forecasts regarding future changes in biogeochemical cycling.

In order to describe the ongoing effects of climate related impacts on Arctic landscapes and to predict the future, it is essential to have access to data regarding chemical concentrations and biotic properties for a wide range of ecosystem domains and functional units (abiotic and biotic pools). There are several studies investigating different aspects of Arctic ecosystems, in West Greenland (e.g., Anderson et al., 2008; Ryves et al., 2009; Sobek et al., 2014; Jessen et al., 2014; Whiteford et al., 2016) and elsewhere (e.g., Chapin et al., 1995: Rysgaard and Glud, 2007; Smol and Douglas, 2007; Tarnocai et al., 2009; Rydberg et al., 2010; Palmtag et al., 2015; Hindshaw et al., 2016: Lundin et al., 2016). These studies provided crucial insights of high-latitude ecosystems that have increased our understanding of the significance and sensitivity of periglacial landscapes. However, no or very few studies have taken a whole-ecosystem approach and looked simultaneously at both the terrestrial and the aquatic ecosystems of the Arctic landscape. Hence, there is a need for studies that take such a holistic approach in order to better understand how biogeochemical processes are connected (Solomon et al., 2015).

In this work, we present biogeochemical data from a detailed whole-ecosystem investigation of the terrestrial and aquatic parts of a lake catchment in the Kangerlussuaq region, West Greenland. The lake is situated close to the Greenland ice sheet and is hereinafter referred to as "Two Boat Lake", or, for brevity, TBL; the lake is also known as SS903 (e.g., Saros et al., 2014; Sobek et al., 2014; Whiteford et al., 2016) or as Lake 26/L26 and "Talik Lake" (e.g., Anderson et al., 2001; Harper et al., 2011). The field investigation reported in this work was performed within the Greenland Analogue Surface Project (GRASP) and was conducted between 2010 and 2015. The overarching objective of the GRASP project was to describe the hydrology and ecosystem functions under periglacial conditions on a landscape scale. Further, GRASP aims at constraining the impact of climate change on such systems over long timescales, including one to several glacial cycles.

The specific aim of the field data collection and methodology presented here was to provide biogeochemical data that – together with hydrological data and carbon age determinations – can be used in mass balance models of various elements for a periglacial catchment. The data will also inform process understanding on element transport within and between catchment ecosystems. The specific objectives of the present work are to describe the study design, sample collection and analytical methods and to make the biogeochemical data from the TBL catchment collected within the GRASP project available to the scientific community.

This paper is structured as follows. First, we provide a short site description that outlines the main site-specific conditions and defines the site in a regional context. This is followed by a section containing a presentation of the general study design and the conceptualisation of the investigated landscape system, followed by a detailed description of all installations, sample collection and in situ measurements for each of the studied ecosystem units. The next section describes all analytical methods used in the chemical analyses, as well as the methods used for data processing. Finally, the results section briefly describes and exemplifies the data made available in the PANGAEA database (doi:10.1594/PANGAEA.860961).

2 Site description

The TBL catchment, described in more detail by Johansson et al. (2015a, b), is situated in close proximity to the Greenland ice sheet and approximately 30 km from the settlement of Kangerlussuaq, West Greenland (lat 67.125940, long −50.180370; Fig. 1). The catchment area is 1.56 km^2, of which the lake occupies 0.37 km^2. Although close to the ice margin, it is not part of the meltwater-driven surface water system (Fig. 2), and surface water runoff to the lake is generated by precipitation only. A digital elevation model (DEM) describing both the topography of the terrestrial part of the catchment and the bathymetry of the lake has been developed by Petrone et al. (2016; PANGAEA, doi:10.1594/PANGAEA.845258).

The Kangerlussuaq region is characterised by a hilly tundra landscape with numerous lakes and is the most extensive ice-free part of Greenland. Continuous permafrost, except where interrupted by through taliks under larger lakes, covers the area. An active layer of 0.15–5 m, depending on soil type and vegetation cover, overlies the permafrost in the area from the settlement of Kangerlussuaq up to the ice sheet margin (van Tatenhove and Olesen, 1994). Measurements in the TBL catchment indicate an active layer depth around 0.4–2.0 m (Petrone et al., 2016). Permafrost depths in the range 300–400 m have been observed from temperature measurements in deep bedrock boreholes 5–6 km from TBL (Harper et al., 2011).

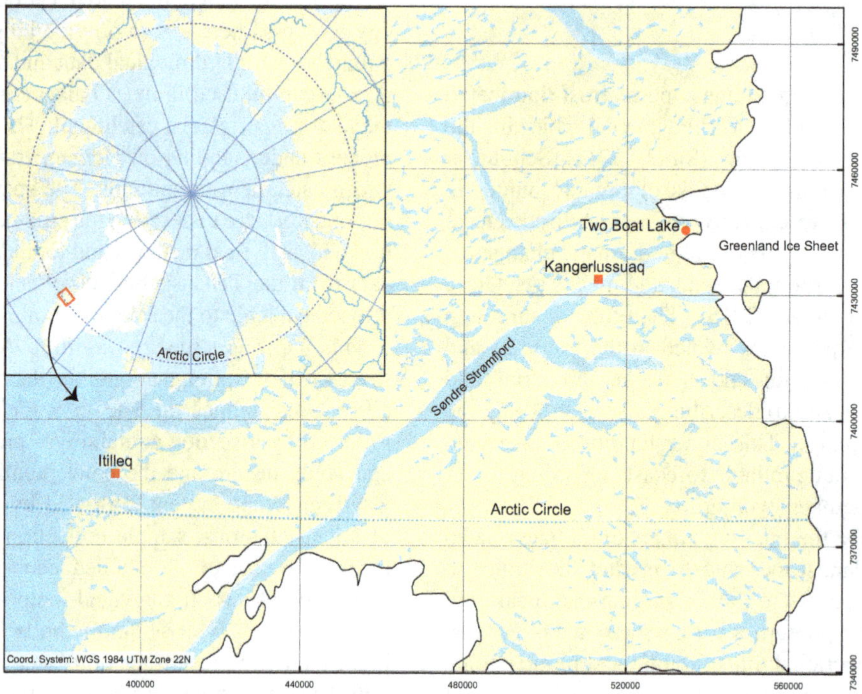

Figure 1. Map of West Greenland and location of Two Boat Lake catchment area. Modified from Johansson et al. (2015a).

Figure 2. Photo of the Two Boat Lake catchment heading north east. To the right the Greenland ice sheet is seen.

The regional climate is dominated by relatively stable high-pressure cells over the ice sheet and can be characterised as continental (Engels and Helmens, 2010). The mean annual air temperature (MAAT) in Kangerlussuaq is −5.1 °C (measured 1977–2015; Cappelen, 2016). The MAAT observed for the available data period 2011–2015 in the TBL catchment is −4.4 °C while the MAAT in Kangerlussuaq for the corresponding period is −5.3 °C, indicating somewhat warmer temperatures in the TBL area. The regional climate is dry with a mean annual precipitation (corrected for wind and adhesion losses) of 173 mm in Kangerlussuaq, with 67 mm falling as snow and 106 mm as rain (Johansson et al., 2015a).

The hydrology of the TBL catchment has been extensively studied during the period 2010–2015. The meteorological and hydrological data collected by GRASP at TBL are presented in Johansson et al. (2015a; doi:10.1594/PANGAEA.836178). Johansson et al. (2015b) developed a detailed water balance model of the TBL catchment. From a decadal perspective precipitation barely exceeds evapotranspiration, and the net annual water budget is alternating between positive and negative. According to these modelling results, the lake has had a surface water outflow in 1 out of 3 years during the last 30 years. Thus, TBL has no continuous surface outflow, and it may take several years between outflow events. Surface outflow from the lake was last observed during the summer of 2009 (Johansson et al., 2015b). There are no permanent streams in the catchment, which means that ice wedges provide the major transport routes of water from the catchment to the lake. The snow melt period (May–June) accounts for almost half of the inflow of water to the lake.

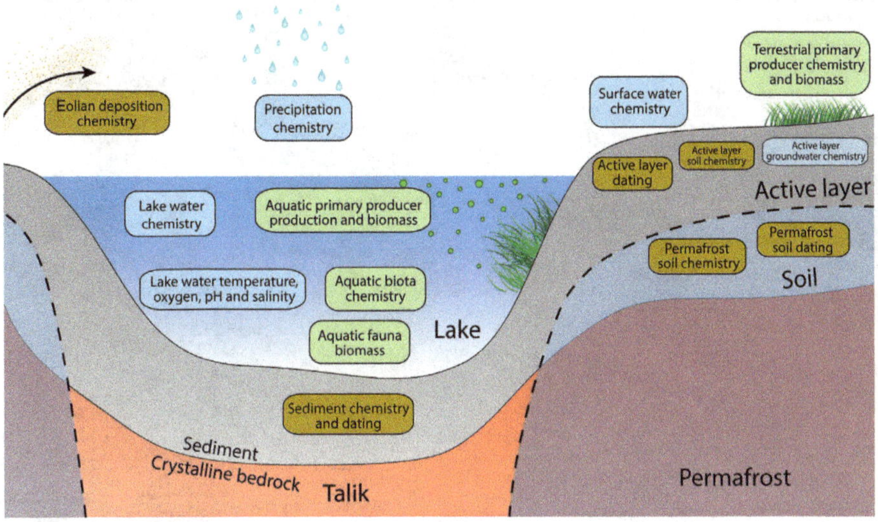

Figure 3. Conceptual model of the TBL catchment describing major ecosystem entities (boxes) in the landscape. Boxes are divided into water (blue), biotic (green) and solid matter (brown).

Maximum and average water depths in the lake are 29.9 and 11.3 m, respectively, and lake volume is 4.29 Mm^3. Secchi depth is about 15 m, indicating a photic depth of around 30 m. A thick mat of *Chara* sp. and Bryophytes dominates the lake bottom down to 14 m depth. From a lake water chemistry and morphometry perspective, TBL is typical for precipitation-driven lakes close to the ice margin in the Kangerlussuaq area (Ryves et al., 2002; Clarhäll, 2011; Andersson et al., 2009; Sobek et al., 2014).

A model describing the spatial distribution of soil depth and active layer thickness was developed by Petrone et al. (2016). Soils in the TBL catchment are dominated by till and glaciofluvial deposits, which to a large extent are overlain by a 0.3–0.4 m thick layer of eolian silt to fine sand (Rydberg et al., 2016; Petrone et al., 2016). With the exception of areas with exposed bedrock or till, most of the catchment has a sparse but continuous vegetation cover. Meadow, dry grass and dwarf-shrub communities dominate the vegetation. There are no trees, and shrubs rarely exceed heights of 0.5 m. The vegetation has been classified based on field surveys and aerial photos, and a vegetation map is available (Clarhäll, 2011).

3 Installations, sampling and in situ measurements

The overall study design and data collection protocol for TBL were based on a conceptual model describing the different functional units of the ecosystem (Fig. 3). By collecting biogeochemical data for all identified ecosystem units in the conceptual model and then combining them with the existing hydrological model (Johansson et al., 2015b) and geometrical information (Petrone et al., 2016), the purpose was to make it possible to develop numerical models describing the fluxes and accumulation of various elements on a catchment scale.

No specific target element was chosen for this work, but rather the aim has been to provide understanding of the general fluxes of dissolved and particulate matter in the system. Therefore, we analysed concentrations of carbon, nitrogen and phosphorus – together with their associated species and isotopic composition – as well as total concentrations of a long list of major and trace elements, which together with the hydrological model can help us to understand the fluxes in the landscape. Through the age determinations of different soil and sediment layers, together with estimates of biomass and primary production in both the terrestrial and aquatic systems, we also wanted to make it possible to capture the accumulation of various elements in different landscape units.

3.1 Water in terrestrial areas

3.1.1 Lysimeters

In August 2011, soil lysimeters were installed in three transects, each with two or five locations (Fig. 4a). At each location, two to five lysimeters – installed at approximately every 10 cm starting from 10 cm below the soil surface – form a depth profile down to the permafrost at about 35 to 70 cm soil depth (Fig. 5a and b). Lysimeters were installed at the same sites as the time-domain reflectometer instruments that were used to measure soil water content in the hydrological studies performed by Johansson et al. (2015a, b). Two of the transects are located in the wetter areas with peaty silt (transect 2 and 3), while the third transect (transect 1) is located in a drier area with eolian silt and fine sand (Fig. 4c; see also Rydberg et al., 2016). The lysimeters (10 cm long, 2 cm in diameter, with an attached hose) were made of porous PTFE

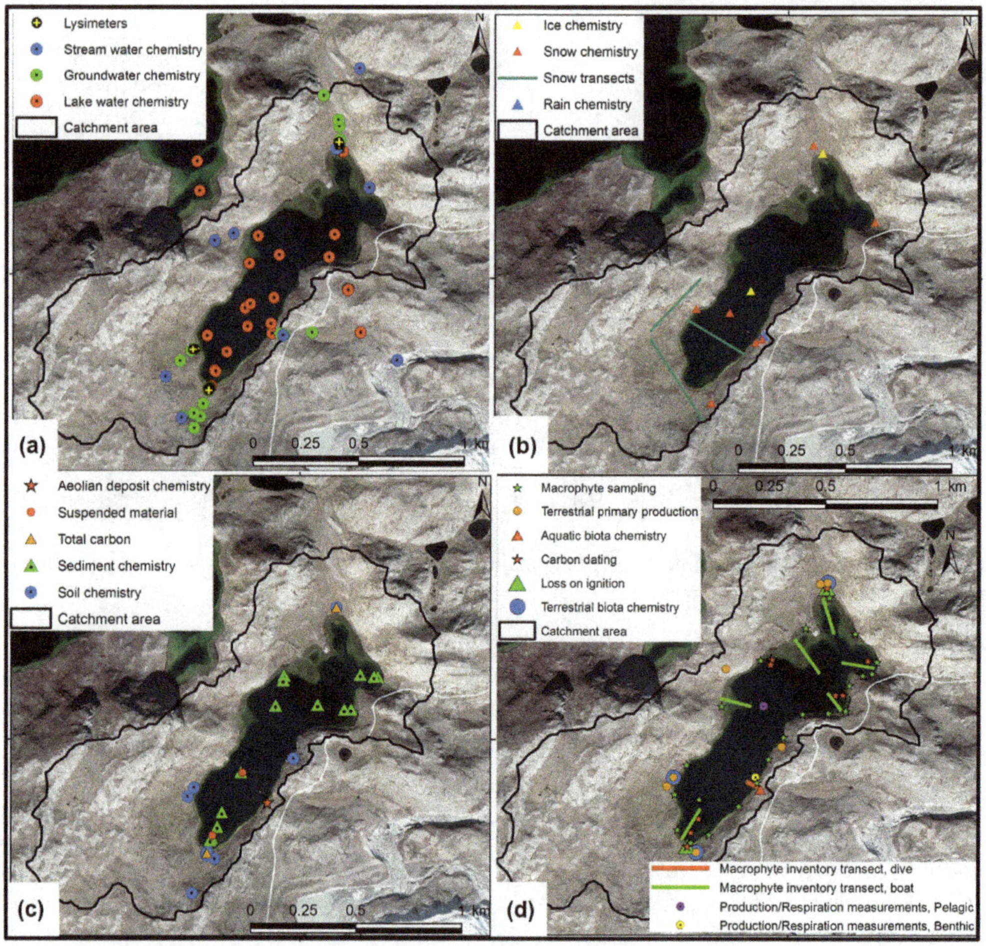

Figure 4. Maps illustrating sampling points, installations and measurement sites in the TBL catchment for groundwater and surface water **(a)**, ice and precipitation **(b)**, soils and sediment **(c)** and biota, carbon dating and soil loss on ignition **(d)**. For information on coordinates and ID for each sampling location, see PANGAEA database (http://doi.pangaea.de/10.1594/PANGAEA.860961).

mixed with silica flour (pore size of 1 μm) similar to Petrone et al. (2007). Installation was done by digging a pit and inserting the lysimeters into the unaffected upslope wall of the pit to ensure a minimum impact on soil and hydrological conditions. The pit was then refilled in stratigraphic order.

In August or September each year from 2011 to 2014, water samples were collected from the lysimeters by attaching glass flasks (Fig. 5a) with vacuum to the plastic tube leading to the lysimeter. The flasks were then left for 12 h to extract water through suction from the lysimeters. No additional filtering was made, except for the filtering effect of the lysimeter itself. Samples were kept frozen awaiting analysis.

3.1.2 Soil water wells and surface water

In August or September from 2011 to 2014, soil water was collected from 15 soil water wells, which are described in detail in Johansson et al. (2015a; see also Fig. 4a). Water was sampled through a plastic tube attached to a pump, and was collected in plastic flasks. In addition, water in small temporary ponds and ice wedges was sampled by hand-filling plastic flasks (Fig. 4a). Both unfiltered and filtered (0.45 μm) samples were collected. Samples were kept frozen awaiting analysis.

3.2 Lake water

3.2.1 Water and ice sampling

In April 2011–2014, June 2013–2014 and August/September from 2011 to 2014, lake water samples were collected using a Ruttner sampler (Figs. 4, 6). Both filtered (0.45 μm) and unfiltered samples were collected in plastic flasks. In April 2011, 13 locations distributed across the lake were sampled at two depths (5 m below the surface and 5 m above the sediment surface) in order to make a first evaluation of the spatial variations in the lake water chemistry (Fig. 4a). This evaluation showed very small differences between sites but considerable differences between the two sampled depths. In

Figure 5. Photos of lysimeter flasks **(a)**, lysimeter installation (left in the pit) **(b)** and respiration chambers for benthic production/respiration measurements **(c, d)**.

Figure 6. Lake water sampling on the lake ice using a Ruttner sampler.

light of this, subsequent lake water sampling was conducted at 5 and 25 m water depth at the centre of TBL (Fig. 4a). In August 2011 and August 2014, additional water samples were collected at temporary lake inlets and at the lake outlet (Fig. 4a). In winter (April) an ice auger was used to penetrate the ice (1.5–1.7 m thick, sampling points shown in Fig. 4b). During drilling, lake ice samples were collected in plastic bags, thawed and transferred to plastic flasks. Samples were kept frozen awaiting analysis.

3.2.2 In situ measurements

In April, June and September from 2012 to 2014, a TROLL 9500 multi-parameter probe was used for in situ measurements of pH, temperature, dissolved oxygen and alkalinity profiles in the water column at the lake centre. The probe was lowered in steps with an average speed of about 1 m per minute, and a field computer (Panasonic Toughbook) was used for online monitoring to ensure that the probe had stabilised before lowering it further. The multi-parameter probe was maintained and calibrated regularly following the instruction in the TROLL 9500 operator's manual.

During the period from 15 June 2013 to 23 August 2014 (except for the period from 22 March 2014 to 26 June 2014), continuous measurements of oxygen saturation and temperature were conducted at two depths close to the lake outlet using HOBO U26-001 loggers. During the first part of the measurement period the loggers were placed at 5 and 17 m water depth, but when they were redeployed after maintenance on the 23 August 2013 the loggers were placed at 8 and 20 m water depth.

Additionally, water temperature was measured at approximately 4.3 m water depth by two Leveltroll 700 transducers (accuracy of $\pm 0.1\,\%$) that were placed directly on the

Figure 7. Photos of snow with inclusions of eolian deposits **(a, b)**, soil sampling in steel cylinders **(c)** and sediment core from the lake bottom **(d)**.

sediment surface. These loggers were also used to monitor changes in lake water level by Johansson et al. (2015a). Small variations in measurement depth were due to slight changes in the positions of the transducers following annual data retrieval and maintenance. Temperature was logged every third hour from August 2011 to August 2015.

3.3 Precipitation and deposition

In June and August 2013 and 2014, i.e., during all precipitation events that coincided with our sampling campaigns, precipitation samples were collected using nine white plastic pans (34 × 22 × 4.5 cm). The pans were placed on the ground close to the lake outlet (Fig. 4b) during precipitation events, and the water from each pan was mixed and transferred unfiltered to plastic flasks after 1 h of collection. Samples were kept frozen while awaiting analysis.

In April 2011, snow samples were collected every 10th metre along three transects using a standardised snow tube sampler (4.5 cm inner diameter; Fig. 4b). The collected samples were pooled and thawed, and a subsample from each transect was transferred unfiltered to a plastic flask. Samples were kept frozen while awaiting analysis.

In April 2012 and 2013, old snow packs in sheltered locations were sampled to get an estimate of the eolian deposition (Fig. 4b). In these snow packs the accumulated eolian material is clearly visible as dark layers (Fig. 7a and b). Samples were collected using a standardised snow tube (inner diameter 4.5 cm), and the samples were thawed and transferred unfiltered to plastic flasks. Samples were kept frozen while awaiting analysis.

3.4 Soils in terrestrial areas

During installation of lysimeters in August 2011, soil samples with known volumes were collected using steel cylinders. Samples were collected at 10 cm depth intervals from 10 cm below the soil surface to the permafrost. In 2011 soil samples were also collected from different soil units using a steel spade (Fig. 4c, Soil chemistry). These soil samples were oven dried and sieved into 0–35, 35–63, 63–125, 125–250 and >250 µm fractions.

During the 2014 campaign, the top soils in areas of each vegetation type were sampled using a soil corer with a diameter of 10 cm (Fig. 7c). Wetland and birch types were sampled as duplicates (approximately 1 m apart). At each location living plants and litter were removed before sample collection, and the top 12 cm of the core was divided into four 3 cm sections and stored in plastic bags. Each sample was dried to constant weight at 35 °C, homogenised and passed through a 2 mm sieve to remove roots >2 mm; this belowground biomass was collected and weighed. In 2015, additional soil profiles were collected from two locations (triplicates) in 10 cm sections down to 40 or 50 cm soil depth. These samples were analysed as bulk samples without the removal of roots.

In August 2014, soil samples extending down into the permafrost were collected at two locations with peaty silt – one at the southern end and one in the northern end of the lake

Table 1. The five main vegetation types and the vegetation-free barren land within the catchment area and their respective areal coverage. The number of investigated plots (n) in August 2011 and short characterisations of the vegetation types are also given.

Vegetation type	Area (m^2)	n	Vegetation
Dry grassland	231 655	1	Mainly tufted species from the genus *Kobresia*/*Carex* with a prominent litter layer found on exposed places.
Meadow	445 814	6	Mainly grasses and herbs almost lacking any woody species. Also some lichens and bryophytes.
Betula	156 056	5	Birch (*Betula nana*) dominating in dense stands with a dominating litter layer on south-facing slopes and a dominating bryophyte layer on north-facing slopes.
Wetland	83 975	3	*Carex* and grasses dominate the field layer and bryophytes cover the bottom layer.
Salix	6278	0	*Salix* dominating but kept close to ground, not reaching more than a height of app. 0.8 m.
Barren	257 431	0	Scares of exposed moraine/till, where the silt layer has been removed by the wind. Void of vegetation.

using a soil coring device in steel that was driven down into the soil using a sledge hammer. At the southern end the profile extended down to 110 cm (not reaching the underlying till), and at the northern location it extended down to 60 cm and into the underlying till.

3.5 Lake sediments

In August 2011, surface sediment cores were collected from the centre of the lake and near the shore in the southern part of the lake using a HTH gravity corer (Renberg and Hansson, 2008). The 25 cm long core from the lake centre was sectioned into 1 cm sections down to 15 cm and then into 5 cm sections, while the 29 cm long core from the other location was sectioned into 5 cm sections throughout its entire length. All samples were placed in plastic bags and stored and transported cold. Upon arriving at the analysis laboratory the samples from the lake centre core were dried and weighted to determine the bulk density.

In August 2012, 13 additional surface sediment cores (23–40 cm long) were collected in four transects using a KC Kajak corer (Figs. 4c, 7d). Sampling locations were selected to provide a good representation of the lakes sediments, in terms of differences in bottom type (vegetated and non-vegetated areas), location in the lake and water depth. All cores were subsectioned in the field as follows: 0–10 cm in 1 cm sections, 10–20 cm in 2 cm sections and below 20 cm in 4 cm sections. The samples were placed in plastic containers, and stored and transported cold. Upon arrival at the laboratory the samples were frozen and freeze dried.

3.6 Biota

3.6.1 Terrestrial biota

In August 2011 – during peak biomass season – the areal extents of the five main vegetation types in the TBL catchment (*Betula*, *Salix*, dry grassland, meadow and wetland) were determined (Table 1). The observed pattern coincided well with an orthophoto taken at the same time of the year. For each of the vegetation types (except *Salix*, due to a low total coverage) a main plot (0.5 × 0.5 m) and replicate plots (0.2 × 0.2 m; not for dry grassland) were selected (Fig. 4d, "terrestrial primary production"). All living above-ground biomass and litter in each plot were sampled through destructive sampling. In the main plots, the living biomass was separated into functional groups (herbs/grasses, shrubs (wood and green tissue), lichens, bryophytes), whereas no separation was done in the replicate plots. Litter was separated into standing or attached to living biomass and litter found on the ground.

In late June 2014, additional biomass samples were collected from the four dominating vegetation types (*Betula*, dry grassland, meadow and wetland) in order to assess respiration by comparing the carbon pools in living biomass, litter and soil organic matter (see Sect. 3.4 regarding the soil sampling). For each vegetation type, three 170 cm^2 circular areas within a 1 m by 1 m plot were sampled destructively for both living biomass and litter; for the *Betula* and wetland vegetation types, one additional replicate plot each was sampled. Similarly to the 2011 sampling, the living biomass was divided into functional groups (herbs/grasses, shrubs (wood and green tissue), lichens, bryophytes); however, in the 2014 sampling campaign litter was retained as a single functional group.

In August 2011 and 2014, a smaller number of terrestrial biota samples were collected for chemical analysis. Litter, wood and living parts of *Betula*, and dead parts of bryophytes were collected from different locations (Fig. 4d). Pooled samples of the dominating field layer species were collected at one location close to the northern lysimeter transect and one close to the weather station (Fig. 4d). Sampling was conducted using a pair of scissors. The northern site was dominated by *Calamagrostis* and other genus within *Poaceae* and *Carex capillaris*, with smaller contribution of *Salix arctica/glauca*, *Vaccinium myrtillus* and *Carex norveica*. The southern site was dominated by two species of *Calamagrostis* and a *Poa* sp., with a small contribution of *Carex scirpoides*. Muscle tissue from muskox (*Ovibos moschatus*) was also provided by a local hunter.

3.6.2 Aquatic biota

Water was collected with a Ruttner sampler at the lake centre for quantitative samples for phytoplankton, bacterioplankton and zooplankton (Fig. 4d). In August 2011, August 2012 and April 2012, phytoplankton and bacterioplankton samples were collected from the epilimnion (at the surface and at 5 m water depth) and from the hypolimnion (20 m water depth). In August 2011, August 2012 and June 2013, zooplankton samples were collected from 5 m water depth with a Ruttner sampler (6 L of water was filtered through a net with 65 µm mesh size). Samples for phytoplankton and zooplankton were preserved with an acidic Lugols solution according to Olrik et al. (1998), and bacterioplankton samples were preserved with formaldehyde.

In August 2011 and August 2012, benthic bacteria were sampled with a HTH-corer (diameter 8.3 cm; Renberg and Hansson, 2008) from six locations: three deep and three shallow (Fig. 4c). The 0–1 cm sections of the cores were mixed, and a 1 mL subsample was preserved with formaldehyde. In August 2011, the areal coverage of aquatic macrophytes was mapped from a boat along six transects and by walking along the shore, and the total biomass of aquatic benthic macrophytes was estimated at 21 locations. The estimates were done by collecting, drying and weighing all macrophytes from 0.12 m^2 plots (Fig. 4d). In June 2013, five additional 0.12 m^2 plots were sampled in the same way from the macrophytes dive transect (see Sect. 3.6.4). In addition, biomass was quantitatively measured in nine production/respiration chambers (see Sect. 3.6.3).

In August 2012 and June 2013, benthic fauna was sampled both qualitatively and quantitatively. For the qualitative sampling, 13 samples were taken in August 2012 and 5 in June 2013 with an Ekman grabber (16 × 16 cm) and sieved through 0.5 mm. In addition to the species found in the quantitative sampling, qualitative sampling was also conducted in shallow areas using a sieve and by turning stones.

In August or September 2011, 2012 and 2014, samples of aquatic biota for chemical analysis were collected. Living tissue of vascular plants and macroalgae (*Chara* sp., *Cladophora* sp.) was collected along the lake shore. *Nostoc* sp. and benthic fauna (Chironomidae, Oligochaeta, Ostracoda) were sampled using an Ekman grabber. *Lymnea* sp., Dytiscidae, Odonata, *Branchinecta paludosa* and *Lepidurus arcticus* were sampled from the lake shore.

All biota samples were frozen before transport to the laboratory.

3.6.3 Aquatic production/respiration measurements

In June and August 2013, benthic production/respiration experiments were performed in plexiglas chambers with a diameter of 0.29 m and a volume of 0.017 m^3 (volume above sediment) placed at the lake bottom. Three chambers, each at 0.7, 6 and 12 m water depth, were placed in June and five chambers, at 0.7 m depth, in August (Figs. 4d, 5c and d). Oxygen and temperature were measured at the start of the incubation, after 2 h and at a few occasions during the following day. Primary production and respiration were calculated based on the oxygen concentration difference before and after the incubation. After the incubation, all vegetation in the chamber was sampled, dried and weighted to allow for biomass quantification (cf. Sect. 3.6.2).

In August 2012 and June 2013, pelagic production/respiration was estimated by incubating 100 mL glass bottles (two transparent and one dark bottle for each depth; Fig. 4d). Production of macroalgae (*Chara* sp. and *Cladophora* sp.) was also measured in 100 mL glass bottles in August 2012. For pelagic production/respiration measurements, glass bottles with unfiltered lake water were incubated at 0.1, 0.25, 0.5, 1, 2, 3, 5, 10, 15 and 20 m water depth. For *Chara* sp. and *Cladophora* sp., pieces of algae were placed in 100 mL glass bottles filled with filtered (0.45 µm) water from the same depths as the incubation, *Chara* sp. was incubated at 1, 2, 5 and 10 m water depth and *Cladophora* sp. at 5 m water depth. Phytoplankton, *Chara* sp. and *Cladophora* sp. were incubated for 4 h around midday. The oxygen concentration and temperature in the bottles were measured before and after the incubation using a HANNA HI 9146 oxygen sensor.

3.6.4 Macrophytes inventory transect

In June 2013 observation of spatial coverage and collection of benthic primary production biomass were conducted at the lake. (Fig. 4d). A dive survey was performed along a transect perpendicular to the depth gradient following the standard of the national environmental monitoring in Sweden (Fig. 8); see Moy et al. (2010) for a detailed description and a comparison with other methodologies. In brief, the divers started from the deeper end of the transect and followed a measuring tape towards the shoreline. Along the way, the divers noted the distance and depth together with bottom type, species composition (macrophytes) and their respective surface coverage on a seven-point scale, i.e., 1, 5, 10, 25, 50, 75 and 100 % coverage. Distance, depth and species coverage were noted every time a change occurred in the bottom substrate, species composition or surface coverage. Estimation of coverage was done for a 6–10 m wide corridor (i.e., 3–5 m on each side of the tape). This resulted in a detailed description of the bottom structure, the coverage of various species and how they changed with depth and distance from the shoreline.

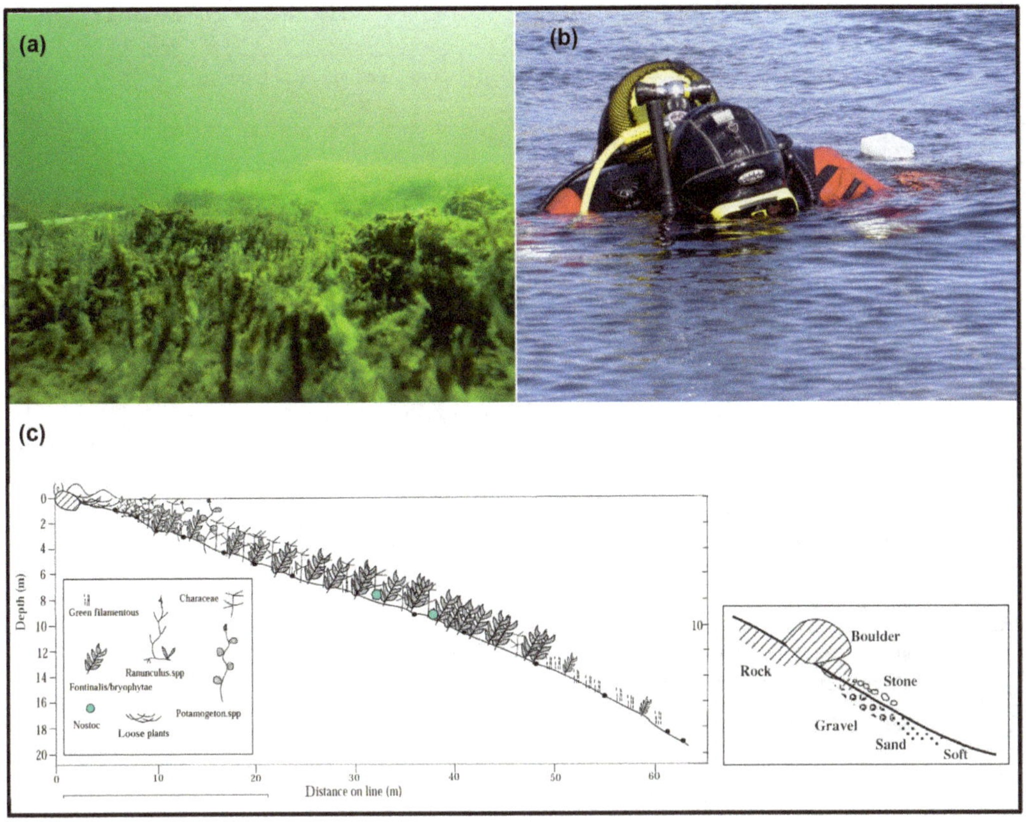

Figure 8. Photos from the macrophyte inventory **(a, b)** and sketches from the resulting inventory transect **(c)**.

4 Chemical analyses and other data processing

4.1 Major and trace elements in water

The elemental concentrations of major and trace elements (Al, Ca, Fe, K, Mg, Mn, Na, P, Si, Ti, Ag, As, B, Ba, Be, Cd, Co, Cr, Cs, Cu, Ga, Ge, Hf, Hg, Li, Mo, Nb, Ni, Pb, Rb, S, Sb, Sc, Se, Sn, Sr, Ta, Th, Tl, U, V, W, Y, Zn, Zr, Ce, La, Pr, Nd, Sm, Eu, Gd, Tb, Dy, Ho, Er, Tm, Yb and Lu) in liquid samples were determined using inductively coupled plasma-sector field mass spectroscopy (ICP-SFMS) (Table 2). All elements were analysed quantitatively on pre-filtered samples. Halogenes (Cl, Br, I) were analysed separately under alkaline conditions. Replicate analysis ($n = 8$) shows that most elements were within either 10 % or 0.01 $\mu g\,L^{-1}$ (calculated as 1 standard deviation). Elements with larger deviations were Cu (0.6 $\mu g\,L^{-1}$), Fe (0.3 $\mu g\,L^{-1}$), P (0.3 $\mu g\,L^{-1}$), Pb (0.03 $\mu g\,L^{-1}$), Ti (0.05 $\mu g\,L^{-1}$), W (0.07 $\mu g\,L^{-1}$) and Zn (0.7 $\mu g\,L^{-1}$).

Silica in water was analysed by allowing the silica to react with a molybdate reagent, reduced by ascorbic acid to form molybdenum blue and quantified using segmented flow analysis (SFA) following the procedure outlined in OI Analytical methodology part no. 319529. In a small number of samples, the concentrations of halogens (Br^-, Cl^-, F^-) and sulfate (SO_4^{2-}) were determined using liquid ion chromatography (DIN EN ISO 10304-1/-2; Table 2).

Elemental concentrations of *Al*, *Ca*, Fe, K, *Mg*, Mn, Na, P, *Si*, Ti, Ag, As, B, Ba, Be, Cd, Co, Cr, Cs, Cu, Ga, *Ge*, Hf, Hg, Li, Mo, Nb, Ni, Pb, Rb, S, Sb, Sc, Se, Sn, Sr, *Ta*, Th, Tl, U, V, W, Y, Zn, *Zr*, Ce, La, Pr, Nd, Sm, Eu, Gd, Tb, Dy, Ho, Er, Tm, Yb and Lu in suspended solids trapped on acid washed filters (0.22 µm) were determined using ICP-SFMS (Table 2). The filters and suspended solids were digested in strong acid (HNO_3 : HF) in closed Teflon vessels (losses might occur for elements denoted in italic). Concentrations in suspended solids were corrected for dissolved substances by including the analysis of filters soaked in pre-filtered water and are reported as $\mu g\,L^{-1}$.

4.2 Major and trace elements in solid samples

Major and trace elements in solid samples were analysed using either wavelength dispersive X-ray fluorescence spectroscopy (WD-XRF) on loose-powder samples, following the method described by Rydberg (2014), or ICP-SFMS, following either digestion (total concentrations) or leaching (extractable concentrations) (Table 2). All samples analysed using WD-XRF were pulverised using a Retsch swing mill (2 min at 30 Hz) prior to analysis.

Table 2. Overview of the laboratories and analytical methods that have been used to produce the presented data set.

Parameter	Sample type	Laboratory – analytical method		
Organic content	Solid	EMG, UmU – LOI	SLU, Uppsala – LOI	ALS Scandinavia – LOI
Tot-C	Solid	SLU, Umeå – EA-IRMS	SLU, Uppsala – EA	ALS Scandinavia – EA
TOC, DOC	Water	DEEP, SU – NPOC	ALS Scandinavia – NPOC/TC-IC	
TIC, DIC	Water	DEEP, SU – IR after acidification	ALS Scandinavia – IR after acidification	
$\delta^{13}C$	Solid Water	SLU, Umeå – EA-IRMS IGV, SU – IRMS	ISO-Analytical – EA-IRMS	
Tot-N	Solid Water	SLU, Umeå – EA-IRMS DEEP, SU – SFA	SLU, Uppsala – EA ALS Scandinavia – CFA	ALS Scandinavia – EA
NH_4^+, NO_2^-, NO_3^-	Solid Water	ALS Scandinavia – IC DEEP, SU – SFA	 ALS Scandinavia – CFA	
$\delta^{15}N$	Solid	SLU, Umeå – EA-IRMS	ISO-Analytical – EA-IRMS	
Tot-P	Water Filters	DEEP, SU – SFA DEEP, SU – combustion SFA		
PO_4^{3-}	Solid Water	ALS Scandinavia – CFA DEEP, SU – SFA		
Chlorophyll *a*	Filters	DEEP, SU – spectrometry		
Major + trace elements	Solid Water	EMG, UmU – WD-XRF ALS Scandinavia – ICP-SFMS	ALS Scandinavia – ICP-SFMS	
Si	Water	DEEP, SU – SFA		
Br^-, Cl^-, F^-	Solid Water	ALS Scandinavia – IC (F only) ALS Scandinavia – IC		
SO_4^{2-}	Solid Water	ALS Scandinavia – IC ALS Scandinavia – IC		
$\delta^{34}S$	Water	IGV, SU – EA-IRMS		
$\delta^{2}H$	Water	SLU, Umeå – liquid water isotope analyzer		
$\delta^{18}O$	Water	SLU, Umeå – liquid water isotope analyzer		
$^{234}U:^{235}U$	Water	FOI – ICP-SFMS		
^{14}C	Solid	^{14}C dating laboratory, LU – AMS		
^{210}Pb, ^{137}Cs, ^{226}Ra etc.	Solid	Flett Research – alpha spectroscopy (^{210}Pb only)	Nutech, DTU – gamma spectroscopy	

EMG – Dept. of Ecology and Environmental Science; LOI – loss on ignition; SFA – segmented flow analysis; SLU – Swedish University of Agricultural Sciences; EA – elemental analysis; CFA – continuous-flow analysis; DEEP – Dept. of Ecology, Environment and Plant Science; IRMS – isotope-ratio mass spectroscopy; IC – liquid ion chromatography; IGV – Dept. of Geological Sciences; NPOC – non-purgeable organic carbon; XRF – X-ray fluorescence spectroscopy; FOI – Swedish defense research agency; TC–IC – total carbon–inorganic carbon; ICP-SFMS – inductively coupled plasma-sector field mass spectroscopy; IR – IR detector; AMS – accelerated mass spectroscopy.

Total concentrations were determined following either lithium borate ($LiBO_2$) fusion (Al, Ba, Be, Ca, Ce, Cr, Dy, Er, Eu, Fe, Ga, Gd, Hf, Ho, K, La, Lu, Na, Nb, Nd, P, Pr, Rb, Sc, Si, Sm, Sr, Ta, Tb, Th, Ti, Tm, U, W, V, Y, Tb, Zr), strong acid ($HNO_3:HF$) digestion (Ag, As, Au, B, Bi, Cd, Co, Cs, Cu, Hg, Ir, Li, Mo, Ni, Os, Pb, Pd, Pt, Re, Rh, Ru, S, Sb, Se, Sn, Te, Tl, Zn) or sintering at 550 °C with ZnO and $NaCO_3$ (Cl, Br, I). For the analysis of extractable metals, the samples were leached in ammonium acetate solution (NH_4Ac-CH_4COO) buffered to pH 4.5 for 16 h. Methane was added to enhance the analytical performance for Ag and Pd, and halogens (Cl, Br, I) were analysed separately in an alkaline environment. For samples analysed in 2011–2013, Al, As, Ba, Be, Br, Ca, Cd, Cl, Co, Cr, Cs, Cu, Fe, I, K, Li, Mg,

Mn, Mo, Na, Nb, Ni, P, Pb, Pd, S, Se, Si, Sn, Sr, Ti, U, V and Zn were analysed quantitatively, while the remaining elements were analysed semi-quantitatively (i.e., the reported data are based on the relative analytical response for the element; however, because the response is not fully calibrated it only represents a relative measure of the concentration). In 2014, all elements were analysed quantitatively.

For the WD-XRF analysis, lower limits of detection range from a few $\mu g\,g^{-1}$ for trace elements to a few tens of $\mu g\,g^{-1}$ for major elements. Accuracy was assessed using 10 different certified reference materials ($n = 5$) and were within $\pm 10\,\%$ (or a few $\mu g\,g^{-1}$) for all elements except S, which had an accuracy of about 20 % (Rydberg, 2014). Precision, based on replicate measurements of TBL samples ($n = 30$), was always better than 5 % (or $\pm 1\,\mu g\,g^{-1}$ for elements present in low concentrations). Fluoride (F^-) and sulfate (SO_4^{2-}) in solid samples were analysed by liquid ion chromatography (Table 2), following the procedure outlined in DIN EN ISO 10304-1.

4.3 Carbon and nutrients in water

Organic carbon (OC), inorganic carbon (IC), nitrogen (total N, NH_4–N, $NO_2 + NO_3$–N) and phosphorus (total P; PO_4–P) in water samples were analysed (Table 2). Total organic carbon (TOC) and dissolved organic carbon (DOC) were determined by thermal combustion (680 °C) following acidification (to remove any IC), while IC was analysed by converting all IC to CO_2 in a reactor vessel filled with 25 % H_3PO_4 (AnalyticJena Multi N/C-3100).

Nitrogen and phosphorus, as well as their different species, were determined using SFA following the procedures outlined in OI Analytical methodology part no. 319526 (NH_4–N), no. 319527 (total N, $NO_2 + NO_3$–N) and no. 319528 (total P, PO_4–P). The reported analytical uncertainty was less than 10 %. Total C, TOC, DOC, total inorganic carbon (TIC) and dissolved inorganic carbon (DIC) were analysed following the procedures outlined in DIN EN ISO 1484-H3. Total N analysis was made after oxidative digestion with peroxydisulfate (DIN EN ISO 11905-1:1997), NO_3–N was analysed through ion chromatography (DIN EN ISO 10304-1/-2-D19) and NO_2–N and NH_4–N were analysed through continuous-flow analysis and spectrometric detection (DIN EN ISO 13395-D28 and DIN EN ISO 11732-E23, respectively).

In addition to the total and dissolved concentrations described above, chlorophyll *a* (Chl *a*) and phosphorus associated with the particulate fraction were analysed (Table 2). In brief, known volumes of water were filtered through 0.45 µm cellulose acetate filters. For Chl *a* analysis the Chl *a* was extracted from the filters using ethanol (96 %) and the Chl *a* concentration in the solution was measured using a spectrophotometer (Helcom Combine manual; Helcom, 2016, Annex C-4) and phosphorus on the filters was determined using SFA following combustion (OI Analytical methodology part no. 319528). The reported analytical uncertainty was 25 % for the Chl *a* measurements and less than 5 % for C, N and P.

4.4 Organic material and nutrients in solid samples

The organic contents of solid samples were estimated by either total carbon analysis or loss on ignition (LOI). A smaller number of soil samples were analysed using both approaches to allow for comparison between the two methods. Total carbon and nitrogen concentrations were determined either separately using an elemental analyser or simultaneously as the analysis of stable C and N isotopes (see further details below; Table 2).

By comparing the carbon contents in acid-treated samples with those in samples not treated with acid, Rydberg et al. (2016) concluded that inorganic carbon contents in the sediments of TBL are negligible, implying that total carbon in the sediment equals total organic carbon. LOI samples were first dried at 105 °C to constant weight, then burnt in a muffle furnace at 550 °C for 4 h and reweighed.

Nitrate and phosphate in solid samples were determined following the procedure outlined in DIN EN ISO 10304-1/-2-D19/20 and DIN EN ISO 15681-2 (D46), respectively.

4.5 Stable isotopes

Carbon, nitrogen and their stable isotopes in solid samples were determined simultaneously using an elemental analyser coupled with an isotope-ratio mass spectrometer (EA-IRMS) (Table 2). The data were corrected for instrumental drift and sample-size effects using internal standards, and $\delta^{13}C$ and $\delta^{15}N$ were calculated with reference to Vienna Pee Dee Belemnite (VPDB) and atmospheric air, respectively. Analytical performance was assessed using different internal standards that are regularly checked against certified reference materials. The standard deviation of all internal standards ($n = 42$ and $n = 10$ for SLU and ISO-Analytical, respectively) were always less than $\pm 0.7\,\%$ for C, $\pm 0.3\,‰$ for $\delta^{13}C$, $\pm 0.1\,\%$ for N and $\pm 0.2\,‰$ for $\delta^{15}N$ for SLU and less than $\pm 2.5\,\%$ for C, $\pm 0.1\,‰$ for $\delta^{13}C$, $\pm 0.4\,\%$ for N and $\pm 0.2\,‰$ for $\delta^{15}N$ for ISO-Analytical Ltd.

In all water samples stable carbon isotopic composition in DIC was determined by injecting 5 mL of water into 12 mL septum-sealed glass vials in the field (Table 2). The glass vials had been prefilled with 100 µL of 85 % H_3PO_4 and flushed with helium gas (100 mL min^{-1}) for 5 min, and the isotopic composition of the head space was determined using a Gasbench II coupled to a continuous-flow IRMS (Thermo scientific MAT 253). The isotopic composition, $\delta^{13}C$, was calculated with reference to VPDB, and the reproducibility of the analysis was reported to be better than 0.1 ‰.

Analyses of hydrogen and oxygen isotopes in water samples were conducted using a liquid water isotope analyser (Picarro L1102-i). The $\delta^{18}O$ and $\delta^{2}H$ values were calculated

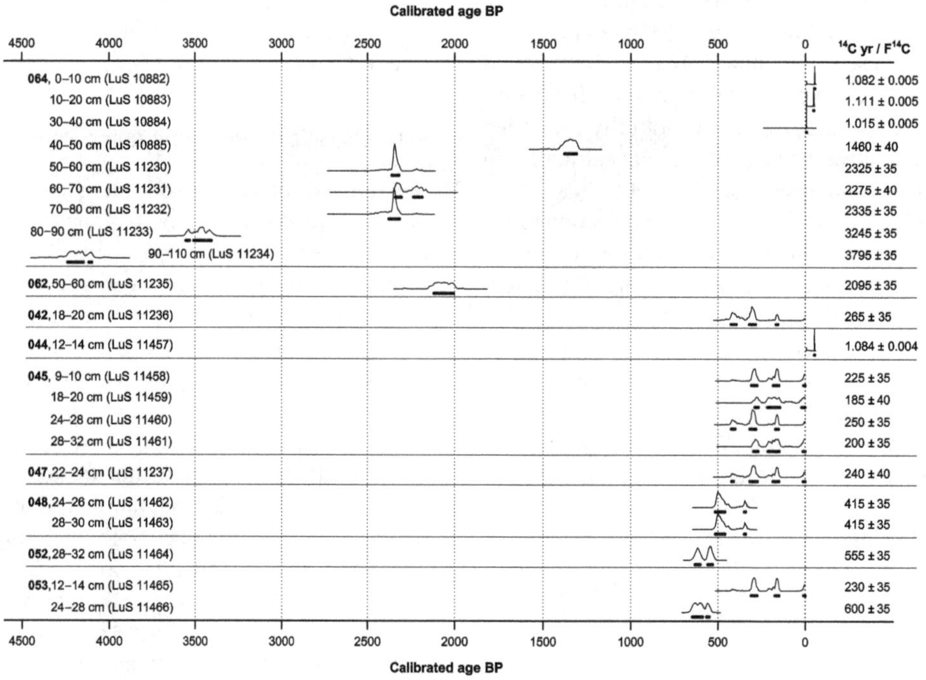

Figure 9. Calibrated age distributions for all ^{14}C dated soil and sediment samples. The thin line represents the probability distribution for each respective sample, and the thick black line shows the 1σ age distributions. The column to the left gives the sample name followed by the laboratory sample number, while the numbers to the right are the uncalibrated ^{14}C dates (or $F^{14}C$ values for samples with a modern carbon composition) with their respective error estimates.

with reference to Vienna Standard Mean Ocean Water (VS-MOW), and the reported analytical uncertainties for $\delta^{18}O$ and $\delta^{2}H$ were 0.065 and 0.0280 ‰, respectively.

Sulfur-34 (^{34}S) and sulfur-32 (^{32}S) were determined. The samples were precipitated as $BaSO_4$ and then analysed using an EA-IRMS system (a Carlo Erba NC2500 connected to a Thermo Delta plus). The isotopic composition, $\delta^{34}S$, was calculated with reference to Vienna-Canyon Diablo Troilite (V-CDT), and the reproducibility of the analysis was calculated to be better than 0.2 ‰.

4.6 Radioisotopes

The soil, permafrost and sediment samples were analysed for carbon-14 (^{14}C) (Table 2). The samples were converted to graphite using an automated graphitisation equipment (AGE) and then analysed for their ^{14}C content using single stage accelerator mass spectrometry (AMS). The values reported in the database are either uncalibrated ^{14}C ages or the fraction of modern carbon ($F^{14}C$) for post-bomb dates. Calibrated ages were obtained using either the InterCal13.14C or the region-1 post-bomb curve (Reimer et al., 2013; Hua et al., 2013) in the online version of Oxcal 4.2 (https://c14.arch.ox.ac.uk; Bronk Ramsey, 2009) and are shown in Fig. 9.

Lead-210 (^{210}Pb) was analysed either measuring the activity of its daughter isotope polonium-210 (^{210}Po) by means of alpha spectrometry or as a direct measurement together with cesium-137 (^{137}Cs), radium-226 (^{226}Ra), thorium-232 (^{232}Th), thallium-208 (^{208}Tl), potassium-40 (^{40}K) and americium-241 (^{241}Am), using gamma spectroscopy (Table 2). Alpha spectrometry measurements were conducted using between 0.25 and 0.5 g of sample and a counting time of 60 000 s. The method detection limit was estimated to 0.1–0.2 DPM g^{-1} (disintegrations/time), the counting errors were generally less than 10 % (except for very low ^{210}Po activities), and replicate analysis ($n = 6$) deviated less than 0.1 disintegrations per minute g^{-1}. The gamma spectrometry was conducted using 5–23 g of sample and a counting time between 83 000 and 340 000 s. The reported uncertainty of the analysis varies between 6 and 30 % depending on the specific radioisotope and the activity of the specific samples (lower activities generally have higher relative uncertainty).

The uranium-234 : 235 ($^{234}U:^{235}U$) isotopic ratio measurements were conducted by allowing 1 L of water to desiccate. The residue was then digested using 2 % HNO_3 and the $^{234}U:^{235}U$ ratio was determined using ICP-SFMS. The analytical performance was assessed using certified reference materials and the reported uncertainties (95 % confidence interval) were below 5 %, except for the Pingo spring sample that had an uncertainty of 30 %. The Pingo spring is located outside of the focus area, close to a meltwater outlet from the ice sheet and not representative for the TBL catchment.

4.7 Biomass quantification

4.7.1 Terrestrial

For the main plots in the 2011 biomass sampling and all plots in the 2014 biomass sampling, the living biomass of each functional group – herbs/grasses, shrubs, lichens, mosses and two litter types (2011 only) – was determined by subtracting the moisture content, determined by drying subsamples at 80 °C, from the total sample weight. The weight of attached mineral grains was also corrected for by analysing LOI (loss on ignition) in subsamples. In the replicate plots from the 2011 biomass sampling, the same approach was employed to obtain the total biomass samples. The litter biomass in the 2011 replicate plots and all 2014 biomass sampling plots was estimated by using the ratio between living biomass and litter from the main plots in the 2011 biomass sampling.

4.7.2 Aquatic

The species composition and biovolumes of phytoplankton and zooplankton were determined using an inverted phase-contrast microscope, and samples were allowed to settle overnight. For phytoplankton, the carbon biomass was estimated by assuming a wet weight of $1\,g\,cm^{-3}$ and using the taxa-specific carbon-content conversion factors presented by Olrik et al. (1998). For zooplankton, carbon biomass was estimated using length-to-weight conversion factors (Johansson et al., 1976; Ruttner-Kolisko, 1977) and assuming a conversation factor of dry weight (DW) to carbon of 0.48.

Both planktonic and benthic bacteria samples were analysed. The benthic samples were first sonicated at 75 W for one minute. All samples were stained with acridine orange, filtered onto 0.2 µm polycarbonate filters and counted using an epifluorescence microscope. A total of at least 400 cells were counted for each sample, and total biovolumes were estimated by measuring at least 100 cells. DW was calculated according to Loferer-Krössbacher et al. (1998) using Eq. (1):

$$\text{dry weight} = 435 \times \text{volume}^{0.86}. \qquad (1)$$

The carbon content was estimated assuming a dry-weight-to-carbon conversion factor of 0.5.

The total mass of aquatic macrophytes was converted to carbon biomass by assuming a dry-weight-to-carbon conversion factor of 0.4. Biomass for benthic fauna was calculated by assuming that the DW was 20 % of the wet weight, and the carbon content was estimated by assuming a dry-weight-to-carbon conversion factor of 0.3 (Kautsky, 1995).

5 Results

5.1 Precipitation and deposition

Concentrations of TOC/DOC (TOC or DOC), total nitrogen and phosphorus in precipitation showed considerable variation depending on precipitation type (Fig. 10). Newly fallen snow, as well as filtered samples, had very low TOC/DOC, nitrogen and phosphorus concentrations (1–3, 0.3–1 and 33–83 $\mu g\,L^{-1}$, respectively), whereas unfiltered old snow – that also contained large amounts of eolian material – had much higher concentrations (40–60, 3–4 $mg\,L^{-1}$ and 1400–2700 $\mu g\,L^{-1}$, respectively). There was also a difference in the $\delta^{18}O$ and $\delta^{2}H$ between rain and snow, with rain having slightly higher values (Fig. 11). As an example of major and trace elements we have chosen to plot the concentrations of aluminium (i.e., a lithogenic element involved in few biological processes), silicon (i.e., a lithogenic element that is an important constituent of, e.g., diatom valves and some sedges) and copper (i.e., a trace element that is an important micronutrient and that binds readily to organic matter). The concentrations of all these elements were low or very low in both rainwater and snow. In rainwater the aluminium, silicon and copper concentrations were 3–8, 0.008–0.01 and 1–7 $mg\,L^{-1}$, respectively, while in snow the concentrations were 0.3–176, 0.01–0.05 and 0.5–4 $mg\,L^{-1}$ for aluminium, silicon and copper, respectively (Fig. 12).

5.2 Soil water

Deep soil water (samples from lysimeters and soil water wells) and surface water (puddles and small inlets) showed relatively similar TOC/DOC, nitrogen and phosphorus concentrations (21–37, 0.9–1.5 $mg\,L^{-1}$ and 10–50 $\mu g\,L^{-1}$, respectively; Fig. 10). Because the TOC/DOC values for the unfiltered samples were virtually identical, it was concluded that all organic carbon was present in the dissolved phase (i.e., TOC equals DOC; Fig. 10). Even if deeper soil water samples showed considerable variation in DIC concentrations, they were generally higher as compared to the DIC concentrations in surface waters (Fig. 10). $\delta^{18}O$ and $\delta^{2}H$ values in soil water overlapped with the values for precipitation, while $\delta^{13}C$ in DIC was more depleted in the heavy isotope in deep soil water as compared to surface water (Fig. 11). The concentrations for aluminium (12–2709 $mg\,L^{-1}$), silicon (0.02–2 $mg\,L^{-1}$) and copper (1–225 $mg\,L^{-1}$) were noticeably higher in soil water as compared to precipitation (Fig. 12).

5.3 Lake water

Lake water temperature at 4 m depth varied between 1 and 14 °C, with maximum temperatures occurring in late July or early August (Fig. 13). Similar to the soil water samples, there were very small differences in TOC/DOC between unfiltered and filtered samples, and there were also only minor differences in TOC/DOC, nitrogen and phosphorus between surface and bottom waters (Fig. 10). TOC/DOC, nitrogen and phosphorus concentrations in lake water samples were lower than those in the soil water samples (8–10, 0.5–0.9 $mg\,L^{-1}$, and 3–11 $\mu g\,L^{-1}$, respectively; Fig. 10), while DIC concentrations in lake water (14–21 $mg\,L^{-1}$) were in

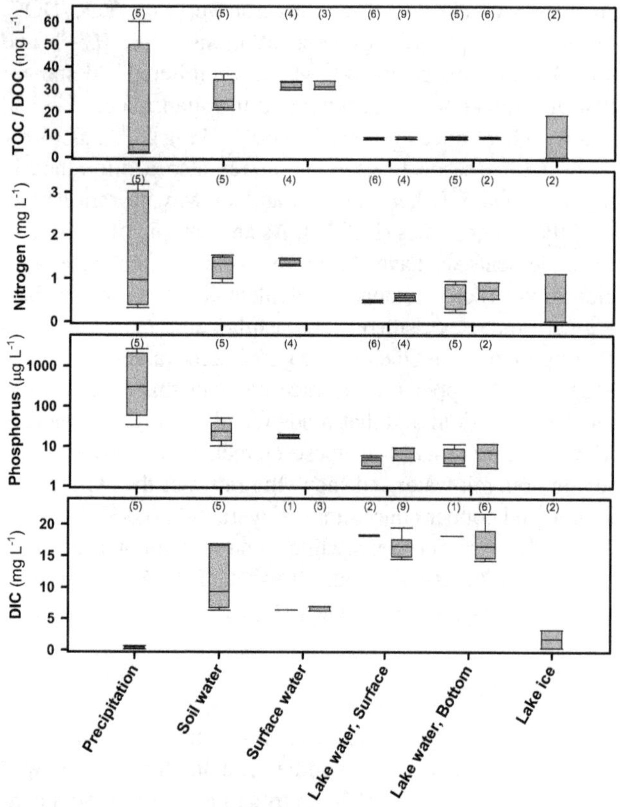

Figure 10. Box plot showing the concentrations of TOC/DOC, total nitrogen, total phosphorus and DIC in water samples from different catchment compartments. For surface and lake water, the box to the left represents unfiltered samples, while the box to the right represents filtered samples. The boxes denote the median and the 25th and 75th percentiles, while the whiskers denote the 10th and 90th percentiles. The numbers within parenthesis represent the number of samples for each sample type.

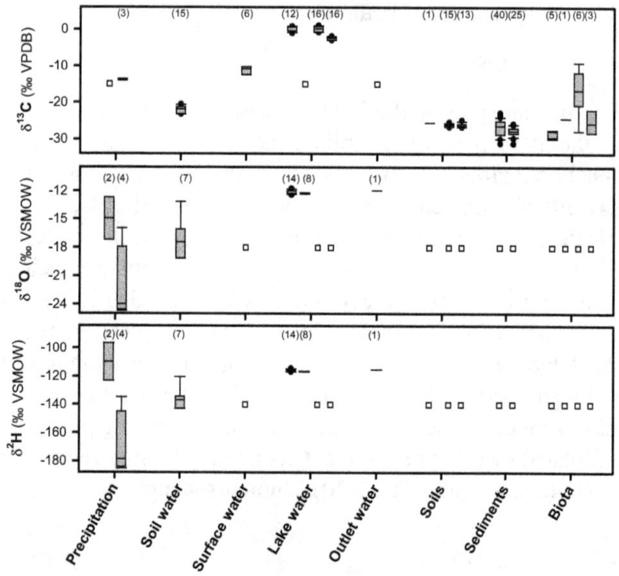

Figure 11. Box plot showing the isotopic composition of carbon, oxygen and hydrogen in different sample types. For water samples the $\partial^{13}C$ is determined on DIC only, while for solid samples the measurements were made on total carbon. The two boxes for precipitation represent rain (left) and snow (right). The four boxes for lake water represent (from left to right) surface and bottom water from locations other than the lake centre and surface and bottom water from the lake centre. The three boxes for soils represent silt (left), peaty-silt active layer (middle) and peaty-silt permafrost (right). The two boxes for sediments represent samples collected from water depths < 14 m (left) and > 14 m (right), respectively. The four boxes for biota represent (from left to right) terrestrial primary producers, terrestrial consumers, aquatic primary producers and aquatic consumers. The boxes denote the median and the 25th and 75th percentiles, while the whiskers denote the 10th and 90th percentiles. Outliers are marked with a black dot, and the open squares mark that no data are available for this sample type. The numbers within parenthesis represent the number of samples for each sample type.

the upper range of those measured in soil water samples (6–16 mg L^{-1}; Fig. 10). $\delta^{18}O$ and δ^2H plot in a very narrow range in the uppermost range of the results for precipitation and soil water, while $\delta^{13}C$ in lake water DIC was considerably less depleted in the heavy isotope than in all other catchment compartments (Fig. 11). Lake water had the lowest concentrations of aluminium, silicon and copper of all sampled ecosystem compartments. Silicon was below the detection limit in all samples, whereas aluminium and copper concentrations were 0.5–3 and 1–2 mg L^{-1}, respectively (Fig. 12).

Results from the in situ measurements of dissolved oxygen and temperature (Fig. 13) show the highest oxygen concentrations in the end of the autumn. During winter, a sharp decrease in dissolved oxygen is observed at 17–20 m depth, whereas the oxygen level at 5–8 m depth is relatively stable the whole winter period. During summer the oxygen concentrations decrease slightly both at shallow and greater depths as the lake temperature rises.

5.4 Soils

Analysis of ^{14}C in the soil profile in the southern part of the catchment (Fig. 4a) shows that the lower most sample is about 4000 years old. It is also apparent that the three samples from the upper 40 cm all have a modern ^{14}C signal and that the three samples from the 50 to 80 cm depth interval all have an age of around 2300 years (Fig. 9).

Using the 24 soil samples in which both LOI and total carbon were analysed and that had carbon concentrations above 1 %, the site-specific carbon : LOI ratio was calculated (0.52 ± 0.05). The highest concentrations of organic matter in soils were found in the areas with peaty silt (LOI of 2–30 %), while upland soils generally contained smaller amounts of organic matter (LOI of 0.2 to 11 %). Soil organic matter showed a narrow range of $\delta^{13}C$ around −26 ‰, i.e.,

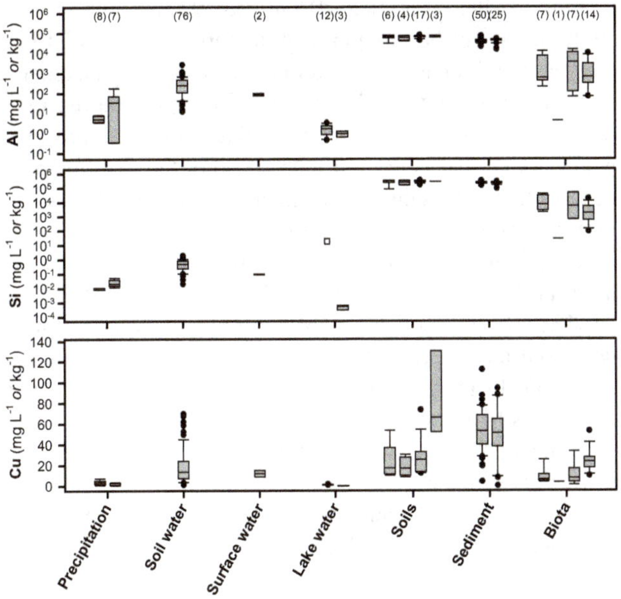

Figure 12. Box plot showing the concentrations of aluminum, silicon and copper in different sample types (the concentrations are expressed in mg L^{-1} for water samples and in mg kg^{-1} for solid samples). The two boxes for precipitation represent rain (left) and snow (right). The two boxes for lake water represent lake water (left) and suspended matter (right). The four boxes for soils represent (from left to right) silt, O horizon, peaty-silt active layer and peaty-silt permafrost. The two boxes for sediments represent samples collected from water depths < 14 m (left) and > 14 m (right). The four boxes for biota represent (from left to right) terrestrial primary producers, terrestrial consumers, aquatic primary producers and aquatic consumers. The boxes denote the median and the 25th and 75th percentiles, while the whiskers denote the 10th and 90th percentiles. Outliers are marked with a black dot, and the open squares mark that all analysed samples were below the detection limit. The numbers within parenthesis represent the number of samples that were analysed for each sample type.

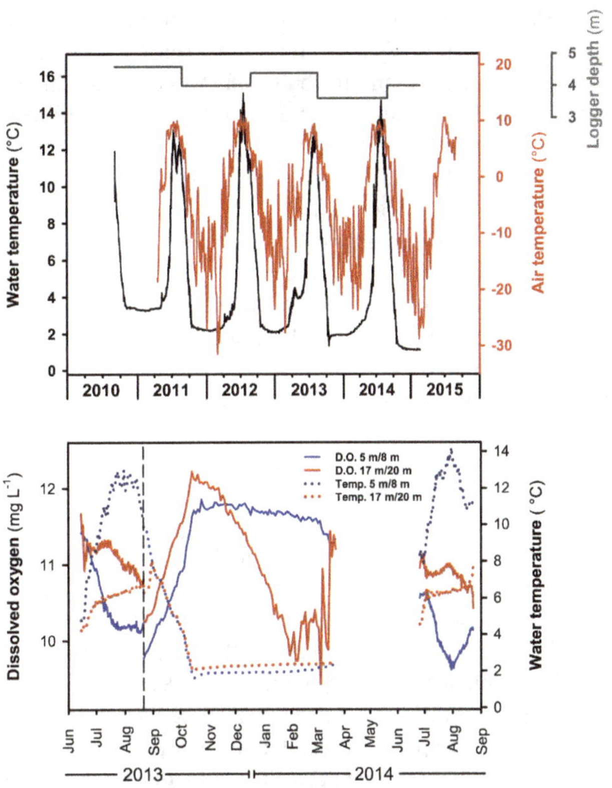

Figure 13. Upper figure: graph displaying the variations in lake water temperature at about 4 m water depth and air temperature during the period from 2011 to 2015. The black line represents lake water temperature, the red line the air temperature, and the grey line represents the exact depth of the lake water temperature logger. Lower figure: graph displaying the variations in dissolved oxygen and temperature during measurements in 2013 and 2014. The dashed line represents the time when the depth of the loggers was changed from 5 and 17 m to 8 and 20 m.

slightly more depleted than the soil water (Fig. 11). The total aluminium, silicon and copper concentrations were, as can be expected, much higher in soils as compared to the water samples. Total concentrations for aluminium generally ranged from 3 to 7 % depending on soil type (silt, O horizon or peaty silt), silicon ranged from 8 to 33 % and copper from 4 to 130 mg kg^{-1} (Fig. 12). The concentrations of aluminium, silicon and copper in the solutions from the ammonium acetate leaching were around 2, 4 and 1 orders of magnitude lower, respectively, when compared to the total concentrations.

5.5 Lake sediments

Even though there are considerable differences in the downcore trends for the 13 surface sediment cores, a couple of general features do exist. All cores show a more or less pronounced increase in organic material (as indicated by the increasing carbon concentrations) in the topmost sediment sample. This is likely to a large extent a result of degradation of organic matter in the most recent sediment layers (Gälman et al., 2008). All cores also show a more long-term increase in organic material in the top 5 to 20 cm of the cores.

Depending on whether the cores were collected in areas with or without benthic vegetation, they can be grouped into two categories. First, cores from locations with benthic vegetation, i.e., cores taken at water depths of less than 14 m, showed carbon concentrations in the top 5 cm ranging from 6 to 16 %. Aluminium, silicon and copper concentrations were 2–7, 14–31 % and 4–112 mg kg^{-1}, respectively, while phosphorus concentrations ranged from 800 to 1300 mg kg^{-1} (Fig. 12). The bulk density of the sediment ranges from 30 to 125 mg cm^{-3}. Second, at deeper locations without benthic vegetation, i.e., deeper than 14 m, the carbon concentration in the top 5 cm of the core ranges from 1 to 11 %. Aluminium, silicon and copper concentrations were 1.5–7,

9–26 % and 0–94 mg kg^{-1}, respectively, while phosphorus concentrations ranged from 1100 to 2600 mg kg^{-1} (Fig. 12). For the deeper locations, the bulk density was a little higher, from 60 to 700 mg cm^{-3}. The $\delta^{13}C$ values of all sediments showed a much larger range than the soil samples (−31 to −23 ‰; Fig. 11), and there was also considerable variation in $\delta^{15}N$ values (−0.8 to 3.5 ‰).

In order to compare the sediment accumulation across the TBL basin, a number of the sediment profiles were analysed for ^{210}Pb. A number of moss macrofossils were also dated through ^{14}C analysis. The sediment depth where the sediment is about 150 years – i.e., the sediment level where atmospherically supplied, unsupported, ^{210}Pb no longer can be detected (Appleby, 2008) – was between 5 and 12 cm in the six cores where ^{210}Pb was analysed (TBL-1, 5, 6, 8, 10, 13). The calibrated ages of the ^{14}C-dated macrofossils are all between 0 and 500 years, which is a period that is very difficult to date accurately due to the large variations in the atmospheric ^{14}C concentrations resulting from the increased burning of fossil fuels (Reimer et al., 2013). This results in very broad calibrated age ranges with many possible dates, which complicate the interpretation (Fig. 9). In one of the sediment cores (TBL-4), it is likely that the modern ^{14}C signature of the sample indicates that the analysed macrofossil was pushed down through the sediment by the corer, indicating that this result should not be used.

5.6 Terrestrial biota

Generally, lichens, bryophytes and litter (i.e., material found on or close to the ground) had more attached mineral particles compared to standing litter and standing living biomass. Assuming that pure plant material gives an ash content of approximately 5 %, standing living biomass had a mineral particle content of almost zero.

The most exposed habitat, ridges with dry grassland, had the lowest biomass, and it was found that biomass increased in the order meadow, wetland and *Betula* vegetation types. This pattern was consistent among both sampling campaigns (mid-August 2011 and late June 2014). Even if it could be expected that standing biomass would be higher later in the season, the opposite pattern was found for both the *Betula* and the wetland vegetation types. A closer examination suggests that, at least for the *Betula* vegetation type, this can be attributed to denser stands being sampled in 2011 than in 2014 (higher wood biomass). Bryophytes were also more dominant in the sampling plots used during the 2014 sampling campaign. For the dry ridges the observed higher biomass in June is consistent with expectations, i.e., that peak biomass should occur when the moisture content still is relatively high.

Both when comparing the main and the replicate plots from the 2011 sampling and when comparing the 2011 and 2014 sampling campaigns, there was a high variability in biomass for the different functional groups. This suggests that the smaller plots (0.2×0.2 m and 170 cm^2) used in this study might not be large enough to capture the spatial heterogeneity within each vegetation type. Hence, it is expected that the larger plots (0.5×0.5 m) should provide the most reliable estimates of terrestrial biomass. Furthermore, it is worth noting that the primary objective of the 170 cm^2 plots used in the 2014 sampling campaign was to relate respiration and production to biomass, litter and organic content in the soil and not to estimate total biomass. Terrestrial primary producers had $\delta^{13}C$ and $\delta^{15}N$ values of -29 ± 1 and -0.4 ± 3 ‰, respectively, while the single sample from a terrestrial consumer had $\delta^{13}C$ and $\delta^{15}N$ of −24 and 4.3 ‰ (Fig. 11). In primary producers the concentrations of aluminium, silicon and copper were 0.02–1.3, 0.2–3.7 % and 4–25 mg kg^{-1}, respectively, while our only sample of a terrestrial consumer showed considerably lower concentrations, i.e., 0.0004, 0.003 % and 4 mg kg^{-1} for aluminium, silicon and copper, respectively (Fig. 12).

5.7 Aquatic biota

The dominant phytoplankton taxon in August 2011 and April 2012 was *Dinophyceae* followed by *Chyptophyceae*. In August 2012, *Chlorophyceae* was the dominant phytoplankton type, followed by *Cryptophyceae*. Phytoplankton biomass varied between 13 and 62 mg C m^{-3}. Bacterioplankton biomass was considerably higher in the samples from August 2011 (on average 0.023 g C m^{-3}) than in the samples from 2012 (average values of 0.012 and 0.010 g C m^{-3} for April and August, respectively). The zooplankton community was dominated by *Copepoda* and *Cladocera* (mainly *Daphnia*), and zooplankton biomass ranged between 0.01 to 0.07 g C m^{-3}.

Contrary to bacterioplankton, benthic bacterial biomass was higher in samples from 2012 (average 1.4 g C m^{-2}) than in samples from 2012 (0.6 g C m^{-2}). The diversity and biomass of benthic fauna were low, and several of the Ekman grabs contained no benthic macrofauna (0–1758 individuals per m^2). The benthic fauna was dominated by *Chironomidae* and *Oligochaeta*.

Fontinalis antipyretica and *Chara* spp. dominated the macrophytes community in the lake but several more species (ca. 20 including vascular plants, bryophytes and epiphytic algae) were identified during the macrophytes mapping of the lake. There were dense stands of macrophytes in wind-sheltered littoral and lower coverage and biomasses of macrophytes in wind-exposed littoral zones. Biomass measured in 2011 varied between less than 1 to 1382 g DW m^{-2}. The dive transect in 2013 did not show correlation of biomass with depth but showed that there were high biomasses of macrophytes also at depths greater than 16 m. Mean biomass at the different depths 1.6, 2.5, 5.2, 9.2 and 16.3 m was 155, 55, 100, 137 and 52 g DW m^{-2}, respectively.

Primary production measured in glass bottles showed that respiration was higher than production in the pelagic

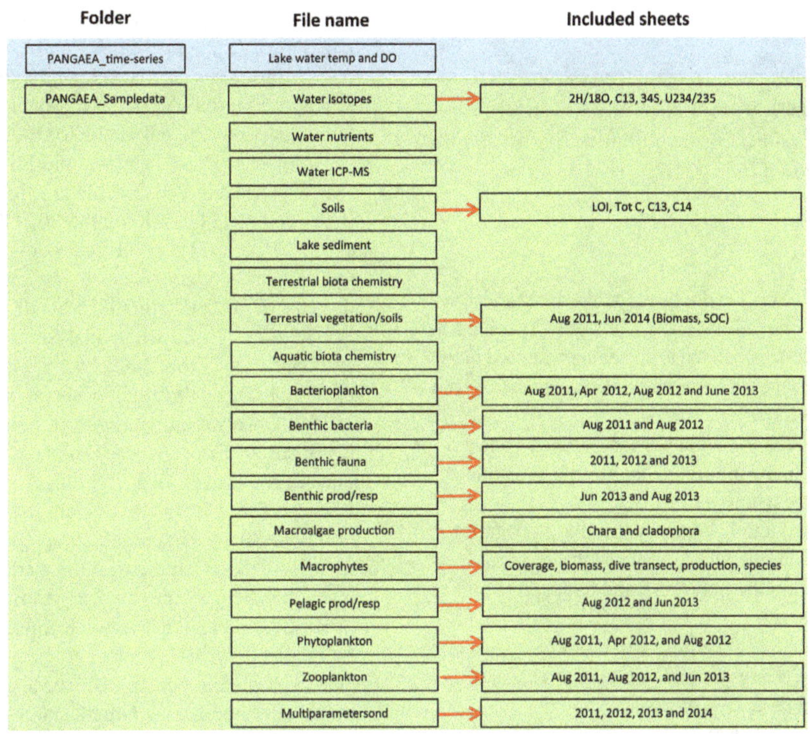

Figure 14. Illustration of folders, resulting data tables and associated sheets in the PANGAEA database.

habitat in both August 2012 and June 2013. Gross primary production in the water column was higher in August (mean of 0.7 $g\,C\,m^{-2}\,h^{-1}$) than in June (mean of 0.2 $g\,C\,m^{-2}\,h^{-1}$). As expected, considering the higher water temperature in late summer, the respiration was also higher in August (0.25 $g\,C\,m^{-3}\,h^{-1}$) than in June (0.006 $g\,C\,m^{-3}\,h^{-1}$). Benthic production and respiration measured in transparent chambers showed that primary production was higher than respiration in both June and August during daytime (increase in oxygen concentrations). When including the dark hours and measuring for 24 h, respiration was higher than primary production in the benthic habitat in August (decrease in oxygen concentration) even though primary production during light hours was higher in August (mean net production at 0.7 m depth: 0.2 $mg\,O_2\,m^{-2}$) than in June (mean net production at 0.7 m depth: 0.03 $mg\,O_2\,m^{-2}$). The production measurements on pieces of macroalgae showed that *Chara* sp. had similar production at 2, 5 and 10 m depth although there was a trend towards lower primary production with depth (mean production 0.038, 0.032 and 0.025 $g\,O_2\,g\,DW^{-1}\,h^{-1}$ at 2, 5 and 10 m, respectively). The production of *Cladophora* sp. at 5 m depth (mean production of 0.022, minimum of 0.013, maximum of 0.035 $g\,O_2\,g\,DW^{-1}\,h^{-1}$) was somewhat lower than the production of *Chara* sp. at corresponding depth (mean production of 0.032, minimum of 0.030, maximum of 0.035 $g\,O_2\,g\,DW^{-1}\,h^{-1}$).

$\delta^{13}C$ and $\delta^{15}N$ in aquatic primary producers varied considerably between species (Fig. 11). *Nostoc* sp. had a $\delta^{13}C$ of −9 ‰ and a $\delta^{15}N$ of −0.4 ‰, while aquatic bryophytes and *Chara* sp. had $\delta^{13}C$ and $\delta^{15}N$ of −28 and 1.24 ‰ and −15 and 1.3 ‰, respectively. Aquatic consumers showed a much narrower range in $\delta^{13}C$ and $\delta^{15}N$, with averages of -25 ± 3 and 3.8 ± 1.8 ‰, respectively (Fig. 11). Concentrations of aluminium, silicon and copper in aquatic primary producers were 0.01–1.5, 0.1–5 % and 1–33 $mg\,kg^{-1}$, respectively. Aquatic consumers had slightly lower concentrations for aluminium and silicon – 0.1–1 and 0.01–2 %, respectively – but slightly higher copper concentrations – 10–52 $mg\,kg^{-1}$ (Fig. 12).

6 Conclusions

This paper presents a large and comprehensive biogeochemical data set covering both the terrestrial and aquatic environments in a periglacial catchment. By the inclusion of data from both liquid and solid phases, as well as abiotic and biotic components, this data set allows for the calculation of whole-ecosystem mass balance budgets for a long list of elements. Together with the hydrological data presented by Johansson et al. (2015a), it also allows for calculations of fluxes between different catchment compartments as well as the development of ecosystem models.

7 Data availability

We encourage anyone interested in using the data set to contact the corresponding author regarding any questions about the data or the field site. A schematic figure of the PANGAEA tables is shown in Fig. 14.

Author contributions. Tobias Lindborg prepared the manuscript with contributions from all coauthors and he has been involved in all fieldwork and data processing related to the data presented in the present paper. Tobias Lindborg was also responsible for the overall field campaign strategy.
Johan Rydberg was involved in the sediment sampling and has written the associated sections in the present paper as well as taking active part in the writing of the whole manuscript.
Mats Tröjbom was involved in the chemistry sampling programme and the data set preparation to the PANGAEA database.
Sten Berglund was involved in the field work and took active part in the writing of the whole paper.
Emma Johansson was involved in the field work and has produced maps and other figures to the present paper. She also prepared the data set for storage in the PANGAEA database.
Johannes Petrone prepared the data set for storage in the PANGAEA database and was involved in parts of the field work described in this paper. Ander Löfgren, Peter Saetre, Eva Andersson and Sara Nordén were involved in the terrestrial and aquatic sampling programmes as well as writing associated texts.
Micke Borgiel was responsible for the macrophyte dive transect inventory. Gustav Sohlenius was involved in the soil sampling.
Ulrik Kautsky and Hjalmar Laudon were involved in field work and different measurement campaigns described in the present paper. They also contributed to the overall field campaign strategy.
Tobias Lindborg (ecology and chemistry) and Emma Johansson (hydrology and meteorology) are responsible for the GRASP field programme.

Acknowledgements. The work was conducted as a part of the Greenland Analogue Surface Project (GRASP) funded by the Swedish Nuclear Fuel and Waste Management Company (SKB). The authors would like to thank the Greenland Analogue Project (GAP) and Kangerlussuaq International Science Support (KISS) for providing logistical support throughout the years.

Edited by: A. Demény

References

Anderson, N. J., Harriman, R., Ryves, D. B., and Patrick, S. T.: Dominant factors controlling variability in the ionic composition of West Greenland lakes, Arct. Antarct. Alp. Res., 33, 418–425, 2001.

Anderson, N. J., Brodersen, K. P., Ryves, D. B., McGowan, S., Johansson, L. S., Jeppesen, E., and Leng, M. J.: Climate versus in-lake processes as controls on the development of community structure in a low-Arctic lake (South-West Greenland), Ecosystems, 11, 307–324, doi:10.1007/s10021-007-9123-y, 2008.

Anderson, N. J., D'Andrea, W., and Fritz, S. C.: Holocene carbon burial by lakes in SW Greenland, Glob. Change Biol., 15, 2590–2598, doi:10.1111/j.1365-2486.2009.01942.x, 2009.

Appleby, P. G.: Three decades of dating recent sediments by fallout radionuclides: a review, Holocene, 18, 83–93, doi:10.1177/0959683607085598, 2008.

Bronk Ramsey, C.: Bayesian analysis of radiocarbon dates, Radiocarbon, 51, 337–360, 2009.

Cappelen, J. (Ed.): Weather observations from Greenland 1958–2015 – Observation data with description, Danish Meteorological Institute, Copenhagen, Denmark, DMI Report 16-08, 2016.

Chapin F. S., Shaver G. R., Giblin A. E., Nadelhoffer K. J., and Laundre J. A.: Responses of arctic tundra to experimental and observed changes in climate, Ecology, 76, 694–711, 1995.

Clarhäll, A.: SKB studies of the periglacial environment – report from field studies in Kangerlussuaq, Greenland 2008 and 2010, SKB P-11-05, Svensk Kärnbränslehantering AB, Stockholm, Sweden, 2011.

Engels, S. and Helmens, K.: Holocene environmental changes and climate development in Greenland, Svensk Kärnbränslehantering AB, SKB R-10-65, Stockholm, Sweden, 2010.

Gälman, V., Rydberg, J., Sjöstedt de-Luna, S., Bindler, R., and Renberg, I.: Carbon and nitrogen loss rates during ageing of lake sediment: Changes over 27 years studied in a varved lake sediment, Limnol. Oceanogr., 53, 1076–1082, 2008.

Harper, J., Hubbard, A., Ruskeeniemi, T., Claesson Liljedahl, L., Lehtinen, A., Booth, A., Brinkerhoff, D., Drake, H., Dow, C., Doyle, S., Engström, J., Fitzpatrik, A., Frape, S., Henkemans, E., Humphrey, N., Johnson, J., Jones, G., Joughin, I., Klint, K. E., Kukkonen, I., Kulessa, B., Londowski, C., Lindbäck, K., Makahnouk, M., Meierbachtol, T., Pere, T., Pedersen, K., Petterson, R., Pimentel, S., Quincey, D., Tullborg, E. L., and van As, D.: The Greenland Analogue Project Yearly Report 2010, Svensk Kärnbränslehantering AB, Stockholm, Sweden, Report SKB R-11-23, 2011.

Helcom: Manual for Marine Monitoring in the Combine Programme of Helcom, available at: http://helcom.fi/Documents/Action areas/Monitoring and assessment/Manuals and Guidelines/Manual for Marine Monitoring in the COMBINE Programme of HELCOM.pdf, last access: 4 April 2016.

Hindshaw, R. S., Lang, S. Q., Bernasconi, S. M., Heaton, T. H. E., Lindsay, M. R., and Boyd, E. S.: Origin and temporal variability of unusually low δ^{13}C-DOC values in two high Arctic catchments, J. Geophys. Res.-Biogeo., 121, 1073–1085, doi:10.1002/2015JG003303, 2016.

Hua, Q., Barbetti, M., and Rakowski, A. Z,: Atmospheric radiocarbon for the period 1950–2010, Radiocarbon, 55, 2059–2072, 2013.

Jessen, S., Holmslykke, H. D., Rasmussen, K., Richardt, N., and Holm, P. E.: Hydrology and pore water chemistry in a permafrost wetland, Ilulissat, Greenland, Water Resour. Res., 50, 4760–4774, doi:10.1002/2013WR014376, 2014.

Johansson, E., Berglund, S., Lindborg, T., Petrone, J., van As, D., Gustafsson, L.-G., Näslund, J.-O., and Laudon, H.: Hydrological and meteorological investigations in a periglacial lake catchment near Kangerlussuaq, west Greenland – presentation of a new multi-parameter data set, Earth Syst. Sci. Data, 7, 93–108, doi:10.5194/essd-7-93-2015, 2015a.

Johansson, E., Gustafsson, L.-G., Berglund, S., Lindborg, T., Selroos, J.-O., Claesson Liljedahl, L., and Destouni, G.: Data evaluation and numerical modeling of hydrological interactions between active layer, lake and talik in a permafrost catchment, Western Greenland, J. Hydrology, 527, 688–703, doi:10.1016/j.jhydrol.2015.05.026, 2015b.

Johansson, J.-Å., Olofsson, H., and Ramberg, L.: Studier av zooplanktons konsumtion i Botjärn, Limnologiska inst., Uppsala universitet, Sweden, Klotenprojektets rapport nr. 5, 27 pp., 1976.

Kautsky, U.: Ecosystem processes in coastal areas of the Baltic Sea, PhD thesis, Department of Zoology, Stockholm University, Sweden, 1995.

Loferer-Krössbacher, M., Klima, J., and Psenner, R.: Determination of bacterial cell dry mass by transmission electron microscopy and densitometric image analysis, Appl. Environ. Microbiol., 64, 688–694, 1998.

Lundin, E. J., Klaminder, J., Giesler, R., Persson, A., Olefeldt, D., Heliasz, M., Christensen, T. R., and Karlsson, J.: Is the subarctic landscape still a carbon sink? Evidence from a detailed catchment balance, Geophys. Res. Lett., 43, 1988–1995, doi:10.1002/2015GL066970, 2016.

McGuire, A. D., Anderson, L. G., Christensen, T. R., Dallimore, S., Guo, L., Hayes, D. J., Heimann, M., Lorenson, T.D., Macdonald, R. W., and Roulet, N.: Sensitivity of the carbon cycle in the Arctic to climate change, Ecol. Monogr., 79, 523–555, doi:10.1890/08-2025.1, 2009.

Moy, F. E., Dahl, K., Karlsson, J., Kautsky, H., Ruuskanen, A., and Carstensen, J.: Nordic intercalibration of hard bottom macroalgae monitoring methodologies, Algamony, Nordic Council of Ministers, Copenhagen, Denmark, TemaNord 2010:543, 2010.

Olrik, K., Blomqvist, P., Brettum, P., Cronberg, G., and Eloranta, P.: Methods for quantitative assessment of phytoplankton in freshwaters, Part 1: Sampling, processing, and application in freshwater environmental monitoring programmes, Naturvårdsverket, Stockholm, Sweden, Rapport 4860, 1998.

Palmtag, J., Hugelius, G., Lashchinskiy, N., Tamstorf, M. P., Richter, A., Elberling, B., and Khury, P.: Storage, landscape distribution, and burial history of soil organic matter in contrasting areas of continuous permafrost, Arct. Antarct. Alp. Res., 47, 71–88, doi:10.1657/aaar0014-027, 2015.

Petrone, J., Sohlenius, G., Johansson, E., Lindborg, T., Näslund, J.-O., Strömgren, M., and Brydsten, L.: Using ground-penetrating radar, topography and classification of vegetation to model the sediment and active layer thickness in a periglacial lake catchment, Western Greenland, Earth Syst. Sci. Data Discuss., doi:10.5194/essd-2016-19, in review, 2016.

Petrone, K., Buffam, I., and Laudon, H.: Hydrologic and biotic control of nitrogen export during snowmelt: a combined conservative and reactive tracer approach, Water Resour. Res., 43, W06420, doi:10.1029/2006WR005286, 2007.

Reimer, P. J., Bard, E., Bayliss, A., Beck, J. W., Blackwell, P. G., Bronk Ramsey, C., Buck, C. E., Cheng, H., Edwards, R. L., Friedrich, M., Grootes, P.-M., Guilderson, T. P., Haflidason, H., Hajdas, I., Hatté, C., Heaton, T. J., Hoffmann, D. L., Hogg, A. G., Hughen, K. A., Kaiser, K. F., Kromer, B., Manning, S. W., Niu, M., Reimer, R. W., Richards, D. A., Scott, E. M., Southon, J. R., Staff, R. A., Turney, C. S. M., and van der Plicht, J.: IntCal13 and Marine13 radiocarbon age calibration curves 0–50 000 years cal BP, Radiocarbon, 55, 1869–1887, 2013.

Renberg, I. and Hansson, H.: The HTH sediment corer, J. Paleolimnol., 40, 655–659, doi:10.1007/s10933-007-9188-9, 2008.

Ruttner-Kolisko, A.: Suggestions for biomass calculations of plankton rotifers, Arch. Hydrobiol. Beih. Ergebn. Limnol., 8, 71–76, 1977.

Rydberg, J.: Wavelength dispersive X-ray fluorescence spectroscopy as a fast, non-destructive and cost-effective analytical method for determining the geochemical composition of small loose-powder sediment samples, J. Paleolimnol., 52, 265–276, doi:10.1007/s10933-014-9792-4, 2014.

Rydberg, J., Klaminder, J., Rosén, P., and Bindler, R.: Climate driven release of carbon and mercury from permafrost mires increases mercury loading to sub-arctic lakes, Sci. Total Environ., 408, 4778–4783, doi:10.1016/j.scitotenv.2010.06.056, 2010.

Rydberg, J., Lindborg, T., Solenius, G., Reuss, N., Olsen, J., and Laudon, H.: The importance of eolian input on lake-sediment geochemical composition in the dry proglacial landscape of western Greenland, Arct. Antarct. Alp. Res., 48, 93–109, doi:10.1657/AAAR0015-009, 2016.

Rysgaard, S. and Glud, R. N. (Eds.): Carbon cycling in Arctic marine ecosystems: Case study Young Sound, the Commission for Scientific Research in Greenland, Meddelelser om Grønland, Bioscience 58, Copenhagen, Denmark, 2007.

Ryves, D. B., McGowan, S., and Anderson, N. J.: Development and evaluation of a diatom-conductivity model from lakes in West Greenland, Freshwater Biol., 47, 995–1014, doi:10.1046/j.1365-2427.2002.00832.x, 2002.

Ryves, D. B., Battarbee, R. W., and Fritz, S. C.: The dilemma of disappearing diatoms: incorporating diatom dissolution data into palaeoenvironmental modelling, Quaternary Sci. Rev., 28, 120–136, doi:10.1016/j.quascirev.2008.08.021, 2009.

Saros, J. E., Strock, K. E., McCue, J., Hogan, E., and Anderson, N. J.: Response of *Cyclotella* species to nutrients and incubation depth in Arctic lakes, J. Plankton Res. 36, 450–460, doi:10.1093/plankt/fbt126, 2014.

Schuur, E. A. G., McGuire, A. D., Schädel, C., Grosse, G., Harden, J. W., Hayes, D. J., Hugelius, G., Koven, C. D., Kuhry, P., Lawrence D. M., Natali S. M., Olefeldt, D., Romanovsky V. E., Schaefer, K., Turetsky, M. R., Treat, C. C., and Vonk, J. E.: Climate change and the permafrost carbon feedback, Nature, 520, 171–179, doi:10.1038/nature14338, 2015.

Smol, J. P.: Climate change: A planet in flux, Nature, 483, S12–S15, doi:10.1038/483S12a, 2012.

Smol, J. P. and Douglas, M. S. V.: From controversy to consensus: making the case for recent climate change in the Arctic using lake sediments, Front. Ecol. Environ., 5, 466–474, 2007.

Sobek, S., Anderson, N. J., Bernasconi, S. M., and Del Sontro, T.: Low organic carbon burial efficiency in Arctic lake sediments, J. Geophys. Res. Biogeo., 119, 1231–1243, doi:10.1002/2014JG002612, 2014.

Solomon, C. T., Jones, S. E., Weidel, B. C., Buffam, I., Fork, M. L., Karlsson, J., Larsen, S., Lennon, J. T., Read, J. S., Sadro, S., and Saros, J. E.: Ecosystem consequences of changing inputs of terrestrial dissolved organic matter to lakes: Current knowledge and future challenges, Ecosystems, 18, 376–389, doi:10.1007/s10021-015-9848-y, 2015.

Tarnocai, C., Canadell, J. G., Schuur, E. A. G., Kuhry, P., Mazhitova, G., and Zimov, S.: Soil organic carbon pools in the northern circumpolar permafrost region, Global Biogeochem. Cy., 23, GB2023, doi:10.1029/2008GB003327, 2009.

van Tatenhove, F. G. M. and Olesen, O. B.: Ground temperature and related permafrost characteristics in West Greenland, Permafrost Periglac., 5, 199–215, 1994.

Whiteford, E. J., McGowan, S., Barry, C. D., and Anderson, N. J.: Seasonal and regional controls of phytoplankton production along a climate gradient in South-West Greenland during ice-cover and ice-free conditions, Arct. Antarct. Alp. Res., 48, 139–159, doi:10.1657/AAAR0015-003, 2016.

12

OCTOPUS: an open cosmogenic isotope and luminescence database

Alexandru T. Codilean[1,2], Henry Munack[1,3], Timothy J. Cohen[1,2], Wanchese M. Saktura[1], Andrew Gray[4], and Simon M. Mudd[5]

[1]GeoQuEST Research Centre, School of Earth and Environmental Sciences, University of Wollongong, Wollongong NSW 2522, Australia

[2]ARC Centre of Excellence for Australian Biodiversity and Heritage, University of Wollongong, Wollongong NSW 2522, Australia

[3]Research Institute for Post-Mining Landscapes (FIB), Finsterwalde 03238, Germany

[4]Information Technology Services, University of Wollongong, Wollongong NSW 2522, Australia

[5]School of GeoSciences, University of Edinburgh, Edinburgh EH8 9XP, UK

Correspondence: Alexandru T. Codilean (codilean@uow.edu.au)

Abstract. We present a database of cosmogenic radionuclide and luminescence measurements in fluvial sediment. With support from the Australian National Data Service (ANDS) we have built infrastructure for hosting and maintaining the data at the University of Wollongong and making this available to the research community via an Open Geospatial Consortium (OGC)-compliant web service. The cosmogenic radionuclide (CRN) part of the database consists of ^{10}Be and ^{26}Al measurements in modern fluvial sediment samples from across the globe, along with ancillary geospatial vector and raster layers, including sample site, basin outline, digital elevation model, gradient raster, flow-direction and flow-accumulation rasters, atmospheric pressure raster, and CRN production scaling and topographic shielding factor rasters. Sample metadata are comprehensive and include all necessary information for the recalculation of denudation rates using CAIRN, an open-source program for calculating basin-wide denudation rates from ^{10}Be and ^{26}Al data. Further all data have been recalculated and harmonised using the same program. The luminescence part of the database consists of thermoluminescence (TL) and optically stimulated luminescence (OSL) measurements in fluvial sediment samples from stratigraphic sections and sediment cores from across the Australian continent and includes ancillary vector and raster geospatial data. The database can be interrogated and downloaded via a custom-built web map service. More advanced interrogation and exporting to various data formats, including the ESRI Shapefile and Google Earth's KML, is also possible via the Web Feature Service (WFS) capability running on the OCTOPUS server. Use of open standards also ensures that data layers are visible to other OGC-compliant data-sharing services. OCTOPUS and its associated data curation framework provide the opportunity for researchers to reuse previously published but otherwise unusable CRN and luminescence data. This delivers the potential to harness old but valuable data that would otherwise be lost to the research community. OCTOPUS can be accessed at https://earth.uow.edu.au (last access: 28 November 2018). The individual data collections can also be accessed via the following DOIs: https://doi.org/10.4225/48/5a8367feac9b2 (CRN International), https://doi.org/10.4225/48/5a836cdfac9b5 (CRN Australia), and https://doi.org/10.4225/48/5a836db1ac9b6 (OSL & TL Australia).

1 Introduction

Cosmogenic radionuclide (CRN) exposure dating and luminescence dating are suites of geochronological techniques that have become important for the studying of Earth surface processes (e.g. Rhodes, 2011; Granger et al., 2013). Both permit quantifying the timing of geological events by dating individual landforms. In addition, CRNs can also be used to measure the rate at which landforms or landscapes are being denuded by physical and chemical erosion processes. Thus, the two suites of techniques have been extensively used among others to quantify basin-wide denudation rates (von Blanckenburg, 2005; Granger and Schaller, 2014), to reconstruct the extent of Quaternary glaciations (Spencer and Owen, 2004; Balco, 2011; Ivy-Ochs and Briner, 2014), to study how rivers have adapted to past climate change via incision and aggradation (Schaller et al., 2004; Lewis et al., 2009; Wallinga et al., 2010), and to study the timing of dune construction (Fitzsimmons et al., 2007; Fujioka et al., 2009; Bristow et al., 2010). Both suites of techniques are costly (both in terms of time and money) and require specialised training, laboratories, and equipment. As such, CRN and luminescence studies are often very focused and involve a relatively small number of samples ($n < 100$). The research questions being addressed by these studies are very specific and study areas are often relatively small. Hence, CRN and luminescence studies will produce small data sets that are unmanaged and that may become *forgotten* once the study has been completed and results are published. Further, despite there being calls for minimum data reporting standards (e.g. Dunai and Stuart, 2009; Frankel et al., 2010), the published work will often not include appropriate levels of metadata to make the raw data reusable with ease. The latter is especially important in the case of cosmogenic nuclides as procedures used to interpret CRN data are regularly revised and updated, requiring denudation rates and/or exposure ages to be recalculated using updated measurement standards and calculation protocols. Such recalculations are also necessary when comparing results produced by different accelerator mass spectrometry (AMS) facilities that happen to normalise results to different AMS standards. Therefore, without periodic recalculation and maintenance of the data, CRN-based age and rate estimates, for example, can become out of date after a few years. A system and framework for managing CRN and luminescence data and metadata are critical to ensuring the longevity and value of such data collections.

Here we present a database of cosmogenic radionuclide and luminescence measurements in fluvial sediment – OCTOPUS. With support from the Australian National Data Service (ANDS) we have built infrastructure for hosting and maintaining the data at the University of Wollongong and making this available to the research community via an Open Geospatial Consortium (OGC)-compliant web service (http://www.opengeospatial.org, last access: 28 November 2018). The CRN part of the database consists of ^{10}Be and ^{26}Al measurements in fluvial sediment samples from across the globe. Sample metadata are comprehensive and include all necessary information for the recalculation of denudation rates using CAIRN, an open-source program for calculating basin-wide denudation rates from ^{10}Be and ^{26}Al data (Mudd et al., 2016). To this end, the database also includes a comprehensive suite of geospatial data layers, both vector (e.g. sample site and basin outline) and raster (e.g. elevation, gradient and flow-routing rasters, atmospheric pressure, and CRN production scaling and topographic shielding factors). The luminescence part of the database consists of thermoluminescence (TL) and optically stimulated luminescence (OSL) measurements in fluvial sediment samples from stratigraphic sections and sediment cores from across the Australian continent. Comprehensive metadata and ancillary vector and raster geospatial data are likewise included and available for download. OCTOPUS can be accessed at https://earth.uow.edu.au (last access: 28 November 2018).

2 CRN and luminescence dating in a nutshell

This section briefly describes the two suites of dating techniques and provides information on CAIRN.

2.1 Inferring denudation rates from cosmogenic ^{10}Be (and ^{26}Al)

Cosmogenic nuclide exposure dating is based on the study of rare isotopes produced by high-energy cosmic radiation breaking up the atoms that make up the minerals and rocks at the Earth's surface. The term "in situ" is used to distinguish these isotopes from those that are produced through the same cosmic-ray-induced nuclear reactions in the atmosphere – termed "meteoric" (Dunai, 2010; Granger et al., 2013). Several of the in situ cosmogenic nuclides, including the stable ^{3}He and ^{21}Ne, and the radioactive ^{10}Be, ^{26}Al, and ^{36}Cl, are now routinely measured and have been used in geomorphological studies for the last three decades (Bierman and Nichols, 2004; von Blanckenburg, 2005; Dunai, 2010; Granger and Schaller, 2014). Of these nuclides, however, ^{10}Be produced in quartz is the workhorse for in situ applications, and most in situ cosmogenic nuclide studies have used ^{10}Be either alone or in conjunction with other cosmogenic nuclides such as ^{26}Al and ^{21}Ne. Given the long half-life of ^{10}Be ($T_{1/2} = 1.387$ Myr, Chmeleff et al., 2010; Korschinek et al., 2010) and the increasingly low analytical backgrounds that can be realised, it is now possible to analyse samples covering a wide range of temporal settings, including historic times (e.g. Schaefer et al., 2009). The rate at which cosmogenic nuclides are produced is extremely low – a few atoms per gram of rock per year (Borchers et al., 2015) and the rapid attenuation of cosmic radiation with depth confines the production of cosmogenic nuclides to the upper few metres of the crust, production rates decreasing roughly exponentially with depth (Argento et al., 2015a, b). Production rates of cosmogenic nuclides are mainly a function of geomagnetic latitude and altitude above sea level (Balco et al., 2008; Lifton et al., 2014). Site-specific cosmogenic nuclide production rates are also subject to several other factors, the most important of these being the geometry of the surrounding topography, which shields part of the incoming cosmic radiation (Dunne et al., 1999; Codilean, 2006; DiBiase, 2018).

The application of ^{10}Be (or any other in situ-produced cosmogenic nuclide) to the study of Earth surface processes is based on the principle that its concentration is directly proportional to the exposure time to cosmic radiation. Cosmogenic nuclides will accumulate in surficial deposits over time such that their concentration will be directly related to not only the exposure age but also the rate at which the surface is eroding (Lal, 1991; Granger et al., 2013; von Blanckenburg and Willenbring, 2014). As a parcel of rock or sediment is brought toward the surface by erosion on a hillslope, its ^{10}Be concentration increases at a rate that depends mainly on the rate of erosion, and the ^{10}Be surface production rate at that locality. When the parcel of rock or sediment reaches the surface, it is transported via hillslope processes to the fluvial system, where it mixes with sediment from other parts of the contributing catchment. Thus, rivers act not only as agents of erosion but also as integrators, collecting sediment from all parts of the catchment in an amount that is proportional to their denudation rate such that, at the outlet of the catchment, the sediment will contain an average concentration of ^{10}Be (and ^{26}Al) that is a measure of the catchment's mean denudation rate (von Blanckenburg, 2005; Granger and Schaller, 2014). The technique of determining basin-wide denudation rates from CRN concentrations in stream sediments was first introduced in the mid-1990s (e.g. Brown et al., 1995; Bierman and Steig, 1996; Granger et al., 1996), and since that time denudation rates have been determined in over 4000 river basins from a wide range of tectonic and climatic settings.

2.2 Luminescence dating of sediment

Luminescence dating provides an estimate of the amount of time elapsed since mineral grains (quartz or feldspar) were last exposed to intense heat or sunlight. The suite of techniques includes thermoluminescence dating (TL), in which the luminescence signal is produced by heating mineral grains in the laboratory during measurement (Aitken, 1985; Huntley et al., 1985), and optically stimulated luminescence dating (OSL), in which the luminescence signal is produced by exposing the mineral grains to an intense light source (Aitken, 1998). The suite of techniques can be used to date events as young as a few decades (e.g. Wolfe et al., 1995; Rustomji and Pietsch, 2007; Pietsch et al., 2015; Croke et al., 2016) to those as old as nearly 1 Ma (Arnold et al., 2015). The basis of both TL and OSL dating resides in measurements of the trapped charge (e.g. electrons) within mineral lattice imperfections which accumulate over time. When electrons are exposed to ionising radiation produced by the decay of radioisotopes contained in the surrounding sediment matrix, and/or via exposure to high-energy cosmic rays, electrons will move from a lower energy level (valence band) to a higher energy level (conduction band). Moving between the two bands, some of the energised electrons will become trapped by defects in the crystal lattice. In a parcel of sediment that is buried and thus shielded from sunlight and/or intense heat, the number of trapped electrons will increase steadily with time in proportion to the intensity of the ionising radiation flux (i.e. dose rate) and water saturation of the sediment. When the irradiated mineral grains are exposed to sunlight (or intense heat) the electrons will escape the traps, and the luminescence clock is zeroed. Thus, TL and OSL provide ages that represent the last time the electron traps were emptied or bleached – either by exposure of the sediment to sunlight (e.g. during sediment transport) or by heating (e.g. during a bush fire or in aboriginal hearths) (Wintle, 2008; Rhodes, 2011).

When luminescence dating techniques are applied to sediments, an often used assumption (when analysing multiple grains) is that the electron traps were completely emptied prior to deposition and so the luminescence clock has been effectively zeroed. In the case of OSL even short exposure to sunlight (< 1 min) is sufficient to bleach the sediment grains and thus zero the luminescence clock; however for TL, a longer exposure to sunlight is required to remove the TL signal. Fine mineral grains ($< 63\,\mu$m) that are transported by wind or as suspended fluvial sediment should be exposed to sunlight while airborne or in the upper parts of the water column. On the other hand, larger grains that travelled as bedload might only be partially bleached. Independent of grain size, when dealing with fluvial sediment, the bleaching characteristics of the sample need to be assessed in order to determine the use of an applicable age model. Possible strategies for determining bleaching characteristics include using geomorphic models that reconstruct mineral grain pathways and thus predict optimal bleaching regimes (e.g. Fuchs and Owen, 2008), analysing recently deposited sediment to assess their residual OSL–TL signals (e.g. Rhodes and Bailey, 1997; Singarayer et al., 2005), pairing with a second independent dating technique such as radiocarbon dating (e.g. Olley et al., 2004), and analysing different grain size fractions or closely spaced samples with different depositional energies under the assumption that these might have behaved differently during sediment transport (e.g. Richards et al., 2000). Alternatively, various age models can be applied to either single or multi-grain data sets (e.g. minimum age model; Galbraith et al., 1999), which statistically differentiate partially bleached grain populations so as to derive the equivalent dose and subsequent age of the depositional event.

2.3 The CAIRN method for calculating CRN-based basin-wide denudation rates

CAIRN is an automated, open-source method for calculating basin-averaged denudation rates so that inferred denudation rates are reproducible: the method ingests topographic data, cosmogenic ^{10}Be and ^{26}Al concentrations and a parameter file and any two users with the same inputs will calculate the same denudation rate (Mudd et al., 2016). CAIRN forward models ^{10}Be or ^{26}Al concentrations at every pixel for a given denudation rate, taking into account latitude and altitude scaling of CRN production rates as well as snow, self, and topographic shielding. The obtained concentrations are averaged to predict a basin-averaged ^{10}Be or ^{26}Al concentration, and Newton's method is then used to find the denudation rate for which the predicted concentration matches the measured concentration and to derive associated uncertainties. CAIRN is also capable of ingesting fixed denudation rates in masked portions of the input raster, allowing users to calculate spatially varying denudation rates in nested basins. In addition, CAIRN outputs spatially averaged CRN production scaling and topographic shielding values that can be used with other available CRN calculators that do not provide spatial averaging, including the online calculators formerly known as the CRONUS-Earth online calculators (Balco et al., 2008) and the Microsoft Excel-based COSMOCALC (Vermeesch, 2007). Because there is no graphical interface and because releases of the software are tagged, CAIRN users can simply publish digital elevation model (DEM) metadata, CRN data files, and CAIRN input files, and denudation rates should be reproducible. The open-source framework means that the code can be modified to include updated methods for production rates and scaling factors. Future users can thus recalculate denudation rates using updated versions of the code. CAIRN includes scripts for producing separate basin rasters for each cosmogenic sample from a regional topographic raster so that the denudation rate calculations can be run on multiple processors, meaning that large regional data sets can be processed simultaneously on compute clusters.

3 Accessing data from OCTOPUS

This section provides a description of the software infrastructure behind OCTOPUS. It also describes the ways in which data can be accessed, interrogated, and downloaded. The software infrastructure behind OCTOPUS consists of a combination of off-the-shelf open-source packages, bespoke code for handling the upload and download of data, and a web interface.

3.1 System architecture

The software architecture behind OCTOPUS is illustrated in Fig. 1. The data are stored in two separate locations. First, tabular data and the point and polygon geometries as-

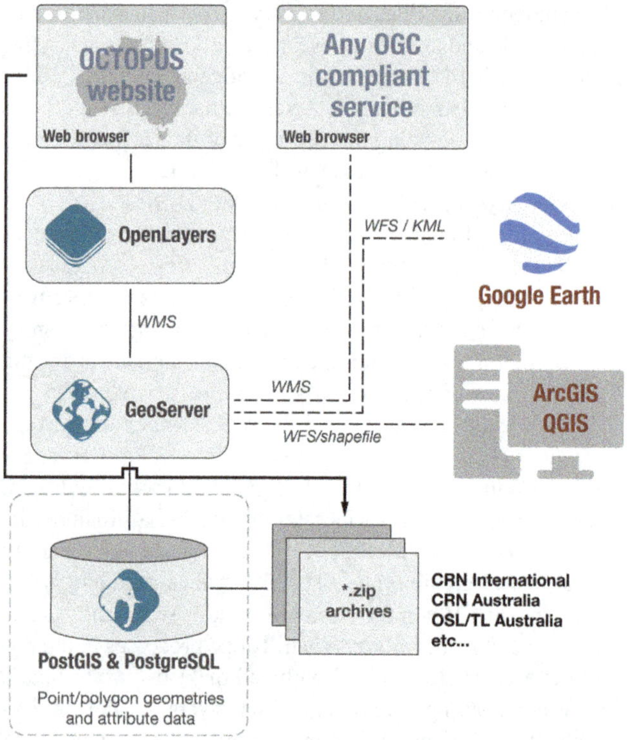

Figure 1. The setup of the OCTOPUS data-storage and -sharing platform. See text for more details.

sociated with each sample site or study (see Sect. 4) are stored in a PostGIS database. PostGIS (https://postgis.net, last access: 28 November 2018) is a spatial database extender for the PostgreSQL object-relational database management system (https://www.postgresql.org, last access: 28 November 2018), adding support for geographic objects and allowing location-based queries to be run in SQL. Second, all data (tabular, vector, and raster) and auxiliary information (e.g. CAIRN input and output files) (see Sect. 4) are also stored in separate zip archives, with one zip file for each study. This hybrid setup was chosen over having all tabular, vector, and raster data together in the PostGIS database because (i) it offered more flexibility regarding the list of files and file formats that could be included for download, and (ii) it made the coding of data upload and download simpler. The PostGIS database is connected to a GeoServer instance (Fig. 1). GeoServer is an open-source server that allows the sharing, processing, and editing of geospatial data (http://geoserver.org, last access: 28 November 2018), and implements a range of OCG data-sharing standards, including the widely used Web Feature Service (WFS) and the Web Map Service (WMS) standards. GeoServer also produces a variety of commonly used geospatial data formats via WFS, including KML and the ESRI Shapefile, and so can export data for using with popular desktop GIS applications such as Google Earth, ArcGIS, and QGIS (Fig. 1). It is also possible to connect to the GeoServer instance directly in QGIS and

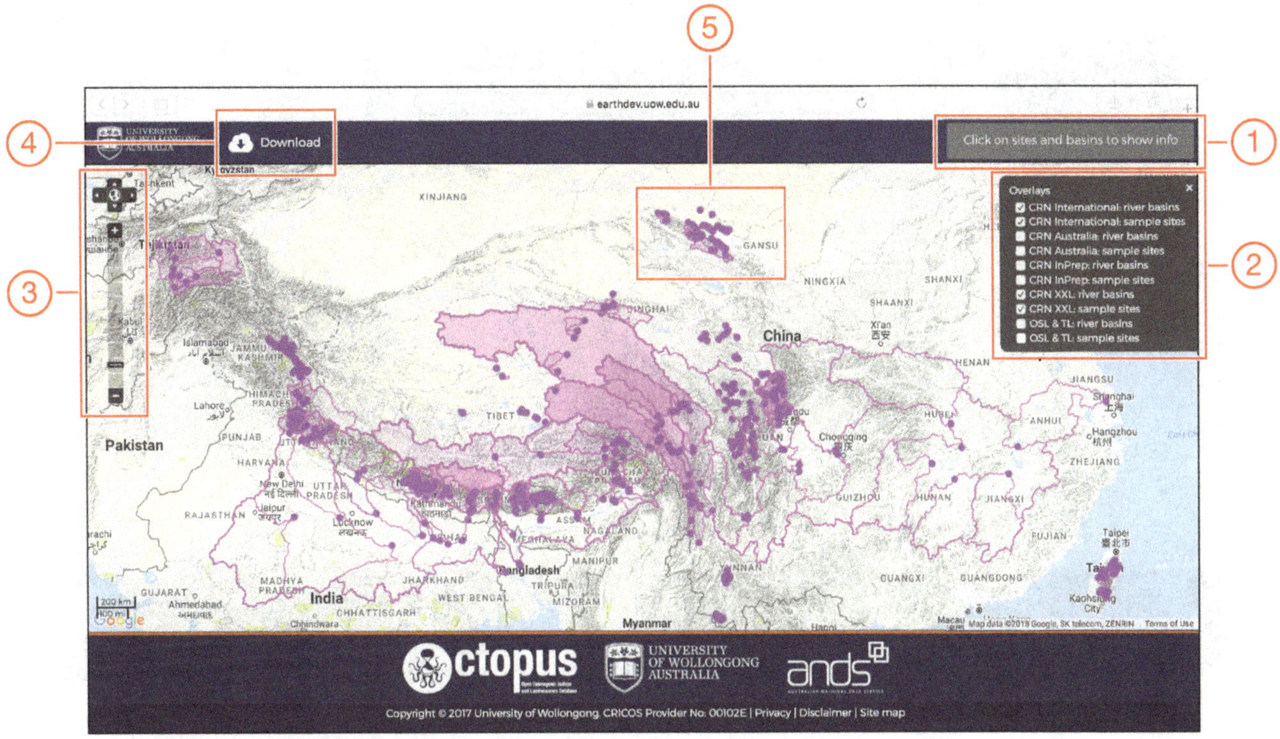

Figure 2. The OCTOPUS web interface with main elements: (1) message box that provides the user with step-by-step help on how to navigate the web page, (2) collapsible panel with a list of all available data layers, (3) navigation buttons, (4) data download button, (5) Google Terrain base map and point and polygon data layers. See text for more details.

interrogate the data via the WFS protocol (see below). The OpenLayers (https://openlayers.org, last access: 28 November 2018) JavaScript library is used to display the geospatial data served by the GeoServer instance in a web browser (Fig. 1). OpenLayers also allows for the data to be queried and a selection to be made for download.

3.2 Accessing data using the web interface

The web interface has a simple design and its sole purpose is to enable users to visualise the various data collections and to select data for download. The web interface includes the following elements (Fig. 2): a message box that provides the user with step-by-step help on how to navigate the web page (Fig. 2, #1); a collapsible panel with a list of all available data layers – these are grouped by data collection (see Sect. 4) and can be toggled on or off (Fig. 2, #2); navigation buttons allowing zooming and scrolling (Fig. 2, #3); and the data download button (Fig. 2, #4). The latter opens a dialogue panel and switches the cursor from panning mode to selection mode, allowing for data layers to be selected and added to a download list. The OpenLayers map frame uses Google Terrain as the base layer and the point and polygon data are displayed using different colours for each collection (Fig. 2, #5).

Figure 3 illustrates a typical user interaction with the web interface. First, the user displays the data collection(s) of interest and navigates to the desired geographical area. This can be achieved by using the navigation buttons or simply by clicking (to zoom) and dragging (to pan) on the map area. To query the data, the user clicks on a point or polygon feature. This action displays an information panel that includes a subset of the records available as part of the attribute table for each point or polygon feature. In the case of overlapping features, the information panel displays records for all features (Fig. 3, #1). Displayed information includes sample ID, publication details, and recalculated ^{10}Be-based denudation rate with uncertainty for CRN data, or published age with uncertainty for OSL–TL data. The dialogue panel closes automatically once the user clicks anywhere outside of the panel in the map display window. The displayed data are only a subset of the available attribute data and are meant to provide the user with basic information about each point or polygon record. To download data, the user clicks on the download button. This action turns the cursor into a selection tool (the user drags a box around desired points and polygons to select) and displays a dialogue panel requesting user information such as name, email address, and the intended use of the data (Fig. 3, #2–3). The user has the option to fine-tune the list of selected studies by toggling on or off each study from the list generated after the selection box is drawn. It is

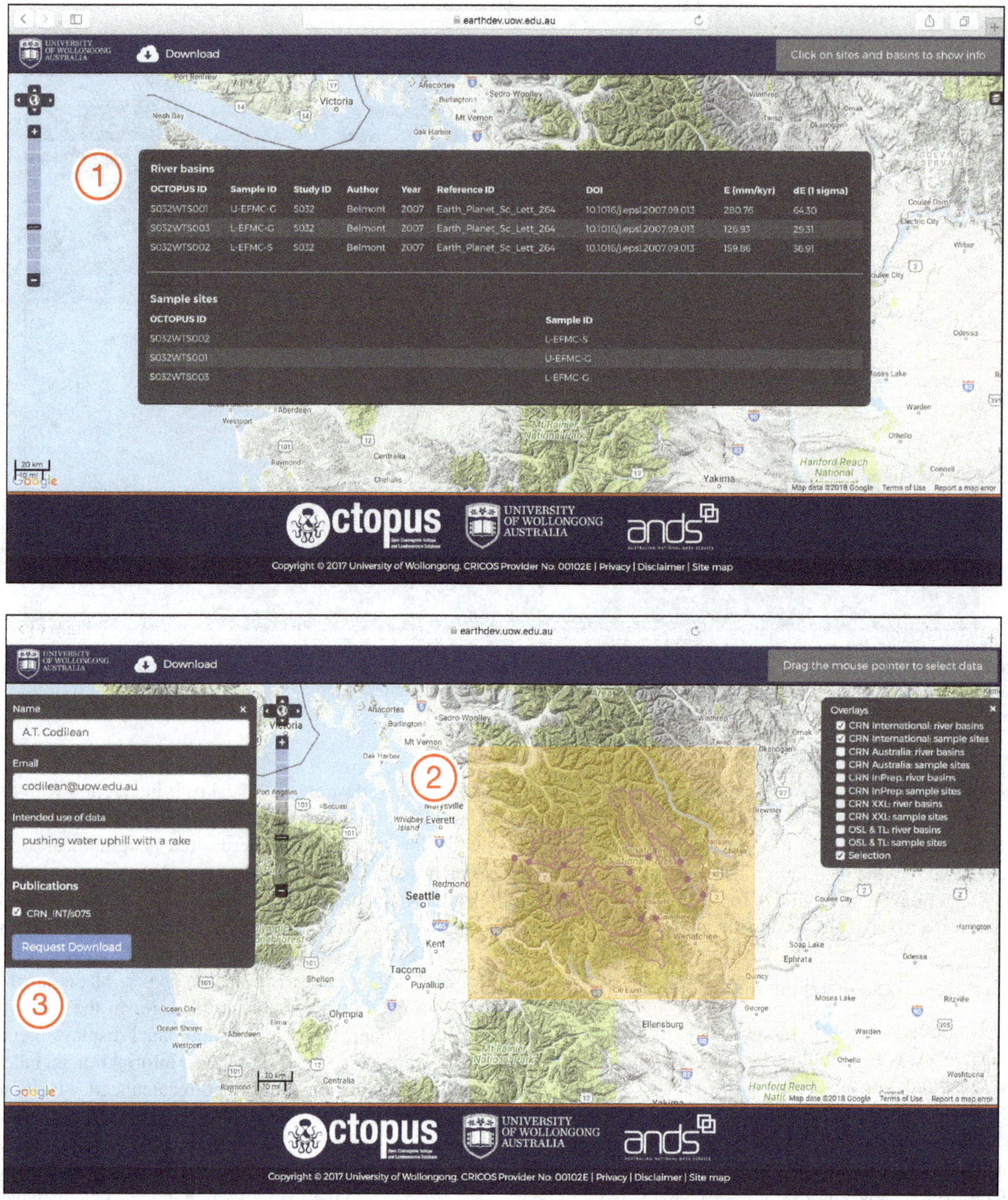

Figure 3. Screenshots of the OCTOPUS web page illustrating a typical user session. See text for more details.

possible to select multiple studies from multiple collections at the same time. A valid email address is required as links to the data are sent to the user via email immediately after the download button is pressed. There is no verification of who the data requestor is or where that person is from; however, none of the fields can be left empty and all entered information is stored in a log file permanently and is used for reporting purposes. Thus, although not mandatory, providing some meaningful information when downloading the data via the web interface will support future efforts to secure funding for OCTOPUS.

3.3 Accessing data using the Web Feature Service (WFS) capability

The web interface allows users to download all of the data: tabular, vector, and raster. The data are organised in studies – each publication is a "study" – with files belonging to each study stored in separated zip archives (see below). The size of these zip archives ranges from as small as 1 MB to just over 2.5 GB, and so the web interface is meant for downloading a small number of studies per session rather than the entire collection – the size of which at the time of writing was just over 165 GB. Users who want access to a subset or to the entire collection but who do not need access to the raster data, however, can download the tabular and vector data using the WFS capability running on the GeoServer instance, instead. WFS allows geospatial data to be interrogated and requested for download using a URL via a web browser or displayed directly in desktop applications such as QGIS (see below). It is beyond the scope of this paper to provide a manual on WFS. Rather, here we provide a series of examples that users can modify to perform basic queries and download data. For a more comprehensive introduction to WFS and GeoServer, the reader is referred to Iacovella and Youngblood (2013) or to the GeoServer documentation web page at http://docs.geoserver.org (last access: 28 November 2018).

The following example WFS request will download all drainage basins belonging to the CRN International collection in the ESRI Shapefile format:

```
http://earth.uow.edu.au:80/geoserver/
        wfs?request=GetFeature&typename
        =be10-denude:crn_int_basins&
        outputformat=SHAPE-ZIP
```

In the above example, `be10-denude:crn_int_basins` is the name of the data layer to be downloaded and `SHAPE-ZIP` refers to the format used to export the data (here, ESRI Shapefile). To obtain a full list of available data layers and export data formats, one should use the following request:

```
http://earth.uow.edu.au:80/geoserver/
        wfs?request=GetCapabilities
```

and look under `FeatureTypeList` and `ows:OperationsMetadata` in the results displayed for layer names and data formats, respectively. It is possible to request only a subset of the data by using the CQL/ECQL query language. For example, the following WFS request will download all drainage basins (with CRN data in the attribute table) from the CRN Australia collection in the ESRI Shapefile format that were published between 2005 and 2010:

```
http://earth.uow.edu.au:80/geoserver/
        wfs?request=GetFeature&typename
        =be10-denude:crn_aus_basins&
        outputformat=SHAPE-ZIP&
        CQL_FILTER=pubyear+BETWEEN+
        2005+AND+2010
```

where `pubyear` is the name of the field containing the publication year (see Table S1, included as part of the Supplement). Similarly, `CQL_FILTER=ebe_mmkyr<10` will download all records with a ^{10}Be denudation rate < 10 mm kyr^{-1}, and `CQL_FILTER=studyid='S066'` will download all records that belong to study S066. Lastly, it is also possible to subset the data by geographic location:

```
http://earth.uow.edu.au:80/geoserver/
        wfs?request=GetFeature&typename
        =be10-denude:crn_int_basins&
        outputformat=SHAPE-ZIP&
        BBOX=-0,20,40,60,EPSG:4326
```

where `0,20,40,60` are the coordinates of the bounding box used to clip the data (following the format: x1,x2,y1,y2) and `EPSG:4326` indicates that the coordinates are WGS84 latitude and longitude in degrees (for a complete list of EPSG references see http://spatialreference.org, last access: 28 November 2018).

For users that wish to display and interrogate the data without actually downloading any files, it is possible to access OCTOPUS from QGIS directly by using the "add WFS layer" function and connecting to

```
http://earth.uow.edu.au:80/geoserver/wfs
```

QGIS is a free and open-source cross-platform desktop GIS application that supports viewing, editing, and analysis of geospatial data (https://www.qgis.org, last access: 28 November 2018). An example QGIS session with a WFS connection to OCTOPUS is shown in Fig. 4. A WFS connection will provide direct access to the data, and so any data accessed remotely is treated in the same way as data stored locally. Thus, it is possible to modify the symbology of the layers, query the data, and run analysis functions on them. However, all the above come at the cost of much more data being transmitted, and so users wanting to perform analyses on the OCTOPUS collections are recommended to instead download a local copy of the data first using one of the methods described above.

4 The OCTOPUS data collections and data structure

The compiled CRN and OSL–TL data are organised in three collections, namely (i) CRN International, including ^{10}Be (and ^{26}Al) measurements in fluvial sediment samples from across the globe but excluding Australia; (ii) CRN Australia, including ^{10}Be (and ^{26}Al) measurements in fluvial sediment samples from Australia; and (iii) OSL & TL Australia, including OSL and TL measurements in fluvial sediment samples from stratigraphic sections and sediment cores from across the Australian continent. The aim of OCTOPUS is to

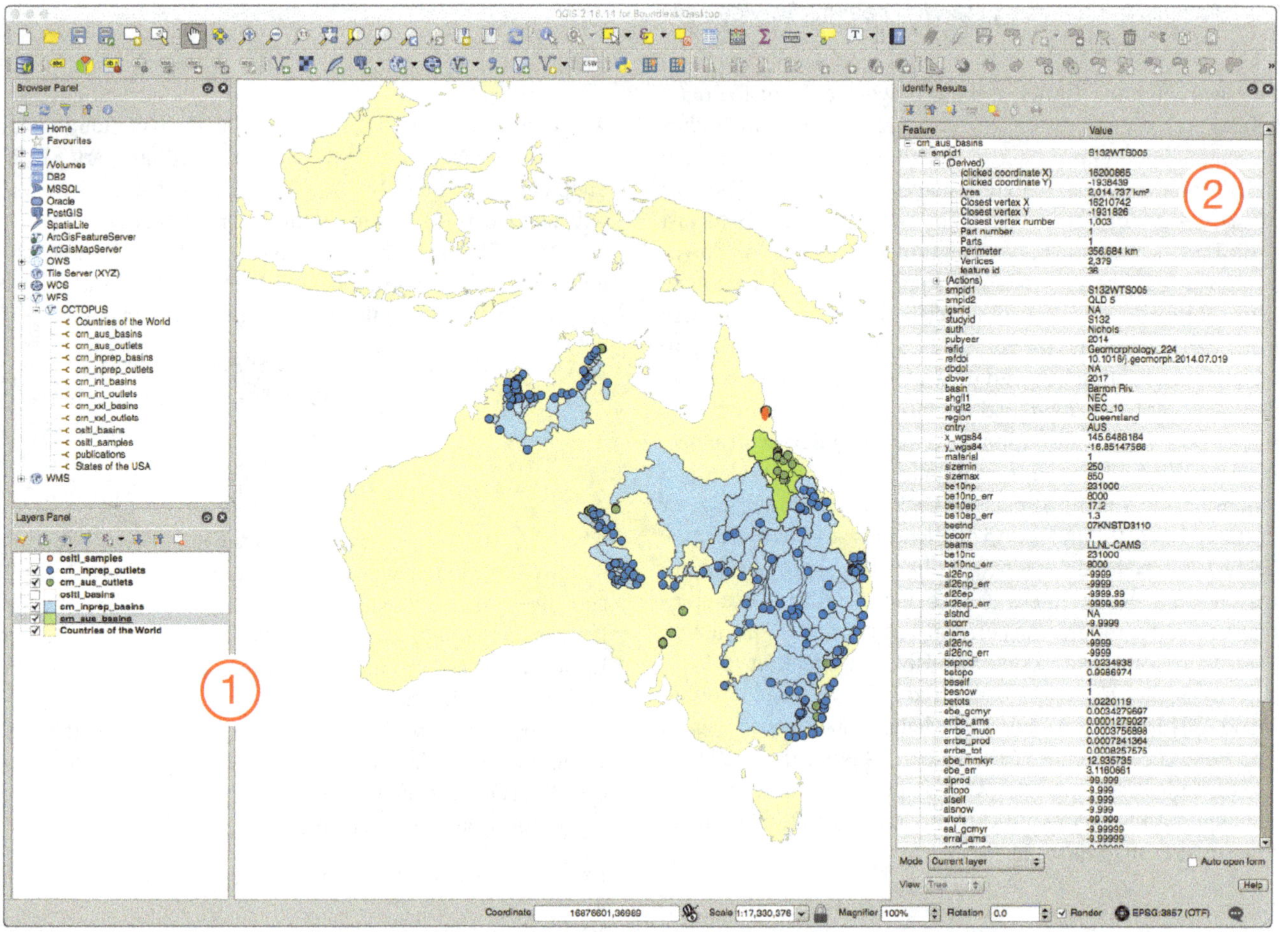

Figure 4. Screenshot of the QGIS application interface displaying data from OCTOPUS accessed through a WFS connection. A WFS connection will provide direct access to the data, meaning that (1) its symbology can be modified, (2) it can be queried, and analysis functions can be run on it.

compile and incorporate all data – both published and unpublished – that is publicly available and we do not think that it is our role to decide on the quality of the data that is already published. However, in some instances, where a publication did not provide sufficient information for the data files to be produced (e.g, insufficient information to be able to confidently locate and delineate drainage basins) and this information could not be obtained from elsewhere, those data were excluded from OCTOPUS. Further, despite our best efforts it is likely that we have missed some studies during our search. Given the above, there are studies that were excluded from the current release of OCTOPUS. This was not an editorial decision except in cases where we had no choice due to a lack of information (see above).

4.1 CRN International and CRN Australia

The CRN International and CRN Australia collections consist of ^{10}Be (and where available, also ^{26}Al) basin-wide denudation rates published in the peer-reviewed literature up to 2018. As already mentioned, the data are organised in studies, with files belonging to each study stored in separated zip archives (Fig. 5). The mean sample number per study is ~ 20 and the ratio of published ^{26}Al to published ^{10}Be measurements is approximately 1 to 10. For each ^{10}Be data point, there is a point geometry file representing the location of the sample site, and a polygon geometry file representing the outline of the drainage basin from which the sampled material is originating. An attribute table including published and recalculated ^{10}Be (and ^{26}Al) data and a comprehensive set of metadata is linked to the polygon geometry file. A complete description of all attribute data entries is provided in Table S1, included as part of the Supplement. For each study, each zip archive also includes seven raster layers: (i) a hydrologically corrected DEM with elevation values in metres (file name suffix: `_demhydro`), (ii) a flow-direction raster calculated using the D8 flow-routing method (Jenson and Domingue, 1988) (`_d8flowdir`), (iii) a flow-accumulation raster calculated with the same D8 method (`_flowacc`), (iv) a slope gradient raster calculated using

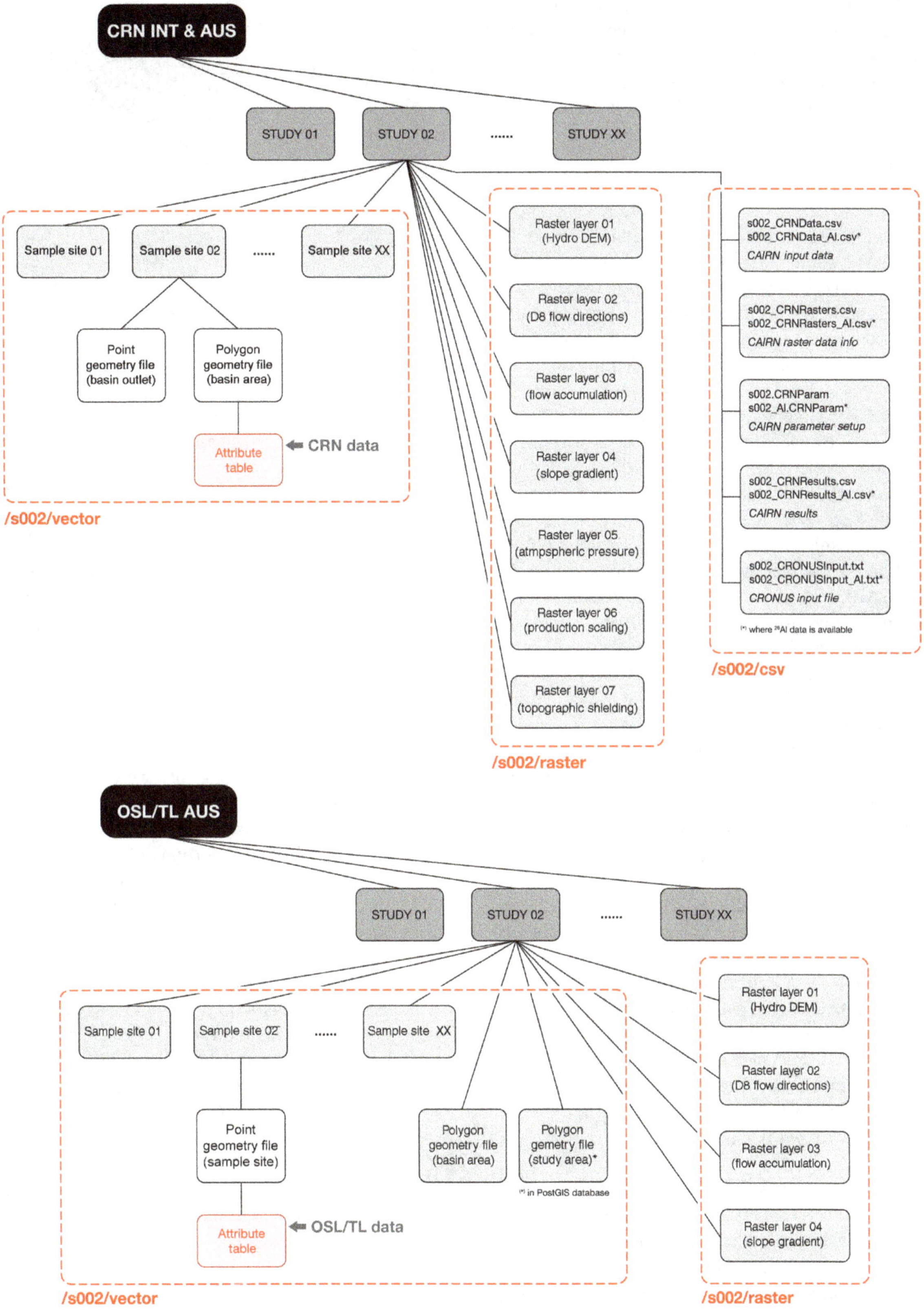

Figure 5. Data organisation diagrams for the CRN and OSL–TL collections. See text for details.

the method described in Horn (1981) with units in m km^{-1} (_gradmkm), (v) an atmospheric pressure raster, showing local atmospheric pressure in hPa calculated based on the NCEP2 climate reanalysis data (Compo et al., 2011) (_atmospres), (vi) a cosmogenic nuclide production scaling raster calculated using the method described in Stone (2000) (_prodscale), and (vii) a cosmogenic nuclide production topographic shielding raster calculated using the method described in Codilean (2006) (_toposhield). All raster layers were derived using the Shuttle Radar Topography Mission (SRTM) 90 m Digital Elevation Database (Farr et al., 2007) and extend 20 km beyond the boundaries of the drainage basins in each study. For two studies, namely Henck et al. (2011) and Reber et al. (2017), due to very large basin areas, all raster layers with the exception of slope gradient were calculated from SRTM data resampled to 500 m resolution. Each zip archive also includes a series of text files representing CAIRN configuration and input data files, and CAIRN output files, including files to be used with the online calculators formerly known as the CRONUS-Earth online calculators (Balco et al., 2008).

Published ^{10}Be concentrations (atoms g^{-1}) were renormalised to the Nishiizumi 2007 ^{10}Be AMS standard (Nishiizumi et al., 2007), and basin-wide denudation rates recalculated with CAIRN. Basin-averaged nuclide production from neutrons and muons was calculated with the approximation of Braucher et al. (2011) and using a sea-level and high-latitude total ^{10}Be production rate of 4.3 atoms g^{-1} yr^{-1} (Mudd et al., 2016). Production rates for catchment-wide denudation rates were calculated at every pixel using the SRTM 90 m DEM, with the time-independent Lal/Stone scaling scheme (Stone, 2000). Atmospheric pressure was calculated via interpolation from the NCEP2 reanalysis data (Compo et al., 2011). Topographic shielding was calculated from the same DEM using the method of Codilean (2006) with $\theta = 8°$ and $\phi = 5°$. Following the submission of this paper, a new study by DiBiase (2018) showed that topographic shielding corrections are inappropriate for calculating basin-wide denudation rates, in most settings, and are only required for steep catchments with non-uniform distribution of quartz and/or denudation rates. For this reason, future iterations of the CRN International and CRN Australia collections will also include ^{10}Be (and where available ^{26}Al) denudation rates calculated without correcting for topographic shielding. All calculations assumed a ^{10}Be half-life of 1.387 ± 0.012 Myr (Chmeleff et al., 2010; Korschinek et al., 2010).

For simplicity and consistency across the global compilation, no corrections were made for lithological differences in quartz abundance, glacier cover, and snow shielding. Performing such corrections in a consistent manner on a global scale is impossible. However, all CAIRN input and configuration files are provided and these corrections can be readily applied by end users to individual studies. Further, CRN International and CRN Australia include both the originally published denudation rates and the ones recalculated using CAIRN, and so detailed comparisons can be made by users. Figure 6 shows a first-order comparison between published and recalculated ^{10}Be denudation rates. With the exception of a small number of data points ($n \sim 10$), there is good agreement between published and recalculated ^{10}Be denudation rates, with no obvious trends related to elevation or basin size. Where large discrepancies exist, these are due either to differences in drainage basins as published versus drainage basins identified on the SRTM DEM during data recalculation or due to corrections that were applied to the data in the original publication that were not appropriately described in the latter. Discrepancies also exist in the case of studies where substantial portions of a drainage basin consisted on non-quartz-bearing lithologies (e.g. Safran et al., 2005; Croke et al., 2015) and where corrections for quartz abundance were applied to the data in the original publications but were not replicated here. The number of such basins is small, however, and will not impact any regional or larger-scale analyses done with the CRN data. For small-scale studies users should compare published with recalculated denudation rates and determine whether a new recalculation that involves corrections for quartz abundance, glacier cover, and/or snow shielding is warranted.

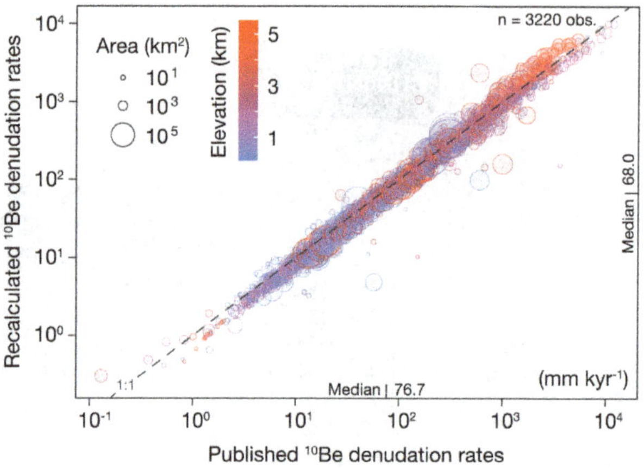

Figure 6. Published versus recalculated ^{10}Be-based denudation rates. Data points are coloured according to average basin elevation and circle sizes are proportional to basin area. Note the good agreement between the two data sets and the lack of obvious trends related to basin elevation and basin area.

Approximately 5 % of compiled ^{10}Be measurements – all of which were published in two highly regarded journals – could not be incorporated into OCTOPUS due to information that is insufficient to reproduce drainage basin extents.

In terms of geographical extent, the global CRN compilation exhibits considerable bias (Fig. 7a). The majority of the ^{10}Be (and ^{26}Al) measurements are from Northern Hemisphere drainage basins, clustering around distinct, mostly tectonically active, topographic regions, such as the Pacific coast of the United States, the Appalachians, the European

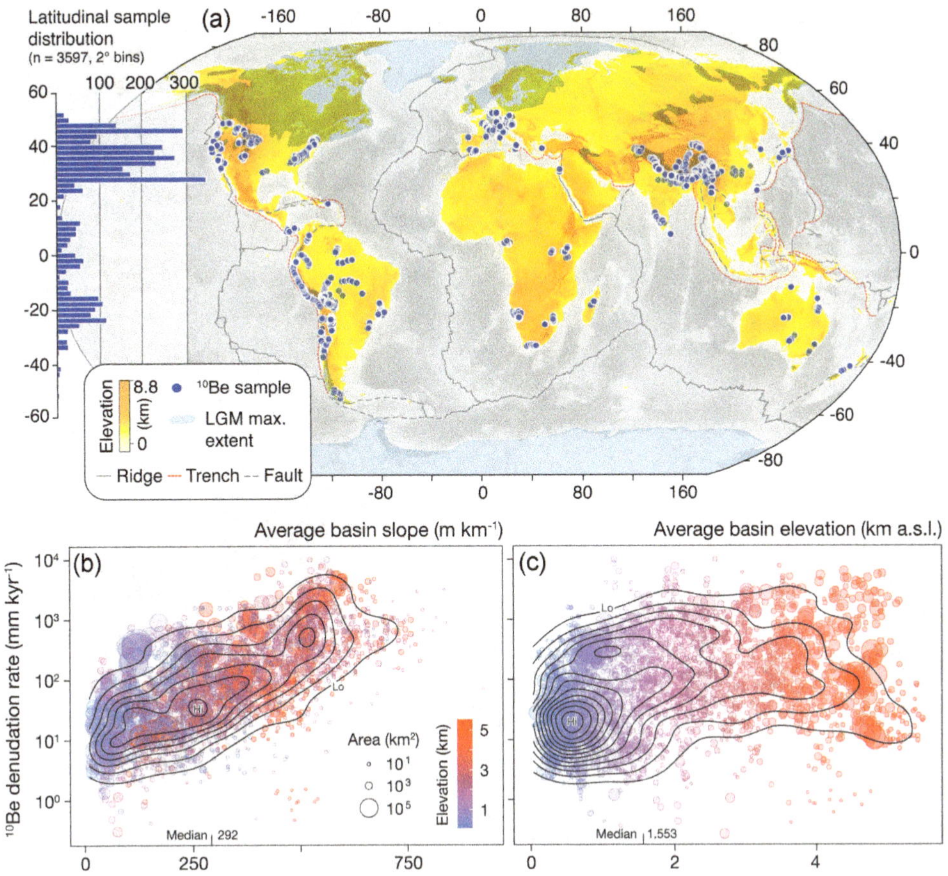

Figure 7. The global CRN data set: **(a)** geographical extent and latitudinal sample distribution, **(b)** average basin slope versus recalculated ^{10}Be denudation rate, and **(c)** average basin elevation versus recalculated ^{10}Be denudation rate. Data points in **(b)** and **(c)** are coloured according to average basin elevation and circle sizes are proportional with basin area. Contour lines show kernel density estimates for the point clouds (arbitrary units).

Alps, and the Tibet-Himalaya region. Due to some recent studies, there is also good coverage of the South American Cordillera. However, there is a considerable lack of data from low-gradient and tectonically passive regions, such as large parts of Australia, most of Africa, and most of Asia less the Tibet-Himalaya region. Further, there are no data from latitudes above $\sim 55°$. The observed geographical bias is a reflection of the intense interest of the geomorphological community in estimating rates of erosion and weathering in tectonically active mountain regions with one of several aims to understand the role of surface processes in the global climate system (e.g. Molnar and England, 1990; Raymo and Ruddiman, 1992; Willenbring and von Blanckenburg, 2010; Herman et al., 2013). Further, the lack of data from high latitudes is partly due to the desire to stay away from formerly glaciated environments. Although the geographical bias does not make the CRN collection less valuable, it may confound studies aiming to infer global-scale trends from these data (cf. Portenga and Bierman, 2011; Willenbring et al., 2013; Harel et al., 2016). Despite the geographical bias, however, the global CRN data sample basins with a wide range of slope gradients, elevations, and basin areas (Fig. 7b–c).

4.2 OSL & TL Australia

The OSL & TL Australia collection consists of thermoluminescence (TL) and optically stimulated luminescence (OSL) measurements in fluvial sediment samples from stratigraphic sections and sediment cores from across the Australian continent. The collection includes data published in the peer-reviewed literature up to 2017 and also previously unpublished data compiled from technical reports and various Honours, MSc, and PhD theses. The majority of the TL data are from sources published from 1986 up to 2005, whereas the majority of the OSL data are from sources less than 10 years old (Fig. 8). In terms of geographical extent both TL and OSL data are concentrated in the south-eastern and eastern parts of the Australian continent, with ~ 500 measurements from Australia's largest river basins – Lake Eyre (LEB) and Murray-Darling (MDB) basins – and with an equal amount from rivers draining the eastern seaboard

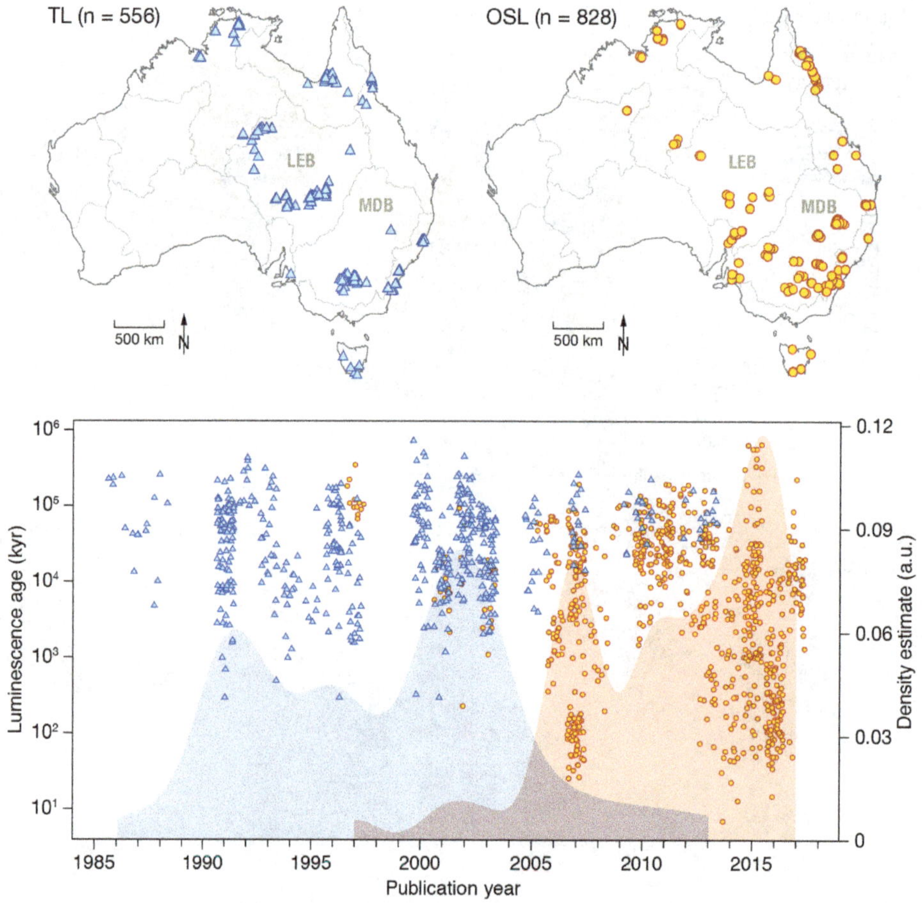

Figure 8. The spatial and temporal extents of the OSL & TL Australia data set. Blue triangles denote TL measurements and yellow circles denote OSL measurements. Grey lines depict major topographic drainage divisions and river regions based on the Australian Hydrological Geospatial Fabric (Geofabric).

(Fig. 8). The western half of Australia is severely understudied, with one single OSL study for the entire region, namely Veth et al. (2009). Focused interest on river systems is proximal to high-population density areas, where floods are a potential threat (e.g. Brisbane River, after major floods in 2011; Croke et al., 2016), or where rivers are of great agricultural importance, such as the MDB. This well-justified bias, however, leaves a gap in the understanding of regions dominated by rivers that are now dry or ephemeral and yet that could hold information on past climatic regimes now buried under the desert sand. The focus on south-eastern coastal river systems draining the Great Dividing Range could be a source of bias in continent-wide interpretations, where the rivers draining the western intracontinental ranges and plains remain underrepresented.

Similarly to the CRN collections, the data are organised in studies – each publication is a study – with files belonging to each study stored in separated zip archives (Fig. 5). For each OSL or TL data point, there is a point geometry file representing the location of the sample site. An attribute table including published OSL or TL ages and a comprehensive set of metadata is linked to the point geometry file (a complete description of all attribute data entries is provided in Table S2, included as part of the Supplement). The zip archive also includes two separate polygon geometry files: one representing the outline of the drainage basin of the most downstream sample and one representing the area extending 20 km beyond the boundaries of this drainage basin. For studies with basin areas up to 100 000 km^2, each zip archive also includes four raster layers: (i) a hydrologically corrected DEM with elevation values in m (file name suffix: `_demhydro`), (ii) a flow-direction raster calculated using the D8 flow-routing method (Jenson and Domingue, 1988) (`_d8flowdir`), (iii) a flow-accumulation raster calculated with the same D8 method (`_flowacc`), and (iv) a slope gradient raster calculated using the method described in Horn (1981) with units in m km^{-1} (`_gradmkm`). All raster layers were derived using the hydrologically enforced SRTM 30 m digital elevation model (DEM-H) obtained from Geoscience Australia (Geoscience Australia, 2011) and were clipped to the extent of the 20 km buffer polygon layer. For studies with basin areas exceeding 100 000 km^2 (e.g. Callen and Nanson,

1992; Bourman et al., 2010; Jansen et al., 2013) raster layers (i) to (iii) were derived using Geoscience Australia's GEODATA 250 m digital elevation model and flow-direction grid (Geoscience Australia, 2008), as the SRTM DEM produced files that were too large to transfer online.

4.3 Other collections

In addition to the CRN and OSL–TL collections described above, the current version of OCTOPUS also includes additional CRN data organised under two collections: CRN XXL and CRN In-Prep. These two collections are not officially supported by the OCTOPUS project, and are included here only for completeness. The first collection consists of five studies with samples from the Yangtse (Chappell et al., 2006), Amazon (Wittmann et al., 2009, 2011), Ganga (Lupker et al., 2012), and Brahmaputra basins (Lupker et al., 2017). These studies focused on very large basins that could only be handled by CAIRN when run using a 500 m resolution DEM that, however, produced drainage basins that were substantially different to what was published, especially in the case of rivers in the Amazon basin. Further, Chappell et al. (2006) do not report denudation rates – suggesting that calculating these might have little meaning for their samples – and both Wittmann et al. (2009, 2011) and Lupker et al. (2012, 2017) perform corrections to the data, some of which (e.g. removing floodplain areas from production rate calculations) we did not wish to replicate. To this end, CRN XXL does not include recalculated values nor does it include any raster layers. CRN In-Prep is an inventory of samples processed at the University of Wollongong where ^{10}Be and ^{26}Al have been measured and the data are not yet published. The collection includes sample metadata and point and polygon geometry files.

5 User contributions to OCTOPUS

User contributions to OCTOPUS are welcome. Those wishing to submit data should download a study and use that as the template for data structure, formats, and naming convention (see also Fig. 5). As a minimum, a contribution should include point and polygon geometry files, and an attribute table with all records listed in Tables S1 and S2 with the exception of those records that are output by CAIRN. Data files should be submitted to the contact address listed in the email received from OCTOPUS when downloading data. The data collections making up OCTOPUS have been assigned digital object identifiers (DOIs), and as a consequence, adding new data needs to follow a versioning scheme, with each new version requiring new DOIs. Thus, data contributed by users will be incorporated in the next release of a given collection, rather than being added to the current one.

6 Data availability

This copy is not supported and is used for testing modifications to the website and data collections before deployment to the official site. Users should refer to the DOIs provided to ensure that they are accessing the current and supported version of the data.

7 Conclusions

We have produced a database of cosmogenic radionuclide and luminescence measurements in fluvial sediment and we have built infrastructure for hosting and maintaining the data at the University of Wollongong and making them available to the research community via an Open Geospatial Consortium (OGC)-compliant web service. The database consists of ^{10}Be, ^{26}Al, TL, and OSL measurements in fluvial sediment samples along with ancillary geospatial vector and raster layers. Sample metadata are comprehensive and include all necessary information for the recalculation of ^{10}Be and ^{26}Al denudation rates using the CAIRN program. The repository and visualisation system enable easy search and discovery of available data. Use of open standards also ensures that data layers are visible to other OGC-compliant data-sharing services. Thus, this project will turn data that were previously invisible to those not within the CRN and luminescence research community into a findable resource. This aspect is of particular importance to industry or local government who are yet to discover the value of geochronological data, for example, in evaluating how human-induced land use practices have accelerated soil erosion and which measures are necessary for restoring these rates to their natural benchmark levels. Our intention is for OCTOPUS to become the default go-to place for CRN and luminescence data. The availability of the repository and its associated data curation framework will provide the opportunity for researchers to store, curate, recalculate and reuse previously published but otherwise unusable CRN and luminescence data. This delivers the potential to harness old but valuable data that would otherwise be lost to the research community. OCTOPUS will enable new research and generate new knowledge by converting a multitude of disconnected data sets into one connected and streamlined database. Current data sets allow local-scale analyses. The streamlined database will allow for regional-scale and even continental- or global-scale analyses. The transparent data reanalysis framework will also reduce research time and avoid the duplication of effort, which will be highly attractive to other researchers. Ultimately, OCTOPUS will ensure that CRN and luminescence data are reusable beyond the scope of the project for which they were initially collected.

Author contributions. ATC, HM, TJC, and WMS compiled the CRN and luminescence data; ATC and HM performed the GIS analyses and the data recalculations using CAIRN with input from SMM; AG designed and built the OCTOPUS platform and web interface with input from ATC. All authors contributed to the writing of the manuscript.

Competing interests. The authors declare that they have no conflict of interest.

Acknowledgements. We acknowledge financial support from the Australian National Data Service (ANDS – Project HVC15) and the University of Wollongong (UOW) to Alexandru T. Codilean and Timothy J. Cohen. Henry Munack benefited from an Endeavour Postdoctoral Research Fellowship granted by the Australian Government. A prototype of OCTOPUS was built while Alexandru T. Codilean worked at the Deutsches GeoForschungsZentrum (GFZ). We thank Monika Oakman, Ben Cornwell, Peter Newnam, Clare Job, Gabriel Enge, Heidi Brown, Alan Glixman, and Keith Russell for their support during this project. We also thank the CRN and luminescence communities for generously sharing their data. We thank Greg Balco, Zsófia Ruszkiczay-Rüdiger, Sebastian Kreutzer, and István Gábor Hatvani for careful reviews of the discussion paper. We also thank Ingrid Ward, Todd Ehlers, and Vincent Godard for providing comments on the discussion paper.

Edited by: Attila Demény

References

Aitken, M. J.: Thermoluminescence dating, Academic Press, London, 1985.

Aitken, M. J.: An Introduction to Optical Dating, Oxford University Press, Oxford, 1998.

Argento, D. C., Stone, J. O., Reedy, R. C., and O'Brien, K.: Physics-based modeling of cosmogenic nuclides part I: Radiation transport methods and new insights, Quat. Geochronol., 26, 29–43, https://doi.org/10.1016/j.quageo.2014.09.004, 2015a.

Argento, D. C., Stone, J. O., Reedy, R. C., and O'Brien, K.: Physics-based modeling of cosmogenic nuclides part II: Key aspects of in-situ cosmogenic nuclide production, Quat. Geochronol., 26, 44–55, https://doi.org/10.1016/j.quageo.2014.09.005, 2015b.

Arnold, L. J., Demuro, M., Parés, J. M., Pérez-González, A., Arsuaga, J. L., de Castro, J. M. B., and Carbonell, E.: Evaluating the suitability of extended-range luminescence dating techniques over early and Middle Pleistocene timescales: Published datasets and case studies from Atapuerca, Spain, Quaternary Int., 389, 167–190, https://doi.org/10.1016/j.quaint.2014.08.010, 2015.

Balco, G.: Contributions and unrealized potential contributions of cosmogenic-nuclide exposure dating to glacier chronology, 1990–2010, Quaternary Sci. Rev., 30, 3–27, https://doi.org/10.1016/j.quascirev.2010.11.003, 2011.

Balco, G., Stone, J. O., Lifton, N. A., and Dunai, T. J.: A complete and easily accessible means of calculating surface exposure ages or erosion rates from ^{10}Be and ^{26}Al measurements, Quat. Geochronol., 3, 174–195, https://doi.org/10.1016/j.quageo.2007.12.001, 2008.

Bierman, P. R. and Nichols, K. K.: Rock to sediment – slope to sea with ^{10}Be – rates of landscape change, Annu. Rev. Earth Pl. Sc., 32, 215–255, https://doi.org/10.1146/annurev.earth.32.101802.120539, 2004.

Bierman, P. R. and Steig, E.: Estimating rates of denudation using cosmogenic isotope abundances in sediment, Earth Surf. Proc. Land., 21, 125–139, https://doi.org/10.1002/(SICI)1096-9837(199602)21:2<125::AID-ESP511>3.0.CO;2-8, 1996.

Borchers, B., Marrero, S., Balco, G., Caffee, M. W., Goehring, B. M., Lifton, N. A., Nishiizumi, K., Phillips, F., Schaefer, J. M., and Stone, J. O.: Geological calibration of spallation production rates in the CRONUS-Earth project, Quat. Geochronol., 31, 188–198, https://doi.org/10.1016/j.quageo.2015.01.009, 2015.

Bourman, R. P., Prescott, J. R., Banerjee, D., Alley, N. F., and Buckman, S.: Age and origin of alluvial sediments within and flanking the Mt Lofty Ranges, southern South Australia: a Late Quaternary archive of climate and environmental change, Aust. J. Earth Sci., 57, 175–192, https://doi.org/10.1080/08120090903416260, 2010.

Braucher, R., Merchel, S., Borgomano, J., and Bourles, D. L.: Production of cosmogenic radionuclides at great depth: A multi element approach, Earth Planet. Sc. Lett., 309, 1–9, https://doi.org/10.1016/j.epsl.2011.06.036, 2011.

Bristow, C., Augustinus, P., Wallis, I., Jol, H., and Rhodes, E.: Investigation of the age and migration of reversing dunes in Antarctica using GPR and OSL, with implications for GPR on Mars, Earth Planet. Sc. Lett., 289, 30–42, https://doi.org/10.1016/j.epsl.2009.10.026, 2010.

Brown, E., Stallard, R., Larsen, M. C., Raisbeck, G., and Yiou, F.: Denudation Rates Determined From the Accumulation of In Situ-Produced ^{10}Be in the Luquillo Experimental Forest, Puerto-Rico, Earth Planet. Sc. Lett., 129, 193–202, https://doi.org/10.1016/0012-821X(94)00249-X, 1995.

Callen, R. and Nanson, G.: Formation and age of dunes in the Lake Eyre Depocentres, Geol. Rundsch., 81, 589–593, https://doi.org/10.1007/BF01828619, 1992.

Chappell, J., Zheng, H., and Fifield, K.: Yangtse River sediments and erosion rates from source to sink traced with cosmogenic ^{10}Be: Sediments from major rivers, Palaeogeogr. Palaeocl., 241, 79–94, https://doi.org/10.1016/j.palaeo.2006.06.010, 2006.

Chmeleff, J., von Blanckenburg, F., Kossert, K., and Jakob, D.: Determination of the ^{10}Be half-life by multicollector ICP-MS and liquid scintillation counting, Nucl. Instrum. Meth. B, 268, 192–199, https://doi.org/10.1016/j.nimb.2009.09.012, 2010.

Codilean, A. T.: Calculation of the cosmogenic nuclide production topographic shielding scaling factor for large areas using DEMs, Earth Surf. Proc. Land., 31, 785–794, https://doi.org/10.1002/esp.1336, 2006.

Codilean, A. T., Cohen, T. J., and Munack, H.: OCTOPUS –CRN International, https://doi.org/10.4225/48/5a8367feac9b2, 2017a.

Codilean, A. T., Cohen, T. J., and Munack, H.: OCTOPUS – CRN Australia, https://doi.org/10.4225/48/5a836cdfac9b5, 2017b.

Codilean, A. T., Cohen, T. J., Munack, H., and Saktura, W. M.: OCTOPUS – OSL/TL Australia, https://doi.org/10.4225/48/5a836db1ac9b6, 2017c.

Compo, G. P., Whitaker, J. S., Sardeshmukh, P. D., Matsui, N., Allan, R. J., Yin, X., Gleason, B. E., Vose, R. S., Rutledge, G., Bessemoulin, P., Brönnimann, S., Brunet, M., Crouthamel, R. I., Grant, A. N., Groisman, P. Y., Jones, P. D., Kruk, M. C., Kruger, A. C., Marshall, G. J., Maugeri, M., Mok, H. Y., Nordli, O., Ross, T. F., Trigo, R. M., Wang, X. L., Woodruff, S. D., and Worley, S. J.: The Twentieth Century Reanalysis Project, Q. J. Roy. Meteor. Soc., 137, 1–28, https://doi.org/10.1002/qj.776, 2011.

Croke, J., Bartley, R., Chappell, J., Austin, J. M., Fifield, K., Tims, S. G., Thompson, C. J., and Furuichi, T.: ^{10}Be-derived denudation rates from the Burdekin catchment: The largest contributor of sediment to the Great Barrier Reef, Geomorphology, 241, 122–134, https://doi.org/10.1016/j.geomorph.2015.04.003, 2015.

Croke, J., Thompson, C., Denham, R., Haines, H., Sharma, A., and Pietsch, T.: Reconstructing a millennial-scale record of flooding in a single valley setting: the 2011 flood-affected Lockyer Valley, south-east Queensland, Australia, J. Quaternary Sci., 31, 936–952, https://doi.org/10.1002/jqs.2919, 2016.

DiBiase, R. A.: Short communication: Increasing vertical attenuation length of cosmogenic nuclide production on steep slopes negates topographic shielding corrections for catchment erosion rates, Earth Surf. Dynam., 6, 923–931, https://doi.org/10.5194/esurf-6-923-2018, 2018.

Dunai, T. J.: Cosmogenic Nuclides: Principles, concepts and applications in the Earth surface sciences, Cambridge University Press, Cambridge, 2010.

Dunai, T. J. and Stuart, F. M.: Reporting of cosmogenic nuclide data for exposure age and erosion rate determinations, Quat. Geochronol., 4, 437–440, https://doi.org/10.1016/j.quageo.2009.04.003, 2009.

Dunne, J., Elmore, D., and Muzikar, P.: Scaling factors for the rates of production of cosmogenic nuclides for geometric shielding and attenuation at depth on sloped surfaces, Geomorphology, 27, 3–11, https://doi.org/10.1016/S0169-555X(98)00086-5, 1999.

Farr, T. G., Rosen, P. A., Caro, E., Crippen, R., Duren, R., Hensley, S., Kobrick, M., Paller, M., Rodriguez, E., Roth, L., Seal, D., Shaffer, S., Shimada, J., Umland, J., Werner, M., Oskin, M., Burbank, D., and Alsdorf, D.: The Shuttle Radar Topography Mission, Rev. Geophys., 45, RG2004, https://doi.org/10.1029/2005RG000183, 2007.

Fitzsimmons, K. E., Rhodes, E. J., Magee, J. W., and Barrows, T. T.: The timing of linear dune activity in the Strzelecki and Tirari Deserts, Australia, Quaternary Sci. Rev., 26, 2598–2616, https://doi.org/10.1016/j.quascirev.2007.06.010, 2007.

Frankel, K. L., Finkel, R. C., and Owen, L.: Terrestrial Cosmogenic Nuclide Geochronology Data Reporting Standards Needed, Eos T. Am. Geophys. Un., 91, 31–32, https://doi.org/10.1029/2010EO040003, 2010.

Fuchs, M. and Owen, L. A.: Luminescence dating of glacial and associated sediments: review, recommendations and future directions, Boreas, 37, 636–659, https://doi.org/10.1111/j.1502-3885.2008.00052.x, 2008.

Fujioka, T., Chappell, J., Fifield, L. K., and Rhodes, E. J.: Australian desert dune fields initiated with Pliocene–Pleistocene global climatic shift, Geology, 37, 51–54, https://doi.org/10.1130/G25042A.1, 2009.

Galbraith, R. F., Roberts, R. G., Laslett, G. M., Yoshida, H., and Olley, J. M.: Optical dating of single and multiple grains of quartz from Jinmium rock shelter, northern Australia: Part I, experimental design and statistical models, Archaeometry, 41, 339–364, https://doi.org/10.1111/j.1475-4754.1999.tb00987.x, 1999.

Geoscience Australia: GEODATA 9 second DEM and D8: Digital Elevation Model Version 3 and Flow Direction Grid 2008. Bioregional Assessment Source Dataset, available at: http://data.bioregionalassessments.gov.au/dataset/ebcf6ca2-513a-4ec7-9323-73508c5d7b93 (last access: 28 November 2018), 2008.

Geoscience Australia: Geoscience Australia, 1 second SRTM Digital Elevation Model (DEM), Bioregional Assessment Source Dataset, available at: http://data.bioregionalassessments.gov.au/dataset/9a9284b6-eb45-4a13-97d0-91bf25f1187b (last access: 28 November 2018), 2011.

Granger, D. E. and Schaller, M.: Cosmogenic Nuclides and Erosion at the Watershed Scale, Elements, 10, 369–373, https://doi.org/10.2113/gselements.10.5.369, 2014.

Granger, D. E., Kirchner, J. W., and Finkel, R. C.: Spatially averaged long-term erosion rates measured from in situ-produced cosmogenic nuclides in alluvial sediment, J. Geol., 104, 249–257, 1996.

Granger, D. E., Lifton, N. A., and Willenbring, J. K.: A cosmic trip: 25 years of cosmogenic nuclides in geology, Geol. Soc. Am. Bull., 125, 1379, https://doi.org/10.1130/B30774.1, 2013.

Harel, M. A., Mudd, S. M., and Attal, M.: Global analysis of the stream power law parameters based on worldwide ^{10}Be denudation rates, Geomorphology, 268, 184–196, https://doi.org/10.1016/j.geomorph.2016.05.035, 2016.

Henck, A. C., Huntington, K. W., Stone, J. O., Montgomery, D. R., and Hallet, B.: Spatial controls on erosion in the Three Rivers Region, southeastern Tibet and southwestern China, Earth Planet. Sc. Lett., 303, 71–83, https://doi.org/10.1016/j.epsl.2010.12.038, 2011.

Herman, F., Seward, D., Valla, P. G., Carter, A., Kohn, B., Willett, S. D., and Ehlers, T. A.: Worldwide acceleration of mountain erosion under a cooling climate, Nature, 504, 423–426, https://doi.org/10.1038/nature12877, 2013.

Horn, B. K. P.: Hill shading and the reflectance map, P. IEEE, 69, 14–47, 1981.

Huntley, D. J., Godfrey-Smith, D. I., and Thewalt, M. L. W.: Optical dating of sediments, Nature, 313, 105–107, https://doi.org/10.1038/313105a0, 1985.

Iacovella, S. and Youngblood, B.: GeoServer Beginner's Guide, Packt Publishing, Birmingham, 2013.

Ivy-Ochs, S. and Briner, J. P.: Dating Disappearing Ice with Cosmogenic Nuclides, Elements, 10, 351–356, https://doi.org/10.2113/gselements.10.5.351, 2014.

Jansen, J., Nanson, G., Cohen, T., Fujioka, T., Fabel, D., Larsen, J., Codilean, A., Price, D., Bowman, H., May, J.-H., and Gliganic, L.: Lowland river responses to intraplate tectonism and climate forcing quantified with luminescence and cosmogenic ^{10}Be, Earth Planet Sc. Lett., 366, 49–58, https://doi.org/10.1016/j.epsl.2013.02.007, 2013.

Jenson, S. K. and Domingue, J. O.: Extracting topographic structure from digital elevation data for geographic information system analysis, Photogramm. Eng. Rem. S., 54, 1593–1600, 1988.

Korschinek, G., Bergmaier, A., Faestermann, T., Gerstmann, U. C., Knie, K., Rugel, G., Wallner, A., Dillmann, I., Dollinger, G., von Gostomski, C. L., Kossert, K., Maiti, M., Poutivtsev, M., and Remmert, A.: A new value for the half-life of ^{10}Be by Heavy-Ion Elastic Recoil Detection and liquid scintillation counting, Nucl. Instrum. Meth. B, 268, 187–191, https://doi.org/10.1016/j.nimb.2009.09.020, 2010.

Lal, D.: Cosmic ray labeling of erosion surfaces: in situ nuclide production rates and erosion models, Earth Planet. Sc. Lett., 104, 424–439, https://doi.org/10.1016/0012-821X(91)90220-C, 1991.

Lewis, C. J., McDonald, E. V., Sancho, C., Peña, J. L., and Rhodes, E. J.: Climatic implications of correlated Upper Pleistocene glacial and fluvial deposits on the Cinca and Gállego Rivers (NE Spain) based on OSL dating and soil stratigraphy, Global Planet. Change, 67, 141–152, https://doi.org/10.1016/j.gloplacha.2009.01.001, 2009.

Lifton, N. A., Sato, T., and Dunai, T. J.: Scaling in situ cosmogenic nuclide production rates using analytical approximations to atmospheric cosmic-ray fluxes, Earth Planet. Sc. Lett., 386, 149–160, https://doi.org/10.1016/j.epsl.2013.10.052, 2014.

Lupker, M., Blard, P.-H., Lavé, J., France-Lanord, C., Leanni, L., Puchol, N., Charreau, J., and Bourlès, D.: ^{10}Be-derived Himalayan denudation rates and sediment budgets in the Ganga basin, Earth Planet. Sc. Lett, 333–334, 146–156, https://doi.org/10.1016/j.epsl.2012.04.020, 2012.

Lupker, M., Lavé, J., France-Lanord, C., Christl, M., Bourlès, D., Carcaillet, J., Maden, C., Wieler, R., Rahman, M., Bezbaruah, D., and Xiaohan, L.: ^{10}Be systematics in the Tsangpo-Brahmaputra catchment: the cosmogenic nuclide legacy of the eastern Himalayan syntaxis, Earth Surf. Dynam., 5, 429–449, https://doi.org/10.5194/esurf-5-429-2017, 2017.

Molnar, P. and England, P.: Late Cenozoic Uplift of Mountain-Ranges and Global Climate Change – Chicken or Egg, Nature, 346, 29–34, https://doi.org/10.1038/346029a0, 1990.

Mudd, S. M., Harel, M.-A., Hurst, M. D., Grieve, S. W. D., and Marrero, S. M.: The CAIRN method: automated, reproducible calculation of catchment-averaged denudation rates from cosmogenic nuclide concentrations, Earth Surf. Dynam., 4, 655–674, https://doi.org/10.5194/esurf-4-655-2016, 2016.

Nishiizumi, K., Imamura, M., Caffee, M. W., Southon, J. R., Finkel, R. C., and Mcaninch, J.: Absolute calibration of ^{10}Be AMS standards, Nucl. Instrum. Meth. B, 258, 403–413, https://doi.org/10.1016/j.nimb.2007.01.297, 2007.

Olley, J., Deckker, P. D., Roberts, R., Fifield, L., Yoshida, H., and Hancock, G.: Optical dating of deep-sea sediments using single grains of quartz: a comparison with radiocarbon, Sediment. Geol., 169, 175–189, https://doi.org/10.1016/j.sedgeo.2004.05.005, 2004.

Pietsch, T., Brooks, A., Spencer, J., Olley, J., and Borombovits, D.: Age, distribution, and significance within a sediment budget, of in-channel depositional surfaces in the Normanby River, Queensland, Australia, Geomorphology, 239, 17–40, https://doi.org/10.1016/j.geomorph.2015.01.038, 2015.

Portenga, E. W. and Bierman, P. R.: Understanding Earth's eroding surface with ^{10}Be, GSA Today, 21, 4–10, https://doi.org/10.1130/G111A.1, 2011.

Raymo, M. E. and Ruddiman, W. F.: Tectonic Forcing of Late Cenozoic Climate, Nature, 359, 117–122, https://doi.org/10.1038/359117a0, 1992.

Reber, R., Delunel, R., Schlunegger, F., Litty, C., Madella, A., Akçar, N., and Christl, M.: Environmental controls on ^{10}Be-based catchment-averaged denudation rates along the western margin of the Peruvian Andes, Terra Nova, 29, 282–293, https://doi.org/10.1111/ter.12274, 2017.

Rhodes, E. and Bailey, R.: The effect of thermal transfer on the zeroing of the luminescence of quartz from recent glaciofluvial sediments, Quaternary Sci. Rev., 16, 291–298, https://doi.org/10.1016/S0277-3791(96)00100-X, 1997.

Rhodes, E. J.: Optically Stimulated Luminescence Dating of Sediments over the Past 200,000 Years, Annu. Rev. Earth Planet. Sc., 39, 461–488, https://doi.org/10.1146/annurev-earth-040610-133425, 2011.

Richards, B. W., Benn, D. I., Owen, L. A., Rhodes, E. J., and Spencer, J. Q.: Timing of late Quaternary glaciations south of Mount Everest in the Khumbu Himal, Nepal, Geol. Soc. Am. Bull., 112, 1621–1632, https://doi.org/10.1130/0016-7606(2000)112<1621:TOLQGS>2.0.CO;2, 2000.

Rustomji, P. and Pietsch, T.: Alluvial sedimentation rates from southeastern Australia indicate post-European settlement landscape recovery, Geomorphology, 90, 73–90, https://doi.org/10.1016/j.geomorph.2007.01.009, 2007.

Safran, E. B., Bierman, P. R., Aalto, R., Dunne, T., Whipple, K. X., and Caffee, M.: Erosion rates driven by channel network incision in the Bolivian Andes, Earth Surf. Proc. Land., 30, 1007–1024, https://doi.org/10.1002/esp.1259, 2005.

Schaefer, J. M., Denton, G. H., Kaplan, M. R., Putnam, A., Finkel, R. C., Barrell, D. J. A., Andersen, B. G., Schwartz, R., Mackintosh, A., Chinn, T., and Schluechter, C.: High-Frequency Holocene Glacier Fluctuations in New Zealand Differ from the Northern Signature, Science, 324, 622–625, 2009.

Schaller, M., von Blanckenburg, F., Hovius, N., Veldkamp, A., van den Berg, M., and Kubik, P. W.: Paleoerosion Rates from Cosmogenic ^{10}Be in a 1.3 Ma Terrace Sequence: Response of the River Meuse to Changes in Climate and Rock Uplift, J. Geol., 112, 127–144, https://doi.org/10.1086/381654, 2004.

Singarayer, J., Bailey, R., Ward, S., and Stokes, S.: Assessing the completeness of optical resetting of quartz OSL in the natural environment, Radiat. Meas., 40, 13–25, https://doi.org/10.1016/j.radmeas.2005.02.005, 2005.

Spencer, J. Q. and Owen, L. A.: Optically stimulated luminescence dating of Late Quaternary glaciogenic sediments in the upper Hunza valley: validating the timing of glaciation and assessing dating methods, Quaternary Sci. Rev., 23, 175–191, https://doi.org/10.1016/S0277-3791(03)00220-8, 2004.

Stone, J. O.: Air pressure and cosmogenic isotope production, J. Geophys. Res.-Sol. Ea., 105, 23753–23759, https://doi.org/10.1029/2000JB900181, 2000.

Vermeesch, P.: CosmoCalc: An Excel add-in for cosmogenic nuclide calculations, Geochem. Geophy. Geosy., 8, Q08003, https://doi.org/10.1029/2006GC001530, 2007.

Veth, P., Smith, M., Bowler, J., Fitzsimmons, K., Williams, A., and Hiscock, P.: Excavations At Parnkupirti, Lake Gregory, Great Sandy Desert: OSL Ages for Occupation before the Last Glacial Maximum, Australian Archaeology, 69, 1–10, https://doi.org/10.1080/03122417.2009.11681896, 2009.

von Blanckenburg, F.: The control mechanisms of erosion and weathering at basin scale from cosmogenic nuclides in river sediment, Earth Planet. Sc. Lett., 237, 462–479, https://doi.org/10.1016/j.epsl.2005.06.030, 2005.

von Blanckenburg, F. and Willenbring, J. K.: Cosmogenic Nuclides: Dates and Rates of Earth-Surface Change, Elements, 10, 341–346, https://doi.org/10.2113/gselements.10.5.341, 2014.

Wallinga, J., Hobo, N., Cunningham, A. C., Versendaal, A. J., Makaske, B., and Middelkoop, H.: Sedimentation rates on embanked floodplains determined through quartz optical dating, Quat. Geochronol., 5, 170–175, https://doi.org/10.1016/j.quageo.2009.01.002, 2010.

Willenbring, J. K. and von Blanckenburg, F.: Long-term stability of global erosion rates and weathering during late-Cenozoic cooling, Nature, 465, 211–214, https://doi.org/10.1038/nature09044, 2010.

Willenbring, J. K., Codilean, A. T., and McElroy, B.: Earth is (mostly) flat: Apportionment of the flux of continental sediment over millennial time scales, Geology, 41, 343–346, https://doi.org/10.1130/G33918.1, 2013.

Wintle, A. G.: Luminescence dating: where it has been and where it is going, Boreas, 37, 471–482, https://doi.org/10.1111/j.1502-3885.2008.00059.x, 2008.

Wittmann, H., von Blanckenburg, F., Guyot, J., Maurice, L., and Kubik, P.: From source to sink: Preserving the cosmogenic ^{10}Be-derived denudation rate signal of the Bolivian Andes in sediment of the Beni and Mamoré foreland basins, Earth Planet. Sc. Lett., 288, 463–474, https://doi.org/10.1016/j.epsl.2009.10.008, 2009.

Wittmann, H., von Blanckenburg, F., Maurice, L., Guyot, J.-L., Filizola, N., and Kubik, P. W.: Sediment production and delivery in the Amazon River basin quantified by in situ-produced cosmogenic nuclides and recent river loads, Geol. Soc. Am. Bull., 123, 934–950, https://doi.org/10.1130/B30317.1, 2011.

Wolfe, S. A., Huntley, D. J., and Ollerhead, J.: Recent and late Holocene sand dune activity in southwestern Saskatchewan, Current Research, Geological Survey of Canada, 1995-B, 131–140, 1995.

13

The GIK-Archive of sediment core radiographs with documentation

Hannes Grobe[1], **Kyaw Winn**[2], **Friedrich Werner**[†], **Amelie Driemel**[1], **Stefanie Schumacher**[1], **and Rainer Sieger**[1]

[1]Alfred-Wegener-Institut, Helmholtz-Zentrum für Polar- und Meeresforschung, 27515 Bremerhaven, Germany

[2]Institut für Geowissenschaften (formerly Geologisch-Paläontologisches Institut und Museum, GIK), Christian-Albrechts Universität, 24118 Kiel, Germany

[†]deceased

Correspondence to: Hannes Grobe (hannes.grobe@awi.de)

Abstract. The GIK-Archive of radiographs is a collection of X-ray negative and photographic images of sediment cores based on exposures taken since the early 1960s. During four decades of marine geological work at the University of Kiel, Germany, several thousand hours of sampling, careful preparation and X-raying were spent on producing a unique archive of sediment radiographs from several parts of the World Ocean. The archive consists of more than 18 500 exposures on chemical film that were digitized, geo-referenced, supplemented with metadata and archived in the data library PANGAEA®. With this publication, the images have become available open-access for use by the scientific community at https://doi.org/10.1594/PANGAEA.854841.

1 Introduction

During the late 1950s, the new field of marine geology was developed in Germany at the Geologisch-Paläontologisches Institut und Museum (Geological-Palaeontological Institute and Museum) at Christian-Albrechts-Universität zu Kiel (GIK), since 1998 known as Institut für Geowissenschaften (Institute of Geosciences). With the commission of the new German research vessel *Meteor* in 1964 and its maiden voyage in the Persian Gulf, GIK developed new techniques and assimilated existing methods to recover sediments from the ocean floor (e.g. Seibold, 1958; Werner, 1998). A simple but efficient gravity corer with a 12 cm diameter barrel and up to 1.5 t lead weight (*Schwerelot*) was constructed by the company Hydrowerkstätten Kiel. Piston coring technology was applied in the 1970s with the Kiel version of the former Kullenberg corer (Kullenberg, 1947). A vibrocorer supplemented the set of devices for sampling harder sediments. High-volume coring technology was performed by the kasten corer (*Kastenlot* by Kögler, 1963) with a rectangular size of 15 cm × 15 cm and length of 6.4 m for clay-like sediments. In the 1970s, a larger version recovered cores of 30 cm × 30 cm and lengths of 12–15 m, weighing up to 3.5 t. Besides the most commonly deployed gravity and piston corers, the kasten corer is also used due to its well-known ability to recover undisturbed and continuous sedimentary sequence for providing sufficient material to fulfil numerous interdisciplinary sampling demands.

To obtain an undisturbed sediment surface from the sea floor for the investigation of the sediment–water interface, the spade box corer was added to the suite of sampling devices (Reineck, 1963). The corer ensures that the pristine sediment succession is collected from the top of the seafloor surface. It was first employed for sampling deep-sea sediments during the *Meteor* cruise 25 in 1971. An extended large version of the spade corer with a box size of 50 cm × 55 cm was developed by Scripps and USNEL (United States Naval Electronic Laboratory) (Farris and Crezée, 1976). This type was modified and rebuilt by Wuttke GmbH (Henstedt-Ulzburg) to become the German *Großkastengreifer* (GKG). Since its first deployment in 1980 on RV

Meteor during cruise M60 (Thiel, 1982), it has been in use on most expeditions of German marine research vessels.

Schwerelot, *Kastenlot* and *Großkastengreifer* became valuable devices recovering large volumes of high-quality sediments, always providing sufficient material for X-ray sample preparation even for multiple sets of radiographies. For more than five decades, doctoral dissertations and publications of GIK have included radiograph interpretations, as described, for example, in Exon (1972), Werner (2002), Winn (1974, 2006), Wetzel (1979), Löwemark (2001), Hinz et al. (1971), Whitaker and Werner (1981), and Winn and Averdieck (1984).

2 Application of X-ray techniques in sedimentology

In radiography, the structural heterogeneity or homogeneity of an object is made visible by the different attenuation of X-rays on a photographic negative film. The resulting image is referred to as a *radiograph*, with the key quality parameters blackening, contrast and resolution. In the late 1990s, positive films were also exposed and interpreted. Applications of radiography are best known from medical and industrial studies, for example to verify the welding quality of steel products.

X-ray imaging on marine sediment cores was initiated at GIK around 1960 by means of a self-constructed device, which simply consisted of an X-ray source located in a shielded cabinet. In the early 1970s, the Faxitron cabinet X-ray system was invented by physicist Joseph Edmonds Henderson at the Applied Physics Laboratory, University of Washington (Faxitron, 2017). In 1974, Hewlett-Packard took over the product for use in manufacturing silicon chips. GIK applied the professional technology of the Faxitron model type 43855 (10 to 110 kV, 3 mA, size 84 cm × 55 cm × 51 cm, weight 176 kg) to the study of sediment slabs taken from marine sediment cores. Its use finally resulted in a comprehensive collection of large-format radiography (Werner, 1998). In marine geology, the Faxitron became the most frequently used device in X-ray imaging. The technology and preparation procedures of GIK as described below were subsequently adopted by various other sedimentology laboratories in Germany.

2.1 Preparation and exposure

Marine sediment cores are archived in segments of 1 m in length for convenient handling and cut longitudinally in two halves for further processing. (This is not required for box and kasten cores, where samples are taken from the outer side.) After photography and a visual lithological description of the sediment sequence (structure, texture, colour), the "work" half is sampled for various analyses, with the preparation of X-ray slabs being the first step of the sampling workflow (Grobe, 1986; Fig. 1). It was common practice first to prepare sediment slices for radiographs and use the exposure as a guide for further sampling, in particular across strongly bioturbated sections. The remaining "archive" half is sealed in airtight D-tubes for future investigation. As a common practice in geological repositories worldwide, the core segments are archived in repositories at +4 °C.

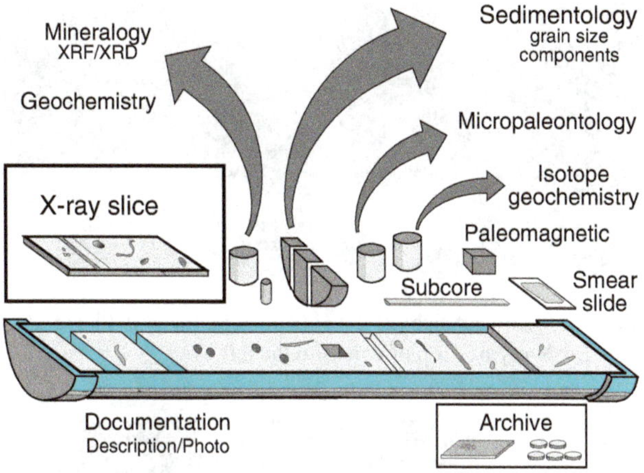

Figure 1. Standard sampling workflow for the investigation of sediment cores as developed at GIK. The first step during the sampling sequence was the preparation of sediment slabs for X-ray imaging along the core profile.

To ensure optimum quality of the X-radiograph, sediment slices have to be prepared with the utmost care so as not to destroy the original sediment structures and to avoid artefacts. The surface of the longitudinal core half is smoothened with a wet glass plate, eliminating the grooves and furrows on the sediment surface usually caused by larger sediment particles. A hard Plexiglas® lid measuring 25 cm × 10 cm × 1 cm is pushed gently into the level surface of the working half to provide support and stability for the sediment slab (Fig. 2). Core label, depth interval and an arrow pointing downwards in the direction of penetration of the coring device are then marked on the lid (in reverse chronological order, in contrast to the convention at other research institutes and programmes, such as the International Ocean Discovery Program). A kasten core typically prepared for the sampling of X-ray slabs is shown in Fig. 3. The slab within the lid is separated from the remaining core by pulling a nylon line up-core behind the plastic lid, thereby using the two sides of the lid orientated parallel to the core axis as guides. The slab is carefully lifted with a cheese knife and removed from the core. The slit may be moistened if the slab tends to adhere to the main sediment core. The slices in the lids are vacuum sealed in polyethylene lay-flat tubing after evacuation of the enclosed air. The equipment used for the preparation is shown in Fig. 4. It is important to note that the X-radiograph technique is non-destructive and that – after X-raying – the sample material could be used for further analyses.

Figure 2. Sediment slab (25 cm × 10 cm) stabilized in Plexiglas lid ready for X-raying. Top of core is to the left.

Figure 3. Sediment core taken with a kasten corer (30 cm × 30 cm) prepared with Plexiglas lids to remove the samples for X-raying.

X-ray film is coated on both sides and thus has higher sensitivity and contrast compared to film used in light photography. For exposure the film type Structurix D 4 manufactured by Agfa-Gevaert was chosen. The film is not sensitive to red light and thus can easily be handled in a darkroom under low-light conditions. The film is cut into 25 × 10 cm strips and stored in black film covers. Each 25 cm long slab is exposed to the X-ray beam with the slab surface not covered by the lid facing the X-ray source (Fig. 5).

The characteristics of the X-radiation determine the quality of the sediment images, with the wavelength being the most important factor. High energy will produce "harder" radiation with shorter wavelength while lower energy will result in radiation with longer wavelength. Due to the soft composition of unconsolidated sediments, a spectrum of longer wavelengths and thus "weaker" radiation is preferred to produce images with a moderate contrast (Werner, 1975).

Exposure times depend on sediment type and are mostly controlled by grain size and compaction, the thickness of the slab and the strength of the radiation. A 1 cm thick slice has to be exposed for a time span from 3 (soft clay) to 20 min (high sand content) at a voltage of 30–35 kV and an electrical current of 3 mA. Identification of the core ID and the depth interval on the negative is assured by putting corresponding lead letters and numbers on the film during exposure. The film is developed in a darkroom for 3 min by using the devel-

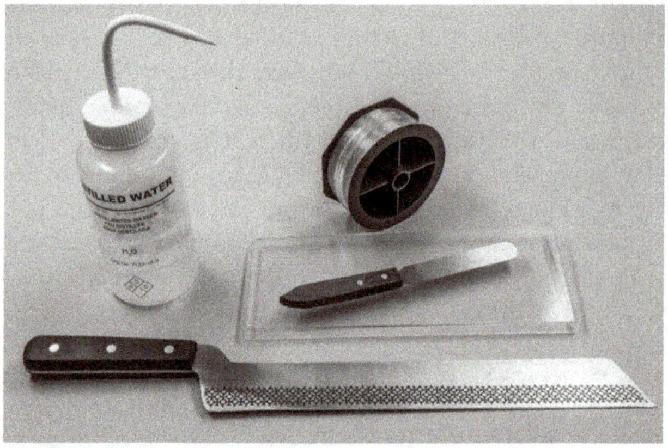

Figure 4. Equipment used for the preparation of sediment slabs for X-ray imaging comprise fishing line, distilled water, cheese knife, spatula and special Plexiglas lids for support of the sediment slab. Lids used for preparation measure 25 cm × 10 cm × 1 cm (Bensberger Kunststoffwerk Lappe GmbH). Lay-flat tubing and sealing device (not shown) are used to protect the sample from drying.

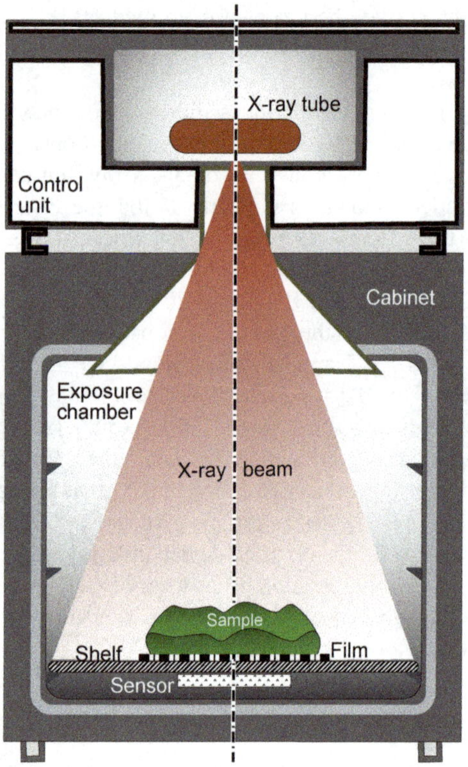

Figure 5. Schematic drawing of an X-ray device of the FAXITRON series (Hewlett-Packard), which has an upper chamber for the X-ray tube with a control unit and an exposure chamber below (Faxitron X-ray Corporation, 1975). X-ray cone is shown in red. Both are fully lead shielded allowing for operation under normal laboratory conditions without special protection of registration. The sensor below the sample shelf can be used to set automatic exposure times. In modern digital operation mode, shelf, film and sensor are replaced by a scanner sensitive to X-ray beams.

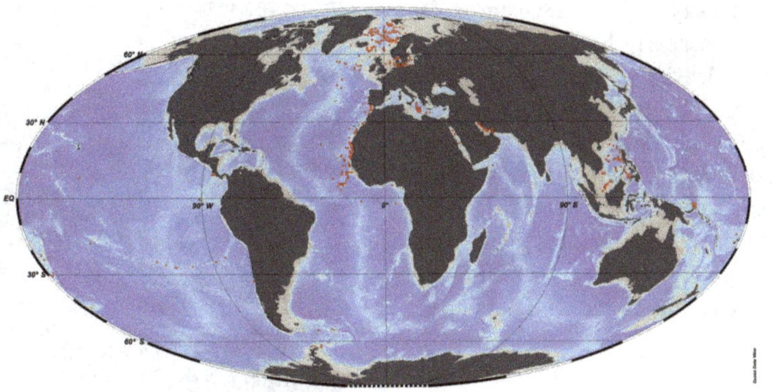

Figure 6. Map showing the locations of marine sediment cores, from which X-radiographs were obtained as part of the GIK-Archive. The cores were collected between 1965 and 2000 on a total of 93 expeditions.

oper G124 and the fixer G335 (Agfa), washed for 20 min in distilled water and dried. A detailed description of the procedure can be found in Werner (1975).

Over 50 years of marine geological research at GIK, a suite of 1355 sediment cores with a total length of 3547 m were investigated (Fig. 6). More than 18 500 sediment slices were prepared for X-ray imaging and exposure. Between 2010 and 2014 the images were transferred to the PANGAEA department at the Alfred Wegener Institute for Polar and Marine Research, Bremerhaven (AWI). They were digitized at a resolution of 600 dpi by a transmitting light scanner (Microtek ScanMaker 9800XL). The X-radiographs, photos and descriptions were supplemented with the metadata of the corresponding cores including position, water depth, sampling device, expedition, date/time, ship, etc. The resulting collection of 1355 datasets was combined into a single parent dataset, which is available at https://doi.org/10.1594/PANGAEA.854841.

2.2 Analysis of images

Besides the documentation by light photography, X-radiography became a standard imaging technique in marine geology to complement and support the visual description of sediment profiles, which comprises the detailed logging of the lithological composition of the sediment, its texture and sedimentary structures (Bouma, 1969). In particular, radiographs reveal details of structures, such as bioturbation and graded bedding, diagenetic modifications and large internal components (e.g. fossils or dropstones) which are not discernible in normal light photography.

The dominant control on beam attenuation is bulk sediment density (Holyer et al., 1996), which in turn is affected by grain size, mineralogical composition, abundance of biogenic components and physical parameters such as water content, porosity and compaction. Thus radiographs can be used to determine a whole range of various sediment properties. During examination of the images, some specific points need to be considered. These are described in brief in the following examples of typical sedimentary textures as shown in Fig. 7.

Structures that may be unrecognizable to the naked eye include boundaries of strata, non-conformities, fine lamination, graded bedding and most importantly – bioturbation (Werner, 1968; Winn, 1974; Wetzel, 1979). *Lebensspuren* are the most common structures in marine sediments which allow the identification of the species and the reconstruction of paleoecological and paleoenvironmental conditions (Löwemark, 2001). The resulting taxonomy of palichnology is the basis for its identification and classification (Bromley, 1999; Seilacher, 2007). Sedimentological sequences formed by distinct processes (e.g. deposition by turbidity or contour currents) and their evolution over time are also part of this structural group. Features such as base and top boundaries, type, thickness, frequency, rhythms and cycles indicate facies differentiation and changes.

Physical properties can be identified by the brightness of the negatives, as well as by internal structures such as layering (e.g. ash layers), lamination, bedding planes, cross-bedding, current ripples or sorting. By using a magnifying glass while investigating the X-ray image, individual grains with a size of > 1 mm can be classified in terms of grain shape and composition. Support of a high-resolution sedimentology is given in the millimetre to centimetre scale, including large components such as mud clasts or gravel grains, e.g. as ice-rafted debris (Grobe, 1987; Principato, 2004). Any gravel fraction reveals itself by the distinct appearance of each individual grain.

Extracted high-resolution greyscale curves were correlated with physical parameters (colour, gamma-ray density, magnetic susceptibility, grain size; St-Onge et al., 2007; Kowalczk, 2005). There is a clear relationship between grain size distribution and brightness/blackening of the film. Sandy to clay-like sediments are composed of just a few minerals (quartz, feldspar, clay minerals, carbonates), all of which have a similar specific grain density of 2.6 to 2.8 g cm^{-3}. Diatomaceous oozes may result in nearly black X-radiograph negatives because of the low density of opal. Thus, differences in brightness result mostly from changes in grain size rather than from a heterogeneous distribution of minerals. Water content, density and porosity are the major factors governing greyscale values; porosity increases from coarse-grained to fine-grained sediments. The sediment density and thus the brightness of the image negative increases with core depth because the compaction results in reduced pore space and water content.

Minerals may result from diagenesis including authigenic pyrite, zeolite, or the rarely formed porcelanite (Gerland et al., 1997). Heavy minerals such as pyrite and other iron sulfide as well as iron oxide minerals can easily be identified by their high brightness/X-ray attenuation and their specific grain shapes and internal structures. However, dark grey features visible in negatives can be areas with an extremely high water content, plant fossils, wood or even small voids. In cases where the samples were stored for a longer period, new minerals may have formed through chemical processes in the sediment (e.g. as per Fig. 6g).

Artefacts reflecting the post-depositional disturbance of the original sedimentary structure must be identified within the core. These effects can have various causes which should always be considered while investigating and interpreting the images (Skinner and McCave, 2003). During the coring process and recovery, especially with a gravity, kasten, piston or vibrocorer, the mostly soft and often "soupy" sediments from near the seafloor surface (i.e. at the core top) may flow, resulting in a loss of the original structure. In some instances it is not even possible to prepare a sediment slab suitable for X-radiography from the upper decimetres of the core. Coring disturbance caused by the piston or gravity coring process may result in "pseudo-tectonic" features (e.g. faults, fractures, sediment mixing, "flow-in") which are predominantly observed at the bases of longer piston and gravity cores. In particular, gravity coring can cause micro-faulting within the sediment and result in an artificial shortening of the sediment column (Fig. 6c). Especially in clay-like sequences, even pseudo-hiatuses can occur when parts of a sediment section succession allow the core barrel to pass but are squeezed out and thus not recovered. In addition, the outer edges of a core segment may show downward bending of layers in close proximity to the core liner, which results from the friction between the liner and the sediment when the core barrel penetrates the seabed.

Further effects visible in radiography might include the following: if the sediment slice is not properly sealed in lay-flat tubing it may dry out and produce drying cracks, which can, however, be clearly identified. If a sediment slice has a variable thickness, the brightness of the X-radiograph will vary throughout the sample. If the sediment contains larger particles, the marginal areas around the particle may be disturbed during the preparation. Regular stripes and patterns

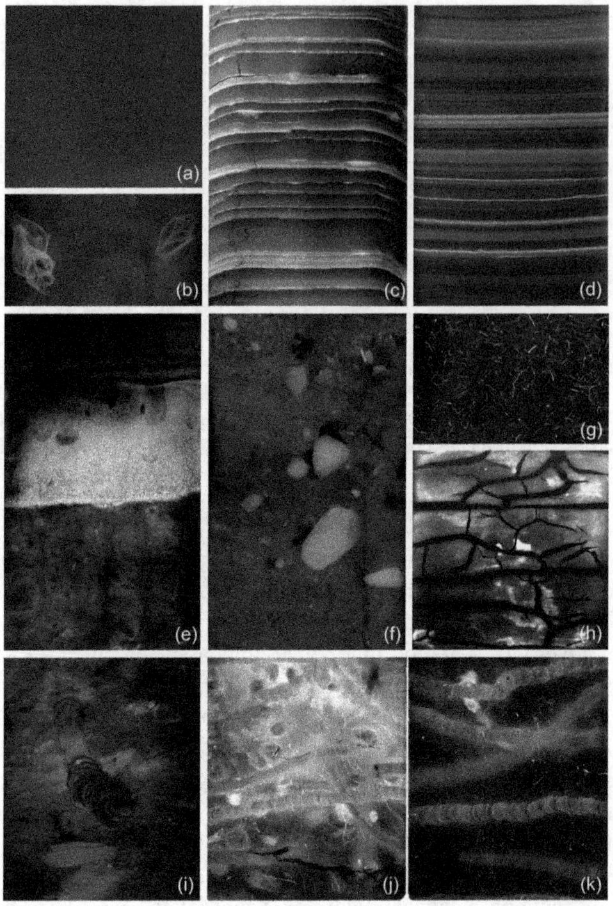

Figure 7. Examples of various X-radiographs from the GIK-Archive. Each has a width of 10 cm. **(a)** Homogeneous clay-like sediment, Mediterranean Sea, https://doi.org/10.1594/PANGAEA.720925; **(b)** fossil molluscs, Persian Gulf, https://doi.org/10.1594/PANGAEA.720253; **(c)** interbedded strata with artificial down-bending of layers and fault lines as a result of gravity coring, Baltic Sea, https://doi.org/10.1594/PANGAEA.690661; **(d)** laminated sediment, Red Sea, https://doi.org/10.1594/PANGAEA.720616; **(e)** turbidite with graded bedding, South China Sea, https://doi.org/10.1594/PANGAEA.720737; **(f)** gravel as ice-rafted debris, Norwegian Sea, https://doi.org/10.1594/PANGAEA.720368; **(g)** pyritized lebensspuren, West Atlantic – off Senegal, https://doi.org/10.1594/PANGAEA.705737; **(h)** artificial cracks from drying out of an unprotected sediment slab, https://doi.org/10.1594/PANGAEA.705491; **(i, j, k)** examples of bioturbation, Baltic Sea off Flensburger Förde, https://doi.org/10.1594/PANGAEA.705626 and African continental slope, https://doi.org/10.1594/PANGAEA.705737.

observed also on an X-radiograph are usually the result of insufficient smoothening of the slab surface and reflect tracks from the cutting wire. The projection also has to be taken into account during analysis: structural elements of a three-dimensional slice are projected in two dimensions onto the film. A 1 cm thick slab sample for X-raying does not work well for gravel- and pebble-rich sediments (such as diamictons), and thus X-raying of half cores works better for those sediments. However, X-raying of half cores has its own problems, especially overlaps and projection issues.

3 Archiving

The X-radiographs archived at GIK were taken from 1355 globally distributed sediment cores recovered on 93 voyages of the German research vessels *Wattenberg*, *Alkor*, *Littorina*, *Poseidon*, *Meteor* and *Sonne*, mostly on expeditions between 1964 and 2000 (Fig. 4). For the list of cores with metadata, please refer to https://doi.org/10.1594/PANGAEA.875415, linked as "Further details" to the parent set. RV *Sonne* cores collected on behalf of the Preussag manganese nodule project and BGR-led (Bundesanstalt für Geowissenschaften und Rohstoffe, Hanover – Federal Institute for Geosciences and Natural Resources) cruises to the equatorial and South Pacific were also sampled and analysed. Most of the remaining material is available in the core storage of GIK and at the Lithothek of GEOMAR Helmholtz Centre for Ocean Research Kiel.

Image digitization and archiving

More than 18 500 exposures of X-radiographs were digitized using two A3-format scanners, model Microtek ScanMaker 9800XL, at a resolution of 600 dpi and stored in JPEG format with moderate compression to generate file sizes for convenient Internet download times. Not all images were post-processed and thus some may still be underexposed. Brightness, lucidity and contrast can be corrected as required for investigation with any image processing software. The full information is in each image due to the high-resolution scan of the fine-grained film. Images were uploaded to PANGAEA and stored in a database. One dataset includes all images of one core. Metadata and additional documentary files including core descriptions and photos were added, if available. The metadata cover core ID, latitude and longitude, water depth, recovery, coring device and date/time when the core was taken. The label of the expedition linking to the cruise report (if published) is also provided. Each dataset starts with a "Citation" tagged line, consisting of the name of the principal investigator(s), a standard title "Documentation of sediment core GIKxxxxx-x", year of electronic storage and thus public availability, and the source institute (in this case always set to GIK). The DOI as a persistent link to the dataset is a mandatory part of any modern citation. If the images were already used in publications, the corresponding references can be found under the "Related to" field. Selected examples of images from this collection are presented in Fig. 7.

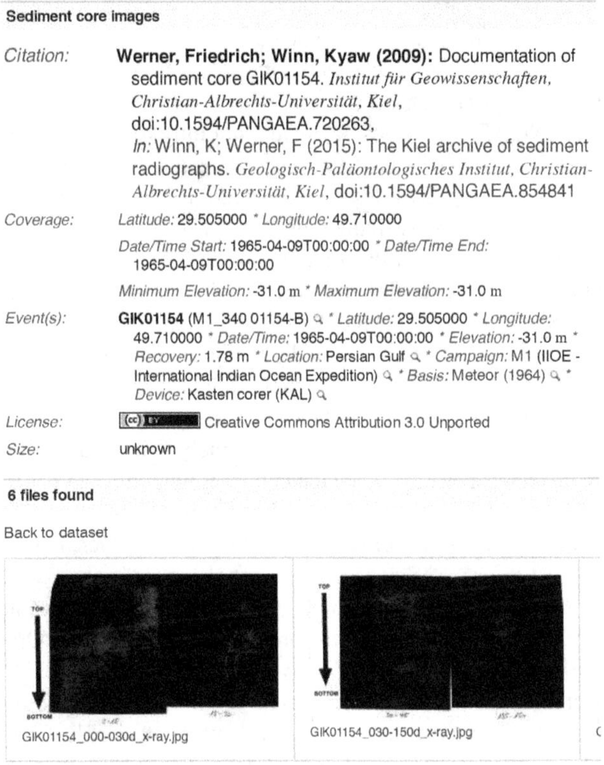

Figure 8. Example of the standard metadata header provided by PANGAEA from dataset https://doi.org/10.1594/PANGAEA.720263. Starting with the citation, comprising author(s), year, title, source and DOI. Citation is followed by the georeference in space and time and links to further references or reports. For this collection, images of each core location have their own data subset. The total of 1355 data subsets of this paper are grouped together in one "parent" dataset (https://doi.org/10.1594/PANGAEA.854841). At the end of the metadata header, each image is shown as a thumbnail and can be downloaded either individually or as part of a compilation of all images in a single zip archive.

With the establishment of polar research through the foundation of the Alfred Wegener Institute for Polar Research (AWI) in 1980 in Bremerhaven, the methods of sediment core sampling and analysis developed at GIK were utilized and adopted by the department of marine geology at AWI (Grobe, 1987). All geological sample material taken aboard RV *Polarstern* was archived in an institutional core repository, administered by a database. Between 1987 and 1997, this system was further developed, in part driven by an initiative of the German National Climate Project of the BMFT in 1994. It finally became an archiving and publishing system for data from earth system research, namely PANGAEA® – Data Publisher for Earth and Environmental Science (Diepenbroek et al., 2002). Technically PANGAEA is a relational database (RDB) with geo-reference in time and space for a consistent storage of analytical and observational data. A storage system (tape robot) for files and binary objects such as images assists the RDB. For single items or collections of files, only the metadata are stored in the relational tables of the data model, including stable links to the image files on tape (Fig. 8).

PANGAEA provides its content not only for direct download from its website (http://www.pangaea.de) but also for data harvesting. Besides standard search engines, the image datasets are also distributed via web services through library catalogues, e.g. WorldCat and a number of portals (listed at http://wiki.pangaea.de/wiki/Portal). Most images of a core can be found easily via the PANGAEA query window or even via an Internet search engine by using the (unique) core label as a search phrase. The requested dataset is usually listed among the first search results.

Metadata of PANGAEA are routinely mirrored in DataCite (reference http://data.datacite.org), which is the central entry portal for citable research datasets on the Internet. This information is also stored in the catalogue of the German National Library of Science and Technology (reference TIB), co-inventor of the data DOI and co-founder of DataCite. Since 2004, PANGAEA has provided its content for OAI-PMH harvesting (Open Archives Initiative – Protocol for Metadata Harvesting). The current operator is OCLC (https://www.oclc.org) with WorldCat (https://www.worldcat.org), which incorporates the content of repositories following the OAI standard and thus also includes the metadata of the PANGAEA content.

4 Conclusion

With this publication, the complete digitized archive of more than 18 500 radiographs from the World Ocean has been made available to the scientific community. This dataset is publicly available under the CC-BY 3.0 licence (https://creativecommons.org/licenses/by/3.0/) with a persistent identifier (https://doi.org/10.1594/PANGAEA.854841) as the Supplement to this publication.

Although X-ray imaging is a method well-suited to supplementing the documentation of sediment cores, this technology has been increasingly neglected in some parts of the scientific community because it is time-consuming and new high-resolution analytical techniques (e.g. multi-sensor core logging, X-ray fluorescence and colour scanning) have been continuously introduced and gained priority over the last couple of decades. In addition, digital X-ray images of sediments can now be taken very quickly both on-board and in the lab (e.g. with the ITRAX XRF scanner; Croudace et al., 2006). For the old FAXITRON models a digital X-ray scanner is now available to fit in (NTB, 2005). The time of analogue X-ray imaging of sediments is over and will now continue with digital X-ray devices and X-ray computed tomography (CT) systems (e.g. Freifeld et al., 2006).

Competing interests. The authors declare that they have no conflict of interest.

Acknowledgements. With this publication the work of Wilma Rehder and Ursula Faber, technicians at GIK, is greatly acknowledged. Both ensured the continuously very high quality of both the sediment slices for X-radiography and of the X-radiographs themselves during many sampling activities on research vessels and at the institute. For nearly four decades they have demonstrated the proper preparation of samples to hundreds of students, as well as guiding them through the exposure and darkroom procedures and archiving the images. We are thankful to Michael Seebeck, who reliably and with great patience digitized those thousands of exposures in a long-lasting workflow. For the improvement of the manuscript the authors are grateful to Michael Sarnthein and Claus-Dieter Hillenbrand.

Edited by: David Carlson

References

Bouma, A. H.: Methods for the study of sedimentary structures, Wiley, New York, 458 pp., https://doi.org/10.1002/iroh.19710560422, 1969.

Bromley, R. G.: Spurenfossilien, in: Biologie, Taphonomie und Anwendungen, Springer, Berlin/Heidelberg, 347 pp., https://doi.org/10.1007/978-3-642-59832-6, 1999.

Croudace, I. W., Rindby, A., and Rothwell, R. G.: ITRAX: description and evaluation of a new multi-function X-ray core scanner, in: New techniques in sediment core analysis, edited by: Rothwell, R. G., Geological Society, London, Special Publications, 267, 51–63, https://doi.org/10.1144/GSL.SP.2006.267.01.04, 2006.

Diepenbroek, M., Grobe, H., Reinke, H., Schindler, U., Schlitzer, R., Sieger, R., and Wefer, G.: PANGAEA – an information system for environmental sciences, Comput. Geosci., 28, 1201–1210, https://doi.org/10.1016/S0098-3004(02)00039-0, 2002.

Exon, N.: Sedimentation in the outer Flensburg Fjord area (Baltic Sea) since the last glaciation, Meyniana, 22, 5–62, https://doi.org/10.2312/meyniana.1972.22.5, 1972.

Farris, R. A. and Crezée, M.: An improved Reineck Box for sampling coarse sand, Int. Rev. Hydrobiol., 61, 703–705, https://doi.org/10.1002/iroh.3510610515, 1976.

Faxitron X-Ray Corporation, Rohde & Schwarz GmbH: Faxitron documentation and manual, Technical Manual, Bedienungsanleitung, 3 documents, http://hdl.handle.net/10013/epic.50709, 1975–1996.

Faxitron: Web site "About us", http://www.faxitron.com/life-sciences-ndt/about, last access April 2017.

Freifeld, B. M., Kneafsey, T. J., and Rack, F. R.: On-site geological core analysis using a portable X-ray computed tomographic system, Geological Society, London, Special Publications, 267, 165–178, https://doi.org/10.1144/GSL.SP.2006.267.01.12, 2006.

Gerland, S., Kuhn, G., and Bohrmann, G.: Physical properties of a porcellanite layer (Southwest Indian Ridge) constrained by geophysical logging, Mar. Geol., 140, 415–426, https://doi.org/10.1016/S0025-3227(97)00046-7, 1997.

Grobe, H.: Spätpleistozäne Sedimentationsprozesse am antarktischen Kontinentalhang vor Kapp Norvegia, östliche Weddell See, Berichte zur Polarforschung, 27, 1–121, https://doi.org/10.2312/BzP_0027_1986, 1986.

Grobe, H.: Determination of IRD in sediment cores, Polarforschung, 57, 123–345, https://doi.org/10.2312/polarforschung.57.3.123, 1987.

Hinz, K., Kögler, F.-C., Richter, I., and Seibold, E.: Reflexionsseismische Untersuchungen mit einer pneumatischen Schallquelle und einem Sedimentecholot in der westlichen Ostsee. Teil II: Untersuchungsergebnisse und geologische Deutung, Meyniana, 21, 17–34, https://doi.org/10.2312/meyniana.1971.21.17, 1971.

Holyer, R. J., Young, D. K., Sandidge, J. C., and Briggs, K. B.: Sediment density structure derived from textural analysis of cross-sectional X-radiographs, Geo-Mar. Lett., 16, 204–211, https://doi.org/10.1007/BF01204510, 1996.

Kögler, F.-C.: Das Kastenlot, Meyniana, 13, 1–7, https://doi.org/10.2312/meyniana.1963.13.1, 1963.

Kowalczk, A.: The application of digital x-radiograph imaging for the determination of bulk density, NSF Research Experience for Undergraduates (REU) Project Report (P. Dickhut and C. Friedrichs advisors), Virginia Institute of Marine Science, Gloucester Point, VA, Final Report, CHSD-2005-04, 8 pp, http://hdl.handle.net/10013/epic.50706.d001, 2005.

Kullenberg, B.: The piston core sampler, Svenska Hydrografisk-biologiska Kommissionen Skrifter, Tredge serien hydrografi, 1, 1–46, http://hdl.handle.net/10013/epic.50738.d001, 1947.

Löwemark, L.: Biogenic traces as paleoceanographic indicators in Late Quaternary sediments from the SW Iberian margin, Berichte-Reports, Institut für Geowissenschaften, Universität Kiel, 14, 138 pp., https://doi.org/10.2312/reports-ifg.2001.14, 2001.

NTB elektronische Geraete GmbH: Digital x-ray scanners of the EZ series, Dickel, Germany, hdl:10013/epic.42410, http://www.ntbxray.com/products/digital_x_ray_scanner.html (last access: April 2017), 2005.

Principato, S. M.: X-ray radiographs of sediment cores: a guide to analyzing diamicton, in: Image analysis, sediments and paleoenvironments, edited by: Francus, P., Kluwer Academic Publishers, Dordrecht, the Netherlands, 165–178, https://doi.org/10.1007/1-4020-2122-4_9, 2004.

Reineck, A.: Bergung von Proben aus dem Meeresboden, Natur und Museum, 93, 331–336, http://hdl.handle.net/10013/epic.50704.d001, 1963.

Seibold, E.: Ein Stoß-Kolbenlot für Flachwasser, Geol. Rundschau, 47, 28–36, https://doi.org/10.1007/BF01802314, 1958.

Seilacher, A.: Trace Fossils Analysis. Springer-Verlag, Berlin/Heidelberg, 226 pp., https://doi.org/10.1007/978-3-540-47226-1, 2007.

Skinner, L. C. and McCave, I. N.: Analysis and modelling of gravity- and piston coring based on soil mechanics, Mar. Geol., 199, 181–204, https://doi.org/10.1016/S0025-3227(03)00127-0, 2003.

St-Onge, G., Mulder, T., Francus, P., and Long, B.: Continuous physical properties of cored marine sediments, Developments in Marine Geology, 1, 63–98, https://doi.org/10.1016/S1572-5480(07)01007-X, 2007.

Thiel, H.: Forschungsschiff "Meteor" Reise Nr. 60, SUBTROPEX '82 Expeditionsprogramm, Deutsche Forschungsgemeinschaft, 23 pp., http://hdl.handle.net/10013/epic.45025.d001, 1982.

Werner, F.: Gefügeanalyse feingeschichteter Schlicksedimente der Eckernförder Bucht (westliche Ostsee), Meyniana, 18, 79–105, https://doi.org/10.2312/meyniana.1968.18.79, 1968.

Werner, F.: Radiographien von Lockersedimenten - Herstellung von Radiographien an Sedimentkernen nach der im GIK üblichen Methode, unpublished report, Geologisch-Paläonotologisches Institut und Museum der Christian-Albrechts-Universität, Kiel (GIK), 9 pp., http://hdl.handle.net/10013/epic.50708.d001, 1975.

Werner, F.: 40 Jahre marin-geowissenschaftliche Forschung am Geologisch-Paläontologisches Institut der Universität Kiel, Meyniana, 50, 7–12, https://doi.org/10.2312/meyniana.1998.50.7, 1998.

Werner, F.: Bioturbation structures in marine Holocene sediments of the Kiel Bay (Western Baltic), Meyniana, 54, 41–72, https://doi.org/10.2312/meyniana.2002.54.41, 2002.

Wetzel, A.: Bioturbation in Spätquartären Tiefwasser Sedimentation vor NW-Afrika, dissertation, Mathematisch-Naturwissenschaftliche Fakultät der Christian-Albrechts-Universität zu Kiel, 111 pp., http://hdl.handle.net/10013/epic.50707.d001, 1979.

Whitaker, M. J. and Werner, F.: Pockmarks: Submarine vents of natural gas or freshwater seeps?, Geo-Mar. Lett., 1, 193–199, https://doi.org/10.1007/BF02462433, 1981.

Winn, K.: Present and postglacial Sedimentation in the Great Belt Channel (Western Baltic), Meyniana, 26, 63–101, https://doi.org/10.2312/meyniana.1974.26.63, 1974.

Winn, K.: Bioturbation structures in marine Holocene sediments of the Great Belt (Western Baltic), Meyniana, 58, 157–178, https://doi.org/10.2312/meyniana.2006.58.157, 2006.

Winn, K. and Averdieck, F.-R.: Postboreal development of the Western Baltic: comparison of two local sediment basins, Meyniana, 36, 35–50, https://doi.org/10.2312/meyniana.1984.36.35, 1984.

Permissions

All chapters in this book were first published in ESSD, by Copernicus Publications; hereby published with permission under the Creative Commons Attribution License or equivalent. Every chapter published in this book has been scrutinized by our experts. Their significance has been extensively debated. The topics covered herein carry significant findings which will fuel the growth of the discipline. They may even be implemented as practical applications or may be referred to as a beginning point for another development.

The contributors of this book come from diverse backgrounds, making this book a truly international effort. This book will bring forth new frontiers with its revolutionizing research information and detailed analysis of the nascent developments around the world.

We would like to thank all the contributing authors for lending their expertise to make the book truly unique. They have played a crucial role in the development of this book. Without their invaluable contributions this book wouldn't have been possible. They have made vital efforts to compile up to date information on the varied aspects of this subject to make this book a valuable addition to the collection of many professionals and students.

This book was conceptualized with the vision of imparting up-to-date information and advanced data in this field. To ensure the same, a matchless editorial board was set up. Every individual on the board went through rigorous rounds of assessment to prove their worth. After which they invested a large part of their time researching and compiling the most relevant data for our readers.

The editorial board has been involved in producing this book since its inception. They have spent rigorous hours researching and exploring the diverse topics which have resulted in the successful publishing of this book. They have passed on their knowledge of decades through this book. To expedite this challenging task, the publisher supported the team at every step. A small team of assistant editors was also appointed to further simplify the editing procedure and attain best results for the readers.

Apart from the editorial board, the designing team has also invested a significant amount of their time in understanding the subject and creating the most relevant covers. They scrutinized every image to scout for the most suitable representation of the subject and create an appropriate cover for the book.

The publishing team has been an ardent support to the editorial, designing and production team. Their endless efforts to recruit the best for this project, has resulted in the accomplishment of this book. They are a veteran in the field of academics and their pool of knowledge is as vast as their experience in printing. Their expertise and guidance has proved useful at every step. Their uncompromising quality standards have made this book an exceptional effort. Their encouragement from time to time has been an inspiration for everyone.

The publisher and the editorial board hope that this book will prove to be a valuable piece of knowledge for researchers, students, practitioners and scholars across the globe.

List of Contributors

Cristina Ponte Lira, Ana Nobre Silva, Rui Taborda and Cesar Freire de Andrade
IDL - Instituto Dom Luiz, Faculdade de Ciências da Universidade de Lisboa, Campo Grande, 1749-016 Lisbon, Portugal

Laura Sánchez, Herbert Arenz and Florian Seitz
Technische Universität München, Deutsches Geodätisches Forschungsinstitut (DGFI-TUM), Arcisstr. 21, 80333 Munich, Germany

Christof Völksen
Bayerische Akademie der Wissenschaften, Erdmessung und Glaziologie, Alfons-Goppel-Str. 11, 80539 Munich, Germany

Alexandr Sokolov
Technische Universität München, Deutsches Geodätisches Forschungsinstitut (DGFI-TUM), Arcisstr. 21, 80333 Munich, Germany
Bayerische Akademie der Wissenschaften, Erdmessung und Glaziologie, Alfons-Goppel-Str. 11, 80539 Munich, Germany

Kurt Henrik Kjær
Centre for GeoGenetics, Natural History Museum, University of Copenhagen, 1350 Copenhagen K, Denmark

Kristian Kjellerup Kjeldsen
Centre for GeoGenetics, Natural History Museum, University of Copenhagen, 1350 Copenhagen K, Denmark
Department of Earth Sciences, University of Ottawa, Ottawa, K1N 6N5, Canada
DTU Space - National Space Institute, Technical University of Denmark, Department of Geodesy, 2800 Kgs. Lyngby, Denmark

Reimer Wilhelm Weinrebe
GEOMAR, Helmholtz-Zentrum für Ozeanforschung, 24148 Kiel, Germany

Jørgen Bendtsen
Arctic Research Centre, Aarhus University, 8000 Aarhus, Denmark
ClimateLab, Symbion Science Park, Fruebjergvej 3, 2100 Copenhagen Ø, Denmark

Anders Anker Bjørk
Centre for GeoGenetics, Natural History Museum, University of Copenhagen, 1350 Copenhagen K, Denmark
Department of Earth System Science, University of California, Irvine, CA 92697, USA
NASA Jet Propulsion Laboratory, Pasadena, CA 91109, USA

Fei Lun
College of Resources and Environmental Sciences, China Agricultural University, Beijing 100193, China

Junguo Liu
School of Environmental Science and Engineering, South University of Science and Technology of China, Shenzhen 518055, China

Philippe Ciais, Jinfeng Chang, Rong Wang and Daniel Goll
IPSL - LSCE, CEA CNRS UVSQ, Centre d'Etudes Orme des Merisiers, 91191 Gif-sur-Yvette, France

Thomas Nesme
Bordeaux Sciences Agro, Université de Bordeaux, UMR 1391 ISPA, CS 40201, 33175 Gradignan CEDEX, France

Jordi Sardans
CREAF, Cerdanyola del Vallès 08193, Catalonia, Spain

Josep Peñuelas
CSIC, Global Ecology Unit CREAF-CSIC-UAB, Cerdanyola del Vallès 08193, Catalonia, Spain

Michael Obersteiner
International Institute for Applied Systems Analysis, 2361 Laxenburg, Austria

Ana Trobec and Marko Vrabec
University of Ljubljana, Faculty of Natural Sciences and Engineering, Department of Geology, Aškerčeva cesta 12, 1000 Ljubljana, Slovenia

Martina Busetti, Fabrizio Zgur, Luca Baradello, Andrea Cova, Emiliano Gordini, Roberto Romeo, Isabella Tomini and Paolo Diviacco
OGS - Istituto Nazionale di Oceanografia e di Geofisica Sperimentale, Borgo Grotta Gigante 42/c, 34010 Sgonico, Trieste, Italy

Alberto Babich
OGS - Istituto Nazionale di Oceanografia e di Geofisica Sperimentale, Borgo Grotta Gigante 42/c, 34010 Sgonico, Trieste, Italy
Università di Trieste, Dipartimento di Matematica e Geoscienze, Via Weiss 2, 34127 Trieste, Italy

Sašo Poglajen
Harpha Sea d.o.o., Čevljarska ulica 8, 6000 Koper, Slovenia

Peter Köhler
Alfred-Wegener-Institut Helmholtz-Zentrum für Polar- und Meeresforschung (AWI) 12 01 61, 27515 Bremerhaven, Germany

Christoph Nehrbass-Ahles, Jochen Schmitt, Thomas F. Stocker and Hubertus Fischer
Climate and Environmental Physics, Physics Institute, and Oeschger Centre for Climate Change Research, University of Bern, Bern, Switzerland

Kamolphat Atsawawaranunt and Sandy P. Harrison
Centre for Past Climate Change and School of Archaeology, Geography & Environmental Sciences, Reading University, Whiteknights, Reading, RG6 6AH, UK

Laia Comas-Bru and Sahar Amirnezhad Mozhdehi
School of Earth Sciences, University College Dublin, Belfield, Dublin 4, Ireland

Michael Deininger
School of Earth Sciences, University College Dublin, Belfield, Dublin 4, Ireland
Institute of Geosciences, Johannes-Gutenberg-University Mainz, Johann-Joachim-Becher-Weg 21, 55128 Mainz, Germany

Andy Baker
School of Biological, Earth and Environmental Sciences, University of New South Wales, Kensington 2052, Australia

Meighan Boyd
Department of Earth Sciences, Royal Holloway University of London, Egham, Surrey, TW20 0EX, UK

Nikita Kaushal and Franziska A. Lechleitner
Department of Earth Sciences, University of Oxford, South Parks Road, Oxford, OX1 3AN, UK

Syed Masood Ahmad
CSIR - National Geophysical Research Institute, Uppal Road, 500 007 Hyderabad, India
Department of Geography, Faculty of Natural Sciences, Jamia Millia Islamia, New Delhi 110025, India

Yassine Ait Brahim
Institute of Global Environmental Change, Xi'an Jiaotong University, Xi'an, Shaanxi, China

Monica Arienzo
Division of Hydrologic Sciences, Desert Research Institute, 2215 Raggio Parkway, 89512 Reno, NV, USA

Petra Bajo
School of Geography, The University of Melbourne, 3010 Victoria, Australia

Kerstin Braun
Institute of Human Origins, Arizona State University, 85287 Tempe, AZ, USA

Yuval Burstyn
Institute of Sciences, Hebrew University of Jerusalem, Edmond J. Safra Campus, Givat Ram, 91904 Jerusalem, Israel
Geological Survey of Israel, 30 Malkhe Israel, 95501 Jerusalem, Israel

Sakonvan Chawchai
MESA Research Unit, Department of Geology, the Faculty of Science, Chulalongkorn University, 10330 Bangkok, Thailand

Wuhui Duan
Institute of Geology and Geophysics, Chinese Academy of Sciences, University of Chinese Academy of Sciences, No. 19 Beitucheng West Road, Chaoyang District, Beijing, China

István Gábor Hatvani and Zoltán Kern
Institute for Geological and Geochemical Research, Research Centre for Astronomy and Earth Sciences, Hungarian Academy of Sciences, Budaörsi út 45, 1112 Budapest, Hungary

Jun Hu
Department of Earth Sciences, University of Southern California, 3651 Trousdale Parkway, 90089 Los Angeles, CA, USA

Inga Labuhn
Institute of Geography, University of Bremen, Bremen, Germany

Matthew Lachniet
Dept. of Geoscience, University of Nevada, Box 4010, 89154 Las Vegas, NV, USA

Andrew Lorrey
Climate, Atmosphere and Hazards Centre, National Institute of Water & Atmospheric Research, 41 Market Place, Central Business District, Auckland, New Zealand

Carlos Pérez-Mejías
Department of Geoenvironmental Processes and Global Change, Pyrenean Institute of Ecology (IPE-CSIC), Avda. Montañana 1005, 50059 Zaragoza, Spain

Robyn Pickering
Department of Geological Sciences, Human Evolutionary Research Institute, University of Cape Town, 7701 Rondebosch, Cape Town, South Africa

Nick Scroxton
Department of Geosciences, University of Massachusetts Amherst, 611 North Pleasant Street, 01003-9297 Amherst, MA, USA

István Gábor Hatvani, Zoltán Kern and Attila Demény
Institute for Geological and Geochemical Research, Research Centre for Astronomy and Earth Sciences, Hungarian Academy of Sciences, Budapest, Budaörsi út 45, 1112, Hungary

Szabolcs Leél-Őssy
Eötvös Loránd University, Department of Physical and Applied Geology, Budapest, Pázmány Péter stny. 1/C, 1117, Hungary

Domenico Di Giacomo and Dmitry A. Storchak
International Seismological Centre (ISC), Pipers Lane, Thatcham, Berkshire, RG19 4NS, UK

E. Robert Engdahl
University of Colorado, Boulder, CO, USA

Marisa Borreggine and Curtis Deutsch
School of Oceanography, University of Washington, 1503 NE Boat Street, Box 357940, Seattle, WA 98195-7940, USA

Sarah E. Myhre
School of Oceanography, University of Washington, 1503 NE Boat Street, Seattle, WA 98195-7940, USA
Future of Ice Initiative, University of Washington, Johnson Hall, Room 377A, Seattle, WA 98195-1360, USA

K. Allison S. Mislan
School of Oceanography, University of Washington, 1503 NE Boat Street, Seattle, WA 98195-7940, USA
eScience Institute, University of Washington, 3910 15th Ave NE, Seattle, WA 98195, USA

Catherine V. Davis
School of Earth, Oceans, and the Environment, University of South Carolina, 701 Sumter Street, Earth and Water Science Building, Room 617, Columbia, SC 29208, USA

Peter Saetre, Sara Nordén, Eva Andersson, Johannes Petrone and Ulrik Kautsky
Swedish Nuclear Fuel and Waste Management Co. (SKB), 101 24 Stockholm, Sweden

Hjalmar Laudon
Department of Forest Ecology and Management, Swedish University of Agricultural Sciences, 901 83 Umeå, Sweden

Tobias Lindborg
Swedish Nuclear Fuel and Waste Management Co. (SKB), 101 24 Stockholm, Sweden
Department of Forest Ecology and Management, Swedish University of Agricultural Sciences, 901 83 Umeå, Sweden

Johan Rydberg
Department of Ecology and Environmental Science, Umeå University, 901 87 Umeå, Sweden

Mats Tröjbom
Mats Tröjbom Konsult AB, Slänningevägen 28, 761 72 Norrtälje, Sweden

Sten Berglund
HydroResearch AB, Stora Marknadsvägen 15, 183 34 Täby, Sweden

Emma Johansson
Swedish Nuclear Fuel and Waste Management Co. (SKB), 101 24 Stockholm, Sweden
Department of Physical Geography, Bert Bolin Centre for Climate Research, Stockholm University, 106 91 Stockholm, Sweden

Anders Löfgren
EcoAnalytica, Slalomvägen 28, 129 49 Hägersten, Sweden

Gustav Sohlenius
Geological Survey of Sweden (SGU), 751 28 Uppsala, Sweden

Micke Borgiel
Sveriges Vattenekologer AB, Marsjö Hvilan, 643 94 Vingåker, Sweden

Wanchese M. Saktura
GeoQuEST Research Centre, School of Earth and Environmental Sciences, University of Wollongong, Wollongong NSW 2522, Australia

Alexandru T. Codilean and Timothy J. Cohen
GeoQuEST Research Centre, School of Earth and Environmental Sciences, University of Wollongong, Wollongong NSW 2522, Australia
ARC Centre of Excellence for Australian Biodiversity and Heritage, University of Wollongong, Wollongong NSW 2522, Australia

Henry Munack
GeoQuEST Research Centre, School of Earth and Environmental Sciences, University of Wollongong, Wollongong NSW 2522, Australia
Research Institute for Post-Mining Landscapes (FIB), Finsterwalde 03238, Germany

Andrew Gray
Information Technology Services, University of Wollongong, Wollongong NSW 2522, Australia

Simon M. Mudd
School of GeoSciences, University of Edinburgh, Edinburgh EH8 9XP, UK

Hannes Grobe, Friedrich Werner, Amelie Driemel, Stefanie Schumacher and Rainer Sieger
Alfred-Wegener-Institut, Helmholtz-Zentrum für Polar- und Meeresforschung, 27515 Bremerhaven, Germany

Kyaw Winn
Institut für Geowissenschaften (formerly Geologisch-Paläontologisches Institut und Museum, GIK), Christian-Albrechts Universität, 24118 Kiel, Germany

Index

CPSIA information can be obtained
at www.ICGtesting.com
Printed in the USA
BVHW051504180820
586704BV00005B/217